FLORA OF IRAQ

VOLUME FIVE

PART ONE

M. STONES

FLORA OF IRAQ

VOLUME FIVE

PART ONE

ELATINACEAE TO SPHENOCLEACEAE

EDITED BY

SHAHINA A. GHAZANFAR

AND

JOHN R. EDMONDSON

With the collaboration of the staff of
the National Herbarium of Iraq of the
Ministry of Agriculture, Baghdad

Published on behalf of the
MINISTRY OF AGRICULTURE
Republic of Iraq
by
ROYAL BOTANIC GARDENS, KEW
2016

First published in 2016 by
Royal Botanic Gardens, Kew,
Richmond, Surrey, TW9 3AB, UK
www.kew.org

Distributed on behalf of the Royal Botanic Gardens, Kew in North America by the University of Chicago Press, 1427 East 60th Street, Chicago, IL 60637, USA

ISBN 978-1-84246-594-3
e-ISBN 978-184246-595-0

British Library Cataloguing in Publication Data
A catalogue record for this book is available from the British Library

Design and page layout by Christine Beard
Publishing, Design and Photography
Royal Botanic Gardens, Kew

Frontispiece: *Dianthus judaicus* Boiss., painted by Margaret Stones 1959.

Printed in the UK by Marston Book Services Ltd
Printed in the USA by The University of Chicago Press

CONTENTS

FOREWORD vii

LIST OF CONTRIBUTORS viii

PREFACE ix

FAMILIES TREATED IN VOL 5(1)

- Elatinaceae 1
- Molluginaceae 3
- Caryophyllaceae 6
- Aizoaceae 124
- Portulaceae 128
- Polygonaceae 130
- Illecebraceae 163
- Chenopodiaceae 164
- Amaranthaceae 256
- Theligonaceae 271
- Sphenocleaceae 273

INDEX TO FAMILIES, GENERA AND SPECIES 275

INDEX TO VERNACULAR NAMES 285

FOREWORD

Our determination to complete the remaining parts of the *Flora of Iraq* under the difficult circumstances that we face in our country is proof that we have the right to live; our persistence reflects a thousand years of civilization that enriches humanity with science and culture.

These efforts, coming through cooperation with our friends at the Royal Botanic Gardens, Kew, are a gift to humanity and contradict those who advocate death to our beloved country. In general, this part covers eleven plant families from Elatinaceae to Sphenocleaceae, and not only gives information about these plant families in Iraq that will facilitate work of researchers both within and outside our country, but it is also an important step in the completion of the exploration phase for investigating plant diversity in Iraq, which will pave the way in the future for developing important strategies, plans and projects to protect our threatened plant species, and regulate the sustainable exploitation of this natural wealth.

I would like to extend my sincere thanks to all those who have contributed to this work, and also express my gratitude to the Royal Botanic Gardens Kew and the staff of the National Herbarium of Iraq for their great efforts to complete this volume of the Flora.

M. Zain Raoof.

Mohammed Zain Al-Abdeen Mohammed Raoof
Director General of SBSTC, Iraqi National Herbarium
Ministry of Agriculture, Abu Ghraib, Iraq

LIST OF CONTRIBUTORS
to VOLUME 5, PART 1

†Paul Aellen, Basel, Switzerland

John Akeroyd, Hindon, UK

H. A. Alizzi, Baghdad, Iraq

John Edmondson, Royal Botanic Gardens, Kew, UK

Shahina A. Ghazanfar, Royal Botanic Gardens, Kew, UK

Ali Haloob, National Herbarium, Baghdad, Iraq

Charles Jeffrey, Komarov Botanical Institute, St. Petersburg, Russia

Ali Kandemir, Erzincan University, Turkey

V. Melzheimer, Philipps University, Marburg, Germany

Alexander P. Sukhorukov, Moscow State University, Russia

Cliff C. Townsend, Royal Botanic Gardens, Kew, UK

PREFACE

to VOLUME 5, PART 1

The present instalment of the Flora of Iraq contains treatments of eleven families traditionally regarded as members of the Centrospermae (and now largely placed in the core Caryophyllales). In Hutchinson's system they comprise three orders: Caryophyllales, Polygonales and Chenopodiales. The families are the Aizoaceae, Amaranthaceae, Caryophyllaceae, Chenopodiaceae, Theligonaceae (Cynocrambaceae), Elatinaceae, Illecebraceae, Molluginaceae, Polygonaceae, Portulacaceae and Sphenocleaceae. Although our understanding of family relationships has advanced considerably since the scheme of the Flora was drawn up we have retained the traditional format as set out in vol. 2, pp. 8–14. The family Elatinaceae is now placed in the order Malpighiales and the Sphenocleaceae is placed in the Solanales following the APGIII system. The family Chenopodiaceae in now subsumed within the Amaranthaceae, the Theligonaceae (Cynocrambaceae) within Rubiaceae, and the Illecebraceae within Caryophyllaceae, but, apart from the Illecebraceae, these are kept separate here.

The editors are pleased to present the work of external contributors of taxonomic accounts to the present volume in addition to Kew staff who were responsible for writing parts of this volume some thirty years ago. The completion of this project will be increasingly dependent on such assistance, since with the exception of vol. 6 (Asteraceae/Compositae) which is already completed and awaiting final editing, many of the accounts of the remaining unpublished families are still unwritten. Since the publication of vol. 5 part 2 we have seen further steady progress with the Flora of Iran (in Persian), and the first volume of a new Flora of Turkey (in Turkish) has appeared. Though no further volumes of the Flora of the Arabian Peninsula & Socotra have become available, the collections in the herbarium of the Edinburgh Royal Botanic Garden provide good coverage of the countries lying to the south of Iraq. An increasing number of specimen images are now available for consultation online from several European herbaria, making the job of establishing overall distributions considerably easier.

In several treatments, notably in the Chenopodiaceae, a fairly radical revision to generic limits has been made in the light of new research. The genus *Salsola,* in particular, has an altered circumscription and there are many nomenclatural changes and even a few newly defined genera. The large genus *Silene* (Caryophyllaceae) has been prepared by separate authors for its annual and perennial species, and in the Polygonaceae the genera *Polygonum* and *Persicaria* are now separated.

Because of the order in which this work was completed, vol. 5 contains separate indexes to its two parts. Our aim is to provide illustrations of at least one species in each genus, and in order to do so we have been granted permission to reproduce drawings from other Floras to complement those which were specially commissioned for this work. A few small adjustments in the spellings of collection sites have been made in the light of further work on the Gazetteer, but we have tried to ensure that transliterations from Arabic script continue to follow the system set out in vol. 1. Kurdish place-names are more difficult to interpret, not only because of the state of flux in these areas but also because the increased adoption of the Roman script has not yet been accompanied by any overarching process of standardisation.

As for vol. 5 part 2, we have had the full support of colleagues and staff from the National Herbarium, Abu Ghraib, Baghdad, for which we are most grateful, and feel encouraged that beyond the completion of this Flora taxonomic research will still carry on there.

Shahina A. Ghazanfar
Royal Botanic Gardens, Kew

John R. Edmondson
Hon. Research Associate, Royal Botanic Gardens, Kew

88. ELATINACEAE

Dumort., Anal. Fam. Pl. 44, 49 (1829) nom. cons.
Niedenzu in Pflanzenreich, ed. 2, 2: 270 (1925)

C.C. Townsend[1]

Annual or perennial herbs, rarely suffrutescent perennials, growing in or near water. Leaves stipulate, opposite or verticillate. Flowers small, hermaphrodite, actinomorphic, axillary, solitary or in fascicles or glomerate. Sepals 2–5, free or shortly connate below, imbricate, persistent. Petals isomerous with the sepals and not or scarcely exceeding them, free, imbricate, hypogynous, persistent. Stamens in 1 or 2 whorls of 2–5 stamens each, free, hypogynous; anthers bilocular, dehiscing by longitudinal slits; disk absent. Ovary superior, 2–5-locular; ovules numerous, placentation axile. Styles 2–5, free, short. Fruit a septicidal capsule. Seeds without endosperm, embryo straight or curved.

Two genera and about 40 species, in temperate and tropical regions.

In the APG III classification, Elatinaceae belongs in the order Malphigiales, sister to Malphigiacae.

1. **BERGIA** L.

Mant. 2: 152 (1771)

Annual or perennial herbs, or rarely subshrubs, growing in wet places, often as weeds of cultivation. Leaves opposite, subsessile or shortly pseudo-petiolate, finely serrate. Stipules persistent. Flowers pentamerous or (not in Iraq) trimerous, in axillary clusters or fascicles or (not in Iraq) solitary. Sepals free, acuminate, ± keeled with a strong midrib and broad membranous margins. Petals ovate-oblong, membranous, obtuse to shortly acuminate. Stamens the same as or up to twice as many as the petals. Ovary 5-locular; stigmas capitate. Capsule globose or ovoid. Seeds oblong, straight or curved, smooth or ± reticulate.

About 25 species in the tropics and subtropics. Two species in Iraq.

Plant glabrous throughout; petals slightly longer than the sepals; seeds with deep and conspicuous parallel rows of quadrate reticulations . . . . . . 1. *B. capensis*
Plant with glandular and 1–3-celled, white, flexuous hairs; petals shorter than the sepals; seeds with shallow and obscure reticulation . . . . 2. *B. ammanioides*

1. **Bergia capensis** *L.*, Mant. 2: 152, 241 (1771); Rawi in Min. Agr. Tech. Bull. 16: 54 (1964); Verdcourt in Fl. Trop. E. Afr. 3 (1968); Boulos, Fl. Egypt 2:133 (2000); Qiner Yang & Tucker in Fl. China 13: 55 (2007).

B. aquatica Roxb., Pl. Coromand. 2: 28, t. 142 (1798); Boiss., Fl. Orient. 1: 782 (1867); Gorshkova in Fl. URSS 15: 260 (1949); Rech.f., Fl. Iranica 16: 3 (1966).
B. verticillata Willd., Sp. Pl. ed. 4, 2: 770 (1799).
Elatine luxuriana Del., Fl. Aegypt.: 72, t. 26 (1813).
E. verticillata (Willd.) Wight et Arnott, Prodr. Fl. Ind. Or.: 41 (1834).

Glabrous annual herb with a succulent creeping stem rooting at the lower nodes, and ascending branches, 10–50 cm; stem usually pinkish or reddish, terete. Leaves subsessile, lanceolate to oblong-lanceolate, 1.25–5 × 0.8–2.2 cm, finely serrulate wtih reddish-mucronate teeth, teeth becoming more remote towards the cuneate or ± attenuate base. Stipules ± 2 mm, erect and ± adpressed, membranous, serrulate to dentate. Flowers 5-merous, in short, dense, axillary clusters. Sepals lanceolate-ovate, 1.5–2 mm, acuminate, reddish-mucronate, midrib distinct. Petals slightly longer than the sepals, pale and delicate, finally spreading, elliptical to ovate to spatulate. Stamens variable in number, mostly 8–10; filaments filiform. Ovary globose, with 5 very short styles. Capsule subglobose, 1.5–2.5 mm in diameter,

[1] Royal Botanic Gardens Kew. Updated by Shahina A Ghazanfar

Fig. 1. **Bergia capensis**. 1, habit × $^{1}/_{2}$. **B. ammannioides**. 2, habit × $^{1}/_{2}$; 3, flower × 4; 4, sepal × 8; 5, petal × 8; 6, flower with sepals and petals removed × 6; 7, seed enlarged. Drawn by Cai Shuqin. Reproduced from Flora of China 13: fig. 39 with permission.

5-furrowed along the lines of the valves. Seeds cylindrical, ± 0.5 mm, brown, slightly curved, ± shining, with parallel rows of deep and conspicuous quadrate reticulations. Fig. 1: 1.

HAB. In channels in fields; alt. ± 700 m; fl. Aug.
DISTRIB. Along channels in paddy fields in the Upper Plain and Foothills Region; not common. **MAM**: 25 km from Zakho to Kasi Masi, *Botany Staff* 43818 (BAG). **FAR**: Arbil, 8 Aug. 1947, 700 m, *Gillett* 9656!

Europe, Tropical Africa, Iran, India, China (Guangdong), Malaysia, Sri Lanka and Thailand; introduced in tropical America.

2. **Bergia ammannioides** *Roxb.*, Hort. Beng.: 34 (1814), nomen ex Roth, Nov. Pl. Sp.: 219 (1821); Boiss., Fl. Orient. 1: 782 (1867); Thisleton Dyer in Fl. Brit. Ind. 1: 251(1874); Rawi in Min. Agr. Tech. Bull. 16: 54 (1964); Rech.f., Fl. Lowland Iraq: 433 (1964); Rech. f., Fl. Iranica 16: 4 (1966); Verdcourt in Fl. Trop. E. Afr. 3 (1968); Qiner Yang & Tucker in Fl. China 13: 55 (2007).

Bergia pentandra Camb. ex Guill. & Perr., Tent. Fl. Seneg. 42. t. 12 (1830).
Elatine ammannioides (Roxb. ex Roth) Wight & Arn., Prodr. Fl. Penin. Ind. Or. 1: 41 (1834).
Bergia ammannioides var. *pentandra* Wight, Ill. Ind. Bot. 54, t. 25a (1840).

Annual herb, prostrate and branched from the base with ascending to erect branches with numerous long, slender, erect branches from near the base upwards, ± 10–50 cm; stem and branches variably clad with a mixed indumentum of flexuous, white, 1–3-celled hairs and shorter stipitate-glandular, red-tipped hairs which apparently become stouter and more prominent with age. Leaves elliptic, elliptic-oblong or oblong-obovate, subsessile and shortly cuneate below, 7–40 × 2–12 mm, finely serrulate with reddish-mucronate teeth, indumentum similar to that of the stem or subglabrous at least on the upper surface. Stipules 2–4 mm, very narrowly deltoid and acute, glandular-dentate and often also with white flexuose hairs, erect, membranous with a greenish midrib, acuminate. Sepals 1–3 mm, whitish hairy, midrib green, margins membranous, white or frequently pinkish. Petals white, elliptic or oblong-elliptic, slightly shorter than the petals, delicate, erect. Stamens variable, usually 5; filaments filiform. Ovary subglobose, with (3–)5 very short styles. Capsule globose to ovoid-globose, 1.5–2.25 mm in diameter, 5-furrowed along the lines of the valves. Seeds comma-shaped or ovoid-cylindrical, pale brown, less than 0.5 mm, very shiny, with rows of shallow and obscure quadrate reticulations. Fig. 1: 2–7.

HAB. Irrigated fields; banks of Euphrates below high flood mark; alt. 20–35 m; fl. Aug.–Nov.
DISTRIB. Occasional in irrigated fields and wet locations. **MAM**: Mindan Bridge on Khazir r. between Mosul and Aqura, *Alizzi & Omar* 35281 (BAG). **LCA**: Faluja, 7 Oct. 1960, *Agnew, Hadač & Haines* W1809; Baghdad, Zafariniya, 15 Aug. 1954, *Al-Rawi* 19938 & ibid., 14 Nov. 1954, *Haines* W42; 30 km from Aziziya to Kut, *Alizzi & Omar* 34639 (BAG).

Widespread in Tropical W and E Africa, Arabia, Iraq, Iran to Afghanistan, Pakistan and Uzbekistan, and from India to Malaysia, the Philippines and Australia.

89. MOLLUGINACEAE

Bartl., Beitr. Bot. 2: 158. (1825) nom. cons.

C.C. Townsend[2]

Mostly herbs, often ± succulent. Leaves simple, alternate, opposite or whorled, sometimes with membranous stipules. Inflorescences cymose, axillary or terminal. Flowers regular, usually hermaphrodite. Calyx of usually 5 free sepals. Petals absent. Staminodes sometimes present, then often petaloid. Stamens 5-many, hypogynous. Ovary superior, of 2–5 usually united carpels; loculi as many as carpels; ovules 1-many per loculus, axile or rarely basal. Fruits usually capsular, loculicidal. Seeds subreniform, embryo usually curved.

The limits of Molluginaceae as currently circumscribed are unclear and require further study. Molluginaceae are included in the order Caryophyllales and in the APG III classification in the core eudicots. Genera included in Molluginaceae have previously been included in the Aizoaceae or Caryophyllaceae by some authors.

---

[2] Royal Botanic Gardens Kew. Updated by Shahina A Ghazanfar

The taxonomic position of *Telephium*, included by some authors in the Molluginaceae, has long been uncertain (see Williams in Journ. Bot. 44: 289 (1906) and Gilbert in Taxon 36: 47–49 (1987) for a discussion of this problem. In this Flora *Telephium* is included in the Caryophyllaceae.

Endress, M. E. & V. Bittrich (1993). Molluginaceae. Pp. 419–426 *in* K. Kubitzki, J. G. Rohwer and V. Bittrich, eds. The families and genera of vascular plants, vol. II, Magnoliid, hamamelid, and caryophyllid families. Springer Verlag, Berlin, Germany.

1. **GLINUS** L.

Sp. Pl. 463 (1753); Gen. Pl. 208 (1754)

Procumbent or prostrate, stellately pubescent herbs. Leaves opposite or apparently whorled. Flowers fascicled at the nodes, in axillary groups, hermaphrodite, regular. Sepals 5, free. Staminodes 0–20 or more, forked or fimbriate at the apex. Stamens 2–20 or more, hypogynous, free. Ovary superior, of 3–5 united carpels, 3–5-locular; ovules axile, many; stigmas free, sessile, as many as the carpels. Fruit a 3–5-valved loculicidal capsule. Seeds subreniform, each with a distinct filiform-appendaged strophiole.

Widespread in the tropics and subtropics; one species in Iraq.

Glinus lotoides *L.*, Sp. Pl. ed. 1, 463 (1753); Boiss., Fl. Orient. 1: 755 (1867); Post, Fl. Syr. Palest. & Sinai 1: 219 (1932); Blakelock in Kew Bull. 3: 431 (1948) & 12: 462 (1957); Zohary in Dep. Agr. Iraq Bull. 31: 50 (1950); Jeffrey in Fl. Trop. E. Afr., Aizoaceae 15, f. 5, 1–7 (1961). Zohary, Fl. Palaest. 1: 73 (1966); Cullen in Fl. Turk. 2: 346 (1966); Y. Nasir in Fl. Pak. 40: 1 (1973); Hedge & Lammond in Fl. Iranica 114: 5 (1975); Miller in Fl. Arab. Penins. & Socotra 1: 159 (1996); Boulos, Fl. Egypt 1: 41 (1999).

G. dictamnoides Burm. f. , Fl. Ind. 113 (1768); Boiss., Fl. Orient. 1: 756 (1867).
Mollugo hirta Thunb., Prodr. Fl. Cap. 24 (1800).
M. lotoides (L.) O.Kuntze, Revis. Gen. 1: 264 (1891).

Diffuse procumbent or prostrate grey-green densely stellate-hairy annual herb. Stems 5–10 cm. Leaves 10–45 × 5–22 mm (including 1–15 mm petiole), elliptic, obovate or suborbicular, rounded above, cuneate at the base. Flowers greenish white, in clusters of 2–10 at the nodes; pedicels 1.5–4 mm. Sepals 5–8 mm. Staminodes 0–9, strap-shaped, apically forked. Stamens 11–30. Ovary of 5 carpels, 5-locular. Fruit a 5-valved capsule. Seeds many, rough, strophiolate. Fig. 2: 1–6.

HAB. Damp silty places on banks of receding rivers, canals and ditches; also sometimes as a weed in marshy fields; alt. up to 500 m; fl. & fr. May–Dec.
DISTRIB. Upper plains and lower Mesopotamian regions, and particularly common along the whole length of the R. Tigris. **FUJ/FNI**: Mosul, *Bornmüller* 950!; Mosul, *Omar & Karim* 36967 (BAG). **FPF**: Saadiya, *Al Kaisi 47221 (BAG)*. **LCA**: Baghdad, *Haussknecht* s.n.!, ibid., *Guest* 13465!; Zafaraniya, *Rechinger & Naib* 33!; Selencia [Seleucia], *Guest* 272!; Na'maniya, *Guest* 3586!; **LEA**: Kut, *Guest* 207!; **LCA/LEA**: Diyala river by Rustamiya, *Guest* 3510!; Jabel Hamrin, *Sutherland* 480!; **LCA**: Baghdad, *Al-Mokhtar* 35844 (BAG); Zafaraniya, Tigris bank, *Gillett* 5868 (BAG). **LSM**: Amara marshes, Majarr canal, *Guest* 1629!; Kabayish (Kubaish), *Rawi* 12551!; Garmat Bani Sa'd, *Rawi* 16601!; **LBA**: Basra, *Haussknecht* s.n.!

Handel-Mazzetti's specimen, cited by Zohary (l.c. p. 50) as from 'Dschesire' is from over the frontier in Turkey (Cizre) – vide Ann. Naturh. Mus. Wien 26 : 144 (1912).

SOTAIH (*Guest* 207), NAIMAH (*Guest* 1629).

Pakistan, Iran, Iraq, Turkey, Syria, Jordan, Palestine, Cyprus, Crete, Balkans, Sicily, Arabian Peninsula; Algeria, Egypt; widespread in the tropics and subtropics.

Fig. 2. **Glinus lotoides**. 1, habit × ²⁄₃. 2, flower × 3; 3, flower with sepals removed × 3; 4, ovary in cross section × 12; 5, dehisced fruit × 3; seed × 10. All from *Guest* 207.

90. **CARYOPHYLLACEAE**[3] A.L. de Jussieu

Gen. Plant.: 299 (1789), nom. cons.
(including 94. Illecebraceae)

C. C. Townsend[4], V. Melzheimer[5], Ali Kandemir[6], Shahina A. Ghazanfar[7] and Ali Haloob[8]

Annual or perennial herbs, rarely subshrubs or shrubs, usually monoecious, rarely gynodioecious or dioecious. Stems and branches usually swollen at nodes. Leaves opposite, decussate, rarely alternate or verticillate, simple, entire, usually connate at base; stipules scarious, bristly, or often absent. Inflorescence mostly dichasial, sometimes clustered or umbellate, rarely solitary. Flowers actinomorphic, bisexual, rarely unisexual, occasionally cleistogamous. Sepals (4–)5, free or united, leaflike or scarious, persistent, sometimes bracteate below calyx. Petals (4–)5, free, rarely absent, often comprising claw and limb; limb entire or dissected, usually with coronal scales. Stamens (2–)5–10 in one or two whorls; filaments free or basally connate, or sometimes additionaly connate with the petals. Ovary 2–5-carpellate, rarely more, superior, rarely weakly semi-inferior; gynophore present or absent; placentation free, central, rarely basal; ovules (1 or) few or numerous, campylotropous. Styles (1–)2–5, free or connate; stigmas as many as carpels. Fruit usually a capsule, with pericarp crustaceous, scarious, or papery, dehiscing by teeth or valves, sometimes indehiscent with 1–seed; seeds 1 to numerous, reniform, ovoid, or rarely dorsiventrally compressed, abaxially grooved, blunt, or sharply pointed, rarely fimbriate-pectinate; testa granular, striate or tuberculate, rarely smooth or spongy; embryo strongly curved and surrounding perisperm or straight but eccentric.

About 86 genera with about 2200 species, mainly distributed in temperate regions of the northern hemisphere with a concentration in the Mediterranean and Irano-Turanian regions.

Bittrich, V. (1993). Caryophyllaceae. In: Kubitzki (ed.), The families and genera of vascular plants 2: 206–236.

Fior, S., Karis, P.O., Casazza, G., Minuto, L. & Sala, F. (2006). Molecular phylogeny of the Caryophyllaceae (Caryophyllales) inferred from chloroplast matK and nuclear rDNA ITS sequences. American Journal of Botany 93: 399–411.

Greenberg, A.K., & Donoghue, M.J. (2011) Molecular systematics and character evolution in Caryophyllaceae. Taxon 60 (6): 1637-1652.

Harbaugh, D.T., Nepokroeff, M., Rabeler, R.K., McNeill, J., Zimmer, E.A. & Wagner, W.L. (2010) A new lineage based tribal classification of the family Caryophyllaceae. International Journal of Plant Sciences 171(2): 185–198.

Pax, F. & Hoffmann, K. (1934). Caryophyllaceae. In: A. Engler & H. Harms eds.), Die natürlichen Pflanzenfamilien ed. 2, 16c: 273–364.

Rabeler, R.K. & Bittrich, V. (1993). Suprageneric nomenclature in the Caryophyllaceae. Taxon 42: 857–863.

Smissen, R.D., Clement, J.C., Garnock-Jones, P.J. & Chambers, G.K. (2002). Subfamilial relationships within Caryophyllaceae as inferred from 5' NDHF Sequences. American Journal of Botany 89(8): 1336–1341.

Caryophyllaceae is often classified into the three subfamilies Alsinoideae, Caryophylloideae and Paronychioideae. According to Fior *et al.* (2006) none of the subfamilies except Caryophylloideae appear monophyletic. Alsinoideae are paraphyletic; there is strong support for the inclusion of *Spergula-Spergularia* in an Alsinoideae-Caryophylloideae clade. Putative synapomorphies for these groupings are twice as many stamens as number of sepals and a caryophyllid-type of embryogeny. Paronychioideae form a basal grade, where tribe Corrigioleae are sister to the rest of the family. Free styles and capsules with simple teeth are possibly plesiomorphic for the family (Fior *et al.* 2006).

[3] Revised and updated by Ali Kandemir and Shahina A Ghazanfar

[4] Royal Botanic Gardens Kew

[5] Philipps-Universität Marburg, Germany. Dianthus, Silene (species 23–36)

[6] Erzincan University, Turkey

[7] Royal Botanic Gardens Kew. Silene (species 1–22)

[8] National Herbarium, Abu Ghraib, Iraq (Agrostemma)

Key to genera

1. Leaves stipulate .... 2 (I. Subfam. Paronychioideae)
   Stipule exstipulate 11
2. Leaves alternate .... 10. **Telephium**
   Leaves opposite or whorled 3
3. Peduncles dilated, oblanceolate or broadly obovate, wing like in fruit... 4. **Pteranthus**
   Peduncles or pedicels not so 4
4. Sepals with 2 bristles arising from near the base on each side .... 7. **Loeflingia**
   Sepals without 2 bristles arising from near the base on each side 5
5. Fruit 1-seeded, indehiscent 6
   Fruit few to many-seeded, capsule dehiscing by valves or teeth 8
6. Inflorescence a spherical, hard spinose head in fruit .... 3. **Sclerocephalus**
   Inflorescence not so in fruit 7
7. Bracts silvery-white, completely scarious, conspicuous .... 1. **Paronychia**
   Stipules inconspicuous .... 2. **Herniaria**
8. Sepals keeled and hooded .... 5. **Polycarpon**
   Sepals not as above 9
9. Styles 3–5 10
   Style solitary .... 6. **Polycarpaea**
10. Stipules free .... 8. **Spergula**
    Stipules united to surround the nodes .... 9. **Spergularia**
11. Sepals free .... 12 (II. Subfam. Alsinoideae)
    Sepals united to a distinct calyx tube .... 22. (III. Subfam. Caryophylloideae)
12. Petals absent, if present then sepals with ending in an awn half as long as or nearly as long as sepals; fruit 1–seeded, indehiscent 13
    Petals present; sepals not as above; fruit a capsule, dehiscing by teeth or valves 14
13. Petals present; sepals with distinct awn .... 12. **Habrosia**
    Petals absent; sepals acute not distinctly awned .... 11. **Scleranthus**
14. Inflorescence a simple umbel .... 18. **Holosteum**
    Inflorescence usually not as above, if umbel then capsule dehiscing by 10 teeth 15
15. Sepals 4, the outer pair short .... 19. **Bufonia**
    Sepals 5, if 4, then ± equal and plant shorter than 8 cm with very slender stems 16
16. Styles 2 17
    Styles 3–5 18
17. Petals entire to retuse; capsule opening by 2 valves .... 15. **Lepyrodiclis**
    Petals deeply bifid; capsule opening by 4 teeth .... 16. **Stellaria**
18. Capsule teeth or valves as many as styles 19
    Capsule teeth or valves twice as many as styles 20
19. Styles 3; capsule opening by 3 valves .... 14. **Minuartia**
    Styles 4–5; capsule opening by 4–5 valves .... 20. **Sagina**
20. Petals entire or emarginated .... 13. **Arenaria**
    Petals divided to at least 1/3, sometimes emarginate then, capsule teeth usually reflexed or revolute 21
21. Capsule oblong, cylindiric or cylindric-conic, opening by (6–)8–10 teeth .... 17. **Cerastium**
    Capsule ovoid, opening by 4 or 6 teeth .... 16. **Stellaria**
22. Calyx strongly 10-ribbed; teeth foliaceous to 35 mm, exceeding the petals .... 30. **Agrostemma**
    Calyx not strongly 10-ribbed; teeth not foliaceous, shorter than the petals 23
23. Calyx tube usually with commissural veins altering with midveins of sepals; styles 3(–5) .... 29. **Silene**
    Calyx tube without commissural veins; styles 2 24
24. Calyx tube with 5 wings; annuals .... 28. **Vaccaria**
    Calyx tube without 5 wings; annuals or perennials 25

25. Calyx with an epicalyx, scales enclosing the base .21. **Dianthus**
    Calyx not so 26
26. Small shrub with needle-like, spiny leaves; bracts spiny . . . . 27. **Acanthophyllum**
    Annual, biennial or perennial herbs; leaves and bracts not as above 27
27. Petals deeply (3–)5-partite; capsule opening irregularly at base . . . 23. **Ankyropetalum**
    Petals not deeply (3–)5-partite; sometimes divided into 3 or 5 linear-triangular lobes then calyx longer than 6 mm; capsule opening with 4 teeth or valves. 28
28. Calyx intervals membranous, hyaline or scarious 29
    Calyx intervals not membranous, hyaline nor scarious 30
29. Calyx campanulate, turbinate, rarely tubular, often with druses; seeds auriculate with a lateral hilum. . . . . .25. **Gypsophila**
    Calyx tubular or campanulate, without druses; seeds peltate, with a facial hilum . . . . 22. **Petrorhagia**
30. Slender, rigid, dichotomously branched annuals; petals without or with small coronal scales; seeds peltate with facial hilum . . . . 23. **Velezia**
    Habit various; petals usually with well developed coronal scales; seeds reniform with lateral hilum. . . . . .24. **Saponaria**

Key to subfamilies

(following Rechinger, Fl Iranica 1988)

1. Stipules present, scarious or rarely setaceous, sometimes caducous; flowers ± perigynous, petals often small or sometimes absent; fruits either capsules or indehiscent, mostly 1–seeded . . . I. SUBFAMILY PARONYCHIOIDEAE
   [*Paronychia, Herniaria, Sclerocephalus, Pteranthus, Polycarpon, Polycarpaea Loeflingia, Spergula, Spergularia, Telephium*]
   Stipules absent; petals always present. 2
2. Flowers ± perigynous or hypogynous, sepals free; petals not clawed; fruit mainly a capsule, rarely indehiscent. . . II. SUBFAMILY ALSINOIDEAE
   [*Scleranthus, Habrosia, Arenaria, Minuartia, Lepyrodiclis, Stellaria, Cerastium, Holosteum, Bufonia, Sagina*]
   Flowers hypogunous; sepals connate; petals often clawed; anthrophore often present; styles free . . . . .III. SUBFAMILY CARYOPHYLLOIDEAE
   [*Dianthus, Petrorhagia, Velezia, Saponaria, Gypsophila, Ankyropetalum, Acanthophyllum, Vaccaria, Silene*]

I. SUBFAMILY PARONYCHIOIDEAE Meisn.

Chaudhri, M.N. (1968). A revision of the Paronychiineae. Mededelingen Van Het Botanisch Museum En Herbarium Van De Rijksuniversiteit Te Utrecht 285: 1–427.

1. **PARONYCHIA** Miller

Gardn. Dict., Abridged ed. 4, 3 (1754).

Annual or perennial herbs. Leaves opposite, sessile with conspicuous scarious stipules. Flowers in axillary and terminal clusters, 5-merous. Bracts silvery, scarious, ± concealing the flowers. Sepals equal or unequal, herbaceous or semi-scarious, plano-concave or cucculate and awned at apex. Petals subulate-filiform, inserted on the perigynous ring. Stamens 5, rarely 1 or 2. Ovary ovoid; style bifid or with two free styles. Fruit 1-seeded, indehiscent, membraneous, rupturing irregularly at base.

About 110 species, nearly cosmopolitan with centres of distributions in the Mediterranean region, Turkey, SE United States, Peru and Bolivia; four species in Iraq.

1. Leaves mucronate at apex; sepals semi-scarious, cucculate with awn or mucro just below the apex 2
   Leaves not mucronate at apex; sepals herbaceous, not cucculate or mucronate or awned 3

2. Hairs on the perigynous zone hooked .... 1. *P. arabica*
Hairs on the perigynous zone straight or partly hooked .... 2. *P. argentea*
3. Sepals ± equal, erect or connivent .... 3. *P. mesopotamica*
Sepals very unequal, recurved at tip .... 4. *P. kurdica*

1. **Paronychia arabica** (*L.*) *DC* in Poir. Encyl. 5; 24 (1804); Zohary, Fl. Iraq & its Phytogeogr. Subdivis.: 53 (1946); Blakelock in Kew Bull. 2: 201 (1957); Rech. f., Fl. Lowland Iraq: 222 (1964); Chaudhri in Meded. Bot. Mus. Herb. Rijks. Univ. Utrecht 285: 199 (1968); Chaudhri in Fl. Iranica 144: 7 (1980); Chamberlain in Fl. Arab. Penins. & Socotra 1: 185 (1996); Boulos, Fl. Egypt 1: 87 (1999); Ghazanfar, Fl. Oman 1: 67 (2003); Dinarvand in Fl. Iran 65: 11 (2009).

Illecebrum arabicum L., Mant. 1: 51 (1767).
Corrigiola albella Forssk., Fl. Aegypt.-Arab. 207 (1775) non L. (1753).
Herniaria lenticulata Forssk., Fl. Aegypt.-Arab. 52 (1775), non L., Sp. Pl. 218 (1753).
Illecebrum longisetum Bertoloni, Fl. Ital. 2: 733 (1835).
Paronychia longiseta (Bertoloni) Webb & Berthelot, Phyt. Canar. 3: 163 (1840).
Paronychia desertorum Boiss., Diagn. Pl. Orient. Nov. ser. 1, 3: 11 (1843).

Annual or perennial herbs. Stems prostrate, 4–30 cm long, much branched from base, shortly pubescent to puberulous. Leaves linear-oblong to narrowly obovate, 4–10 mm, attenuate to base, mucronate at apex, puberulous to glabrous, ± ciliate on margins. Stipules narrowly ovate to lanceolate, glabrous or ciliate on margins, shorter or longer leaves, sometimes densely imbricate. Glomerules pseudo-axillary and terminal, often densely congested along the short lateral branches. Bracts ovate, up to 3 mm. Flowers with hooked hairs on perigynous zone. Sepals ± 1.5 mm long, hairy or glabrous on herbaceous part, broadly scarious margined, cucculate; awn equalling or shorter than sepals. Petals equalling to filaments. Ovary ovoid, papillose at top; style bilobed. Fruit ovoid-oblong, papillose on upper part. Fig. 3: 1–6.

HAB. Sandy and gravel soils; alt. 0–450 m; fl. Feb.–May.
DISTRIB. Common in desert areas of Iraq. **DLJ**: Between Rawa and Haditha, *Omar et al.* 45148!; 12 km W of Rawa, *Omar et al.* 44453!; **DWD**: Wadi Ghadaf, *Rawi* 14843!; 48 km W of Rutba, *Kaisi* & *Hamad* 46593!; 140 km W of Ramadi, *Rawi* & *Nuri* 27029!; 30 km W of Haditha, *Hamad* 50032!; Between Ramadi and Rutba, *Rechinger* 9815!; Rutba, *Alizzi* & *Omar* 36186!; **DSD**: Jumaima police station, *Kaisi et al.* 48259! & *Weinert et al.* 47580!; Jabal Sanam, *Kaisi, Hamad* & *Hamid* 48600!; 5 km E of Jumaima, *Kaisi, Hamad* & *Hamid* 48204!; Ansab, *Botany Staff* 42235!; 60 km S of Salman, *Botany Staff* 42214!; 5 km SE by E of Zubair, *Guest, Rawi* & *Rechinger* 16791!; **LCA**: nr. Falluja, *Guest* & *Rawi* 13638!; 10 km S of Zubair, *Gillett* & *Rawi* 6075!

A variable species with differences in leaf, stipules and sepals characters. Chaudhri (l.c.) recognized 6 subspecies and many varieties. According to Chaudhri and Rechinger (l.c.) subsp. *arabica* var. *arabica* and subsp. *breviseta* (Aschers. et Schweinf.) Chaudhri var. *breviseta* occur in Iraq. According to Chaudhri, in subsp. *breviseta* the stipules are shorter than the leaves during the normal flowering and fruiting period, and awns are about 0.4 mm; and in subsp. *arabica*, the stipules at the basal parts of the stems exceed the leaves and awns by 0.25–0.35 mm. These differences are not clear in our specimens.

N Africa, E and S Turkey, SW and S Iraq, Arabian Peninsula (Yemen, Oman, Qatar, Bahrain, Kuwait), S Iran.

2. **Paronychia argentea** Lam., Fl. Fr. 3: 230 (1778); Zohary, Fl. Iraq & its Phytogeogr. Subdivis.: 53 (1946); Blakelock in Kew Bull. 2: 201 (1957); Rech. f., Fl. Lowland Iraq 222 (1964); Chaudhri in Fl. Turk. 2: 252 (1967); Chaudhri in Meded. Bot. Mus. Herb. Rijks. Univ. Utrecht 285: 211 (1968); Artelari in Fl. Hellenica 1: 222 (1997); Boulos, Fl. Egypt 1: 88 (1999).

Illecebrum paronychia L., Sp. Pl. 206 (1753).
Paronychia nitida Gaertn., Fruct. Sem. Pl. 2: 218, t. 128 (1791).
P. glomerata Moench., Meth. 315 (1794).
P. hispanica DC in Lam., Encycl. 5: 24 (1804).

Perennial or occasionally biennial herbs. Stems 3–50 cm, prostrate much branched from base, retrorsely pubescent and sometimes glabrous. Leaves glabrous, ovate to oblong-lanceolate, 5–15 mm, acute. Stipules ovate-lanceolate, mostly shorter than leaves. Glomerules 10–15 mm in diameter, lateral and terminal, dense, intermixed with leaves. Bracts ovate, acute, at least twice as long as the flowers. Flowers 2–2.5 mm, shortly pubescent with straight

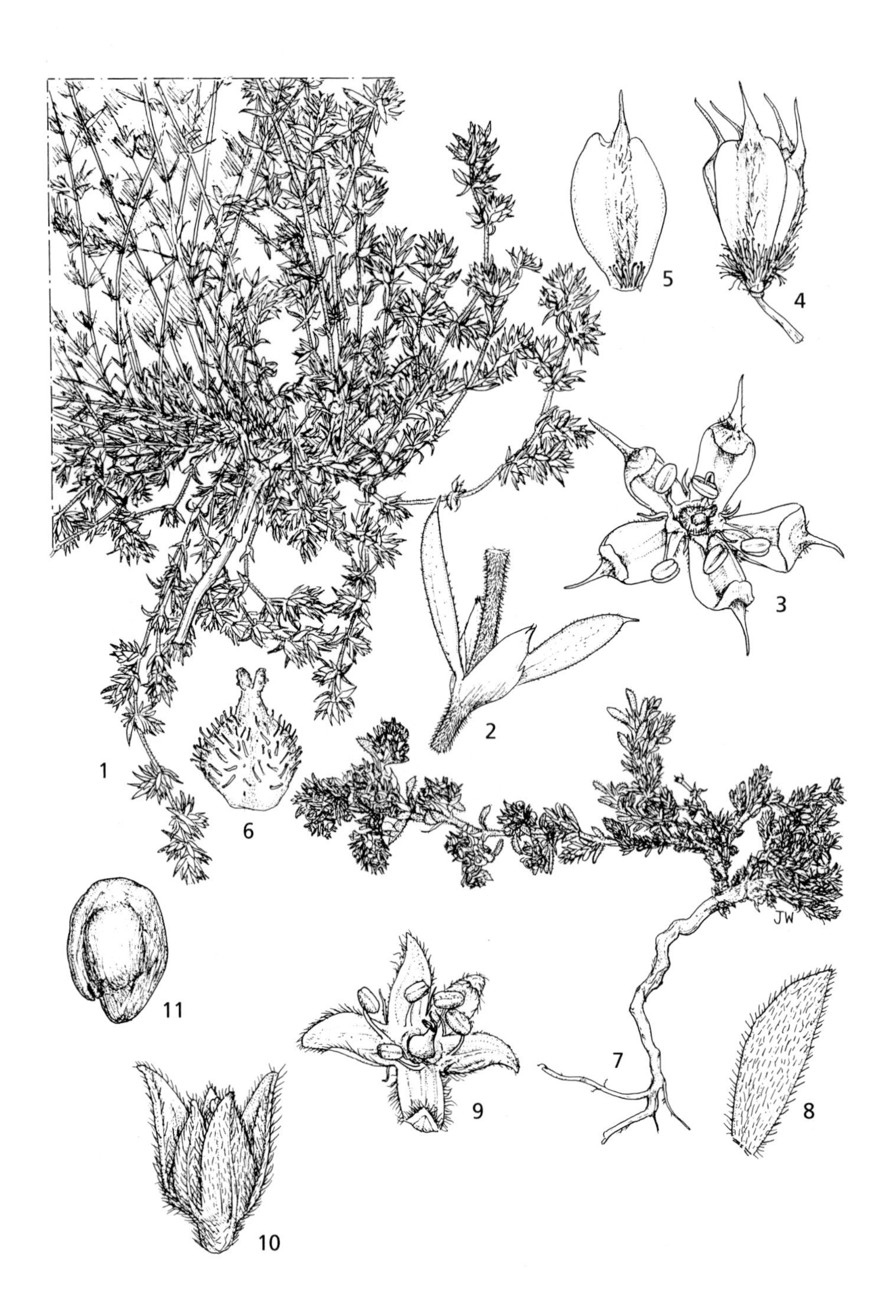

Fig. 3. **Paronychia arabica**. 1, part of habit × ²/₃; 2, pair of leaves with stipules × 4; 3, flower, open × 12; 4, flower, side view × 12; 5, calyx lobe, abaxial view × 12; 6, ovary × 24. **Paronychia kurdica**. 7, part of habit × 1; 8, leaf × 8; 9, flower open × 10; 10, flower side view × 10; 11, seed × 20. 1, 2 from *Rawi* 14799; 3, 4, 5, 6 from *Rawi* 20875; 7, 8 from *Guest* 1975; 9, 10, 11 from *Rawi* 23209. © Drawn by Juliet Beentje.

hairs on outside. Sepals oblong, scarious margins narrower than the herbaceous part, cucculate, awned. Petals slightly shorter than filaments. Ovary ovoid, papillose at top; style nearly half-way split into 2 thick stigmas. Fruit papillose on upper part.

Hab. Sandy banks rocky slopes; fl. Mar.–Jun.
Distrib. Scattered in desert and steppe areas. **dwd**: Jabal Jidra Wa Jundan, 389 km W of Baghdad; Wadi Amji to Wadi Muhammadi, 320 km W of Baghdad; Wadi Muhammadi, 45 km W of Baghdad; **dsd**: Zubair (*Anthony*).

No specimens at K; records from Zohary, Flora of Lowland Iraq (1964).

S Europe, N Africa, SW Asia.

3. **Paronychia mesopotamica** *Chaudhri* in Meded. Bot. Mus. Herb. Rijks-Univ. Utrecht 285: 262 (1968); Chaudhri in Fl. Iranica 144: 9 (1980). — Type: N Iraq: Sarsang, on rocks besides shady gorge, 1300 m, 5.vii. 1955, *Haines* W452 (holo. E!).

Perennial herb, caespitose. Stems 3–7 cm, prostrate, much branched from base, short retrorsely pubescent. Leaves 2.5–5 mm, velutinous-pubescent, oblong or oblanceolate-spatulate, obtuse. Stipules ovate or lanceolate, 3–5 mm. Glomerules 10–15 mm in diameter, terminal and subterminal. Bracts 6–7 mm, ovate, longer than flowers. Flowers ± 2.5 mm, densely appressed pubescent. Sepals herbaceous, oblong, subequal, without hood or awn, connivent to erect. Petals equal to filaments. Ovary subovoid, glabrous; styles paired. Fruit smooth and glabrous.

Hab. On rocks beside shady gorge, on hillsides in Quercus forest; fl. Jun.
Distrib. **mam**: Sarsang, *Haines* W452 (type); **msu**: Melakawa, *Rawi* 22441!

Iraq and Syria.

4. **Paronychia kurdica** *Boiss.*, Diagn. Pl. Orient. Nov. Ser. 1,3: 10 (1843). — Type: Iraq; Mt. Gara Kurdistan, *Kotschy* 334 (holo. G!; iso. HAL, JE, S); Zohary, Fl. Iraq & its Phytogeogr. Subdivis.: 53 (1946); Blakelock in Kew Bull. 2: 201 (1957); Rech. f., Fl. Lowland Iraq 222 (1964); Chaudhri in Fl. Turk. 2: 254 (1967) & in Meded. Bot. Mus. Herb. Rijks. Univ. Utrecht 285: 267 (1968) & in Fl. Iranica 144: 10 (1980); Dinarvand in Fl. Iran 65: 16 (2009).

Illecebrum cephalotes M. Bieb., Fl. Taur.-Cauc., Suppl. 169 (1819) p.p quoad pl. Caucas.
Paronychia capitata (L.) Lam. var. *pubescens* Fenzl in Ledeb. Fl. Ross. 2: 163 (Isus 3) (1843).
Paronychia kurdica Boiss. var. *imbricata* Chaudhri in Meded. Bot. Mus. Herb. Rijks. Utrecht 285: 91 (1968). — Type: Kirkuk, Jarma, *Wheeler-Haines* W277 (E).

Perennial herbs with a woody base. Stems 3–15 cm, procumbent, much branched from base, shortly pubescent. Leaves 3–7 mm, narrowly oblong to oblanceolate, ± obtuse, somewhat succulent, pubescent, sometimes imbricate. Stipules ovate-lanceolate, 4 per node, mostly shorter than leaves. Glomerules (4–)10–15 mm in diameter, profuse and congested. Bracts obliquely ovate to orbicular, acute to acuminate, almost 2× longer than the flowers, sometimes not quite concealing flowers. Flowers 3–3·5 mm, pubescent. Sepals unequal, 3 long, 2 (inner) short, narrowly oblong-lanceolate, obtuse to acute, the outer recurved at tips. Petals equalling to filaments. Ovary ovoid-conical, narrowed to the top, glabrous; styles long and erect. Fruit ovoid-oblong, glabrous. Fig. 3: 7–11.

Hab. Rocky mountain slopes, serpentine rocks, crevices in rock, dry silt ridges; alt. 90–1600; fl. May–Jul.
Distrib. **mam**: 5 km E of Kani Masi, *Omar & Kaisi* 45410!; Bamerny, 20 km NW of Sarsang, *Kaisi & Hamad* 45927!; Sarsang, *Haines* W452!; Ser Amadiya, Dabbagh, *Kaisi & Hamid* 45983!; Bikhair Mt., nr. Zakho, *Rawi* 23148!; Khantur Mt., nr. Zakho, *Rawi* 23364!; Gali Mazurka, *Dabbagh & Jasim* 46867! **mro**: Pushtashan, 15 km NE of Rania lower slope of Qandil range, *Rawi* 23834 (BAG!). **msu**: Jarmo, *Haines* W277! (type of var. imbricata); *Helbaek* 1893!; Kajan Mt., nr. Penjwin, *Rawi* 22734!; Zewiya, Pira Magrum, *Rawi* 12063!; Chemchemal, *Gillett & Rawi* 11608!; Khurmal, *Rawi* 8895!; **mjs**: Jabal Sinjar, *Gillett* 11114!; **fni**: 30 km from Mosul to Aqra, *Kaisi* 49720!; Zab, nr. Eski Kellek, *Gillett* 8208!; **fki**: 13 km N of Kirkuk, *Gillett & Rawi* 10561!; **fpf**: Koma Sank Police Station, nr. Mandali, *Rawi* 20639!; Khanaqin, *Guest* 1834!; 20 km N of Mandali, *Kaisi & Yahya* 45209!; nr. Mandali, *Guest* 864!

P. kurdica is a very variable species. Some specimens are very close to *P. sinaica* Fresen. According to Chaudhri, var. *imbricata* Chaudhri is different from var. *kurdica* in its imbricate leaves, smaller glomerules (± 5 mm in diameter) and almost orbicular bracts. Isotype of var. *imbricata* seen in K shows the bracts oblique as in var. *kurdica*, and some specimens with finely imbricate leaves have glomerules up to 10 mm in diameter.

SW Asia.

2. HERNIARIA L.

Sp. Pl. 218 (1753)

Annuals or short-lived perennials. Stems prostrate or procumbent, branched from base. Leaves opposite or apparently alternate on flowering branches, sessile or subsessile, entire, mostly obovate or broadly oblanceolate to spatulate to nearly elliptic. Stipules minute, scarious. Flowers 4- or 5-merous, very small, in leaf-opposed or apparently axillary, sometimes terminal clusters, slightly perigynous. Bracts inconspicuous. Sepals free, equal or unequal. Petal inconspicuous, subulate, filiform. Stamens 2–5; staminodes 4 or 5, filiform, sometimes absent. Stigmas 2, sessile or on a short style. Fruit 1-seeded, indehiscent.

The genus has about 45 species worldwide, but mainly distributed in Eurasia and N Africa. Represented by four species in Iraq.

F.N. Williams (1896) A Systematic revision of the genus *Herniaria*. Bull. Herb. Boiss. 1 (4): 556–570.
M.N. Chaudhri (1968) A revision of the Paronychiineae in Mededelingen Van Het Botanisch Museum En Herbarium Van De Rijksuniversiteit Te Utrecht 285: 1–427 (1968).

1. Calyx 4-merous 2
   Calyx 5-merous 3
2. Inner sepals distinctly unequal; petals absent; stamens 2. . . . . 3. *H. hemistemon*
   Sepals slightly unequal; petals ± 0.8 mm, linear-lanceolate with broad base; stamens 4. . . . . 4. *H. arabica*
3. Leaves glabrous; flowers glabrous, occasionally slightly puberulous on perigynous region . . . . 1. *H. glabra*
   Leaves and flowers distinctly hairy . . . . 2. *H. hirsuta*

1. **Herniaria glabra** L. Sp. Pl. 1: 218 (1753); Brummitt in Fl. Turk. 2: 246 (1967); Chaudhri in Meded. Bot. Mus. Herb. Rijks. Univ. Utrecht 285: 315 (1968) & in Fl. Iranica 144: 15 (1980); Hartvig in Fl. Hellenica 1: 230 (1997); Lu Dequan & Gilbert in fl. China 6: 3 (2001); Dinarvand in Fl. Iran 65: 23 (2009).

Herniaria glabra L. var. *glaberrima* sensu Chaudhri (1968: 318), non Fenzl in Ledebour, Fl. Ross. 2: 159 (1843).

Annual or perennial herbs. Stems up to 15 cm, prostrate, glabrous or very shortly pubescent. Leaves sessile to subsessile, up to 6 mm, elliptic to obovate or spatulate, glabrous. Stipules ovate, up to 1.5 mm, ciliate to fimbriate on margins. Flower clusters mostly leaf-opposed with 6–10 flowers; flowers sessile, usually glabrous. Sepals 5, 0.6–0.8 mm, narrowly oblong, obtuse with narrow white margins. Petals ± 0.5 mm. Style bilobed. Fruit ovoid to globose, exceeding sepals.

HAB. Field margins, *Quercus* forest, roadsides, open, stony grounds; alt. 1200–2600; fl. May–Jul.
DISTRIB. Very rare. **MRO**: Halgurd Dagh, *Rechinger* 11358; *Gillett* 9533 (BAG). **MSU**: Penjwin, *Rawi* 22580!

Europe, NW Africa, temperate and subtropical Asia.

2. **Herniaria hirsuta** L. Sp. Pl. 1: 218 (1753); Rech. f., Fl. Lowland Iraq: 224 (1964); Brummitt in Fl. Turk. 2: 248 (1967); Chaudhri in Meded. Bot. Mus. Herb. Rijksuniv. Utrecht 285: 337 (1968); Ghafoor in Fl. Pak. 47: 6 (1973); Chaudhri in Fl. Iranica 144:16 (1980); Chamberlain in Fl. Arab. Penins. & Socotra 1: 189 (1996); Hartvig in Fl. Hellenica 1: 230 (1997); Boulos, Fl. Egypt 1: 84 (1999); Gilbert in Fl. Ethiop. & Eritr. 2(1):199 (2000); Ghazanfar, Fl. Oman 1: 68 (2003); Dinarvand in Fl. Iran 65: 26 (2009).

H. cinerea DC in Lam., Fl. Fr. ed. 3, 6: 375 (1815).

Anual herbs. Stems 10–20 cm high, prostrate, branched from base, covered with patent hairs. Leaves opposite, alternate on flowering branches, narrowly elliptic to obovate, 4–8 mm, covered with stiff white patent to antrorse hairs. Stipules ovate-triangular with ciliate margins. Flower clusters leaf-opposed with 7–12-flowered, congested on short lateral branches; flowers sessile. Sepals 5, ovate-oblong, ± 1 mm long, equal to unequal, covered with stiff, whitish, hooked and straight hairs. Petals 0.5–0.6 mm. Stamens 2–5. Style bilobed. Fruit ovoid to globose, enclosed by sepals. Fig. 4: 1–3.

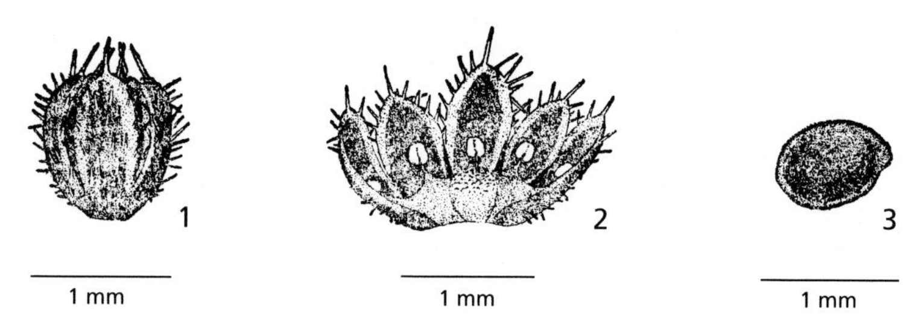

Fig. 4. **Herniaria hirsuta**. 1, flower; 2, flower, open; 3 seed. Reproduced with permission from Flora of Pakistan 47: 1973. Drawn by Asad.

Hab. Sandy soils, dry and stony places, fallow fields, riversides; alt. 0–2200 m; fl. Mar.–Jun.
Distrib. Widespread in the mountains, foothills and alluvial plains. **MJS**: Karsi, Jabal Sinjar, *Gillett* 10909! Karsi NW slope, *Omar, Al-Khayat & Al-Kaisi* 52476 (BAG!); **FUJ**: Ani Ghazal, *Guest* 4075!; **FPF**: Sa'diya, *Kaisi* 44204!; Chilat, Police Station, nr. Persian border, *Rawi & Haddad* 25687!; Koma Sank police station nr. Mandali on Persion border, *Rawi* 20666 (BAG!). **DLJ**: 70 km NW of Falluja, *Chakravarty & Rawi* 30374!; 15 km N of Falluja, *Rawi* 10151!; 10 km W of Tikrit, *Alizzi & Husain* 33730!; **DGA**: 4 km E of Simarra, *Rawi* 20330!; **DWD**: 13 km W of Ukhaidhir, *Rawi* 30858!; 50 km from Rutba, *Jenan* 33376!; 4 km NW of K3, *Alizzi* & *Husain* 34034!; Wadi of Ghadaf, *Rawi* 14846!; 4 km SW of Rutba, *Rawi* 21016!; **DSD**: Jabal Sanam, *Kaisi, Hamad* & *Hamid* 48605!; 70 km W of Nukhaila, *Rawi* 31026!; nr. Jal al-Rawaq, *Guest, Rawi* & *Rechinger* 18916!; 5 km E of Jumaima, *Kaisi et al.* 48193!; 60 km S of Samawa, *Gillett* & *Rawi* 6170!; **LEA**: Basra, *Guest, Rawi* & *Rechinger* 16739; **LCA**: Falluja, *Chakravarty* & *Rawi* 30271!; 25 km Chabbab's Bridge, S of Kut al-Imara, *Rawi* 17955!; Baghdad, *Bornmüller* 948!; Baghdad, *Lazar* 190! & 3890!; 4 km SE of Baghdad, *O.Polunin* 5009; Daltawa (Khalis), *Roger* 328!; Zafariniya, *Gillett* 10186! **LSM**: nr. Azair (between Hor and R. Tigris), *Guest, Rawi & Rechinger* 17399 (BAG).

Eurasia (except North) and N Africa.

3. **Herniaria hemistemon** J. Gay in Duchartre, Rev. Bot. 2: 371 (1847); Rech. f., Fl. Lowland Iraq: 224 (1964); Chaudhri in Meded. Bot. Mus. Herb. Rijks. Univ. Utrecht 285: 393 (1968) & in Fl. Iranica 144: 21 (1980); Chamberlain in Fl. Arab. Penins. & Socotra 1: 189 (1996); Boulos, Fl. Egypt 1: 86 (1999); Dinarvand in Fl. Iran 65: 33 (2009).

H. fruticosa L. var. *hemistemon* (J. Gay) Barratte in Bonn. & Barr., Cat. Tun. 65 (1896).

Perennial (rarely annual to biennial) herbs with a woody long tap root. Stems 3–12 cm, prostrate, shortly retrorsely pubescent to puberulous. Leaves mostly opposite, oblong-elliptic, 2.5–5 mm, subsessile, obtuse, pubescent. Stipules ovate-triangular with ciliate to fimbriate margins; upper ones purple or white. Flowers sessile, in axillary clusters with long hooked hairs on perigynous region. Bracts similar to upper stipules. Sepals 4, papillose or hirtellous on back, unequal; outer pair deltoid-spatulate, ± 1.5 mm long, thickened; inner ovate-oblong, $^1/_3$ as long as the outer ones. Petals absent. Stamens 2; staminodes absent. Stigmas 2, filiform, pale-brown. Fruit ovoid-ellipsoid.

Hab. Sandy gypsaceous soils, stony hillsides, riversides; alt. 0–400 m; fl. Feb.–Apr.
Distrib. **DGA**: 50 km NE of Samara, *Rawi & Shahwani* 25488!; 70 km NW of Falluja, *Chakravarty & Rawi* 30373!; **LCA**: 10 km E of Falluja, *Gillett & Rawi* 6765!; **DWD**: Falluja Desert, *Haines* 192!; Kahan Baghdadi, *Hamid* 36955!; 24 km E of Ana, *Barkley & Brahim-Mohamed* 604!; 15 km W of Ramadi, *Gillett & Rawi* 6803!; N of Lake Habbaniyah, *Guest & Haines* 15993A; **DSD**: 6 km E of Shabicha, *Rawi* 13838!; Al-Samah, *Kaisi et al.* 48442!; 6 km S of Salman, *Rawi et al.* 29218!; 12 km ESE of Salman, *Guest, Rawi & Rechinger* 18853! & 18821!; Jabal Sanam, *Kaisi, Hamad & Hamid* 48635!; 50 km WSW of Karbala, *Rawi* 30814!; 41 km N of Samah, *Botany Staff* 42320!; nr. Zubair, *Gillett & Rawi* 6018!; *Guest, Eig & Zohary* 5047!; **LCA**: on Falluja road, west of Abu Ghraib, *Polunin et al.* s.n!

N Africa, Sinai, Palestine, Jordan, Syria and S Iran.

4. **Herniaria arabica** *Hand.- Mazz.* in Ann. Natur Hist. Hofmus, Wien 26: 145 (1912); Rech. f., Fl. Lowland Iraq:, 224 (1964); Chaudhri in Meded. Bot. Mus. Herb. Rijks. Univ. Utrecht 285: 395 (1968). — Type: Iraq: Abukemal to Ramadi, at Kaijim, below Abukemal, *Handel-Mazzetti* 649 (holo. WU!).

Perennial (rarely annual to biennial) herbs with a long woody tap root. Stems 4–10 cm, prostrate to ascending with patent or retrorsely hairs. Leaves mostly opposite, linear-obovate, up to 6 mm, entire or incurved at margins, often hispidulous. Stipules ovate-triangular with ciliate to fimbriate margins; younger ones usually purple, sometimes white. Flowers sessile in axillary clusters with long hooked hairs in the pergynous region. Sepals 4, subequal to unequal, hirsute; outer cucullate, rounded at apex, attenuate at base, ± 1.5 mm; inner sepals ovate-oblong, ± 1 mm long with membranous margins. Petals present, ± 0.8 mm, linear-lanceolate. Stamens 4. Styles short distinct. Fruit ovoid.

HAB. Sandy gypsaceous soils, calcareous hills; alt. 50–600 m; fl. Apr.–Jun.
DISTRIB. **DJL**: Haditha, *Chakravarty, Rawi, Khatib & Alizzi* 32937!; *Gillett & Rawi* 6905!; **DWD**: 10 km SW of Rutba, *Rawi* 21075!; 15 km W of Ramadi, *Gillett & Rawi* 6799!; 102 km W of Ramadi, *Rawi & Nuri* 27020!; N of Lake Habbaniya, *Guest et al.* 15993!; 24 km SW of Ukhaidhir, *Rawi & Hamad* 33495! **DSD**: S of Zubair, *Rawi* 25882 (BAG); 22 km E of Shabicha, *Chakravarty, Rawi, Khatib & Tikrity* 30116 (BAG).

Syria.

SPECIES DOUBTFULLY RECORDED

Herniaria incana Lam., Encycl. 3: 124 (1789); Blakelock in Kew Bull. 2: 197 (1957); Brummitt in Fl. Turk. 2: 248 (1967); Chaudhri in Meded. Bot. Mus. Herb. Rijks. Univ. Utrecht 285: 369 (1968); Chaudhri in Fl. Iranica 144: 20 (1980); Hartvig in Fl. Hellenica 1: 230 (1997).

H. macrocarpa Sm., Fl. Graec. Prodr. 1: 167 (1806).
H. incana Lam. var. *angustifolia* Fenzl in Ledeb. Fl. Ross. 2: 160 (1843).
H. densiflora Williams in Bull. Herb. Boiss. 4: 562 (1896)

Perennial herbs with a stout woody stock. Stems up to 30 cm, prostrate, often spreading, densely pubescent with stiffly spreading or somewhat deflexed hairs. Leaves up to 12 × 3 mm, oblanceolate to spatulate, sometimes elliptic-obovate, densely hispid, older ones glabrescent or adpressed pubescent, or becoming glabrous with age; stipules usually hairy on the outer surface. Flowers 1·5–2 mm, with up to 12 flowers in a clusters. Sepals 5, ovate-oblong, obtuse, covered with stout, whitish, spreading hairs. Stamens 5. Fruit shorter than, and completely enclosed by the sepals, with a subsessile stigma of 2 widely divergent lobes.

No Iraqi specimens at K. Only one record from Jabal Hamrin (*Anthony* s.n.) in Kew Bull. 2: 197 (1957). This species is common in E and S of Turkey, Iran & Syria. It probably grows in some localities in Iraq.

3. **SCLEROCEPHALUS** Boiss.

Diagn. Pl. Orient. Nov. Ser.1, 3: 12 (1843)

Annual herbs. Stems much branched, prostrate, glabrous. Leaves linear-terete, opposite or fasciculate, fleshly or subcoriaceous, spine-tipped. Stipules scarious. Flowers in compact terminal or axillary spherical heads, becoming accrescent with spiny and indurate (floral) leaves in fruit. Bracts fused with sepal bases. Sepals 5, linear-lanceolate, spine-tipped. Petals absent. Stamens 5. Style solitary, 2–3-lobed at apex. Fruit 1-seeded, indehiscent, ovoid, contained within the indurate receptacle.

The genus has one species, distributed from S Iran to Arabia, N Africa and Cape Verde Islands.

Sclerocephalus arabicus *Boiss.* Diagn. Pl. Orient. Nov. Ser.1, 3: 12 (1843); Rech. f., Fl. Lowland Iraq: 225 (1964); Chaudhri in Meded. Bot. Mus. Herb. Rijks. Univ. Utrecht 285: 62 (1968); Chaudhri in Fl. Iranica 144: 22 (1980); Gilbert in Fl. Somalia 1: 96 (1993); Chamberlain in Fl. Arab. Penins. & Socotra 1: 187 (1996); Boulos, Fl. Egypt 1: 89 (1999); Gilbert in Fl. Ethiop. & Eritr. 2(1): 199 (2000); Ghazanfar, Fl. Oman 1: 67 (2003); Dinarvand in Fl. Iran 65: 35 (2009).

Paronychia sclerocephala Decne. in Ann. Sci. Nat. Bot. Ser. 2 (3): 262 (1835).

Annual herb. Stems 2–16 cm tall, much branched, prostrate, stout, stiff, glabrous. Leaves sessile, opposite, often in axillary clusters, 4–12 × 1 mm, linear-terete or subcylindrical, acute and strongly mucronate, glabrous. Stipules lanceolate, up to 3.5 mm, acute. Sepals 2.5–3 mm, slightly unequal, narrowly oblong with membranous margins, glabrous above, densely lanate below. Fruit 3.5–3 × 1.5–2 mm, hard with a membranous pellucid pericarp, slightly papillose at top. Fig. 5: 1–7.

Fig. 5. **Sclerocephalus arabicus**. 1, habit × 1 1/3; 2, habit showing lower part of plant × 1; 3, flower × 10; 4, sepal × 12; 5, sepal and stamen × 12; 6, inflorescence at fruiting stage × 6; 7, seed × 12. 1, from *Guest, Rawi & Rechinger* 16921; 2, 4–7 from *Guest et al.* 14374; 3 from *Gillett & Rawi* 6152. © Drawn by Juliet Beentje.

HAB. Dry sandy and stony soils; alt. 10–300 m; fl. (Jan.–)Feb.–May(–Jun.).
DISTRIB. Southern Desert area, local. **DSD**: 7 km WNW of Um Qasr, *Guest, Rawi* & *Rechinger* 16921!; *Khatib* & *Alizzi* 33444!; Jabal Sanam, *Chakravarty et al.* 29922! & *Kaisi, Hamad* & *Hamid* 48629! & *Haines* 942! & *Gillett* & *Rawi* 6152! & *Rawi* & *Schwan* 14374!; nr. Chilawa, *Guest, Rawi* & *Rechinger* 17221!

N Africa, Sinai, Arabian Peninsula (Oman, UAE, Qatar, Bahrain, Kuwait) and W Iran.

4. PTERANTHUS Forsskål

Fl. Aegypt.-Arab. 36 (1775)

Annual herbs. Leaves opposite or fasciculate, fleshly, papillose-scabrid. Stipules scarious. Inflorescences of 3-flowered cymes on flattened, leaf like oblanceolate or broadly obovate peduncles; outer flowers sterile; inner flower fertile. Some bracts indurate in fruit and becoming broadly spatulate with a long cusp and pink in colour. Sepals 4, hooded, inner 2 membranous, ovate-lanceolate; outer 2 linear-oblong, keeled. Petals absent. Stamens 4. Style 1, 2-lobed at the tip. Fruit 1-seeded, indehiscent

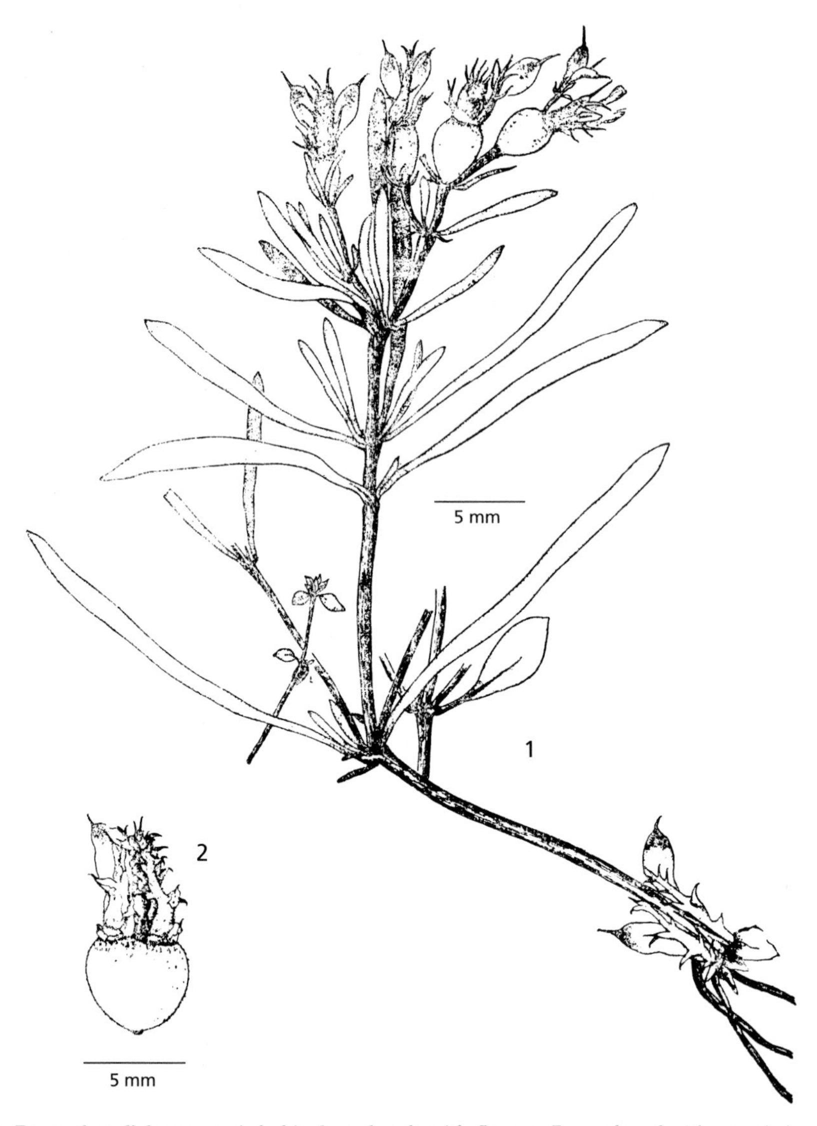

Fig. 6. **Pteranthus dichotomus**. 1, habit; 2, peduncle with flowers. Reproduced with permission from Flora of Pakistan 47: 1973. Drawn by Asad.

Pteranthus dichotomus *Forssk.*, Fl. Aegypt.-Arab. 36 (1775); Zohary, Fl. Iraq & its Phytogeogr. Subdiv.: 53 (1950); Blakelock in Kew Bull. 2: 202 (1957); Rech.f., Fl. Lowland Iraq: 226 (1964); Ghafoor in Fl. Pak. 47: 11 (1973); Rech.f., Fl. Iranica 144: 24 (1980); Chamberlain in Fl. Arab. Penins. & Socotra 1: 187 (1996); Boulos, Fl. Egypt 1: 90 (1999).

Comphrosma pteranthus L., Mantissa 1: 41 (1767).
Pteranthus echinatus Desf., Fl. Atlant. 1: 144 (1789), nom. illeg.
Pteranthus trigynus Caball., Bot. Soc. Esp. Hist. Nat. 13: 88 (1913).

Annual herbs. Stems 4–18 cm tall, procumbent to ascending, glabrous. Leaves linear 7–15 × 1–1.5 mm, fleshly, acute. Stipules lanceolate, 3–4 mm, sometimes irregularly toothed. Peduncles minutely papillate-pubescent. Bracts up to 9 mm, spirally arranged, minutely pubescent, with broad membranous wings with hooded spines. Sepals 4, 4–6 mm, slightly united at base, papillate-hairy, acuminate; the 2 outer, stiff, oblong-linear, dorsally keeled, broadly winged; the 2 inner, ovate-lanceolate, with membranous margins. Fruit enclosed by calyx and hooded spiny processes of bracts. Fig. 6: 1–2.

HAB. Sandy clay soils, stony hillsides alt. 50–630 m; fl. Mar.–Apr.
DISTRIB. Widespread in the alluvial plains and desert areas. **MJS**: Jabal Sinjar, 28 May 1979, *Sharief & Hamad* 50225 (BAG); **FUJ**: Qaiyara, *Bayliss* 111!; **FKI**: Baba Gurgur, *Guest, Eig* & *Zohary* 5107!; 31 km N of Kirkuk, *Rawi, Nuri* & *Kass* 27983!; **FPF**: 16 km S of Barda, *anon.* 18222!; Garaqhan, *Rogers* 100! & 130!; Jabal Hamrin, *Guest* 422! & 671!; 25 km E of Barda, *Rawi* 20762!; Tib Al-Sharhani, *Kaisi* & *Khayat* 50594!; **DLJ**: nr. Thirthar Lake, *Bharucha* & *Abbas* s.n!; Jabal al-Muwaila, *Guest, Rawi* & *Rechinger* 17555!; **DWD**: Rutba, *Omar* & *Hamid* 36725!; 70 km N of Rutba, *Chakravarty et al.* 32927!; 95 km NE of Rutba, *Rawi* & *Khatib* 32360!; 22 km W of Falluja, *Chakravarty* & *Rawi* 30252!; 20 km W of Nukhaib, *Rawi* 30991!; Bottom of Wadi Hauran, *Chakravarty et al.* 31693!; 6 km S of Shithatha, *Gillett* & *Rawi* 6459!; **DSD**: Jabal Sanam, *Alizzi* & *Khatib* 33469!; *Rawi* & *Schwan* 14372!; *Guest, Rawi & Rechinger* 16958! & 17039!; Ma'niya, *Alizzi & Omar* 35542!; Salman, *Rawi* 14904!; 13 km W by N of Shabicha, *Guest, Rawi & Rechinger* 19274 (BAG); 40 km from Aujah to Khidr al-Mai, *Al-Kaisi, Hamad & Hamid* 48478 (BAG). **LEA**: Faka, *Rawi* 25314!; **LCA**: Daltawa, *Rogers* 327!

N Africa, SE Spain, Sinai, Palestine, Arabia (Kuwait), Iran and Pakistan.

5. **POLYCARPON** Loefl. ex L.

Syst. Nat. ed. 10: 881 (1759).

Low annual or perennial herbs. Leaves obovate or oblong, opposite or whroled. Stipules scarious. Inflorescence compound cyme. Bracts scarious. Sepals 5, free, keeled. Petals 5. Stamens 3–5, inserted on a perigynous disk. Style solitary, obscurely 3-lobed. Capsule 3-valved, dehiscing to base.

Eighteen species; two in Iraq.

Stipules and bracts triangular-lanceolate, acuminate, all scarious;
sepals ovate, acuminate . . . . 1. *P. tetraphyllum*
Stipules and bracts ovate-oblong, obtuse, herbaceous with scarious
margins; sepals oblong, obtuse . . . . 2. *P. succulentum*

1. **Polycarpon tetraphyllum** (L.) L. Syst. Nat. ed. 10: 881 (1759); Blakelock, Kew Bull. 2: 202 (1957); Rech.f., Fl. Lowland Iraq:, 226 (1964); Cullen in Fl. Turk. 2: 96 (1967); Rech. f. in Fl. Iranica 144: 26 (1980); Chamberlain in Fl. Arab. Penins. & Socotra 1: 198 (1996); Hartvig in Fl. Hellenica 1: 232 (1997); Boulos, Fl. Egypt 1: 82 (1999); Gilbert in Fl. Ethiop. & Eritr. 2(1): 203 (2000); Ghazanfar, Fl. Oman 1: 71 (2003); Dinarvand in Fl. Iran 65: 43 (2009).

Mollugo tetraphylla L., Sp. Pl. ed. 1: 89 (1753).

Annual or perennial herbs. Stem prostrate to procumbent or ascending, 2–12 cm tall, usually divaricately branched, rarely unbranched, glabrous or papillose. Leaves opposite or in whorls of 4, oblong-spatulate to elliptic, 4–15 mm, glabrous to papillose, sometimes slightly succulent. Stipules triangular-lanceolate, acuminate, scarious. Cymes many-flowered, congested to lax. Bracts similar to stipules. Sepals ovate, mucronate, 1.5–2 mm, keeled with scarious margins. Petals oblanceolate, hyaline, shorter than sepals. Stamens usually 3. Styles shorter than ovary. Capsule ovoid, slightly shorter than sepals. Seeds tuberculate. Fig. 7: 1–7.

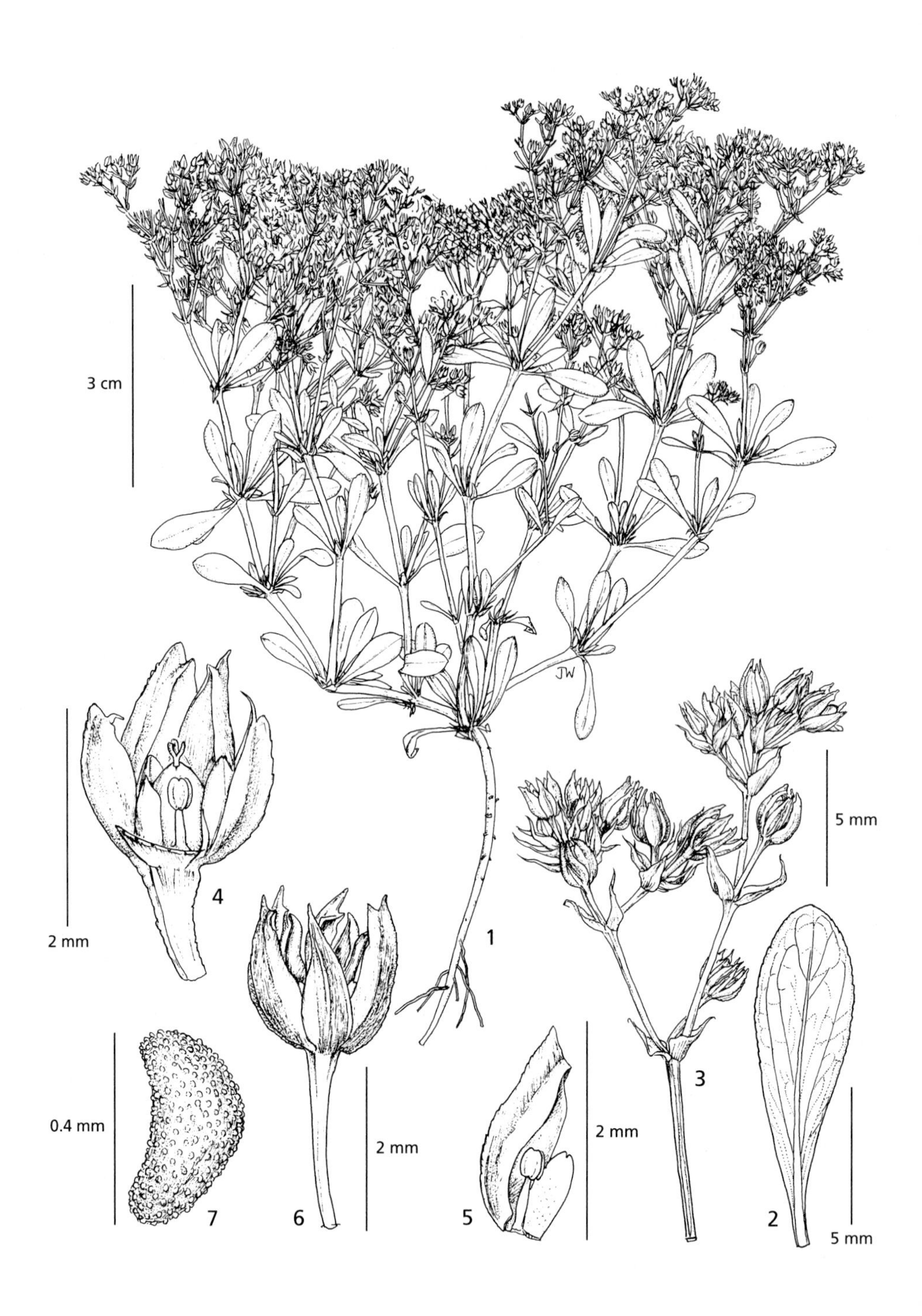

Fig. 7. **Polycarpon tetraphyllum**. 1, habit; 2, leaf upper surface; 3, part of inflorescence; 4, flower with one sepal removed; 5, sepal, petal and stamen × 16; 6, capsule with calyx × 12; 7, seed × 70. 1–3 from *Hamad* 52345; 4, 5 from *Haines* W136; 6, 7 from *Alizzi* 34228. Scales bars: 1 = 3 cm; 2, 3 = 5 mm; 4–6 = 2 mm; 7 = 0.4 mm. © Drawn by Juliet Beentje.

HAB. Sandy clay soils, river-banks, gardens, cliffs; alt. 0–800 m; fl. May.
DISTRIB. **LCA**: Baghdad, Waziriyah, Tigris river-bank, *Haines* 136!; near Baghdad, *Alizzi* 34228 (BAG). **DWD**: *Hamad* 52345!

W & C Europe, Georgia, Mediterranean, Sinai, Syria, Iraq, Arabian Peninsula, N Iran to India. Introduced in S Africa, America and W Australia.

2. **Polycarpon succulentum** (Del.) J. Gay in Rev. Bot. Bull. Mens. 2: 373 (1846); Zohary, Fl. Iraq & its Phytogeogr. Subdivis.: 52 (1946); Blakelock in Kew Bull. 2: 202 (1957); Rech.f., Fl. Lowland Iraq: 227 (1964); Chamberlain in Fl. Arab. Penins. & Socotra 1: 199 (1996); Boulos, Fl. Egypt 1: 83 (1999); Ghazanfar, Fl. Oman 1: 71 (2003).

Alsine succulenta Delile, Descr. Egypte, Hist. Nat. 211 (1814).
Polycarpon arabicum Boiss., Diagn. Pl. Orient. Ser. 1, 10: 13 (1849).

Annual herbs. Stems 2–10 cm high, prostrate, many times 2–3-forked from a diffuse base. Leaves oblong-spatulate to linear-oblong, 4–10 × 2–6 mm, those of stems 4-ranked. Cymes loose, few-flowered. Sepals oblong, obtuse, the interior longer. Petals oblong-ovate, as long as the calyx. Style longer than ovary. Seeds smooth.

HAB. Sandy ground; alt. 0–900 m.

No Iraqi specimens at K; description from Fl. Lowland Iraq (1964).

Egypt, Sinai, Lebanon, Palestine, S Arabia, Oman, Bahrain and Kuwait.

6. **POLYCARPAEA** Lam.

Journ. Hist. Nat. Par. 2:3, t. 25 (1792); Boiss., Fl. Orient. 1: 738 (1867).

Annual or perennial herbs (in Iraq), sometimes woody-based or shrublets. Stems prostrate to erect. Leaves opposite or clustered. Stipules scarious. Flowers sessile or subsessile in terminal or axillary cymes. Bracts scarious. Sepals 5, free, somewhat flattened, not keeled, acute, midrib broad, narrow (in Iraq) or absent, margins scarious. Petals broadly ovate to oblong, shorter than sepals. Stamens 5. Style solitary, 3–lobed at apex. Capsule dehiscing by 3 valves.

About 50 species usually distributed in tropical and subtropical regions. Two species grow in Iraq.

Plant hairy .... 1. *P. repens*
Plant glabrous .... 2. *P. robbairea*

1. **Polycarpaea repens** (Forsskal) Aschers. & Schweinf. in Oesterr. Bot. Z. 39: 126 (1889); Zohary, Fl. Iraq & its Phytogeogr. Subdivis.: 52 (1946); Rech.f., Fl. Lowland Iraq: 227 (1964); Chamberlain in Fl. Arab. Penins. & Socotra 1: 192 (1996); Ghazanfar, Fl. Oman 1: 69 (2003).

P. fragilis Del. Fl. Egypte. 209 (1814) nom. illegit.
Corrigiola repens Forsskal, Fl. Aegypt.-Arab. 207 (1775).

Perennial herbs somewhat woody at base. Stems 10–25 cm high, decumbent to erect, appressed crispy-white tomenteose. Leaves linear to lanceolate, 3–7 mm long, white tomentose, with revolute margins, mucronulate at apex. Flowers in dense terminal and axillary cymes. Stipules ovate or lanceolate up to 5 mm long, sometimes ciliate at margin. Bracts similar to stipules. Sepals ± 2 mm, ovate with broad scarious margin, usually hairy on green midrib, mucronulate. Petals ± 1 mm, ovate, acute, shorter than sepals. Styles as long as the ovary; stigma capitate. Capsule ovoid. Seeds 8–10.

HAB. Sandy soils, rocky areas ; alt. ± 200 m; fl. Mar.–Apr.
DISTRIB. **DSD**: 105 km S of Basra, *Guest, Rawi* & *Rechinger* 17247!; 6 km W of Safwan, *Guest, Rawi* & *Rechecinger* 16950B!; Jabal Sanam, *Haines* 943!; *Guest, Rawi* & *Rechinger* 16977!; *Watson* s.n!; Chilawa, *Guest, Rawi* & *Rechinger* 17218!

N Africa and Sudan eastwards to Pakistan.

Two specimens from Jabal Sanam (G.A.Watson, s.n) and Chilawa (Guest *et al.* 17218) have leaves with 10–14 mm long and sepals almost glabrous on midrib.

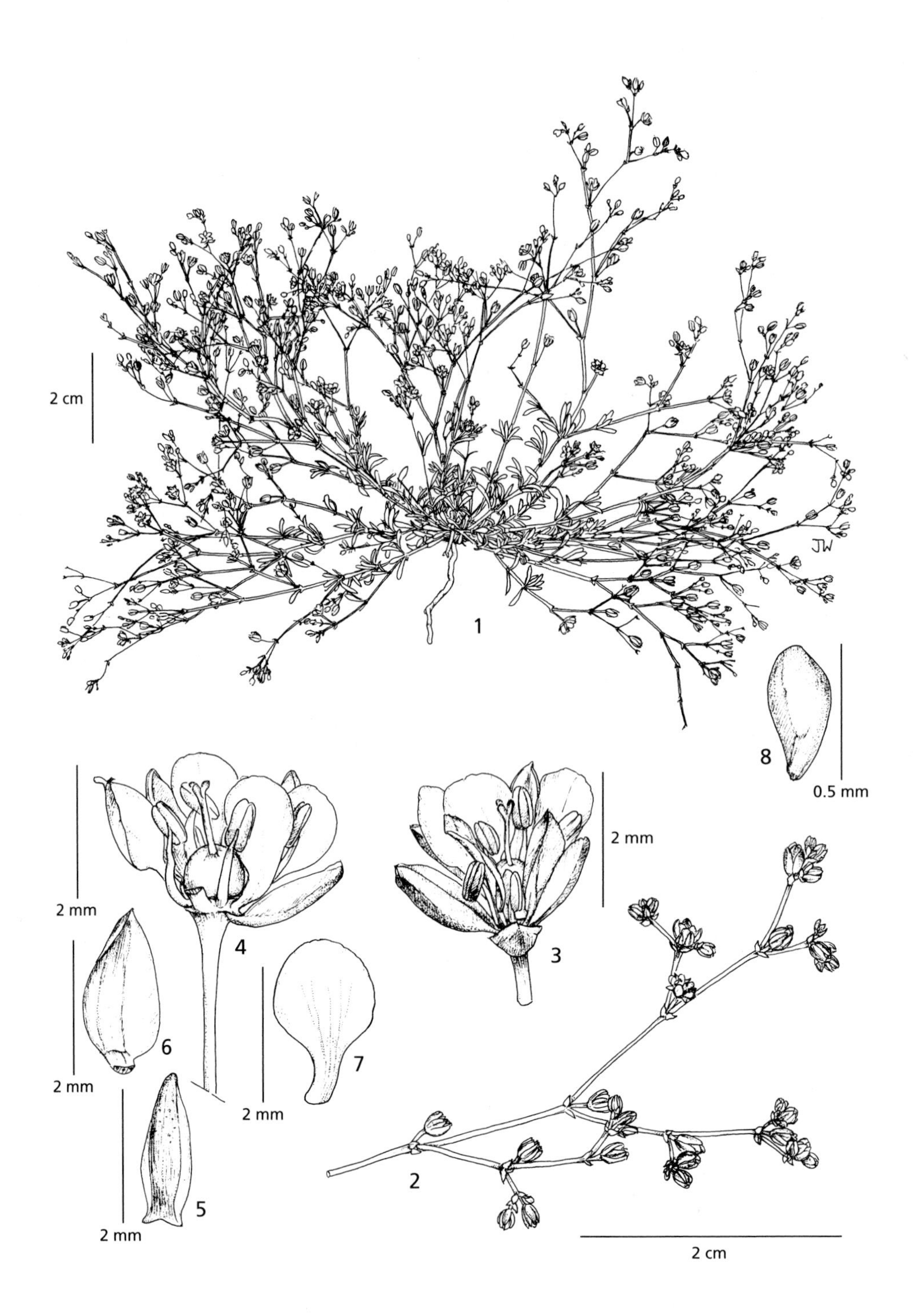

Fig. 8. **Polycarpaea robbairea**. 1, habit; 2, inflorescence detail; 3, flower; 4, flower with front sepals and petal removed; 5, outer sepal, abaxial view; 6, inner sepal, abaxial view; 7, petal; 8, seed. 1, 3–7 from *Alizzi & Omar* 35908; 2, 8 from *Guest, Rawi & Rechinger* 18868. © Drawn by Juliet Beentje.

2. **Polycarpaea robbairea** (Kuntze) Greuter & Burdet in Willdenowia 12: 189 (1982); Chamberlain in Fl. Arab. Penins. & Socotra 1: 197 (1996); Gilbert in Fl. Somalia 1: 100 (1993); Boulos, Fl. Egypt 1: 82 (1999); Gilbert in Fl. Ethiop. & Eritr. 2(1): 206 (2000); Ghazanfar, Fl. Oman 1: 70 (2003).

Polycarpon robbairea Kuntze, Revis. Gen. Pl. 1: 51 (1891).
Robbairea delileana Milne Redhead in Kew Bull. 3: 452 (1948); Rech.f., Fl. Lowland Iraq: 228 (1964).
R. confusa Maire, Mém. Inst. Fr. Afr. Noire 8: 34 (1950).
R. major (Asch. & Schweinf.) Botsch., Nov. Sist. Vyssh. Rast. 1: 370 (1964).

Annual or perennial glabrous herbs. Stems 5–25 cm high, numerous, prostrate to ascending, branched. Leaves 5–8 mm long, linear to narrowly obovate. Stipules 0.7–1.5 mm long, triangular. Flowers in lax cymes; pedicels to 2 mm long. Bracts ovate with green midrib and white scarious margins. Sepals ovate-elliptic, 1.5–2.5 mm long with narrow scarious margins. Petal white to pinkish, ovate to cordate, equalling to slightly exceeding the sepals. Capsule subglobose. Seeds smooth, glossy. Fig. 8: 1–8.

HAB. Sandy calcareous soils, cultivated fields; alt. 0–400 m; fl. Feb.–Apr.
DISTRIB. **DSD**: 40 SSE of Salman, *Guest, Rawi* & *Rechinger*18868!; nr. Salman, *Guest, Rawi* & *Rechinger* 18831!; *Alizzi* & *Omar* 35908!; 5 km NW of Sharar, *Guest, Rawi* & *Rechinger* 19225!; between Wagsa and Shibajcha, *Rechinger* 145!; Southern Desert, *Rawi* & *Gillett* 14140!; Bakur, *Guest, Rawi* & *Rechinger* 18892!; 125 km N of Neutral Zone, *Rawi, Khatib* & *Tikrity* 29204!

N Africa, Sudan, Palestine, S Arabia, Yemen, Oman, UAE, Qatar and Kuwait.

7. **LOEFLINGIA** L.

Sp. Pl. 35 (1753); Gen. Pl. ed. 5: 52 (1754).

Annual herbs. Leaves opposite, linear. Stipules scarious. Flowers small, sessile, in compound cymes, often spicate or aggregated into heads. Bracts leaf-like. Sepals free, the outer 3 larger than the inner 2, with 2 bristles arising from near the base on each side. Petals 3 or 5. Stamens 3 or 5. Style 1, obscurely 3-lobed. Capsule dehiscing by 3 valves. Seeds numerous.

Seven species distributed in Saharo-Arabian region, Mediterranean, N America in mainly dry and sandy habitats; one species in Iraq.

Loeflingia hispanica L., Sp. Pl. 35 (1753); Zohary, Fl. Iraq & its Phytogeogr. Subdivis.: 52 (1946); Rech.f., Fl. Lowland Iraq: 228 (1964); Cullen in Fl. Turk. 2: 96 (1967); Rech. f. in Fl. Iranica 144: 27 (1980); Chamberlain in Fl. Arab. Penins. & Socotra 1: 199 (1996); Dinarvand in Fl. Iran 65: 47 (2009).

Annual herbs. Stems to 12 cm tall, procumbent to ascending or erect, diffusely branched, glandular-puberulous. Leaves linear, 3–5 mm, mucronulate, sparsely hairy. Stipules ± 2 mm, connate with leaf base. Flowers sessile, minute, arranged in dense, leafy, dichotomously branched fascicles. Sepals ovate-lanceolate, 3–4.5 mm, sparsely hairy, mucronulate, with 2 bristles arising from near the base on each side. Petals much shorter than sepals. Style half as long as the ovary. Capsule ovoid, shorter than the sepals Fig. 9: 1–9.

HAB. Sandy stony slopes, sandy shores, fields, waste places; alt. 0–2000 m; fl. Mar.–May.
DISTRIB. Scattered, mainly in the desert and alluvial areas. **DSD**: W of Zubair, *Guest et al.* 5027!; nr. Zubair, *Haines* 1017; Al-Baniya, *Guest, Rawi* & *Rechinger* 17310!; Jabal Sanam, *Guest, Rawi* & *Rechinger* 16978! & 17046!; Samah, *Kaisi et al.* 48457!; between Karbela and Najaf, *Rawi* 20167 (BAG). **LEA**: Fakka, *Rawi* 1496!; **LCA**: Baghdad, *Lazar* 79!; Falluja, *Chakravarty* & *Rawi* 30274! & 30292!; **LBA**: Basra, *Haines* 922!

S Europe, N Africa, S Turkey, Syria, S Iran, Arabia, (Oman, Bahrain and Kuwait).

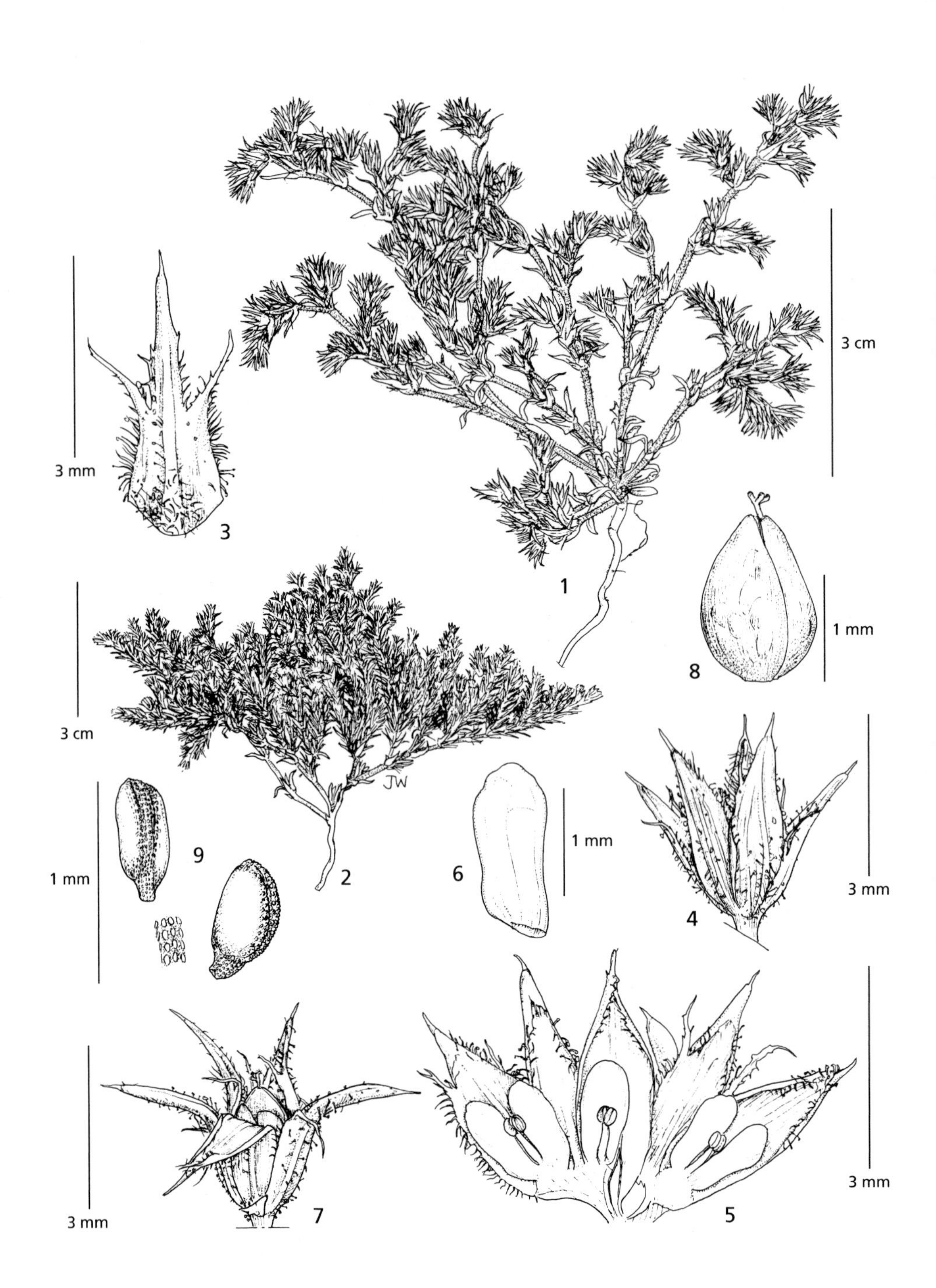

Fig. 9. **Loeflingia hispanica**. 1, habit; 2, habit (compact); 3, leaf with connate stipules; 4, flower side view; 5, flower opened with overy removed; 6, petal; 7, capsule with calyx; 8, capsule (young); 9, seeds. 1, 3–6, 9, 8 from *Guest, Rawi & Rechinger* 7310; 2, 7 from *Guest, Ali & Rawi* 14226. Scales bars: 1, 2 = 3 cm; 3–5, 7 = 3 mm; 6, 8–9 = 1 mm. © Drawn by Juliet Beentje.

8. **SPERGULA** L.

Sp. Pl. 440 (1753); Gen. Pl. ed. 5: 199 (1754)

Annual (rarely perennial) herbs. Stems ascending, often decumbent and much branched at base. Leaves linear, obtuse, decussate; stipules scarious, not united to surround the node; leaf fascicles (short leafy lateral branches) borne on both sides at each node. Perianth 5-merous. Sepals free, green, with scarious margins. Petals white, entire. Stamens 5–10. Styles 3–5. Capsule ovoid to subglobose, dehiscing by 3 or 5 valves. Seeds often winged.

Five species in Europe, Mediterranean area and temperate Asia, one species in S America. Two species in Iraq.

Leaves canaliculate beneath; capsule dehiscing by 5 valves . . . . 1. *S. arvensis*
Leaves not canaliculate beneath; capsule dehiscing by 3 valves. . . . . 2. *S. fallax*

1. **Spergula arvensis** *L.*, Sp. Pl. 440 (1753); Boiss., Fl. Orient. 1: 731 (1867); Rech. f., Fl. Lowland Iraq: 229 (1964); Ratter in Fl. Turk. 2: 92 (1967); Ratter in Fl. Iranica 144: 28 (1980); Ghazanfar & Nasır in Fl. Pak. 175: 47 (1986); S Snogerup & B. Snogerup, Fl. Hellenica 1: 234 (1997); Gilbert in Fl. Ethiop. & Eritr. 2(1): 207 (2000); Lu Dequam & Gilbert in Fl. China 6: 4 (2001).

S. vulgaris Boenn., Prodr. Fl. Monast. 135 (1824).

Annual herbs. Stems 5–70 cm, with ascending stems ± branched at the base, moderately to densely glandular-hairy above. Leaves 1–3(–8) cm, linear, canaliculated beneath. Sepals 3–5 mm, ovate, obtuse, glandular pilose. Petals white, obovate, obtuse, slightly exceeding the sepals. Stamens 5–10. Styles 4. Capsule 4–5 mm, somewhat exceeding the sepals, dehiscing by 5 valves. Seeds 1–2 mm, subglobose, keeled or with a very narrow wing, grey-black, papillose or not.

HAB. Sandy soil, weed of cultivation; alt. up to 200 m; fl. Mar.–Apr.
DISTRIB. Rare in lower Mesopotamian region or most likely under-collected. **LCA**: Nasiriya, *Anthony* s.n; **LEA**: Basra, *Whitehead* s.n.

Cosmopolitan weed. No specimens at K.

2. **Spergula fallax** (*Lowe*) *E.H.L. Krause* in Sturm, Fl. Deutschl. Flora ed. 3, 5: 19 (1901); Rech. f., Fl. Lowland Iraq: 229 (1964); Ratter in Fl. Iranica 144: 28 (1980); Ghazanfar & Nasır in Fl. Pak. 175: 46 (1986); Chamberlain in Fl. Arab. Peninsula & Socotra 1: 199 (1996); Gilbert in Fl. Somalia 1: 102 (1993); Boulos, Fl. Egypt 1: 80 (1999); Ghazanfar, Fl. Oman 1: 72 (2003); Dinarvand in Fl. Iran 65: 49 (2009).

S. pentandra L. var. *intermedia* Boiss., Pl. Orient. Nov. Diagn. ser. 2, 1: 93 (1853).
Spergularia fallax Lowe in Hook. Kew Journ. Bot. 8: 289 (1856).
S. flaccida (Roxb.) Aschers. in Verh. Bot. Ver. Brand. 30: 43 (1889), nom. illegit.

Annual herbs. Stems 5–40 cm, with ascending stems ± branched at the base, glabrous throughout. Leaves 1–3.5 cm, linear, fleshy, not canaliculated beneath. Sepals 3–5 mm, oblong-lanceolate. Petals white, equalling or somewhat shorter than the sepals. Stamens 10, rarely less. Styles 3. Capsule 4–5 mm, somewhat exceeding the sepals, dehiscing by 3 valves. Seeds 0.8–1 mm (excluding wing), compressed, grey-black, encircled by a broad scarious wing. Fig. 10: 1–3.

HAB. Sandy soil in wadis, alluvial flats, roadsides, fields etc. alt. 5–550 m; fl. & fr. Feb.–May.
DISTRIB. Widespread in the deserts and lowlands of Iraq: **LCA**: Abu Ghraib, *Gillett* 6540!; *Alizzi, Thaib & Janan* 32587!; **LEA**: Kut to Aziziya, *Gillett* 6746!; **LSM**: nr. Azair, *Guest, Rawi & Rechinger* 17388!; **DSD**: 8 km E of As Salman, *Gillett & Rawi* 6219 !; Shabicha, *Gillett & Rawi* 6272 !; Jal al-Rawaq, *Guest, Rawi & Rechinger* 18908!; Zubaida's Wells on road to Negev (Najaf), *Rawi, Agnew & Haines* W1652 ! 160–190 km NW of Ramadi to Rutba, *Rawi* 31344!

Madeira, Canary Is., N Africa, SW Asia, NW and C India.

Fig. 10. **Spergula fallax**. 1, habit × 1; 2, capsule × 6; 3, seed × 20. **Sagina apetala**. 4, habit × 1; 5, capsule × 15. Reproduced with permission from Flora of Pakistan 175: f. 6, 1986. Drawn by S. Hameed.

9. **SPERGULARIA** (Pers.) J. & C Presl

Fl. Čechia 94 (1819)

Herbs, sometimes woody at the base. Stems erect, decumbent or procumbent, dilated at the nodes, somewhat flattened. Leaves linear, decussate, often fleshy, with pale scarious stipules united to surround the node, forming ± triangular structures on either side of the stem; leaf-fascicles (short leafy lateral branches) when present borne on one side only at each node. Perianth 5-merous. Sepals free, green, with scarious margins. Petals entire. Stamens 1–10. Styles 3, free. Capsule dehiscing by 3 valves. Seeds often winged.

About 25 species cosmopolitan, especially some halophytes nearly worldwide distribution. Four species occur in Iraq.

1. Capsule 7–9 mm; stamens 10; sepals usually more than 4 mm. . . . . 1. *S media*
   Capsule less than 7 mm; stamens often fewer than 10; sepals usually less than 4 mm 2
2. Capsule usually more than 4 mm, exceeding the sepals; seeds light brown, 0.6–0.7 mm, unwinged or both winged and unwinged; stipules on young shoots connate for about ½ their length, forming a sheath . . . . . . 2. *S marina*
   Capsule usually less than 4 mm, ± equalling the sepals; seeds blackish or if light brown then less than 0.5 mm, unwinged; stipules on young shoots connate for considerably less than ½ their length. 3
3. Inflorescence few-flowered; capsule globose, valves often blackish at maturity; seeds black or brownish. . . . . .3. *S diandra*
   Inflorescence usually many-flowered; capsule ovoid; valves of capsule pale at maturity; seeds light grey-brown . . . . 4. *S Bacconii*

1. **Spergularia media** (*L.*) *C Presl*, Fl. Sic. 161 (1826); Fl. Pal. ed. 2, 1: 207 (1932); Ratter in Fl. Turk. 2: 93 (1967); Ratter in Fl. Iranica 144: 28 (1980); Ghazanfar & Nasır in Fl. Pak. 175: 48 (1986); Lu Dequam & Rabeler in Fl. China 6: 5 (2001).

Arenaria media L., Sp. Pl. ed. 2: 606 (1762).
S marginata (DC.) Kittel, Taschenb. ed. 2, 1003 (1844); Boiss., Fl. Orient. 1: 733 (1867).

Perennial herb with stout ± woody stock. Stems 5–40 cm, glabrous or glandular-hairy in the inflorescence. Leaves fleshy, mucronate, slightly fasciculate; stipules broadly triangular, not acuminate. Sepals 4–6 mm. Petals white or pink, equalling or somewhat exceeding the sepals. Stamens 10. Capsule 7–9 mm, much exceeding the calyx. Seeds 0.7–1.0 mm (excluding wing), dark brown, smooth or tuberculate, usually winged; margin of wing entire or only slightly divided. Fig. 11: 1–4.

HAB. Sea shores, salt-marshes and inland saline areas; alt. 0–1700 m; fl. Mar.–May.
DIST. **FKI**: Minjan, *Sutherland* 199!

This species not common in Iraq. Widespread in suitable habitats in both hemispheres.

2. **Spergularia marina** (*L.*) *Griseb.*, Spic. Fl. Rumel. 1: 213 (1843); Zohary, Fl. Pal. ed. 2, 1: 206 (1932); Ratter in Fl. Turk. 2: 94 (1967) & in Fl. Iranica 144: 30 (1980); Ghazanfar & Nasır in Fl. Pak. 175: 50 (1986); Leonard, Contrib. Fl. et Veg. Deserts d'Iran 6: 20 (1986); Chamberlain in Fl. Arab. Penins. & Socotra 1: 201 (1996); Gilbert in Fl. Ethiop. & Eritr. 2(1): 208 (2000); Lu Dequam & Rabeler in Fl. China 6: 5 (2001); Ghazanfar, Fl. Oman 1: 73 (2003).

Arenaria rubra L. var. *marina* L., Sp. Pl. 423 (1753).
Spergularia salina J. & C.Presl, Fl. Čechia 95 (1819); Zohary, Fl. Iraq and its Phytogeogr. Subdiv. 52 (1950), & Fl. Lowland Iraq: 230 (1964) & Fl. Palaest. 1: 124 (1966).
S. media Wahl., Fl. Goth. 45 (1820–1824); Boiss., Fl. Orient. 1: 733 (1867).

Annual, biennial or rarely perennial herb, with slender or slightly fleshy stock. Stems 3–35 cm, prostrate, procumbent or suberect. Leaves fleshy, mucronate or somewhat armed, not or only slightly fasciculate, stipules short, obtuse, forming a sheath. Sepals 2.5–4 mm. Petals pink above and white near the base, rarely entirely white, not exceeding the sepals. Stamens 1–5(–8). Capsule (3–)4–6 mm, usually considerably exceeding the sepals. Seeds 0.6–0.7 mm, light brown, smooth or densely tuberculate, unwinged or mixed winged and unwinged; wing of seed when present erose to laciniate. Fig. 11: 5–7.

HAB. Fields, sandy soil, moist saline places; alt. s.l.–130(–200) m; fl. & fr. (Dec.–)Feb.–Apr.(–May).
DISTRIB. Widespread in desert and lowland steppe areas of Iraq: **FKI**: Kirkuk, *Rogers* s.n.!; Tuz, *Guest* s.n.!; **FPF**: 10 km E of Mandali, *Rawi* 20688!; **DWD**: Wadi Fuhaimi SE of Ana, *Gillett & Rawi* s.n.!; Shithatha, *Barkley & Hikmat Abbas* 188!; **DSD**: Najaf, *Rawi* 19973!; **DLJ**: 70 m NW of Rawa, *Khatib & Alizzi* 31946!; **LEA**: nr. Basra, island in Shatt al Arab, *Guest, Rawi & Rechinger* 16709!; nr. Shahraban (Muqdadiya), *Rawi* 19996!; Basra, *Bornm.* 112!; **LCA**: nr. Baghdad, *Lazar* 563!; 26 km NE of Nasiriya, *Rawi* 26044!; Abu Ghraib, H. Ahmad s.n.!; Baghdad, Karrada Mariam, *H. Ahmad* s.n.!; Karrada, *Graham* s.n.!; **LSM**: Mesaida nr. Amara, Field & Lazar s.n.!; 30 km E of Amara, *Bharucha, Rawi & Tikriti* 20306!; nr. Azair, *Guest, Rawi & Rechinger* 17390!; 20 km N of Amara, *Rawi* 15012!

Widespread in N Hemisphere; also in S Hemisphere, but there often introduced.

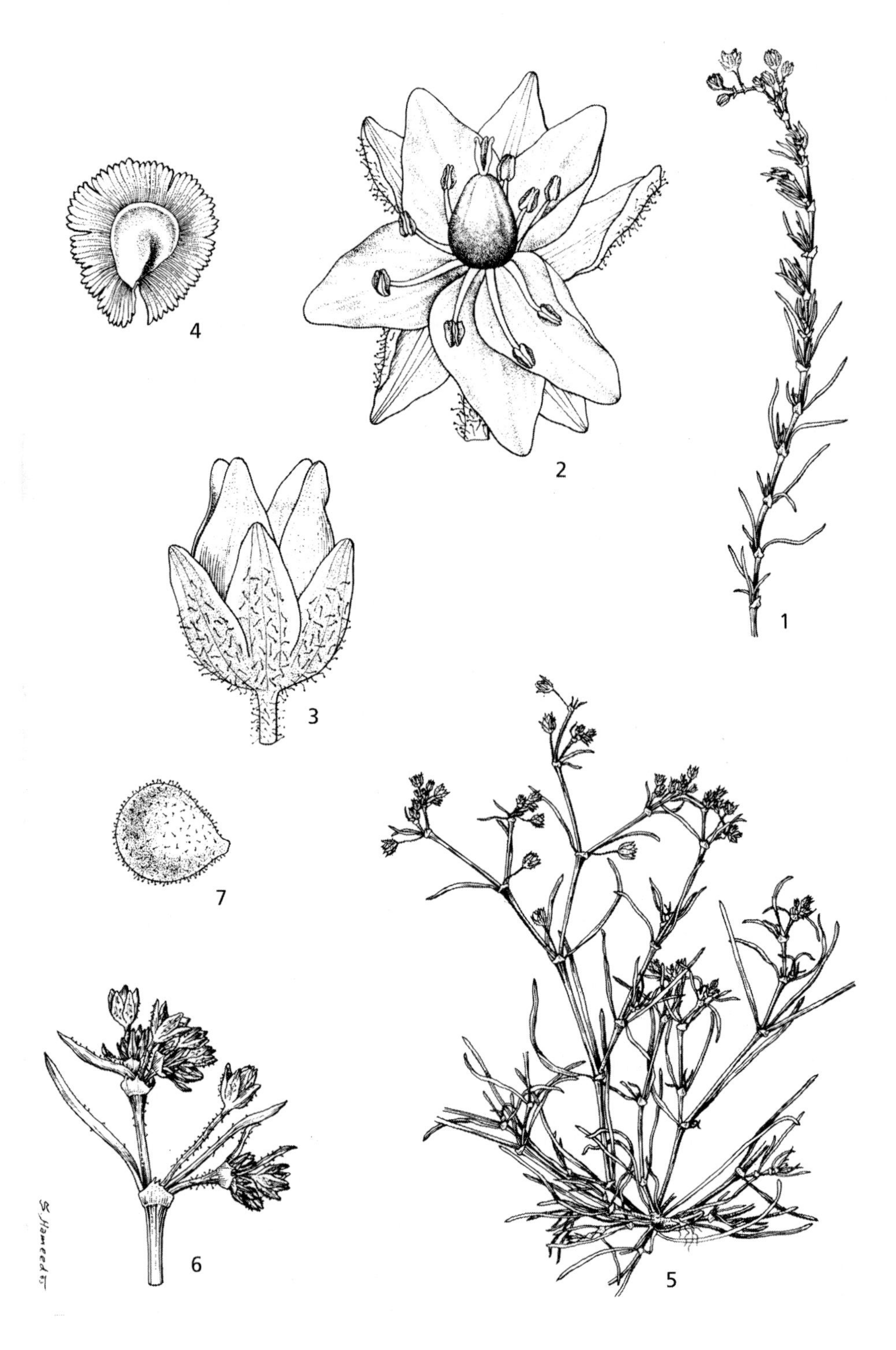

Fig. 11. **Spergularia media**. 1, habit × ½; 2, flower × 7; 3, capsule × 5; 4, seed × 25. **Spergularia marina**. 5, habit × ½; 6, part of shoot showing stipules; 7, seed × 30. Reproduced with permission from Flora of Pakistan 175: f. 7, 1986. Drawn by S. Hameed.

3. **Spergularia diandra** (*Guss.*) *Boiss.*, Fl. Orient. 1: 733 (1867); Ratter in Fl. Turk. 2: 95 (1967) & in Fl. Iranica 144: 31 (1980); Ghazanfar & Nasır in Fl. Pak. 175: 51 (1986); Leonard, Contrib. Fl. et Veg. Deserts d'Iran 6: 20 (1986); Chamberlain in Fl. Arab. Penins. & Socotra 1: 202 (1996); S & B Snogerup, Fl. Hellenica 1: 236 (1997); Lu Dequam & Rabeler in Fl. China 6: 5 (2001); Ghazanfar, Fl. Oman 1: 73 (2003); Dinarvand in Fl. Iran 65: 56 (2009).

Arenaria diandra Guss., Fl. Sic. Prodr. 1: 515 (1827).
Spergularia diandra (Guss.) Heldr. et Sint. in sched. ad Herb. Graec. no. 492; Zohary, Fl. Iraq and its Phytogeogr. Subdiv. 52 (1950); Rech. f., Fl. Lowland Iraq: 230 (1964).

Annual to biennial herb with slender taproot. Stems 3–30 cm, slender, ascending, sparsely to moderately glandular above. Stipules short, triangular (rarely lanceolate). Inflorescence much branched, very slender, without upper bracts. Sepals 2–3 mm, oblong, obtuse. Petals lilac (rarely white), narrowly elliptical, equalling the sepals. Stamens 2–3. Capsule 1.5–3 mm, subglobose, about equalling the sepals; valves often dark at maturity. Seeds 0.6–0.7 mm, unwinged, dark brown to black, rugulose or bristling with rigid papillae.

Hab. Sandy or silty soils; alt. 10–560 m; fl. Feb.–Jun.
Distrib. Widespread in the desert and lowland areas of Iraq: **FKI**: Kirkuk, *Rogers* s.n.!; **DWD**: nr. Ana, *Barkley, Agnew & Palmatier* s.n.!; Wadi Ghadaf, *Rawi* 14841!; 18 km S of Rutba, *Rawi* 14606 !; 22 km W of Ramadi, *F. & E Barkley & Faud Safwat* 5024!; 6 km S of Shithatha, *Gillett & Rawi* 6451!; **DSD**: 20 km SE of Salman, *Rawi* 14894!; Shabicha, *Gillett & Rawi* 6314!; **LEA**: Kut to Aziziya, *Gillett* 6745!;1 km N of Kut, *F. & E Barkley* s.n.!; **LEA/LSM**: between Fakka and Usaieda, *Rawi* 14988!; **LCA**: Za'faraniya, *Gillett* 10189!; Baghdad, *Wheeler-Haines* s.n.!; 3 km S of Baghdad, *O. Polunin* 5001!

S Europe, N Africa, SW and C Asia.

4. **Spergularia bocconii** (Scheele) Aschers. & Graebn., Syn. Mitteleur. Fl. 5 (1): 849 (1919); Ratter in Fl. Turk. 2: 95 (1967) & in Fl. Iranica 144: 33 (1980); Chamberlain in Fl. Arab. Penins. & Socotra 1: 202 (1996); S & B Snogerup, Fl. Hellenica 1: 237 (1997); Boulos, Fl. Egypt 1: 80 (1999); Ghazanfar, Fl. Oman 1: 72 (2003); Dinarvand in Fl. Iran 65: 59 (2009).

Alsine bocconii Scheele in Flora 26: 431 (1843).
Spergularia bocconii (Sol. ex Scheele) Fouc. ex Merino in Mem. Soc. Esp. Hist. Nat. 2: 496 (1904); Täckh., Stud. Fl. Egypt 398 (1956).

Annual or biennial herb with slender tap-root. Stems 5–25 cm, slender, densely glandular-hairy in the inflorescence. Leaves mucronate, not or slightly fleshy, not fasciculate; stipules triangular, not acuminate. Upper bracts much reduced. Sepals 2–3.5 mm. Petals pink with white base, or entirely white, somewhat shorter than or equalling the sepals. Stamens (0–)2–5(–8). Capsule 2–3.5 mm, equalling or shorter than the sepals. Seeds 0.35–0.45 mm, unwinged, light grey-brown, finely tuberculate.

Hab. Sandy soils, waste places, moist depressions. alt. 2–200 m; fl. (Dec.–) Feb.–Jun.
Distrib. Occasional on the Mesopotamian plain of Iraq.
DSD: Shabicha, *Rawi* 14831!; 30 km W by N of Busaiya, *Guest, Rawi & Long* 14168!; **LCA**: Baghdad, *Paranjpye* 132!; Nasiriya, *Abd-ar-Riza la Barbuti* 2535!; **LBA**: 40 km SE of Basra to Fao, *Rawi* 25896!

Mediterranean area, S Europe, SW and C Asia; introduced elsewhere.

10. **TELEPHIUM** L.

Sp. Pl.: 271(1753) & Gen. Pl. ed. 5, 377 (1754)

Perennial, glabrous herbs with woody stock. Stems procumbent to ascending. Leaves alternate. Stipules scarious. Inflorescences terminal, capitate-corymbose cymes. Sepals 5, free, green with white margins. Petals 5, entire. Stamens 5, hypogynous; staminodes 5, petaloid. Ovary incompletely 3–4-locular. Styles 3, recurved. Fruit a capsule, pyramidal, beaked, dehiscing by 3(–4) valves. Seeds numerous, globular-reniform.

Five species distributed in S Europe and the Mediterranean region, Sinai, Arabia to Afghanistan and Madagascar. It is closely related to Molluginaceae and there is some argument as to its inclusion in the Caryophyllaceae.

M.G. Gilbert (1987) The taxonomic position of the genera *Telephium* and *Corrigiola*, Taxon 36 (1): 47–49.

Fig. 12. **Telephium oligospermum**. 1, habit; 2, detail of stipules; 3, 4, inflorescence with mature fruit; 5, capsule; 6, capsule with calyx removed; 7, seed. 1, 3–7 from *Rawi, Tikriti & Nuri* 29023; 2 from *Rawi & Serhang* 18310. Scale bar: 1 = 3 cm; 2, 5–6 = 3.3 cm; 3 = 1 cm; 4 = 5 mm; 7 = 1.6 mm. © Drawn by Lucy T. Smith.

Stems 10–30 cm; leaves oblong, elliptic-obovate or oblong-lanceolate 2–4(–6) × longer than broad; seeds 11–18. . . . . 1. *T. imperati* subsp. *orientale*
Stems (20–)30–50 cm: leaves elliptic-oblanceolate, 7–15 × longer than broad; seeds 6–10 . . . . 2. *T. oligosperma*

1. **Telephium imperati** L. subsp. **orientale** (Boiss.) Nyman, Consp. 254 (1879); Demiriz in Fl. Turk. 2: 97 (1967); Rech.f., Fl. Iranica 144: 34 (1980); Ghazanfar & Nasir in Fl. Pak. 175: 54 (1986); Hartvig in Fl. Hellenica 1: 238 (1997); Boulos, Fl. Egypt 1: 43 (1999); Dinarvand in Fl. Iran 65: 63 (2009).

Telephium orientale Boiss., Diagn. Pl. Orient. Nov. Ser. 1, 10: 11 (1849).
T. imperati L. var. *orientale* (Boiss.) Boiss., Fl. Orient. 1: 754 (1867); Blakelock in Kew Bull. 2: 217 (1957).

Stems 10–30(–35) cm high, procumbent to ascending. Leaves (5–)7–20 × 2–6 mm, leaves oblong, elliptic-obovate or oblong-lanceolate 2–4(–6) × longer than broad, somewhat acute, midrib prominent beneath. Flowers in subcapitate cymes with short pedicels. Sepals 4–5 mm, oblong-linear, obtuse, carinate with scarious margins. Petals as long as the calyx. Styles brownish. Capsule 5–6 × 3 mm, brownish green, shiny, exceeding the persistent sepals. Seeds 11–15(–18).

HAB. Dry stony slopes, rocky screes, *Quercus* scrub; alt. 500–2200 m; fl. May–Jul.
DISTRIB. Rare. **MJS**: Jabal Sinjar, *Handel-Mazzetti* 1477; ibid, *Omar, Kaisi* & *Khayat* 52567!

Greece, Turkey, Cyprus, Egypt, Syria, NW Iran, Caucasia, Pakistan.

2. **Telephium oligospermum** Steud. ex Boiss., Boiss., Fl. Orient. 1: 754 (1867). — Type: Iraq in lapidosis montis Gara Dagh Kurdistaniae, *Kotschy* 320 (syn. G, BM!, K!). Zohary in Dep. Agr. Iraq Bull. 31: 52 (1950); Blakelock in Kew Bull. 2: 217 (1957); Demiriz in Fl. Turk. 2: 98 (1967); Rech.f., Fl. Iranica 144: 35 (1980); Dinarvand in Fl. Iran 65: 64 (2009).

Similar to *T. imperati* L. subsp. *orientale* (Boiss.) Nyman, but with stems (20–)30–50 cm long. Leaves 16–40 × 1.8–3.5 mm, elliptic-oblanceolate, 7–15 × longer than broad. Sepals broadly linear. Styles brownish. Capsule 6–7 × 2 mm, exceeding the persistent calyx. Seeds 6–10. Fig. 12: 1–7.

HAB. Rocky and stony slopes, near streams; alt. 1000–2000 m; fl. Jun–Jul.
DISTRIB. **MAM**: Khantur Mt. Rawi, *Tikriti & Nuri* 29023!; Sarsang, *Haines* W1319!; Bamerny, 20 km NW of Sarsang, *Kaisi & Hamad* 45945! **MRO**: NE of Rania, Qandil Range, *Rawi & Serhang* 18310!; Kanirush, N of Pishtashan, *Serhang & Rawi* 26654!; Pishtashan, *Serhang & Rawi* 24244! & *Rechinger* 11215!

E Turkey, Armenia, NW Iran.

II. SUBFAMILY ALSINOIDEAE Fenzl.

11. SCLERANTHUS L.

Sp. Pl. 406 (1753); Gen. Pl. ed. 5: 190 (1754)

Small annuals and perennial herbs. Stems procumbent to erect, pubescent on two opposite sides. Leaves opposite, linear, sessile, expanded and connate at base into scarious sheath; stipules absent. Flowers in lax or dense, terminal or axillary cymes. Bracts not markedly different from the leaves. Perigynous zone inconspicuous in flower but increasing in size and thickness with development of fruit, and about equalling the sepals in length at maturity. Sepals 5, equal or slightly unequal. Petals absent. Fertile stamens 10 or fewer. Style 2, free with filiform styles at apex. Fruit indehiscent, surrounded by the much thickened, hard wall of the expanded perigynous zone, the flower falling as a whole at maturity.

Ten species distributed in temperate Eurasia, Mediterranean region, Ethiopia and Australasia; one species in Iraq.

Scleranthus uncinatus *Schur* in Verb. Siebenb. Ver. Naturw. 1: 107 (1850); Schischk. in Kom., Fl. URSS 6: 549 (1936); Brumitt in Fl. Turk. 2: 256 (1967).
S. annus L. var. *uncinatus* Boutigny in Bull. Soc. Bot. France II: 768 (1855).

Fig. 13. **Scleranthus uncinatus**. 1, habit; 2, detail of stipules; 3, flower; 4, developing fruit. 1 from *Jacobs* 6500; 2–4 from *Davis & Hedge* 29805. Scale bar: 1 = 1.5 cm; 2 = 2.2 cm; 3, 4 = 1.3 mm. © Drawn by Lucy T. Smith.

Annual or biennial. Stems several, branched from base, erect or ascending up to 10 cm, usually with pubescence confined to one side. Leaves linear, acuminate, up to 10 mm, with ciliate margins below. Inflorescence lax to fairly dense, mostly terminal; flowers pubescent or puberulous on perigynous cup. Calyx 3–4 mm; sepals lanceolate, tapering from base to acute, uncinately curved apex, equalling or slightly exceeding perigynous cup and divergent in fruit; scarious margin much narrower than the green central part. Fig. 13: 1–4.

HAB. Cultivated hillsides, open stony places; alt. 1000– 3000 m; fl. May–Jul.
DISTRIB. Rare. **MSU**: Penjwin, *Rawi* 22558!; Hauraman mountain, *Chakravarty, Rawi, Nuri* & *Alizzi* 19837 (BAG).

S Europ, Caucasus, W and E Mediterranean, Balkans, Turkey (E Anatolia).

12. **HABROSIA** Fenzl

Bot. Zeit. 1: 322 (1843)

Annual glabrous herbs. Stems filiform, branching dichotomously. Leaves setaceous, connate at base. Flowers somewhat perigynous, small. Sepals 5, with apical awn. Petals 5, white, minute. Stamens 5, filaments very short. Styles 2, free. Ovary with two ovules. Nutlet 1–seeded; seeds globose.

Monotypic; distributed in Syria, SE Turkey, Iraq & W Iran.

Habrosia spinuliflora (*Ser.*) Fenzl, Bot. Zeit. 1: 322 (1843); Rech.f., Fl. Lowland Iraq: 235 (1964); Rech.f., Fl. Iranica 163: 6 (1988).

Arenaria spinuliflora Ser. in DC, Prodr. 1: 406 (1824).

Annual. Stems filiform, 5–10 branching dichotomously, glabrous or glandular at nodes. Leaves setaceous, 3–10 mm, connate at base, glabrous to glandular at base. Flowers 1.5–2 mm, perigynous, small. Bracts lanceolate, 1–2 mm with ± glandular scarious margins. Pedicels up to 4 mm. Sepals 5, oblong, carinate, 1-veined with broad scarious margins, long awned at apex. Petals 5, white, distinctly shorter than sepals. Stamens 5, filaments very short. Styles 2, free. Ovary with two ovules. Nutlet 1-seeded; seeds globose.

HAB. Limestone scree, disturbed calcareous stepe, open *Quercus* forest; alt. 1000–1200 m; fl. Apr.–May.
DISTRIB. **MAM**: 8 km S of Zakho, *Rechinger* 12166; Bikhair Mountain, nr. Zakho, *Rawi* 23144!; Aqra, *Rawi* 11305!; **MRO**: Salah ad-Din, *Barkley* & *Haddad* 5722!; Safin, *Bornmüller* 949!; 7 km NW of Rania, *Rawi, Nuri* & *Kass* 28533!; **MSU**: Jarmo, *Helbaek* 357! & 702!; Khurmal, *Rawi* 8857!; **MJS**: Jabal Sinjar, *Kaisi* & *Wedad* 52224!; *Gillett* 1109!; **FNI**: Jabel Makbul, *Haines* W1425!

Syria, SE Turkey, W Iran.

13. **ARENARIA** L.

Sp. Pl. 423 (1753); Gen. Pl. ed. 5: 193 (1754)

Annual or perennial herbs, often caespitose or mat-forming. Leaves exstipulate, suborbicular to linear or setaceous. Inflorescence cymes or cymose panicles or flowers in clusters. Sepals 5, free, herbaceous, scarious or coriacous, 1–3-veined. Petals 5, entire to retuse. Stamens 10. Styles 3. Capsule opening by 6 teeth. Seeds numerous, not arillate, reniform to subreniform, smooth, tuberculate or papillose.

About 150 species mainly in the northern temperate regions, also in the arctic; few on the mountains of S America and NE Africa; five species in Iraq.

Ref.: F.N. Williams (1898) A Revision of the genus *Arenaria* L., Journal of the Linnean Soc. (Bot.) 33: 326–437.

J. McNeill (1962) Taxonomic Studies in the Alsinoideae I. Generic and infra-generic groups, Notes RBG Edinb. 24(2): 79–155.

1. Tufted or mat-forming perennials 2
   Annuals. 3
2. Stems erect, leaves linear-setaceous, longer than 30 mm . . . .
   . . . . 1. *A. gypsophiloides* var. *gypsophiloides*

Stems spreading, leaves elliptic, up to 6 mm long. . . . . 2. *A.balansae*
3. Pedicels deflexed in fruit; sepals about as long as petals . . . . 3. *A. kurdica*
Pedicels not deflexed in fruit; sepals clearly longer than petals 4
4. Inflorescence usually dichasial; capsule conical, gradually tapering towards the apex . . . . 5. *A. leptoclados*
Inflorescence usually an irregular monochasium; capsule ovoid, abruptly narrowing towards apex . . . . 4. *A. serpyllifolia*

1. **Arenaria gypsophiloides** *L.* Mant. 1: 71 (1767); McNeill, Notes R.B.G. Edinb. 24, 3: 295 (1963); McNeill in Fl. Turk. 2: 32 (1967); Rech. f. in Fl. Iranica, 163: 18 (1988).

var. **gypsophiloides**

Alsinanthus gypsophiloides (L.) Desvaux in J. Bot. Desv. 3 (5): 221 (1816).
Eremogone gypsophiloides (L.) Fenzl, Versuch Verbreit. Vertheil. Alsin. T. ad p. 57 (1883).
Arenaria mirdamadii Rech. f., Bot. Jahrb. 75: 342 (1951).

Tufted perennials. Stem 15–45 cm, erect, glabrous or glandular-pubescent above. Leaves linear-setaceous, ± ciliate on margins; basal leaves 1–10 cm long; cauine leaves (18–)35–75 × 0.5–1 mm, longer than internodes; sheaths less than 2 mm. Inflorescence many-flowered, cymules 5–15, 1–8-flowered. Bracts scarious, 6–9 mm long, shorter than sepals, lanceolate, acuminate, margins ciliate or not. Pedicels 5–14 mm. Sepals 2.5–4.5 mm, glandular pubescent. Petals 8–9 mm, 2–2.5 × longer than sepals, obovate to oblong, base attenuate.

HAB. Stony slopes, meadows; alt. 1800–2900 m; fl. Jul.
DISTRIB. Thorn-cushion or alpine zone; rare. **MRO**: Halgurd Dagh, *Rechinger* 11405.

The species close to *A. cucubaloides* Smith. According to Flora of Turkey *Arenaria gypsophiloides* L. var. *glabra* Fenzl occurs in N Iraq; it differs from var. *gypsophiloides* in having glabrous peduncles, pedicels and sepals. No material has been seen or is recorded of this variety from Iraq.

Bulgaria, Transcaucasica, N, W and C Iran.

2. **Arenaria balansae** Boiss., Fl. Orient. 1: 700 (1867); McNeill, Notes RBG Edinb. 24(3): 261 (1963) & in Fl. Turk. 2: 22 (1967); Rech. f., Fl. Iranica 163: 11 (1988).

Mat-forming perennial herbs, glandular-pubescent or glabrous. Stems up to 14 cm, spreading. Leaves 2–6 × 2.5–3.5 mm, elliptic or obovate, sessile, attenuate at base, apex obtuse, glabrous or glandular-pubescent. Inflorescence 1–2(–3)-flowered, terminal and lateral. Pedicels up to 6 mm long, glandular or glabrous, much longer than bracts. Sepals 3.5–4.5 mm, ovate-lanceolate, obscurely 1-veined, glabrous, as long as or shorter than petals. Petals obovate-oblong. Capsule ovoid, shorter than calyx. Seeds, rugose-reticulate.

Plant glandular-pubescent at least in some parts . . . . a. var. *balansae*
Plant entirely glabrous. . . . . b. var. *glaberrima*

a. var. **balansae**

A. neelgherense Wight & Arnott var. *glanduloso-pubescens* Fenzl, Diagn. Pl. Orient. 12 (1860).

HAB. Mountain slopes, on rocks ; alt. ± 3100 m; fl. Aug.–Sep.
DISTRIB. Alpine zone. **MRO**: Qandil Range, *Rawi & Serhang* 24489!

b. var. **glaberrima** (Fenzl) McNeill, Notes RBG Edinb. 24(3): 261 (1963).

A. neelgherense Wight & Arnott var. *glaberrima* Fenzl, Diagn. Pl. Orient. 12 (1860).

HAB. On rocks ; alt. 2800–3200 m; fl. Aug.–Sep.
DISTRIB. Alpine zone. **MRO**: Halgurd Dagh, *Guest* & *Husham* 15832!; *Gillett* 9602!; Ser Kurawa, *Gillett* 9726!

Turkey and W Iran.

3. **Arenaria kurdica** *McNeill* in Notes RBG. Edinb. 23: 510 (1961) & in Notes RBG Edinb. 24(3): 282 (1963). — Type: Iraq: Sefin Dagh, above Shaqlawa, 1200 m, *Gillett* 8076 (holo. K!).

Annual herbs. Stem slender, to 4 cm long, hairy with glandular and eglandular hairs. Leaves glabrous to hairy, up to 5 mm long; basal leaves spatulate; upper ones oblanceolate. Inflorescence 1–9-flowered. Bracts foliaceous, hairy. Pedicels 4–7 mm, glandular, sharply deflexed in fruit. Sepals 2–2.5 mm, ovate-lanceolate, hairy, 3-nerved with scarious margins. Petals lanceolate, as long as sepals or slightly longer, bifid. Capsule subglobose.

HAB. On limestone; alt. ± 1200 m; fl. May.
DISTRIB. Alpine zone, rare. **MRO**: Sefin Dagh, above Shaqlawa, *Gillett* 8076! (type).

Endemic.

4. **Arenaria serpyllifolia** *L.*, Sp. Pl. 423 (1753); McNeill, Notes R.B.G. Edinb. 24, 3: 285(1963); McNeill in Fl. Turk. 2: 28 (1967); Ghazanfar & Nasır in Fl. Pak. 175: 6 (1986); Rech. f. in Fl. Iranica 163: 13 (1988); Chamberlain in Fl. Arab. Peninsula and Socotra 1: 206 (1996); Phitos, Fl. Hellenica 1: 165 (1997); Boulos, Fl. Egypt 1: 72 (1999); Wu Zhengyi, Zhou Lihua & Wagner in Fl. China 6: 45 (2001).

Alisine serpyllifolia (L.) Crantz, Instit. 2: 406 (1766).
Stellaria serpyllifolia (L.) Scop. Fl. Carn. Ed. 2,1: 319 (1772).
Alisanthus serpyllifolius (L.) Desv. in J. Bot. Desv. 3 (5): 222 (1816).
Alsinella serpyllifolia (L.) SF. Gray, Nat. Arr. Br. Pl. 2: 664 (1821).
Euthalia serpyllifolia (L.) Rupr., Fl. Cauc. 220 (1869).

Annual herbs. Stems slender, 4–25 cm high, scabrid, puberulent to glandular-pubescent, erect, ascending to decumbent. Leaves scabrid, glandular or glabrous; basal leaves spatulate, withering at anthesis; upper ovate-triangular, acute to acuminate, sessile, 3–7 mm long. Inflorescence usually monochasial or laxly dichasial with 5–25-flowered. Bracts foliaceous, hairy. Pedicels 2–11 mm, glandular-pubescent, ± thickened at base in fruit. Sepals 3–4 mm, ovate-lanceolate, acute to acuminate, glandular, scabrid, or glabrous, 3–5-veined with scarious margins. Petals ovate, shorter than sepals. Capsule subglobose, flask-shaped, slightly shorter or longer than calyx. Seeds tuberculate. Fig. 14: 1–3.

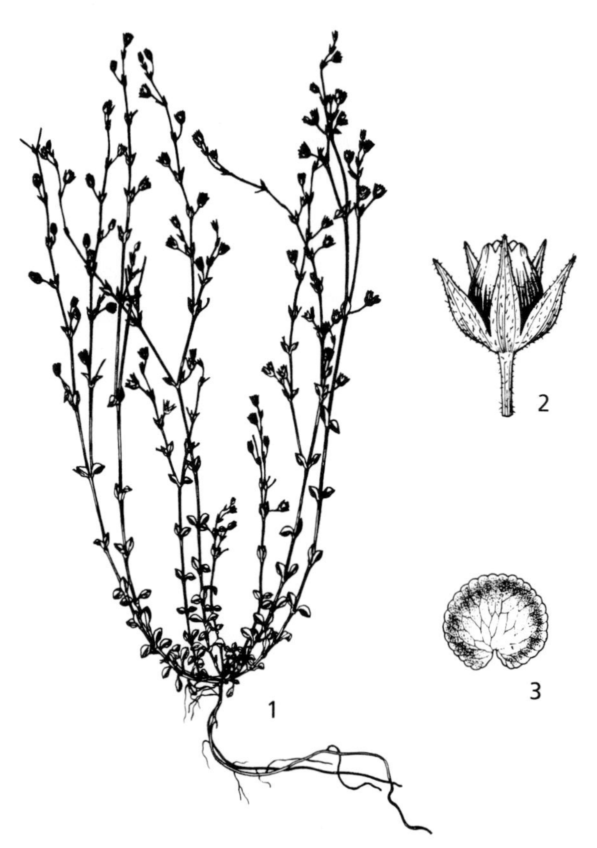

Fig. 14. **Arenaria serpyllifolia**. 1, habit × ½; 2, capsule × 4; 3, seed × 35. Reproduced with permission from Flora of Pakistan 175: f. 1, 1986. Drawn by S. Hameed.

HAB. Plains, rocky places, under trees, open places on mountains; alt. 600–3000 m; fl. May–Jun.
DISTRIB. **MRO**: Hisar-i Sakran, *Hadač* 5622; Halgurd Dagh, *Hadač* 2093; **MSU**: Penjwin, *Hadač* 4976.

No Iraqi specimens at K. Records from Flora Iranica. In addition, var. *macrosepala* Rech.f. with calyx 5 mm long, is recorded from Iraq by Rechinger; the record needs confirmation.

Temperate Eurasia, mountains in NW Africa, Arabian Peninsula, Ethiopia. Introduced in N America and Australia.

5. **Arenaria leptoclados** (*Reichenb.*) *Guss.*, Fl. Sic. Syn. 2: 824 (1845); Blakelock, Kew Bull. 2: 182 (1957); McNeill, Notes R.B.G. Edinb. 24, 3: 286 (1963); McNeill in Fl. Turk. 2: 28 (1967); Ghazanfar & Nasır in Fl. Pak. 175: 8 (1986); Rech. f. in Fl. Iranica, 163: 14 (1988); Gilbert in Fl. Somalia 1: 104 (1993); Phitos, Fl. Hellenica 1: 166 (1997; Gilbert in Fl. Ethiop. & Eritr. 2(1): 213 (2000).

A. serpyllifolia L. var. *leptoclados* Reichenb., Icon. Fl. Germ. Helv. 5: 32 (1841).
A. serpyllifolia L. subsp. *leptoclados* (Reichenb.) Neyman, Consp. Fl. Eur. 112 (1878).

Annual herbs. Stems slender, up to 12 cm high, pubescent to puberulent, glandular haris also present above. Leaves ovate to spatulate, hairy to glabrousu, to 4 mm long, usually mucronulate. Inflorescence usually dichasial. Bract foliaceous, usually separsely glandular hairy. Pedicels 2–7(–9) mm long. Sepals 2.5–3 mm long, lanceolate, glandular, margins scarious. Petals shorter than sepals. Capsule conical, gradually tapering towards, equalling to sepals or shorter.

HAB. Dry steppes, stony hills; alt. 100–1400 m; fl. Mar.
DISTRIB. **MSU**: Jarmo, *Haines* 361! *& Helbaek* 449!; **FPF**: between Sangura and Tanura, *Poore* 481!; Barda, *Gillett* 6604! **DLJ**: Rawa, *Gillett & Rawi* 2041!

Blakelock mentioned from Gillett's specimen (Gillett 8076: not seen) from Sefin Dagh above Shaqlawa that it might be intermediate between *A. leptoclados* and *A. sabulinea* in leaf-shape and length of the sepals. The fruting pedicels are sharply deflexed as in *A. sabulinea*.

Temperate Eurasia, N Africa, introduced in N America and Australia.

IMPERFECTLY KNOWN SPECIES

Arenaria sabulinea *Griseb.* ex *Fenzl*, III. Pl. Syr. Tauri 47 (1843); Zohary, Fl. Iraq & Its Phytogeogr. Subdivis.: p. 51 (1946); McNeill, Notes R.B.G. Edinb. 24, 3: 284 (1963); McNeill in Fl. Turk. 2: 27 (1967).

Annual herbs. Stems slender up to 10 cm high, hairy with glandular and eglandular hairs. Leaves glabrous or hairy on margins; basal leaves withering at anthesis; upper leaves linear-subulate, up to 3 mm long. Inflorescence 3–35–flowered. Bracts subulate, 1–3 mm long, shorter than pedicels. Pedicels 3–13 mm long, longer than sepals, glandular, spreading or deflexed in fruit. Sepals 2–2.5 mm long, ovate, acute, ± glandular with prominently 3-veined. Petals oblong, up to 3.5 mm long, clearly longer than sepals. Capsule subglobose, shorter or equalling to calyx. Seeds tuberculate.

Zohary mentioned that this species was found in the Rowanduz area. His record is not clear. But the species grows in some locations near Iraq in Turkey. It probably extends into Iraq. In addition to this, Blakelock, Kew Bull. 2: 183 (1957), and McNeil mentioned from Bornmüller's record related to *A. sabulinea* Gris. ex Fenzl var. *brevipes* Bornm. in Beih. Bot. Zbl. 28 (2): 149 (1911). This variety was recorded from Shaqlawa (*Bornmüller* 941) and Kirkuk & Sulaimaniya (*Bornmüller* 941b) and described as 'pedicels not more than twice as long as calyx'.

14. **MINUARTIA** L.

Sp. Pl. 89 (1753); Gen. Pl. ed. 5: 39 (1754)

Annual or perennial herbs, often caespitose, sometimes forming cushions or mats. Leaves narrowly lanceolate, linear-subulate or setaceous, exstipulate. Inflorescence terminal, 1-many-flowered cymes or clusters. Sepals 5, free to base, 1–9-veined. Petals 5, free, entire or rarely retuse. Stamens (3–)10, the outer whorl with obsolete, single or bifid basal glands. Styles 3. Capsule opening by 3 teeth. Seeds usualy many, smooth or tuberculate.

About 120 species, found throughout the northern hemisphere, one species (*M. groenlandica* Retz) in Brazil and one species native to Chile; 14 species in Iraq.

McNeill, J. (1962). Taxonomic Studies in the Alsinoideae I. Generic and infra-generic groups. Notes RBG Edinb. 24 (2): 79–155.
McNeill, J. (1963). Taxonomic Studies in the Alsinoideae II. A revision of the species in the Orient. Notes RBG Edinb. 24(3): 311–400.

1. Perennial 2
   Annual 4
2. Outer sepals (4–) 5–9–veined, inner 3–veined; petals shorter than sepals. . . . . 2. *M. recurva* subsp. *oreina*
   Sepals 3–veined; petals clearly longer than sepals. 3
3. Stem nodes prominent swollen; leaves linear-setaceous to subtrete, spinose. . . . . 4. *M. juniperina*
   Stem nodes not prominent swollen; leaves linear, not spinose . . . . 3. *M. kashmirica*
4. Stem leaves in the lower ½ only; petals pinkish. . . . . 1. *M. picta*
   Stems leafly throughout; petals white. 4
5. Bracts with basal membranous margin c. 1 mm broad, strongly recurved in fruit. . . . . 10. *M. hamata*
   Bracts without basal membranous margin, if present than not more than 0.5 mm broad, straigth or slightly curved in fruit. 6
6. Flowers sessile or subsessile, pedicels shorter than 2 mm long 7
   Flowers distinctly pedicellate not sessile or subsessile, at least some pedicels longer than 2.5 mm long 10
7. Calyx rounded at base . . . . 7. *M. montana* subsp. *wiesneri*
   Calyx truncate at base 8
8. Sepals glabrous or very sparsely glandular-pubescent near margin. 9
   Sepals very finely and sparsely glandular-pubescent. . . . . 9. *M. decipiens* subsp. *decipiens*
9. Inflorescence dichotomously corymbose cymes; petals absent . . . . 11. *M. sclerantha*
   Inflorescence not dichotomously corymbose cymes; petals present, very short . . . . 8. *M. intermedia*
10. Leaves 5–7–veined. 11
    Leaves 1–3–veined. 12
11. Lowest flower usually long-pedicelled, remainder densely clustered. . . . . 5. *M. meyerii*
    Flowers ± equally distant, not densely clustered . . . . 6. *M. multinervis*
12. Leaves and bracts 1–veined, sometimes weakly 3–veined an base . . . . 13. *M. subtilis*
    Leaves and bracts 3–veined throughout. 13
13. Petals oblong-ovate, cuneate at base, always shorter than sepals . . . . 14. *M. hybrida*
    Petals ovate to ovate-lanceolate, contracted into short claw usually as long as or longer than sepals, rarely shorter . . 12. *M. mesogitana* subsp. *turcomanica*

1. **Minuartia picta** (*Sibth.* & *Sm.*) *Bornm.* in Beih. Bot. Centr. 28 (2): 148 (1911); McNeill, Notes R.B.G. Edinb. 24, 3: 324 (1963); Rech.f., Fl. Lowland Iraq: 234 (1964); McNeill in Fl. Turk. 2: 32 (1967); Ghazanfar & Nasır in Fl. Pak. 175: 12 (1986); Leonard, Contrib. Fl. et Veg. Deserts d'Iran 6: 16 (1986); Rech. f. in Fl. Iranica 163: 32 (1988); Chamberlain in Fl. Arab. Penins. & Socotra 1: 203 (1996); Boulos, Fl. Egypt 1: 73 (1999).

Arenaria picta Sibth. et Smith, Fl. Graec. Prodr. 1: 304 (1806).
A. filiformis Labill., Icon. Plant. Syr. Rar. 4: 8, t. 3, f. 2 (1812).
Alsine picta (Sibth. & Sm.) Fenzl, Vers. Verbreit. Vertheil. Aslin. t. ad. p. 57 (1853).

Annual herbs. Stem 5–15 cm, erect or divergent, usually branching at base, glabrous or glandular-pubescent above, leafly in the lower ½ only. Leaves setaceous, mucronate, 9–20 mm, often in axillary fascicles, usually glabrous, ± ciliate on margin near base. Inflorescence lax, 5–30-flowered dichasium. Bracts minute, scarious. Pedicels 3–12 mm, glabrous to glandular-pubescent. Sepals 2–2.5 mm, ovate-orbicular, rounded at apex, obscurely nerved, glabrous with broad scarious margin. Petals pink or pink markings, up to 5 mm, clearly longer than sepals, oblong, truncate or emarginate at apex. Staminal glands 5, prominent, flap-like. Capsule ovoid, about twice the length of the calyx. Fig. 15: 1–5.

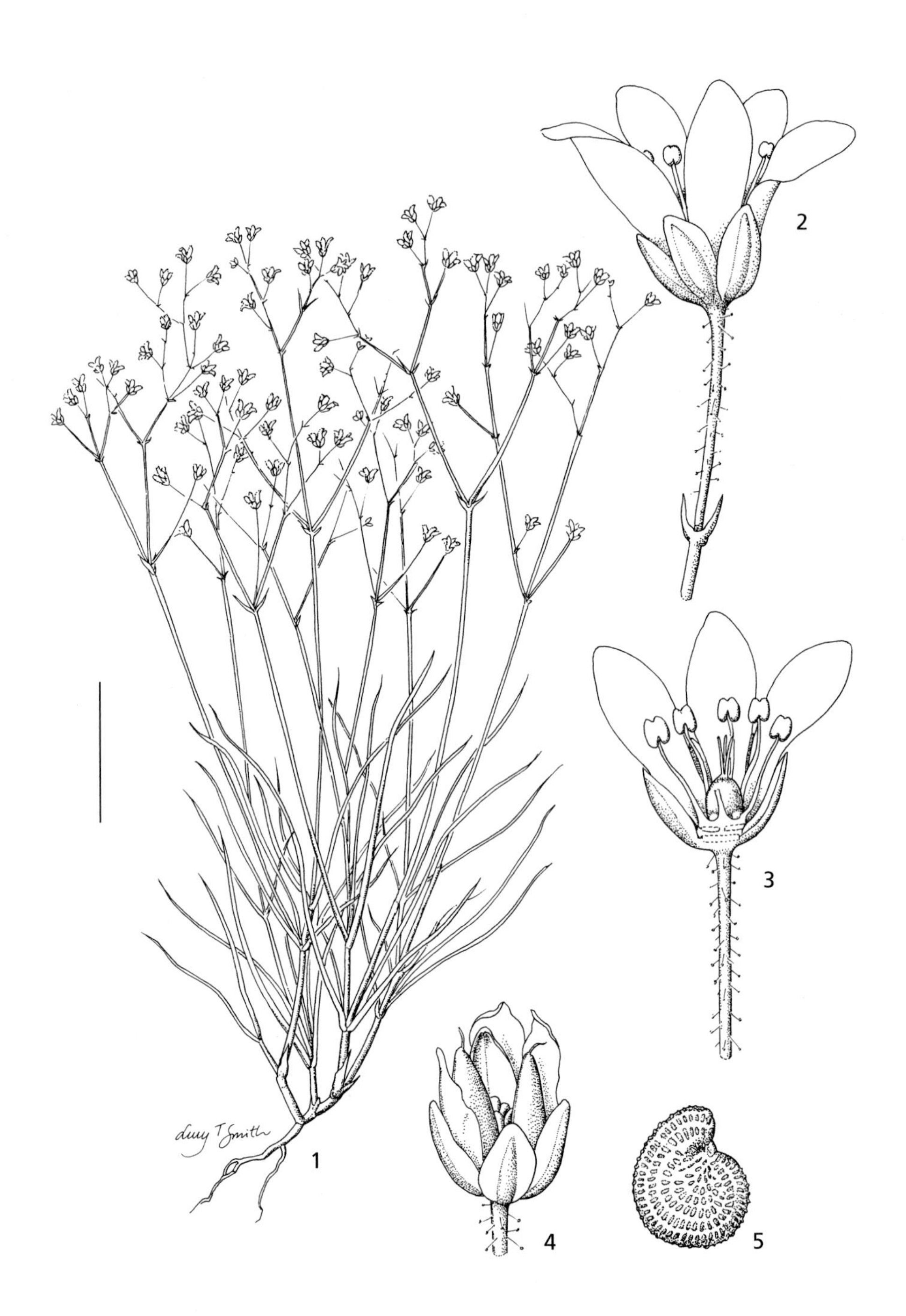

Fig. 15. **Minuartia picta**. 1, habit; 2, flower; 3, flower with front petals and sepals removed; 4, capsule; 5, seed. 1 from *Alizzi* 31992; 2–5 from *Barkley & Haddad* 5717. Scale bar: 1 = 2 cm; 2–4 = 3.3 mm; 5 = 0.8 mm. © Drawn by Lucy T. Smith.

HAB. Cultivated fields, sandy soil, flow fields, gravely lands, cracks of rocks; alt. 200–930 m; fl. Mar.–Apr.
DISTRIB. **MAM**: Dohuk, *Guest* 1319!; **MRO**: Below Salah ad-Din, *Barkley* & *Haddad* 5717!; **MJS**: Jabal Sinjar, *Kaisi* & *Wedad* 52123!; **FUJ**: 5 km N of Sha'bani to Sinjar, *Kaisi* & *Hamad* 48978!; Mosul, *Kaisi* & *Wedad* 51982!; between Balad Sinjar and Tal Afar, *Guest* 4147!; **FAR**: 10–15 km E of Abril, *Shahwani* 25288!; **FKI**: 36 km W of Kirkuk, *A. Barkley* & *E.D. Barkley* 4567!; **FPF**: Kani Kirmaj, *Poore* 473!; **DLJ**: 15 km E of Rawa in Jazira, *Omar et. al.* 45124!; **DWD**: 5 km SW of Ana, *Omar et. al.* 45003!; Abu-Amad. 110 km N E of Rawa, *Khayat* & *Hamad* 51825!; 2 km SW of H2, *Chakr. et al.* 31467!; Wadi Ghadaf, *Rawi* 14730!

Turkey, Cyprus, Syria, Palestine, Sinai, Arabian Peninsula, Iran and Pakistan.

2. **Minuartia recurva** (All.) Schinz & Thellung, Bull. Herb. Boiss. Ser. 2 (7): 404 (1907).
Arenaria recurva All., Fl. Pedem. 2: 113 (1785).
Alsine recurva (All.) Wahlenb., Veg. Clim. Helvet. 87 (1813).

subsp. **oreina** (Mattf.) McNeill, Notes R.B.G. Edinb. 24, 3: 338 (1963); McNeill in Fl. Turk. 2: 47 (1967); Rech. f. in Fl. Iranica 163: 34 (1988).
Alsine recurva (All.) Wahlenb var. *nivalis* Boiss., Boiss., Fl. Orient. 1: 674 (1867).
Minuartia hirsuta (Bieb.) Hand.-Mazz. subsp. *oreina* Mattf. in Feddes Rep. Beih. 15: 114 & 118 (1922).
M. oreina (Mattf.) Schischk. in Izv. Tomsk. Univ. 81: 443 (1928).

Perennial, caespitose to pulvinate with numerous non-flowering shoots. Stems 2–9 cm, glabrous to glandular-pubescent. Leaves 5–12 mm, usually flaccid, slightly flattened, linear, acute, glabrous; leaf-fascicles ± curved. Inflorescence lax 3–15-flowered. Bracts 4–5 mm, glabrous with scarious margins, acuminate, 3-veined. Pedicels up to 7 mm, glabrous to glandular-pubescent. Sepals 3–4.5 mm, glabrous, lanceolate; outer ones (4–)5–9-veined; innermost 3-veined. Petals shorter than sepals. Capsule equalling to calyx.

HAB. Stony mountain slopes; alt. 2900–3800 m; fl. Jul.–Aug.
DISTRIB. **MRO**: Halgurd Dagh, *Rechinger* 11793! & *Gillett* 9593!; Ser Kurawa, *Gillett* 9743!; Qandil range, *Rawi* & *Serhang* 24509!

Caucasus, Turkey and Iran.

3. **Minuartia kashmirica** (*Edgew.*) *Mattf.*, Bot. Jahrb. 57, Beibl. 126: 32 (1921); Ghazanfar & Nasır in Fl. Pak. 175: 14 (1986); Rech. f. in Fl. Iranica 163: 40 (1988); Lu Dequan & McNeill in Fl. China 6: 31 (2001).
Arenaria kashmirica Edgew. ex Edgew. & Hook. f., Fl. Brit. Ind. 1: 236 (1874).
Minuartia lineata (CA.Mey.) Bornm. in Beih. Bot. Centrbl. 27, 2: 318 (1910).
M. sublineata Rech. f. in Fl. Iranica 163: 36 (1988), syn. nov.

Caespitose, perennial herbs with stout tap root and sterile shoots with fasciculate leaves. Stems glabrous to glandular-pubescent, 5– 12(–18) cm; nodes scarcaly swollen. Leaves linear, ± flat, 6–12 mm, acute, glabrous to glandular-pubescent, 3-veined; cauline leaves decreasing in size. Inflorescence 3–7-flowered dichasia. Bracts linear-lanceolate, acuminate. Pedicels 2–11 mm, glabrous or glandular. Sepal 3–4.5 mm, acuminate, glabrous or glandular hairy, 3-veined with scarious margins. Petals oblong, longer than sepals.

HAB. Stony mountain slopes; alt. 2500–3880 m; fl. Jul.–Sep.
DISTRIB. **MRO**: Ser Kurawa, *Gillett* 9746!; Qandil Range, *Rawi* & *Serhang* 24076!; Chiya-i Mandau, *Guest* 3060!; Halgurd Dagh, *Guest* 3073!

Iraq specimens were treated as a new species, *M. sublineata* by Rechinger. Blakelock (Kew Bull. 2: 198, 1962), placed specimens from Iraq under *M. aucheriana* (Boiss.) Bornm. (1910), a species considered endemic to Iran by Rechinger, but also noted that some specimens were intermediate between *M. aucheriana* and *M. lineata*. Specimens at K that I have seen show variations with characters of typical *M. lineata* and *M. sublineata* (both treated here as synonyms of *M. kashmirica*). In our opinion (AK & SAG), Iraq material comes closest to *M. kashmirica*, but more taxonomic work is required to acertain whether Iraq specimens fall under *M. aucheriana* (recorded by Rechinger to be endemic to Iran) and whether these four species be treated as a single polymorphic taxon (*M. aucheriana*) distributed from Azerbaijan, Turkey, Iraq, Iran, Afghanistan and Pakistan to Nepal and Tibet.

Caucasus, Turkey, Iran to Nepal and Tibet.

4. **Minuartia juniperina** (L.) Maire & Petitmengin, Étude Pl. Grèce Vasc. 4: 48 (1908); McNeill, Notes RBG Edinb. 24, 3: 348 (1963); McNeill in Fl. Turk. 2: 50 (1967); Rech.f. in Fl. Iranica 163: 39 (1988); Kamari, Fl. Hellenica 1: 184 (1997).

Arenaria juniperina L. Mantissa 1: 72 (1767).
Alsine juniperina (L.) Wahlenb., Fl. Lapp. 129 (1812), non Cherleria juniperina D. Don.

Tufted to cushion-forming perennial herbs. Stems numerous, up to 25 cm, glabrous with swollen nodes below. Leaves linear-subulate, 7–20 mm, rigid and rather pungent, glabrous, 1 or obscurely 3-veined. Inflorescence 3–10-flowered cyme or cymose clusters. Bracts 1–2.5 mm with narrow scarious margins. Pedicels 5–12 mm, glabrous. Sepals 4–5 mm, ovate-lanceolate to lanceolate, glabrous, prominently 3-veined. Petals ovate to oblanceolate, obtuse to truncate at apex, longer than sepals. Capsule cylindrical, exceeding calyx. Fig. 16: 1–5.

HAB. Stony mountain slopes; alt. 2000–2700 m; fl. Jul.–Aug.
DISTRIB. Very rare in N Iraq. **MAM**: Ser Amadiya, *Haines* W2060!

Greece, W Syria, Turkey, NW and C Iran.

5. **Minuartia meyeri** (*Boiss.*) *Bornm.* in Beih. Bot. Centr. 27 (2): 318 (1910); McNeill, Notes R.B.G. Edinb. 24, 3: 356 (1963); McNeill in Fl. Turk. 2: 53 (1967); Ghazanfar & Nasır in Fl. Pak. 175: 15 (1986); Rech.f. in Fl. Iranica 163: 44 (1988); Chamberlain in Fl. Arab. Penins. & Socotra 1: 204 (1996); Boulos, Fl. Egypt 1: 73 (1999).

Arenaria globulosa Labill. var. *nana* CA. Mey., Verz. Pfl. Cauc. 219 (1831).
Arenaria meyeri (Boiss.) Edgew. & Hook. in Hooker, Fl. Brit. Ind. 1: 236 (1874), non Fenzl (1842).
Alsine meyeri Boiss., Diagn. Ser. 1 (8): 96 (1849).
Alsine brevis Boiss., Diagn. Ser. 1 (8): 96 (1849).

Annual herbs. Stems erect to ascending, 4–20 cm, simple or branched, glandular-pubescent. Leaves 10–25 × 0.5–1.5 mm, linear, 5–7-veined, acute, erect or spreading, margin glabrous or puberulent. Flowers in dichasial clusters. Pedicels 1–5 mm. Bract herbaceous, similar to leaves. Sepal 5.5–8 mm, lanceolate, acuminate, inequal, long sepal 5-veined, short sepal 3-veined. Petal 1–2(–2.5) mm, $1/3$ of calyx length, oblong.

HAB. Dry sandy places; alt. 100–3100 m; fl. Apr.–Jun.
DISTRIB. Rare and local in Iraq. **MRO**: Kani Mazi Shirin, *Hadač* 6032; Sakri-Sakran, *Hadač* 5516; **MSU**: Avroman, *Rechinger* 10298.
No Iraqi specimens at K; description and records taken from Flora Iranica. This species is close to *M. multinervis.*

W Syria, Sinai, Palestine, Syrian Desert, S Arabia, Transcaucasica, Iran, Pakistan, Afghanistan, C Asia.

6. **Minuartia multivervis** (Boiss.) Bornm. in Beih. Bot. Centr. 33 (2): 279 (1915); McNeill, Notes RBG Edinb. 24, 3: 354 (1963); McNeill in Fl. Turk. 2: 54 (1967).

Alsine multinervis Boiss., Fl. Orient. 1: 683 (1867).
Alsine akinfiewii Schmalh. in Ber. Dtsch. Bot. Ges. 10, 278 (1892).
Minuartia akinfiewii (Schmalh.) Woronow in Fomin & Woronow, Opered. Rast. Kavk. Krima (keys Pl. Cauc. Crim.) 2: 176 (1914), syn. nov.

Like *M. meyeri* but more slender. Flowers ± equally distant, (not dense clusters in *M. akinfiewii*). Inflorescence overlaping, the lowest flower is usually long-pedicelled, the pedicels of the rest of flowers in clusters are much shorter in *M. meyeri.* However, some small specimens can be confused.

HAB. Quercus forests, open grounds; alt. 900–2000 m; fl. Jun.–Aug.
DISTRIB. Rare in mountainous areas. **MAM**: Sarsang, *Haines* 1344!; **MRO**: Kani Mazi Shirin, *Agnew et. al.* 2005!

Turkey, Georgia and Armenia.

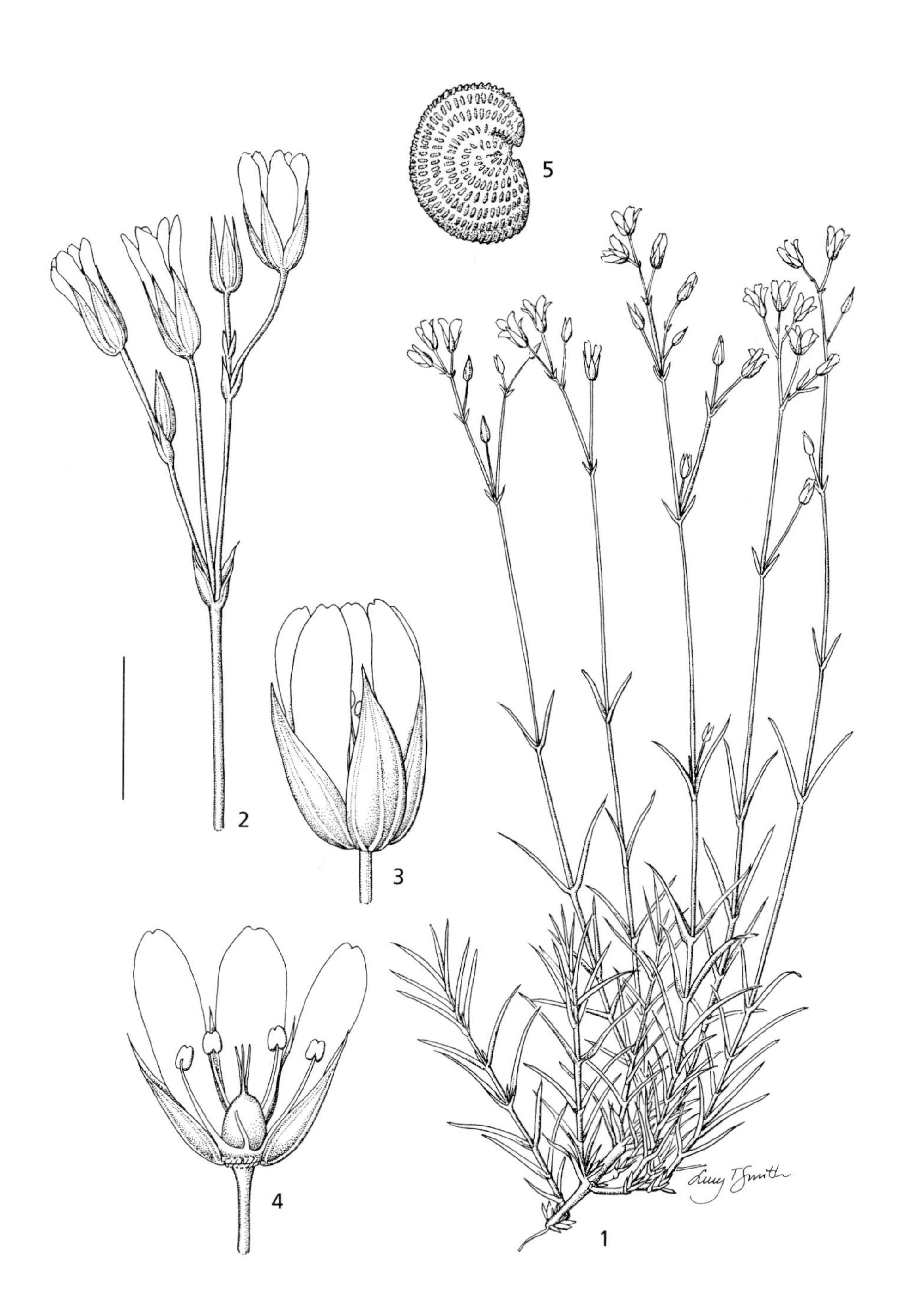

Fig. 16. **Minuartia juniperina**. 1, habit; 2, inflorescence; 3, flower; 4, flower with front petals and sepals removed; 5, seed. All from *Haines* W2060. Scale bar: 1 = 2 cm; 2 = 7 mm; 3, 4 = 3.3 mm; 5 = 0.8 mm. © Drawn by Lucy T. Smith.

7. **Minuartia montana** L., Sp. Pl. 90 (1753).

subsp. **wiesneri** (Stapf) McNeill, Notes R.B.G. Edinb. 24, 3: 359 (1963); McNeill in Fl. Turk. 2: 55 (1967); Rech. f. in Fl. Iranica 163: 46 (1988).

Alsine wiesneri Stapf, Denkschr. Akad. Wiss. Wien Math.-Nat. Kl. 51: 20 (1886).
A. montana sensu Boiss., Fl. Orient. 1: 684 (1867) non L.
A. montana var. *caucasica* (Boiss.) Boiss., Fl.Orient. 1: 685 (1867).

Annual herbs. Stem 2–6 cm, simple or branching, with white crisped hairs. Leaves linear-setaceous 3–5-veined, hairy with few white hairs, expanded at base. Inflorescence ± overlapping, forming dense clusters. Bracts similar to leaves, exceeding clusters. Pedicels less than 2 mm long. Calyx rounded at base; sepals inequal, outer one 6–7 mm, linear-lanceolate, acute to acuminate, glabrous or sparsely glandular-pubescent on margins, prominently 3-veined; inner one 5–7 mm. Petals absent. Staminal glands 5, in the form of a single groove in front of the outer stamens. Seeds more than 6 per capsule.

HAB. Rocky slopes, on hillsides; alt. 300–1400 m; fl. Apr.–Jun.
DISTRIB. **MSU**: 10 km W of Tawila, *Rawi* 22132!; **FKI**: Mirjani, *Sutherland* 196!; **FPF**: Khanaqin, *Haines* W896!
Two specimens from Diyala (Poore 370!) seem to be intermadiate between *M. montana* subsp. *wisneri* and the next species, *M. intermedia.*

Transcaucasia, Turkey, W and S Iran.

8. **Minuartia intermedia** (Boiss.) Hand.-Mazz. in Ann. Nat. Hofm. Wien 26: 148 (1912); McNeill, Notes R.B.G. Edinb. 24, 3: 361 (1963); McNeill in Fl. Turk. 2: 55 (1967).

Alsine intermedia Boiss., Boiss., Fl. Orient., 1: 685 (1867).
Arenaria intermedia (Boiss.) Fernald in Rhodora 21: 6 (1919).

Similar to *M. montana*, but stems 3–9 cm long. Leaves linear-lanceolate. Inflorescence sometimes overlapping. Bracts longer than lower clusters but not usually exceeding upper clusters. Calyx truncate at base; sepals linear-triangular, 4.5–7 mm, veins very distinct, glabrous or sparsely glandular-pubescent near margins. Petal present, short. Staminal glands bifurcate appearing as 10 finger-like strucuctures, a pair lying on each side of the outer stamens. Seeds (1–)2–5 per capsule.

HAB. Limestone mountain with coppiced *Quercus* forest, sandy and gravelly soils; alt. up to 1300 m; fl. Apr.–Jun.
DISTRIB. **MAM**: Occasional in upper forest zone. Dohuk, *Haines* W1443!; **MRO**: Between Karokh and Dargala Village, *Kaisi & Nuri* 27690!; **MJS**: Jabal Sinjar, *Kaisi & Wedad* 52121! & *Gillett* 1110A!

Armenia, Turkey, Cyprus, Syria.

9. **Minuartia decipiens** (Fenzl) Bornm. in Beih. Bot. Centr. 31 (2): 193 (1914); McNeill, Notes R.B.G. Edinb. 24, 3: 361 (1963); McNeill in Fl. Turk. 2: 56 (1967).

Alsine decipiens Fenzl, Pug. 12 (1842).

subsp. **decipiens**

Annual herbs. Stem 5–17 cm, pruinose-velutinous with slender hairs. Leaves linear-lanceolate, up to 28 mm, glandular-pubescent, 3–7-veined. Inflorescence loose or dense clusters. Bracts foliaceous, ± glandular-pubescent, not exceeding clusters. Calyx truncate at base, glandular-pubescent; sepals linear-triangular, 4.5–5.5 mm, veins very distinct. Petals 1/1$^{3}/_{4}$ as long as sepals. Staminal glands single, appearing as a swelling at base of each of the 5 outer stamens. Seeds (1–)2–5 per capsule.

HAB. Limestone mountain with coppiced *Quercus* forest, sandy and gravelly soils; alt. up to 1300 m; fl. Apr.–Jun.
DISTRIB. Very rare and local. **MJS**: Jabal Sinjar, *Gillett* 1110!

Turkey, Cyprus, Syria, Lebanon, Israel, Jordan.

10. **Minuartia hamata** (Hausskn.) Mattf. in Bot. Jahrb. 57 Beibl. 126: 29 (1921); McNeill, Notes R.B.G. Edinb. 24, 3: 364 (1963); McNeill in Fl. Turk. 2: 56 (1967); Rech.f. in Fl. Iranica 163: 47 (1988); Kamari, Fl. Hellenica 1: 172 (1997).

Queria hispanica L., Sp. Pl. 90 (1753).
Scleranthus hamatus Hausskn. in Mitt.Thür. Bot. Ver. 9: 17 (1890).

Annual herbs. Stems 2.5–14 cm, usually branched from base, eglandular-pubescent. Leaves linear-setaceous, 5–11 mm, expanded at base with membranous ciliate margins, 3–5-veined. Inflorescence dense spherical clusters with sessile flowers; the uppermost flowers usually sterile. Bracts 3–7-veined, almost as long as the clusters with broad membranous margins at base, narrowed abruptly toward apex, enclosing flowers, strongly recurved in fruit. Sepals 2–4 mm, lanceolate to linear-lanceolate, 3–5-veined, eglandular-pubescent at base. Petals absent or minute to 1 mm, subulate. Staminal glands 2–5, undivided with an elongate nectar furrow in front of the outer stamens. Capsule 1-seeded.

HAB. Limestone slopes with relic Quercus forest, dry sandy soils, on mountains; alt. 300–2500 m; fl. Apr.–Jun.
DISTRIB. **MRO**: Between Kujar and Kani Gawra, *Kass, Nuri & Serhang* 22635!; Gali Ali Beg, *Barkley & Haddad* 8398!; **MSU**: Penjwin, *Rawi* 42246!; Kani Witman, 20 km SW of Dukan, *Kaisi & Hamad* 49336!; Kanjan Mountain, nr Penjwin, *Rawi* 22735!; **MJS**: Jabal Sinjar, *Gillett* 10970!; *Sharif & Hamad* 50258!; Kursi, *Kaisi & Wedad* 52272!

Mediterranean area, extending to eastern Iran, C Asia and W China.

11. **Minuartia sclerantha** (Fisch. & Mey.) Thellung in Mem. Soc. Sci. Nat. Math. Cherbourg 38: 231 (1912); Rech. f. Lowl. Fl. Iraq 234 (1964); McNeill, Notes R.B.G. Edinb. 24, 3: 365 (1963); McNeill in Fl. Turk. 2: 56 (1967); Rech. f. in Fl. Iranica 163: 48 (1988).

Alsine sclerantha Fisch. & Mey. in Bull. Soc. Nat. Mosc. 33:400 (1838).
Alsine rudbaransis Stapf, Denkschr. Akad. Wiss. Wien.-Nat. Kl. 51: 288 (1886) p.p.

Annual herbs. Stem 5–15 cm, erect, simple or dichotomously branched, woolly. Leaves linear-subulate, 1–1.2 cm long, subsetaceous, somewhat 3–5-veined at base, woolly on margins. Flowers nearly sessile in dense, dichotomously corymbose cymes. Bracts with very narrow membranous margins. Calyx oblong-conic, truncate ± gibbous at base; sepals narrowly linear-lanceolate, 4–6 mm, acute. Petal absent. Staminal glands bifurcate, appearing as 10-finger-like structures between stamens. Seeds 2–3 per capsule.

HAB. In sand or clay; alt. 300–1400 m; fl. May–Jun.
DISTRIB. Rare. **FPF**: Jabal Hamrin, *Sutherland* 198.
No Iraqi specimens at K; record citation from Flora Iranica.

Transcaucasia, Turkey, N and NW Iran, C Asia and W China.

12. **Minuartia mesogitana** (Boiss.) Hand.-Mazz. in Ann. Nat. Hofm. Wien 27: 148 (1912).

Alsine mesogitana Boiss., Diagn. ser. 1(1): 45 (1842).
A. tenuifolia sensu Boiss., Fl. Orient. 1: 686 (1867) p.p.

subsp. **turcomanica** (Schischk.) McNeill in Rech. f. in Fl. Iranica 163: 50 (1988).

Minuartia turcomanica Schischk. in Acta Inst. Bot. Acad. Sci. URSS ser. 1, 3: 168 (1937).
M. mesogitana (Boiss.) Hand.-Mazz. subsp. *lydia* (Boiss.) McNeill var. *turcomanica* (Schischk.) McNeill, Notes RBG Edinb. 24, 3: 388 (1963).

Annual herbs. Stems 5–15 cm, ascending to erect, diffuse, usually branching from base, glandular-pubescent, sometimes glabrous below. Leaves linear-subulate or subulate-setaceous, up to 8 mm, 3-veined. Inflorescence lax, many-flowered dichasia, sparsely glandular-pubescent. Bracts up to 2 mm with scarious margins. Pedicels 5–8 mm, glandular. Sepals 2–2.5 mm, ovate to ovate-lanceolate, acute to acuminate, glandular, 3-veined with scarious margins. Petals ovate-lanceolate, abruptly contracted at base into a very short claw, usually as long as or longer than sepals, sometimes shorter. Capsule not or scarcely exeeding calyx.

HAB. Dry saline soils, slopes; alt. 200–900 m; fl. Apr.–May.
DISTRIB. **MRO**: Kani Mazi Shirin, *Hadač* 6039; **FAR**: Erbil, *Bayliss* 92!

Iraq, Iran, Afghanistan, C Asia and W China.

13. **Minuartia subtilis** (Fenzl) Hand.-Mazz. in Ann. Nat. Hofm. Wien 26: 148 (1912); McNeill, Notes RBG Edinb. 24, 3: 391 (1963); McNeill in Fl. Turk. 2: 66 (1967); Rech. f. in Fl. Iranica 163: 51 (1988).

Alsine tenuifolia var. *subtilis* Fenzl in Tchihat., Asie Min. Bot. 1: 226 (1860).

Similar to *M. mesogitana* but ± erect and always very slender, up to 10 cm long, ± glabrous. Leaves and bracts 1-veined (sometimes 3-veined at base); sepals 2–2.5 mm, ovate-lanceolate or lanceolate, weakly 3-veined. Petals cuneate at base, shorter than sepals. Capsule ovoid, exceeding calyx.

HAB. Stony slopes; alt. 900–2700 m; fl. Jun.–Aug.
DISTRIB. **MRO**: Sakri, *Hadač* 5485.
No specimens at K; citation record from *Flora Iranica.*

Eastern Turkey, Iran, Afghanistan.

14. **Minuartia hybrida** (*Vill.*) *Schischk.* in Fl. URSS 6: 488 (1936); McNeill, Notes R.B.G. Edinb. 24, 3: 393 (1963); McNeill in Fl. Turk. 2: 66 (1967); Ghazanfar & Nasır in Fl. Pak. 175: 16 (1986); Rech. f. in Fl. Iranica 163: 51 (1988); Chamberlain in Fl. Arab. Penins. & Socotra 1: 203 (1996); Kamari in Fl. Hellenica 1: 176 (1997); Boulos, Fl. Egypt 1: 75 (1999).

Arenaria tenuifolia L., Sp. Pl. 526 (1753).
Alsine tenuifolia (L.) Crantz, Instit. 2: 407 (1766).
Arenaria hybrida Vill., Prosp. Pl. Dauph. 48 (1779).
Alsine tenuifolia var. *genuina* sensu Boiss., Fl. Or. 1: 686 (1867).
Minuartia tenuifolia (L.) Hiern in J. Bot. 37: 32 1(1899) non Nees ex Mart. (1814).
M. tenuifolia subsp. *hybrida* (Vill.) Mattf. in Bot. Jahrb. 57 Beibl. 126: 29 (1921).

Annual herbs. Stem 4–18 cm, usually erect, often branched from base, glabrous. Leaves linear to subulate, 5–11(–20) mm, 3-veined, acute, glabrous. Inflorescence in lax, many-flowered cymes. Bracts linear-lanceolate, 3–5 mm, glabrous or sometimes eglandular-pubescent. Pedicels 3–8 mm, glabrous or glandular-pubescent near calyx. Sepals 2–5 mm, linear to linear-lanceolate, acute to acuminate, 3-veined with scarious margins, glandular. Petals oblong-ovate, cuneate at base, shorter than sepals. Capsule cylindrical or ovoid, shorter or longer than calyx.

HAB. Calcarious sandstone in steppes, rocky slopes with Quercus forest, fields alt. 250–1000 m; fl. Feb.–Jun.
DISTRIB. Occasional. **MAM**: Aqra, Rawi 11306A!; Great Zab, nr. Kani Gossik, *Gillett & Rawi* 10534!; **MRO**: Rowanduz, *Guest* 2136!; *Gillett* 8328!; Rowanduz gorge, *Gillett* 8327!; **MSU**: Jarmo, *Haines* 359!; Palegawra, *Helbaek* 638!; Qaranjir, *Gillett & Rawi* 7544!; **MJS**: Kursi, Jabal Sinjar, *Gillett* 10905! & 11120!; **FPF**: Kani Kirmaj, *Poore* 466!

Europe, Mediterranean Region and SW Asia.

15. **LEPYRODICLIS** Fenzl in Endl.

Gen. Pl. 13: 966 (1840)

Annual herbs with weak, sprawling stem. Leaves lanceolate to linear-lanceolate, acute. Flowers in compound cymes. Sepals 5, free, herbaceous. Petals 5, entire to toothed. Stamens 10 or more on a glandular disc. Styles 2(–3). Ovary with 4 ovules. Capsule globose, membranous, opening to base by 2(–3) valves. Seeds globose, 2–4, tuberculate, back canaliculate.

Three species from Turkey to W Tibet and Himalayas. The members of the genus are potential problems in cultivated lands as weed. Two species occur in Iraq.

Ref. G. Wagenitz (1957) Zur Gattung Lepyrodiclis Fenzl (Caryophyllaceae), *Annalen des Naturhistor-ischen Museum in Wien* 61: 74–77.

Sepal 4–5 mm long; petals obovate, entire or retuse at apex . . . . 1. *L. holosteoides*
Sepal 5–6 mm long; petals oblanceolate, toothed at apex . . . . 2. *L. stellarioides*

Fig. 17. **Lepyrodiclis holosteoides**. 1, habit × ½; 2, flower × 3; 3, petal × 5; 4, stamen, detail; 5, ovary with styles, detail. **Lepyrodiclis stellarioides.** 6, flowering branch × ½; 7, flower × 3; 8, petal × 5; 9, stamen, detail; 10, ovary with styles, detail. Reproduced with permission from Flora of China 6: f. 22. Drawn by Li Zhimin.

1. **Lepyrodiclis holosteoides** (*C.A.Meyer*) *Fenzl* ex *Fisch.* & *Mey.*, Enum. Pl. Nov. Schrenk 1: 93 & 110 (1841); Blakelock, Kew Bull. 2: 197 (1957); Cullen in Fl. Turk. 2: 68 (1967); Ghazanfar & Nasır in Fl. Pak. 175: 18 (1986); Rech. f. in Fl. Iranica 163: 54 (1988). Lu Dequan & Rabeler in Fl. China 6: 31 (2001).

Gouffeia holosteoides C.A.Meyer, Verz. Pfl. Cauc. 217 (1831).
Arenaria holosteoides (C.A.Meyer) Edgew. in Hook. f., Fl. Brit. Ind. 1:241 (1874).

Annual herbs. Stem prostrate or ascending, 30–100 cm high, simple or branched, weak, striate, glabrous or papillose-pubescent. Leaves 18–65 × 3–8 mm, lanceolate, sessile, glabrous to subglabrous on surface, glandular or scabrid on veins, ± scabrid at margins, apex acute. Bracts similar to leaves but decreasing in size upwards. Inflorescence paniculate, spreading. Pedicels up to 12 mm long, papillose. Sepals 4–5 mm in anthesis, acute, glandular-pilose. Petals obovate, about as long as the sepals, entire or retuse at apex. Capsule globose, included in the persistent calyx. Seeds globose, tuberculate. 2n=34, 48. Fig. 17: 1–5.

HAB. Near streams, fields, muddy stony soils, hedges; alt. 500–1600 m; fl. Jun.–Aug.
DISTRIB. Local in NE Iraq. **MAM**: Zawita, *Hadač* 5989; **MRO**: Sei-Waka Village, nr S foot of Karoukh, *Kass & Nuri* 27567!; between Rowanduz & Agoyan Village, *Nuri & Kass* 27254!
Closely elated to *L. stellarioides* Schrenk ex Fisch. & CA. Mey and sometimes confused with it, but distinguished by the characters given in the key.

Caucasus, Turkey, Iraq, Iran to NW China.

2. **Lepyrodiclis stellarioides** *Schrenk* ex *Fisch.* & *C.A.Mey.*, Enum. Pl. Nov. Schrenk 1: 93 (1841); Ghazanfar & Nasır in Fl. Pak. 175: 19 (1986); Rech.f., Fl. Iranica 163: 56 (1988); Lu Dequan & Rabeler in Fl. China 6: 31 (2001).

Arenaria holosteoides Edgew. var. *stellarioides* Williams, Journ. Linn. Soc. Bot. 33: 427 (1898).
Lepyrodiclis cerastioides Kar. & Kır., Bull. Soc. Imp. Nat. Mosc. 15: 167 (1842).

Similar to *L. holosteoides* in facies, but with stems glabrous below, glandular above; leaves up to 14 mm wide, sometimes densely glandular. Inflorescence and pedicels densely glandular. Sepals oblong to lanceolate, 5–6 mm in anthesis, up to 9 mm long in fruit. Petals oblanceolate, toothed at apex. Fig. 17: 6–10.

HAB. Shaded limestone rocks in *Quercus* forest; alt. 1200–1800 m; fl. May–Jun.
DISTRIB. Local in the upper forest zone. **MRO**: Sefin Dagh, above Shaqlawa, *Gillett* 8166!; **MSU**: Penjwin, *Rechinger* 1049!; *Rawi* 22538!; Kajan Mountain, nr. Penjwin, *Rawi* 22686!; Kopi Qara Dagh, *Haines* 1084!

Transcaucasia, Iran, Afghanistan, Pakistan & C Asia.

16. STELLARIA L.

Sp. Pl. 421 (1753)

Annual or perennial herbs. Stems usually weak and slender. Inflorescence of dichasial cymes. Sepals 5. Petals absent or 4–5, usually deeply bifid, white. Stamens (2–7–)10. Styles 2–3. Capsule ovoid; teeth twice as many as styles. Seedes numerous, reniform or suborbicular, usually tuberculate.

About 200 species, mainly Eurasian, with centre of distribution in the mountains of east Central Asia and some in Africa; three species in Iraq.

1. Perennials with 2 styles ..... 3. *St. kotschyana*
   Annuals with 3 styles 2
2. Petals present, at least 3 mm long; stamens 3–7 ..... 1. *St. media*
   Petals usually absent or very small; stamens 2–5 ..... 2. *St. pallida*

1. **Stellaria media** (*L.*) *Vill.*, Hist. Pl. Dauphiné 3: 615 (1789); Rech. f., Fl. Lowland Iraq: 232 (1964); Coode in Fl. Turk. 2: 69 (1967); Ghazanfar & Nasır in Fl. Pak. 175: 22 (1986); Rech. f. in Fl. Iranica 163: 63 (1988); Chamberlain in Fl. Arab. Penins. & Socotra 1: 209 (1996); Strid, Fl. Hellenica 1: 195 (1997); Boulos, Fl. Egypt 1: 76 (1999); Chen Shilong & Rabeler in Fl. China 6: 15 (2001).

Alsine media L., Sp. Pl. 272 (1753).

Annuals. Stems 10–35 cm, decumbent or prostrate to ascending, usually with single line of eglandular hairs, sometimes glabrous. Lower leaves petiolate, cordate to ovate, lamina 3–10 × 3–8 mm; petiole glabrous to sparsely ciliate; upper leaves larger, petiolate to sessile. Inflorescence usually lax. Pedicels filiform, to 20 mm, glabrous to sparsely ciliate, longer than sepals. Sepals 3–7 mm, lanceolate, glabrous or hairy with scarious margins. Petals deeply bilobed, shorter than sepals. Stamens 3–7. Styles 3. Capsule narrowly ovoid. Seeds dark brown, tubarculate. Fig. 18: 1–3.

HAB. Cultivated fields, ruderal places, stream sides, loam soil in woodlands; alt. 300–2700 m; fl. Mar.–May.

DISTRIB. Alluvial plain and lower forest zones. **MAM**: 11 km S of Dohuk, *Barkley & Brahim* 46!; Amadiya, *Guest* 1230!; **MRO**: 16 km NW of Rania, *anon.* 28602!; Shaqlawa, *Barkley* 7116!; **LCA**: 70 km S of Baghdad, *Barkley & Brahim* 6393!; Baghdad, *Guest* 1111!

Cosmopolitan in temperate regions.

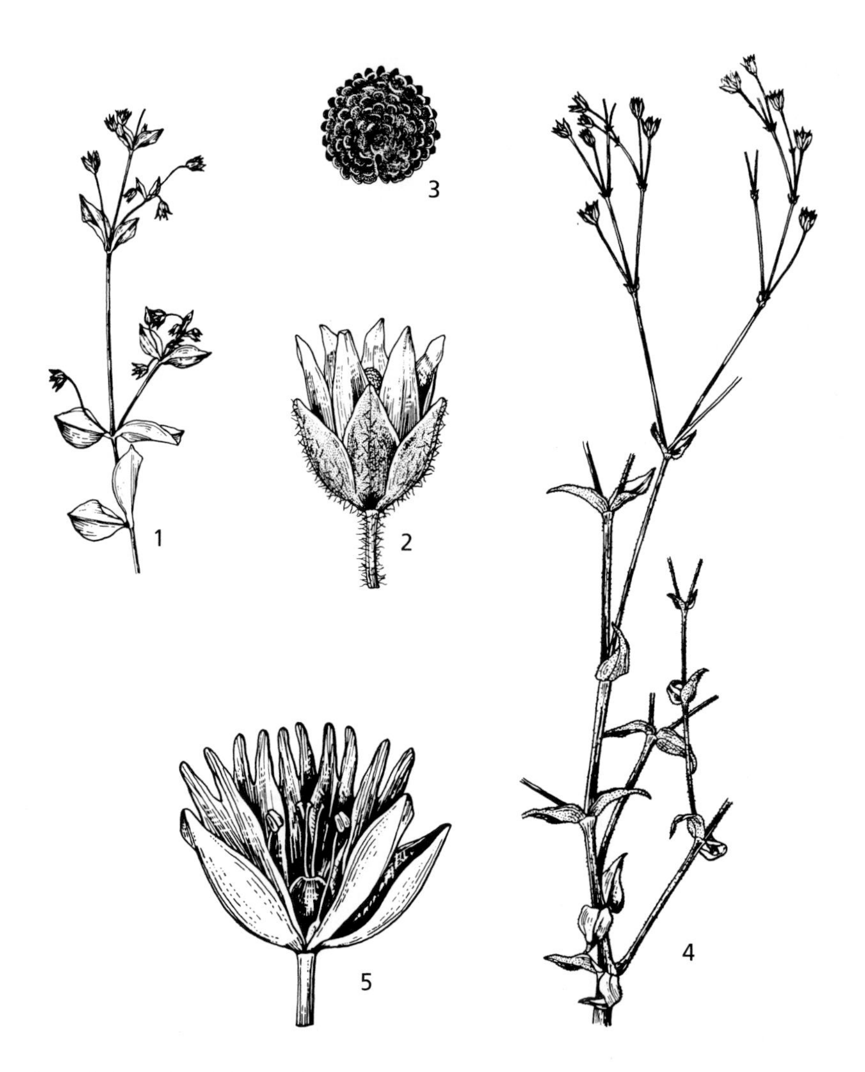

Fig. 18. **Stellaria media**. 1, habit × ½; 2, capsule × 5; 3, seed × 15. **Stellatria kotschyana**. 4, habit × ½; 5, flower × 5. Reproduced with permission from Flora of Pakistan 175: f. 3 & 4, 1986. Drawn by S. Hameed.

2. **Stellaria pallida** (Dumort) Piré in Bull. Soc. Roy. Bot. Belgique 2: 49 (1863); Rech. f. in Fl. Iranica 163: 65 (1988); Chamberlain in Fl. Arab. Penins. & Socotra 1: 209 (1996); Strid, Fl. Hellenica 1: 196 (1997); Boulos, Fl. Egypt 1: 76 (1999); Chen Shilong & Rabeler in Fl. China 6: 16 (2001).

Alsine pallida Dumort., Fl. Belg.: 109 (1827).
Stellaria media (L.) Vill. subsp. pallida (Dumort) Asch. & Graebner, Fl. Nordostd. Flachl. 310 (1898); Coode in Fl. Turk. 2: 69 (1967).
Stellaria apetala auct., non Ucria in Arch. Bot. (Leipzig) 1, 1: 68 (1796).

Similar to *St. media* Vill, but with the inflorescence rather congested, the petals usually absent or very small, the stamens 2–5 and seeds pale brown.

HAB. Cultivated fields, stream sides, rocky mountain sides; alt. 400–2200 m; fl. Feb.–May(–Aug.).
DISTRIB. Forest zone. **MRO**: Shaqlawa, *Haines* W604!; 80 km S W of Haji Umran, *Shahwani* 25268!; **FUJ**: K-2, *F.A. Barkley & E.D. Barkley* 4227!; **FKI**: Kirkuk, *Rogers* 486!; **FPF**: Khanaqin, *Haines* W859!; Makatu, *Guest* 4276!

N Europe, Mediterranean, extending to C Asia.

3. **Stellaria kotschyana** *Fenzl* ex *Boiss.*, Fl. Or. 1: 705 (1867); Coode in Fl. Turk. 2: 72 (1967); Ghazanfar & Nasır in Fl. Pak. 175: 29 (1986).

Mesostemma kotshyana (Fenzl ex Boiss.) Vved., Bot. Nat. Herb. Bot. Inst. Uzhek Fil. Akad. Nauk. SSSR. 3: 4 (1941); Rech.f., in Pl. Syst. Evol. 137: 137 (1981); Rech. f. in Fl. Iranica 163: 78 (1988).
Lepyrodiclis paniculata Stapf, Denkschr. Akad. Wiss. Wien Math.-Nat. Kl. 51: 287 (1886).

Perennial herbs. Stems erect to 45 cm, diffusely branched, glabrous (in Iraq). Leaves 10–40 × 4–6(–10) mm, ovate to ovate-lanceolate, glabrous, sessile, sometimes base slightly cordate in upper leaves; apex acute. Bracts to 4.5 mm, linear-lanceolate. Pedicels up to 16 mm in dichasium, glabrous. Calyx cylindrical; sepals 5–6 m, ovate-lanceolate with scarious margins, glabrous. Petals 5–6 mm, bilobed. Styles 2. Capsule teeth 4. Seeds 1–2, tuberculate. Fig. 18: 4–5.

HAB.Mountain sides, rocky slopes in *Quercus* forests ; alt. 1200–3100 m; fl. Jun.–Aug.
DISTRIB. Upper forest and thorn-cushion zones. **MAM**: Khantur Mountain, NE of Zakho, *Rawi* 23465!; **MRO**: Halgurd Dagh, *Gillett* 9561! & 1375!; Rowanduz, *Bornmüller* 966! (?type of var. *glabra*); Kani Mazu Shirin, *Haines* W2145!; Chiya-i Mandau, *Guest et al.* 2696!; **MSU**: Penjwin, *Rawi* 12257!; mt. Hauraman (Avroman), *Rawi et al.* 19776!; Kamarspa, between Halabja & Tawela, *Rawi* 22242!; Zewiya, *Rawi* 11526!

Turkey, NN Iran, Afghanistan and Pakistan.

All Iraq specimens seen are glabrous. Some specimens from Iran are completely glabrous whilst others subglabrous to entirely pubescent.

17. **CERASTIUM** L.

Spec. Pl. 437 (1753); Gen. Pl. ed, 5: 199 (1753)

Annual or perennial herbs, often glandular hairy. Leaves linear-subulate, ovate to lanceolate, elliptic or obovate to oblong. Inflorescence of lax cymes or of terminal cymose heads. Sepals 5, free. Petals (4–)5, white, emarginate or bifid to lobed. Stamens 5 or 10. Styles 3 or 5. Capsule oblong, cylindiric or cylindric-conic, opening by twice as many as teeth as styles, usually exceeding the sepals; teeth flat, revolute or with revolute margins. Seeds tuberculate.

Cosmopolitan with about 100 species, mainly distrubuted in the temperate regions of the Old World; eleven species in Iraq.

Refs. S. Murbeck (1898) Studier Öfver Kritikse Kärläxtformer III. De Nordeuropeiske Formerna Af Slägtet Cerastium, Botaniska Notiser 1898: 241–268.
F.N. Williams, Critical Notes on Some Species of *Cerastium*, J. Linn. Soc. (Bot.) 37: 116–124; 209–216; 310–315; 474–477 (1899)
W. Möschl (1966) De Cerastis Florae Iranicae, Sitzungsberichte der Österrechischen Akademie der Wissenschaften, Math.-Naturw. Klasse 175: 159–216.

1. Styles 3; capsule opening by 6 teeth 2
   Styles 5; capsule opening by 10 teeth 3
2. Diffuse perennials; stems almost glabrous, ascending to erect, rooting at lower nodes . . . . . 1. *C. cerastioides*
   Annuals: stems distinctly hairy at least above, erect, not rooting at nodes . . . . . 2. *C. dubium*
3. Plants glaucous, entirely glabrous . . . . . 3. *C. perfoliatum*
   Plants hairy 4
4. Perennials, ± caespitose 5
   Annuals 6
5 Petals longer than 10 mm, exceeding calyx with ciliate claws; filaments ciliate . . . . . 5. *C. purpurescens* var. *elburense*
   Petals shorter than 9 mm, not exceeding calyx with glabrous claws; filaments glabrous . . . . . 4. *C. fontanum* subsp. *vulgare*
6. Capsule teeth flat . . . . . 11. *C. dicohotomum*
   Capsule teeth revolute 7
7. Upper bracts scarious in upper $^1/_3$–$^1/_2$ . . . . . 7. *C. semidecandrum*
   Bracts not scarious in upper $^1/_3$–$^1/_2$, with only tips and/or margins scarious or not 8
8. Cauline leaves broadly elliptic to orbicular; inflorescence congested; pedicels shorter than sepals . . . . . 8. *C. glomeratum*
   Cauline leaves elliptic to oblanceolate or linear lanceolate; inflorescence lax; at least some pedicels longer than calyx 9
9. Petal claws pilose; capsule 10–15 mm long . . . . . 9. *C. longifolium*
   Petal claws glabrous; capsule 6–9 mm long 10
10. Some eglandular hairs exceeding the apex of sepals; petals with small auricle at base . . . . . 10. *C brachypetalum* subsp. *roeseri*
   Hairs not exceeding the apex of sepals; petals without auricle . . . . . 6. *C. fragilimum*

1. **Cerastium cerastioides** (*L.*) *Britton*, Mem. Torrey Bot. Club 5: 150 (1891); Zohary, Fl. Iraq & Its Phytogeogr. Subdivis.: 51 (1946); Möschl, Sitz.-Ber. Akad. Wiss Wien, Math.-Nat. Kl. Abt. 1, 175: 167 (1966); Cullen in Fl. Turk. 2: 75 (1967); Ghazanfar & Nasır in Fl. Pak. 175: 36 (1986); Möschl in Fl. Iranica 163: 91, t. 53 (1988); Strid, Fl. Hellenica 1: 200 (1997); Lu Dequan & Rabeler in Fl. China 6: 33 (2001).

Stellaria cerastioides L. Sp. Pl. 422 (1753).
Cerastium trigynum Will., Prosp. Pl. Dauph. 48 (1779).
Dichodon cerastioides (L.) Reichenb., Icon. Fl. Germ. 5: 34 (1842).
Cerastium argeum Boiss. & Bal. in Boiss., Diagn. Ser. 2, 6: 38 (1859).
Cerastium intermedium Williams in Bull. Herb. Boiss. 6: 899 (1898).
Cerastium cerastioides (L.) Britton var. *lalesarense* Bornm., Beih. Bot. Centrbl. 28, 2: 150 (1911).

Perennials. Stems creeping or ascending, 4–10 cm, rooting at lower nodes, glabrous. Leaves (2–)4–7 × 1.2–1.9 mm, oblong-lanceolate, glabrous. Bracts herbaceous, ± 2 mm, glabrous. Cymes lax, 1–6-flowered. Pedicels 5–20 mm, glabrous or glandular, reflexed in fruit. Sepals lanceolate, 4–5 mm, glabrous, often tinged purple, with scarious margins. Petals white, 7–8 mm, obcordate. Capsule exceeding sepals; teeth rolled circinately outwards. Seeds yellowish brown, subreniform, papillate.

Hab. Damp screes, slopes, nr. melting snows in rocky places; alt. 2700–3340 m; fl. July–Aug.
Distrib. Alpine zone of Iraq. **mro**: Qandil Range, *Rawi & Serhang* 26772! & 24514!; Halgurd Dagh, *Guest* 3075!; *Gillett* 9582! & 12327!; Ser Kurawa, *Gillett* 9726A!

On mountains in Europe, Asia and NW Africa.

2. **Cerastium dubium** (Bastrad) Guepin, Fl. Maine et Loire 267 (1830); Möschl, Sitz.-Ber. Akad. Wiss Wien, Math.-Nat. Kl. Abt. 1, 175: 176 (1966); Möschl in Fl. Iranica 163: 93, t. 54 (1988); Strid, Fl. Hellenica 1: 200 (1997).

Setellaria dubia Bastrard, Suppl. à l'Essai sur la Flore du Département de Maine-et-Loire 24 (Angers 1812).
Cerastium anomalum Waldst. & Kit. ex Willd., Sp. Plant. ed. 4: 812 (1799) non Schrank (1795); Zohary, Fl. Iraq & Its Phytogeogr. Subdivis.: 51 (1946); Cullen in Fl. Turk. 2: 75 (1967).

Arenaria anomala (Waldst. & Kit.) Skinners, Sida 1: 50 (1962).
Dichodon dubium (Bastrad) Ikonn., Nov. Syst. Pl. Vasc. Leningrad 10: 141 (1973).

Annuals. Stems 10–18 cm, glabrous to sparsely pubescent, usually glandular above. Leaves linear to lanceolate, (10–)15–35 × 1–4(–6) mm, glabrous, sometimes lower leaves sparsely pubescent on margins. Bracts herbaceous, glandular on margins. Pedicels erect, spreading, up to 15 mm long, longer than calyx in old flowers, glandular-pubescent. Sepals 4–6 mm long, obtuse to subacute, glandular-pilose with scarious margins. Petals white, equalling to slightly exceeding calyx, bifid. Styles 3. Capsule 7–10 mm, longer than calyx; teeth erect or rolled circinately outwards.

HAB. Meadows near stone walls, fields; alt. 1300–3500 m; fl. Mar.–Jun.
DISTRIB. Rare and possibly under-recorded. **MAM**: Amadiya, *Guest* 1255!; Mesopotamia, *Graham* s.n!

Europe, S Russia, Caucasica, Mediterranean, Iran and Afghanistan.

3. **Cerastium perfoliatum** L., Sp. Pl. 437 (1753); Zohary, Fl. Iraq & Its Phytogeogr. Subdivis.: 51 (1946); Möschl, Sitz.-Ber. Akad. Wiss Wien, Math.-Nat. Kl. Abt. 1, 175: 194 (1966); Cullen in Fl. Turk. 2: 78 (1967); Möschl in Fl. Iranica 163: 96, t. 58 (1988).

Annual, glabrous, glaucous herbs. Stems 15–50 cm high, erect. Lower leaves obovate; median leaves oblong or oblong-ovate or elliptic-lanceolate, 10–60 × 5–18 mm, connate at base. Bracts foliaceous; ovate, connate in pairs, shorter than pedicels. Pedicels erect, to 18 mm long. Sepals 7–9 mm long, lanceolate, glabrous, with scarious margins. Petals 5–6 mm, shorter than sepals. Styles 5. Capsule 11–25 mm long; narrowing toward above, teeth circinately revolute.

HAB. Cultivated grounds, damp places, shade places; alt. 900–1200 m; fl. Mar.–May.
DISTRIB. **MAM**: Sharifa nr. Amadiya, *Polunin* 5106!; *Guest* 1235!; Sarsang, *Haines* W958!

Spain, NW Africa, Hungary, Bulgaria, S Russia, Caucasia, Turkey, Syria, Iran and Turkmenistan.

4. **Cerastium fontanum** Baumg., Enum. Stirip. Transsily. 1: 425 (1816).

subsp. **vulgare** (Hartman) Greuter & Burdet in Wildenowia 12: 37 (1982); Strid, Fl. Hellenica 1: 204 (1997).

Cerastium vulgare Hartman, Handb. Skand. Fl.: 182 (1820).
C. vulgatum L., Sp. Pl. ed. 2, 267 (1762), non (1755).
C. caespitosum Gil. Fl., Lith. 5: 159 (1781).
C. holosteoides Fries, Novit Fl. Svec.: 52 (1817).
C. trivale Link, Enum. Hort. Berol. Alt. 1: 143 (1821), nom. illeg.
C. fontanum Baumg. subsp. *trivale* (Link) Jalas in Arch. Soc. Zool. Fenn. 'Vanamo' 18(1): 63 (1963); Cullen in Fl. Turk. 2: 80 (1967); Ghazanfar & Nasır in Fl. Pak. 175: 39 (1986).
C. holosteoides Fries subsp. **trivale** (Link) Möschl in Mém. Soc. Brot. 17: 11 (1964); Möschl in Fl. Iranica 163: 98 (1988).

Lax caespitose, perennials. Stems 6–25(–40) cm, spreading pilose, ascending to erect. Leaves pilose; lower leaves oblanceolate; upper one oblanceolate or oblong-ovate or lanceolate, 3–12(–28) × 3–4(–8) mm. Inflorescence rather dense, many-flowered. Bracts, herbaceous, pilose. Pedicels up to 10 mm long, setose-pilose. Sepals lanceolate, 5–7 mm long, setose-pilose, with scarious margin. Petals equalling to sepals. Styles 5. Capsule 9–11 mm long, longer than calyx opening by 10 with revolute teeth.

HAB. In grass, moist places nr. streams, marshy fields; alt. 1300–2500 m; fl. May–Aug.
DISTRIB. Montane zone. **MRO**: Halgurd Dagh, *Hadač* 2130 & 20701.
No specimens in K. Record from *Flora Iranica*.

Caucasica, Turkey, Iraq, Iran, Afghanistan, Pakistan, Himalaya, China, Taiwan, Korea and Japan.

5. **Cerastium purpurescens** Adams in Weber & Mohr, Beitr. Naturk. 1: 60 (1805).

var. **elburense** (Boiss.) Möschl, Sitz.-Ber. Akad. Wiss Wien, Math.-Nat. Kl. Abt. 1, 175: 198 (1966); Möschl in Fl. Iranica 163: 100, t. 62 (1988).

C. elburense Boiss., Fl. Orient. 1: 729 (1867); Blakelock, Kew Bull. 2: 185 (1957).

Caespitose perennials. Stem ascending to erect, up to 11 cm high, glandular-pilose. Leaves oblong, linear spatulate, 7–11 × 2–3 mm, obtuse to subacute, glandular-pilose. Infloescence usually ± umbellate. Bracts herbaceous, glandular pilose, shorter than pedicels. Pedicels up to 18 mm long, glandular-pilose. Sepals, lanceolate, (5–)7–8 mm long, sparsely pilose, with scarious margins. Petals 10–15 mm long, ciliate near base. Filaments ciliate at base. Styles 5. Capsule about 2× longer than calyx, opening by 10 teeth with revolute margins.

HAB. On mt. slopes; alt. above 3000 m; fl. Jul.–Aug.
DISTRIB. Rare and local in NE Iraq. MRO: Halgurd Dagh, *Guest* 3074!; *Rawi* 24924!

Iran and Pakistan.

6. **Cerastium fragillimum** Boiss., Diagn. Pl. Or. Nov. Ser. 1,1: 54 (1842); Zohary, Fl. Iraq & Its Phytogeogr. Subdivis.: 52 (1946); Blakelock, Kew Bull. 2: 186 (1957); Möschl, Sitz.-Ber. Akad. Wiss Wien, Math.-Nat. Kl. Abt. 1, 175: 179 (1966); Cullen in Fl. Turk. 2: 84 (1967); Möschl in Fl. Iranica 163: 102, t. 64 (1988); Strid, Fl. Hellenica 1: 213 (1997).

C. tmoleum Boiss., Diagn. Pl. Orient. Ser. 1 (5): 86 (1844).

Annual herbs. Stems erect, 8–20 cm high, glandular-pubescent. Basal leaves spatulate, glabrous to eglandular-ciliate; upper leaves narrowly elliptic, pubescent. Inflorescence lax, wide branching. Bracts 1–2.5 mm long, pubescent. Pedicels up to 20 mm long, patent glandular-pubescent, refracted in fruit. Sepals oblong, obtuse, 3.5–5 mm long, patent glandular-pubescent with narrow scarious margins. Petals slightly shorter than sepals, shortly bifid. Styles 5. Capsule 7–8 mm long, exceeding sepals; teeth with recurved margins. Fig. 19. 1–7.

HAB. Screes and forests; alt. 1100–1300 m; fl. Apr.–May.
DISTRIB. Rare and local in NE Iraq. MRO: Shaqlawa, *Haines* W602!

Greece, Cyprus, Lebanon, Turkey, W Iran.

7. **Cerastium semidecandrum** L., Sp. Pl. 438 (1753); Blakelock, Kew Bull. 2: 186 (1957); Möschl, Sitz.-Ber. Akad. Wiss Wien, Math.-Nat. Kl. Abt. 1, 175: 200 (1966); Cullen in Fl. Turk. 2: 83 (1967); Möschl in Fl. Iranica 163: 104, t. 66 (1988); Strid, Fl. Hellenica 1: 211 (1997).

C. pendantrum L., Sp. Pl. 438 (1753); Rech. f., Fl. Lowland Iraq: 233 (1964); Möschl in Fl. Iranica 163: 103 (1988).
C. balearicum Hermann in Verh. Bot. ver. Prov. Brand. 54: 247 (1913); Möschl in Fl. Iranica 163: 102 (1988).
C. dentatum Möschl, Österr. Bot. Zeitschr. 82: 230 (1933).

Annuals. Stems decumbent to erect, 4–20 cm high, glandular-pubescent. Leaves ovate-oblong, 4–10 × 2–5 mm, pilose, sometimes lower leaves glabrescent beneath; lower leaves attenuate at base; upper ones sessile. Inflorescences lax, many-flowered. Bracts up to 2 mm long, glandular-pubescent, upper half glabrous and scarious. Pedicels up to 10 mm long, glandular-pubescent, refracted in fruit. Sepals narrowly lanceolate, 3–4 mm long, glandular-pubescent, with scarious tip and margins. Petals as long as or slightly shorter than calyx. Styles 5. Capsule 7–8 mm long, up to 2× longer than sepals; teeth with revolute margins.

HAB. Sandy soils, cliffs; alt. 25–1100 m; fl. April–May.
DISTRIB. Rare and local in NE Iraq. MRO: Shaqlawa, *Haines* 758!; FPF: Diyala, Khanaqin, *Haines* W894!

Europe, Caucasus, Turkey, Cyprus, Lebanon, Palestine, N Iran, Afghanistan and C Asia.

8. **Cerastium glomeratum** *Thuill.*, Fl. Env. Paris ed. 2: 226 (1799); Möschl, Sitz.-Ber. Akad. Wiss Wien, Math.-Nat. Kl. Abt. 1, 175: 180 (1966); Cullen in Fl. Turk. 2: 82 (1967); Ghazanfar & Nasır in Fl. Pak. 175: 38 (1986); Möschl in Fl. Iranica 163: 104, t. 67 (1988); Chamberlain in Fl. Arab. Peninsula and Socotra 1: 207 (1996); Strid, Fl. Hellenica 1: 210 (1997); Boulos, Fl. Egypt 1: 76 (1999).

C. viscosum L., Sp. Pl.: 437 (1753) p.p., nom. ambig. (cf. Lonsing 1939: 162–163).
C. vulgatum L., Fl. Suec., ed. 2: 158 (1755) p.p., nom. ambig. (cf. Lonsing 1939: 162–163).

Annuals. Stems 6–24 cm high, ascending to erect, hirsute-pilose. Leaves 8–18 × 3–10 mm, eglandular-pilose; lower leaves spatulate to obovate, obtuse; middle ones broadly elliptic to

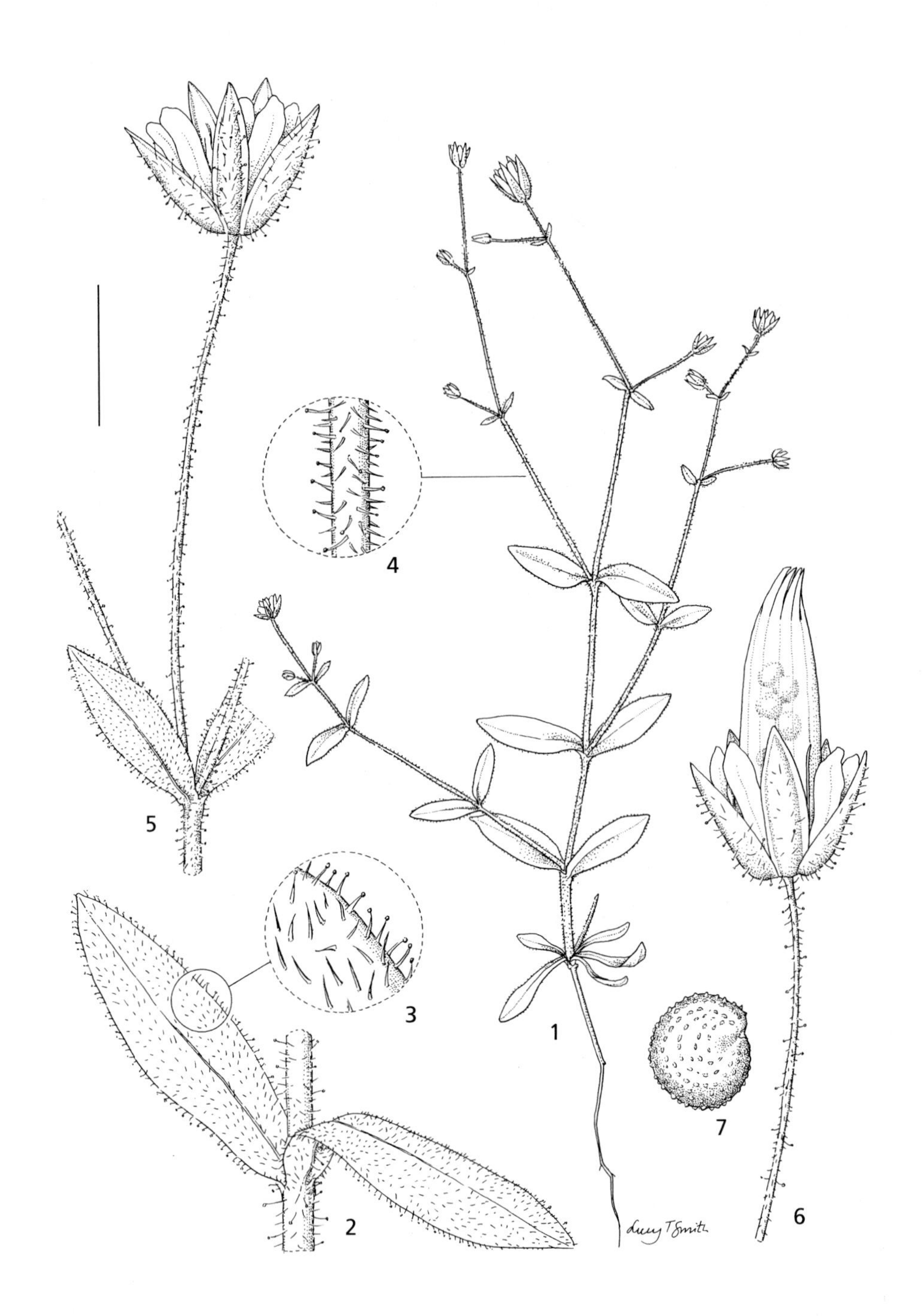

Fig. 19. **Cerastium fragillimum**. 1, habit × 1; 2, stem and leaves × 5; 3, detail of leaf hairs; 4, details of stem hairs; 5, flower × 5; 6, capsule × 5; 7, seed × 12. 1, 4 from *Kennedy* 374; 2, 3, 5–7 from *Haines* W607. Scale bar: 1 = 2 cm; 2, 5–6 = 4 mm; 3, 4, 7 = 1.6 mm. © Drawn by Lucy T. Smith.

orbicular, obtuse to subacute. Inflorescence congested, many-flowered. Bracts herbaceous, up to 3 mm, pilose. Pedicels usually shorter than calyx, glandular- or eglandular-pilose. Sepals lanceolate, 3–4 mm, with scarious margins, glandular- or eglandular-pilose, some eglandular hairs exceeding apex. Petals shorter or equalling the calyx, usually bifid. Styles 5. Capsule 6–8 mm; teeth with revolute margins.

HAB. Sandbank, wet waste lands, steppes with calcarious sandstones; alt. 0–800 m; fl. Mar.–May.
DISTRIB. Occasional in the steppe zone of Iraq. **MAM**: Sarsang, *Haines* 937!; **MRO**: Shaqlawa, *Haines* 601!; 11 km NW of Rania, *Rawi, Nuri & Kass* 28577!; **MSU**: Palegawra, W of Sagirma Dagh, *Helbaek* 685!; Sulaimaniya, *Graham* 722!; Jarmo, *Helbaek* 481!; **FUJ**: Mosul, *Guest* 3388; **FAR**: Zab, below Kau Gossik, *Gillett* 10524!; **FPF**: between Kani Kirmaj and Haji Lar, *Poore* 478!

Mediterranean area, C and W Europe, Caucasus, Iraq, Iran, Arabia, Pakistan; introduced in N America, S and W Australia, New Zealand.

9. **Cerastium longifolium** *Willd.*, Sp. Pl. ed. 4, 2,1: 814 (1799); Möschl, Sitz.-Ber. Akad. Wiss Wien, Math.-Nat. Kl. Abt. 1, 175: 188 (1966); Cullen in Fl. Turk. 2: 81 (1967); Möschl in Fl. Iranica 163: 105, t. 68 (1988).

C. blepharostomum Fisch. & CA. Mey. ex Hohen., Enum. Pl. Talysch in Bull. Soc. Imp. Nat. Mosc. 11: 403 (1838); *C. blepharophyllum* Fisch. & C A. Mey. ex Fenzl in Ledeb., Fl. Ross. 1: 403 (1842).

Annuals. Stems 8–25 cm high, dichotomously branched, pubescent below, glandular-pubescent above. Leaves linear-lanceolate 15–30 × 2–6 mm, glandular-pilose, sometimes glabrescent beneath. Inflorescence lax, many flowered. Bract herbaceous, glandular-pilose, to 4 mm long. Pedicels 5–14(–35) mm, with mixture of glandular and eglandular hairs. Sepals lanceolate, 7–9 mm long, glandular-pilose with scarious margins. Petals equalling sepals, claws ciliate. Filaments ± ciliate. Styles 5. Capsule 10–15 mm long; teeth with revolute margins.

HAB. Screes; alt. 1700–2400 m; fl. Jun.–Jul.
DISTRIB. Rare. **MRO**: N of Rost, *Thesiger* 937.
No specimens at K; record from Flora Iranica.

Armenia, Turkey, NNW Iran.

10. **Cerastium brachypetalum** Pers., Synops. Pl. 1: 520 (1805).

subsp. **roeseri** (Boiss. & Heldr.) Nyman, Consp. Fl. Eur.: 109 (1878); Cullen in Fl. Turk. 2: 82 (1967); Strid, Fl. Hellenica 1: 210 (1997).

C. roeseri Boiss. & Heldr. in Boiss., Diagn. Pl. Orient., ser 2, 3(1): 93 (1854).
C. atticum Boiss. & Heldr. in Boiss., Diagn. Pl. Orient., ser 2, 3(1): 93 (1854).
C. brachypetalum Pers. var. *luridum* Boiss., Boiss., Fl. Orient. 1: 723 (1867).
C. brachypetalum Pers. subsp. *luridum* (Boiss.) Nyman, Consp. Fl. Eur.: 109 (1878).
C. luridum (Boiss.) Lonsing in Repert. Spec. Nov. Regni Veg. 46: 159 (1938), non Guss., Fl. Sicul. Syn. 1: 510 (1843), nom. illeg.

Annuals. Stems 4–15 cm high, ascending to erect, pilose below, glandular-pubescent above. Leaves elliptic to oblanceolate, pilose. Inflorescence lax. Bracts 1 –2 mm long, hairy. Pedicels exceeding calyx, up to 9 mm long, glandular-pubescent with patent hairs. Sepals 3–5 mm long, with mixture of glandular and eglandular patent hairs, some eglandular hairs exceeding apex; margins scarious. Petals as long as sepals or slightly shorter with small auricle at base. Styles 5. Capsule 6–9 mm long; teeth with revolute margins.

HAB. Rocky places; alt. 900–1200 m; fl. Mar.–Jun.
DISTRIB. Rare; upper forest zone. **MRO**: Shaqlawa, *Haines* W603!

Mainly C and E Mediterranean area, extending to Iran.

11. **Cerastium dichotomum** *L.*, Sp. Pl. 438 (1753); Blakelock, Kew Bull. 2: 185 (1957); Cullen in Fl. Turk. 2: 81 (1967); Ghazanfar & Nasır in Fl. Pak. 175: 37 (1986); Möschl, Sitz.-Ber. Akad. Wiss Wien, Math.-Nat. Kl. Abt. 1, 175: 174 (1966); Leonard, Contrib. Fl. et Veg. Deserts d'Iran 6: 13 (1986); Möschl in Fl. Iranica 163: 107 (1988); Chamberlain in Fl. Arab. Peninsula and Socotra 1: 208 (1996); Strid, Fl. Hellenica 1: 201 (1997); Boulos, Fl. Egypt 1: 77 (1999); Lu Dequan & Rabeler in Fl. China 6: 34 (2001).

Fig. 20. **Cerastium dichotomum**. 1, habit × 1; 2, stem and leaves × 3; 3, detail of stem hairs; 4, flower with front petal removed × 3; 5, capsule × 3; 6, seed × 14. 1–3 from *Poore* 344; 4 from *Barkley & Barkley* 8068; 5–6 from *Rawi* 21878. © Drawn by Lucy T. Smith.

Annuals. Stems erect, dichotomously and divaricately branched, 8–25 cm, pilose to pubescent, often glandular. Leaves pubescent with mixture of glandular and eglandular hairs, ± obtuse; basal ones oblanceolate, middle cauline leaves linear-oblong, 10–30 × 3–10 mm. Inflorescence lax to condense, many-flowered. Bracts herbaceous. Pedicels shorter than calyx, rarely equalling, densely patently glandular-pilose. Calyx somewhat inflated in fruit. Sepals lanceolate, 7–13 mm, sharply acute to acuminate, patent glandular-pilose. Petals as long as or shorter than calyx, bilobed. Styles 5. Capsule 14–23 mm, opening by flat and narrowly triangular teeth. Fig. 20: 1–6.

Inflorescence dense; calyx mostly less than 5 mm in diameter near base in fruit. . . . . a. var. *dichotomum*
Inflorescence rather lax; calyx more than 6 mm diameter near base in fruit. . . . . b. var. *inflatum*

a. var. **dichotomum**

HAB. Mountain slopes, open places in Quercus forest, cultivated grounds; alt. 1200–2000 m; fl. Apr.–Jun.
DISTRIB. Occasional in the upper forest zone of Iraq. **MAM**: Dohuk, *A. Barkley & E.D. Barkley* 8068!; **MRO**: Haji Umran, *Guest & Kass* 18571!; Kawreish, E side of Karokh Mt., *Kass & Nuri* 27684!; **MSU**: Kamarspa, between Halabja & Tawela, *Rawi* 22255!; Kopi Qara Dagh, *Haines* W1097!; Penjwin, *Rawi* 8549!

b. var. **inflatum** (Link) Kandemir comb. et stat. nov.

C. inflatum Link in Desf., Cat. Hort. Par. 462 (1829); Möschl in Fl. Iranica 163: 108 (1988).
C. dichotomum L. subsp. *inflatum* (Link) Cullen in Notes R.B.G. Edinb. 27: 211 (1967); Cullen in Fl. Turk. 2: 81 (1967).

HAB. Mountains, dry hillsides, near streams in mountains, stony places; alt. 1200–2100 m; fl. Apr.–Jun/Sep.
DISTRIB. Occasional in the upper forest zone of Iraq. **MAM**: 3 km N E of Amadiya, *A. Barkley* 8202!; nr. Sarsang, *Guest & Kass* 18698!; **MRO**: Rowanduz, *Guest* 2095!; Dergala, *Rawi, Nuri & Kaisi* 28903!; **MSU**: nr. Halabja, *Omar et al.* 37451!; Tawela, *Rawi* 21878; Azmir, *Omar & Kaisi* 38079! & *Poore* 344!; Qara Dagh, *Gillett* 7910!; top of Sinjar, *Omar, Khayat & Kaisi* 52533!; Karsi, *Omar, Khayat & Kaisi* 52451!; *Gillett* 11044!

Distribution of the species: Mediterranean, Caucasus, N Iraq, Iran, Turkestan, Afghanistan and Pakistan.

18. **HOLOSTEUM** L.

Sp. Pl. 88 (1753); Gen. Pl. ed. 5: 376 (1754)

Small annual herbs. Leaves lanceolate to spatulate, opposite; stipules absent. Inflorescence terminal umbellate. Sepals 5, free to base. Petal 5, entire or irregularly toothed. Stamens 3–5; nectaries present. Stamens 3–5 or 8–10. Styles 3(–5). Capsule cylindrical, dehiscing by 6 revolute teeth. Seeds asymmetrical, reniform, tuberculate.

About four species in temperate Eurasia, one record from Ethiopia; two species in Iraq.

Stamens 3–5. . . . . 1. *H. umbellatum*
Stamens 8–10 . . . . 2. *H. glutinosum*

1. **Holosteum umbellatum** *L.*, Sp. Pl. 88 (1753); Rech. f., Fl. Lowland Iraq: 231 (1964); Coode in Fl. Turk. 2: 86 (1967); Ghazanfar & Nasır in Fl. Pak. 175: 33 (1986); Rech. f. in Fl. Iranica 163: 110 (1988); Strid, Fl. Hellenica 1: 198 (1997); Lu Dequan & Rabeler in Fl. China 6: 40 (2001).

Annual herbs. Stems erect to ascending, 5–25(–30) cm, glaucous, glabrous or glandular hairy, sometimes eglandular hairs present. Leaves glabrous; basal leaves spatulate to linear, acute at apex; upper ones linear to ovate, sessile. Inflorescence 2–16-flowered. Pedicels 3–18 mm, glabrous to glandular hairy, erect at first, soon becoming sharply deflexed, erect or deflexed in fruit. Sepals 4–5 mm, ovate, glabrous or glandular. Petals white as long as or longer than sepals. Stamens (2–)3–5. Capsule ± exceeding the sepals. Seeds tuberculate. Fig. 21: 1–5.

HAB. Limestone slopes, open places, road sides; alt. 160–2000 m; fl.Apr.–May.

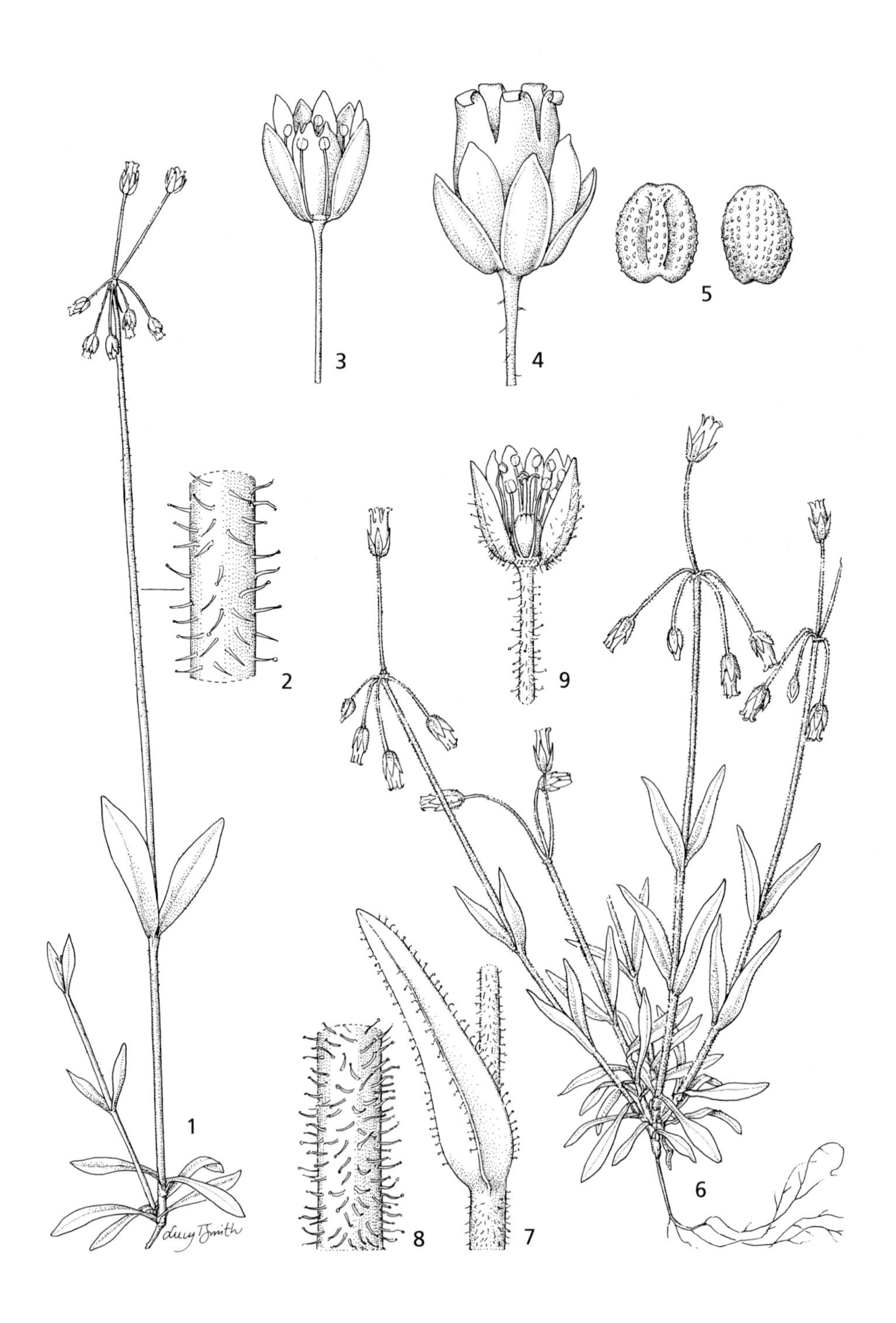

Fig. 21. **Holosteum umbellatum**. 1, habit × 1; 2, detail of stem hairs; 3, flower with front sepal and 2 petals removed × 6; 4, capsule × 6; 5, seed × 14. **Holosteum glutinosum**. 6, habit × 1; 7, stem and leaf × 4; 8, detail of stem hairs; 9, flower with sepal and 2 petals removed × 6. 1–5 from *Roger* 578; 6–9 from *Chapman* 11928. © Drawn by Lucy T. Smith.

DISTRIB. **MAM**: Aqra, *Polunin et al.* 35!; Amadiya, *Guest* 1389!; Zakho, *Guest* 2275!; **MRO**: Shaqlawa, *Haines* W606!; **MSU**: Jarmo, *Helbaek* 356!; Sulaimaniya, *Graham* s.n.!; **FUJ**: Shargat, *Guest* 400!; **DWD**: Ana, *Omar et al.* 45105!; **LCA**: Rustamiya, *Roger* 578!

Europe, Mediterranean region to Pakistan.

2. **Holosteum glutinosum** (M.B.) Fisch. & CA. Mey., Ind. Sem. Horti Petrop. 6: 52 (1839); Rech. f., Fl. Lowland Iraq: 231 (1964); Rech.f. in Fl. Iranica 163: 111 (1988); Chamberlain in Fl. Arab. Penins. & Socotra 1: 204 (1996).

Holosteum liniflorum Fisch. & CA. Mey., Ind. Sem. Horti Petrop. 3: 39 (1839).
Arenaria glutinosa M.B., Fl. Taur.Cauc. 1: 344 (1808).
Holosteum polyganum C Koch. Linnaea 15: 708 (1841).
H. umbellatum L. var. *glutinosum* (M. B). Gay in Ann. Sci. Nat. ser. 3, 4: 33 (1845).

Similar to *H. umbellatum* L. but the indumentum is denser and stamens 8–10. Fig. 21. 6–9.

DISTRIB. **MAM**: Amadiya, *Guest* 1 269!; Zawita, *Khudairi & Mohammed* 34!; **MSU**: 15 km W of Chemchemal, *Gillett & Rawi* 10595!; **MRO**: Haji Umran, *Chapman* 11194! & *Mooney* 4279!; Selahaddin Forestry, *Gillett & Rawi* 10425!; **FUJ**: Qaiyara, *Bayliss* 28!; **FPF**: Garaghan, *Roger* 355!; **DWD**: Wadi Fuhaimi, 40 km S of Ana, *Omar et. al.* 44391!

Caucasus, E Europe, Turkey to Pakistan.

19. **BUFONIA** L.

Sp. Pl. 123 (1753)

Annual or perennial herbs or small subshrubs. Leaves linear-subulate, shorter than internodes. Sepals free, 4, the outer pair shorter than the inner pair. Petals 4, white, entire, or erose or bifid. Stamens 2 or 8, inserted on a perigynous disk. Styles 2. Capsule opening to base by 2 valves, compressed.

A genus of about 20 species mostly on mountains in the Canary Islands, the Mediterranean and Irano-Turanian regions; three species in Iraq.

1. Annual; at least some flowers in racemiform inflorescence . . . . 3. *B. oliveriana*
   Perennial, suffrutescent; inflorescence not so. 2
2. Upper stem leaves minutely scabrous to ciliate on scarious margins; at least outer sepals ciliate on margins . . . . 1. *B. calycina*
   Upper stem leaves glabrous; sepals not ciliate on margins. . . . . 2. *B. leptoclada*

1. **Bufonia calycina** *Boiss. & Hausskn.* in Boiss., Fl. Orient. Suppl. 110 (1888). — Type: Iraq, M. Avroman, *Haussknecht* 218 (G!, G-Boiss., P!); Rech. f. in Fl. Iranica, 163: 120 (1988).

Suffrutuscent. Stems ascending to erect, 20–30 cm high, glabrous. Leaves rigid, subulate, up to 14 mm long, sheats ± 2 mm long, scabrous to ciliate on scarious margin; cauline leaves ± adpressed to stem. Flowers 1–3 in cymes. Bracts up to 2.5 mm long with broad scarious margin. Pedicels usually shorter than calyx, sometimes longer, glabrous or shortly hairy. Sepals ovate-lanceolate to lanceolate, glabrous to papillose, with white broad ciliate scarious margin; inner pair longer than outer, ± 5 mm long. Petals oblong, clearly shorter than calyx. Style subequal to ovary. Seeds 4.

HAB. In *Astragalus* thorn cushion zone on igneous or metamorphic rock; alt. 2600–3000 m; fl. July.
DISTRIB. **MRO**: Eastern slopes of Halgurd (Arl Gird) Dagh, *Gillett* 9612!; **MSU**: Avroman, *Haussknecht* 218!
The species is close to *B. kotschyana* Boiss, but the sepals are subequal in *B. kotschyana*.

NW Iran.

2. **Bufonia leptoclada** *Rech.f.* in Fl. Iranica, 163: 121 (1988). — Type. Iraq: Agnew, *Haines & Hadač* 2153 (syn. E!, G; K!, PR).

Suffrutex, caespitose. Stems numerous, ascending to erect, 20–40 cm high, branched, glabrous, rarely minutely puberulent below the nods. Leaves filiform-subulate, up to 22 mm long, sheats 1–1.5 mm long, flaccid, glabrous, rarely scabrous to ciliate on scarious margin

Fig. 22. **Bufonia leptoclada**. 1, habit × ½; 2, leaves × 5; 3, inflorescence × 1½ ; 4, flower with sepal and 2 front petals removed × 8; 5, flower bud, dissected × 8; 6, flower bud showing sepals × 8. All from *Agnew et al.* 2153. © Drawn by Lucy T. Smith.

in lower leaves. Inflorescence lax, diffuse. Bracts up to 3 mm long with scarious margins. Pedicels 2–5 mm long, glabrous. Sepals 3–3.5 mm long, ovate-lanceolate, acute with broad scarious margins, glabrous; inner pair usually longer than the outer. Petal broadly ovate, shorter than calyx. Style longer than ovary. Fig. 22: 1–6.

HAB. Rocky slopes near *Quercus libani* forests; alt. ± 1900 m; fl. Jun.
DISTRIB. **MRO**: Kani Mazu Shirin, nr. Shirwan Mazin, *Agnew, Haines & Hadač* 2153! (type); *M. Potin, Hadač et al.* 6106.

Endemic.

3. **Bufonia oliveriana** Ser. in DC, Prodr. 1: 388 (1824); Cullen in Fl. Turk. 2: 89 (1967); Ghazanfar & Nasır in Fl. Pak. 175: 45 (1986); Rech.f., Fl. Iranica, 163: 122 (1988).

Annual herbs usually branched from base. Stems ascending to erect, 12–35 cm high, glabrous or minutely papillose. Leaves linear-subulate, 12–25 mm long, 3-veined, with scarious margins at base; margins ciliate or not. Flowers in paniculate or racemiform inflorescence. Bracts lanceolate, acute, ± 2.5 mm with scarious margins. Pedicel up to 2 mm long, glabrous to subglabrous at base, densely pubescent above. Sepals 3.5–4 mm long, lanceolate, acuminate, 3-nerved, glabrous with scarious margins, subequal. Petals linear, shorter than calyx. Style very short. Seeds elongate, the back minutely tuberculate, the face ± smooth.

HAB. Mountain slopes, under forests, near streams, on serpentinite rocks; alt. 300–2600 m; fl. Jun.–Aug.
DISTRIB. Forest zone of NE Iraq. **MAM**: Mt. Khantur, *Rawi, Tikriti & Nuri* 29007!; Sarsang, *Haines* 473!; **MRO**: Pishtashan, *Rawi* 23912!; Haj Omran, *Rawi* 24265!; Kani Rash, *Serhang & Rawi* 26658!; **MSU**: Penjwin, *Rawi* 22531!; *Guest* 12936!; Avroman, *Rawi et al.* 19832!; Sulaimaniya, *Rechinger* 12328!

Caucasus, SW Turkey, Iran, Afghanistan, Pakistan, Turkestan.

20. **SAGINA** L.

Sp. Pl. 128 (1753); Gen. Pl. ed. 5: 62 (1754)

Annual or perenial herbs with very slender stems. Leaves very narrow, slightly connate at base. Flowers usually solitary, sometimes in few-flowered cymes, 4–5-merous. Sepals free to base, very small or absent. Stamens 4–5 or 10. Styles 4–5. Capsule opening to base by 4–5 valves. Seeds numerous, small, orbicular, reniform, tuberculate.

About 25 species distributed in the northern temperate regions, arctic, mountains from S to E Africa, Himalayas, New Guinea and Andes; two species in Iraq.

Perennial; flowers 5–merous .... 1. *S saginoides*
Annual; flowers 4–merous .... 2. *S apetala*

1. **Sagina saginoides** (*L.*) *Karst.*, Deutsch. Fl. Pharm.-med. Bot. 539 (1882); Cullen in Fl. Turk. 2: 90 (1967); Ghazanfar & Nasır in Fl. Pak. 175: 43 (1986); Rech. f., Fl. Iranica 163: 124 (1988); Strid in Fl. Hellenica 1: 216 (1997).

Spergula saginoides L., Sp. Pl. 441 (1753).
Sagina linnaei Presl, Rel. Haenk. 2: 14 (1835–36), nom illeg.

Small caespitose perennial. Stems 1–6 cm, procumbent to ascending ± glabrous. Leaves linear-lanceolate, 5–18 mm, base connate with scarious margins, acute, very shortly mucronate at apex. Flowers usually solitary. Pedicels up to 40 mm, glabrous. Sepals 5, ovate, 1.5–3 mm with hyaline margins, obtuse. Petals 5, white, somewhat shorter than to equalling the sepals. Stamens 5–10. Ripe capsule exserted beyond the persistent sepals. Seeds less than 0.5 mm.

HAB. Damp alpine slopes, igneous rock; alt. 2000–3100 m; fl. Jul.–Aug.
DISTRIB. In thorn-cushion zone of NE Iraq. **MRO**: Ser Kurawa, *Gillett* 9723! & 9724!; Halgurd Dagh, *Hadač* 2182.

On mountains of Europe, NW Africa, Asia, N America and Mexico.

2. **Sagina apetala** *Arduino*, Animadv. Bot. Spec. 2: 22 (1964); Cullen in Fl. Turk. 2: 91 (1967); Ghazanfar & Nasır in Fl. Pak. 175: 43 (1986); Rech.f., Fl. Iranica 163: 126 (1988); Strid in Fl. Hellenica 1: 217 (1997); Boulos, Fl. Egypt 1: 71 (1999).

Sagina procumbens L. var. *apetala* (Ardunio) Huds., Fl. Angl. Ed. 2: 73 (1778).
S. ciliata Fries in Liljebl., Utk. Svensk. Fl. ed. 3, 713 (1816).
S. reuteri Boiss., Diagn. Pl. Orient., Ser. 2,3 (1): 82 (1854).

Annual. Stems 3–8 cm, ± ascending, usually branched from base, lateral stems and branches decumbent to ascending, not rooting at nodes. Leaves linear, ciliate near the base, long-mucronate. Flowers 4-merous, solitary, axillary or terminal. Pedicels filiform 4–18 mm, glabrous or glandular-hairy in the upper part. Sepals oblong-ovate, cucculate at apex, outer two mucronate. Petals white, minute, often quickly caducous. Capsule equalling or exceeding the spreading to adpressed persistent sepals. Seeds minutely and irregularly rugulose. Fig. 10: 4–5.

HAB. On damp limestone rocks near spring, bare open grounds on sandy soil igneous rock; alt. up to 2000 m; fl. Mar.–May.
DISTRIB. **MSU**: Qarachitan, *Gillett* 7710!

Europe, Mediterranean, SW Asia and N America.

III. SUBFAMILY CARYOPHYLLOIDEAE

21. **DIANTHUS** L.

Sp. Plant. ed. 1: 409 (1753); Gen. Pl. ed. 5: 191 (1754)

Perennial, rarely annual or biennial herbs, or subshrubs (not in Iraq). Leaves opposite and decussate, frequently linear and parallel-nerved, connate basally and forming a sheath. Flowers 5- numerous, solitary, clustered, or capitate and surrounded by bracts; each flower subtended by 1- several pairs of imbricate epicalyx scales. Calyx tube ± cylindrical with many parallel nerves, without scarious commissures; teeth 5, usually triangular-lanceolate, acute to acuminate. Petals 5, long-clawed, limb seldom entire, usually dentate and also laciniate, sometimes bearded, pink or reddish-purple, seldom white; coronal scales lacking. Stamens 10; Styles 2. Ovary 1-locular. Fruit a capsule dehiscing by 4 teeth. Seeds dorsiventrally compressed, reticulate, black.

A large genus with about 300 species distributed in North temperate regions, mostly in Asia and Europe and especially concentrated in the Mediterranean region. A few species grow in Africa and America. Eleven native species found in Iraq.
D. barbatus L., *D. chinensis* L. and *D. caryophyllus* L. are grown as ornamental in gardens.

In this account, measurements of leaf sheaths refer to lower part of stems and diameter of stem is measured just below the node. Lengths of internodes are measured from the middle part of stems. Bracts occur in species with capitate inflorescences, and should be distinguished from epicalyx scales as they subtend more than one flower.
The length of the inner and outer epicalyx scales is ± equal but the lamina of the outer is shorter and the cusp is longer; the inner vice-versa. The uppermost pair of cauline leaves just below the epicalyx scales is not included because their size differs markedly. Flower measurements refer only to hermaphrodite flowers at flowering time. Female fertile and male sterile flowers occur sporadically and add to the difficulty in identification. Late flowers, particularly if borne on lateral branches produced by damaged main stems, are possibly abnormal in size and in the number of epicalyx scales.

Ref. F.N. Williams (1893). A Monograph of the Genus Dianthus, *Journ. Linn. Soc. Bot.* 29: 346–478

Key to species

1. Flowers congested in heads, sessile or with 1–2 mm pedicels. 2
   Flowers distinctly pedicellate, not in heads, sometimes in groups of 3–4, then calyx more than 32 mm and some pedicels at least 5 mm long 3
2. Stems and calyx short hairy . . . . 12. *D. hymenolepis*
   Stems and calyx glabrous . . . . 13. *D. persicus*
3. Petal limb entire, subentire or dentate, lacerations less than ¼ of limb length 4
   Petal limb fimbriate, lacerations more than ¼ of limb length 8

4. Calyx verruculose, covered in plain hemispherical protruberances (visible only at a magnification of ×10 5
Calyx not verruculose 6
5. Annual; epicalyx scales ± equalling length of calyx.......... 2. *D. cyri*
Perennial; epicalyx scales at most $^2/_5$ of length of calyx .... 1. *D. strictus*
6. Petal limbs barbellate, sometimes minutely so .... 11. *D. zonatus* var. *hypochlorus*
Petal limbs ebarbellate 7
7. Calyx 13–14 mm, tube not wider than 4 mm.......... 3. *D. siphonocalyx*
Calyx longer than 18 mm, tube wider than 4 mm.......... 4. *D. judaicus*
8. Epicalyx scales densely imbricate, usually more than 16; calyx teeth 4–5 mm .... 7. *D. pendulus*
Epicalyx scales up to 12, if more than calyx teeth longer than 6 mm 9
9. Calyx up to 25 mm.......... 10
Calyx longer than 25 mm 11
10. Sheaths of leaves longer than 4–6 mm; calyx 21–24 mm .... 5. *D. longivaginatus*
Sheaths of leaves mostly shorter 3 mm ; calyx 17–21(–24) mm .... 6. *D. orientalis*
11. Petal limb barbullate; calyx tube glabrous.......... 10. *D. libanotis*
Petal limb ebarbullate; calyx tube hairy, sometimes glabrous in *D. macranthus*.......... 12
12. Epicalyx scales (4–)6–8; calyx shorter than 34 mm.......... 8. *D. basianicus*
Epicalyx scales usually more than 8; at least some calyces more than 34 mm .... 9. *D. macranthus*

1. **Dianthus strictus** *Banks* & *Soland.* in Russell, Nat. Hist. Aleppo ed. 2, 2: 252 (1794); Zohary, Fl. Iraq & Its Phytogeogr. Subdivis.: 53 (1946); Rech.f., Fl. Lowland Iraq: 245 (1964); Zohary, Fl. Palaest. 1: 108 (1966); Reeve in Fl. Tur.2: 106 (1967); Rech.f., Fl. Iranica 163: 135 (1988).

Perennials, with ± stout woody rootstock. Stems many, 20–60 cm, simple or branched, glabrous or short hairy; internodes 2.5–4 cm, nodes sometimes oblique. Leaves linear-lanceolate, 15–30 × 1–2.5 mm, margin shortly ciliate or scabrous, margin hyaline at base, 3-veined; sheaths of cauline leaves as long as stem diameter. Flowers solitary, sometimes also secondary flowers below with shorter pedicels or sessile. Epicalyx scales 4–6 (secondary flowers sometimes have 1–3 pairs of scale-like cauline leaves a short distance below the epicalyx scales which are not included), enclosing to $^1/_3$ of length of calyx, short- to long-cuspidate, seldom caudate, 4–6(–8) mm; awn usually 0.5–1.5 mm, upper margin of lamina of scales broadly hyaline, tips ± weakly striate. Calyx cylindrical-conical (10–)11–13(–15) mm, verruculose, wholly or partially striate; teeth lanceolate, 3.5–5 mm, with narrow, hyaline, ciliate margin, regularly striate, sometimes with a violet tinge. Petals up to 25 mm, obovate-cuneate; limb 4–7(–10) mm, pink to pinkish, sometimes or frequently with dark striations, yellowish on abaxial surface, deeply dentate, barbellate. Capsule included in calyx. Fig. 23: 1–9.

HAB. In fallow fields, edges of cultivated ground, stony slopes in *Quercus* forest, also in damp fields and on sandy soil; alt. 400–2600 m; fl. May–Aug.
DISTRIB. E Aegean Isles, Iran, Jordan, Lebanon, Palestine, Syria, Turkey.
The ± 12 mm calyx is diagnostic for this species, along with the hyaline margins of the calyx teeth and calyx scales, the awn of the calyx scales, and the bearded petal limb.

In my opinion it is preferable not to follow the treatment of Reeve (1967) and Rechinger (1988) in distinguishing also var. *axilliflorus* (Fenzl) Reeve. The arguments of Boissier (1867), Lemperg (Acta Horti Gothob. 11: 71–136, 1936), and Weissmann-Kollmann (Israel Journ. Bot. 14: 141–170, 1965) are very convincing, namely that the late summer flowers (located in the axils) are the result of an interrupted flowering period e.g. by overgrazing or sudden increase in humidity. In fact material named as *D. strictus* var. *axilliflorus* contains also elements of *D. strictus* subsp. *strictus* var. *strictus*, subsp. *strictus* var. *gracilior* and subsp. *strictus* var. *subenervis*.

1. Stem and leaves short hairy .... b. subsp. *velutinus*
Stem and leaves glabrous 2
2. Whole calyx striate.......... a. subsp. *strictus* var. *strictus*
Whole calyx not striate 3

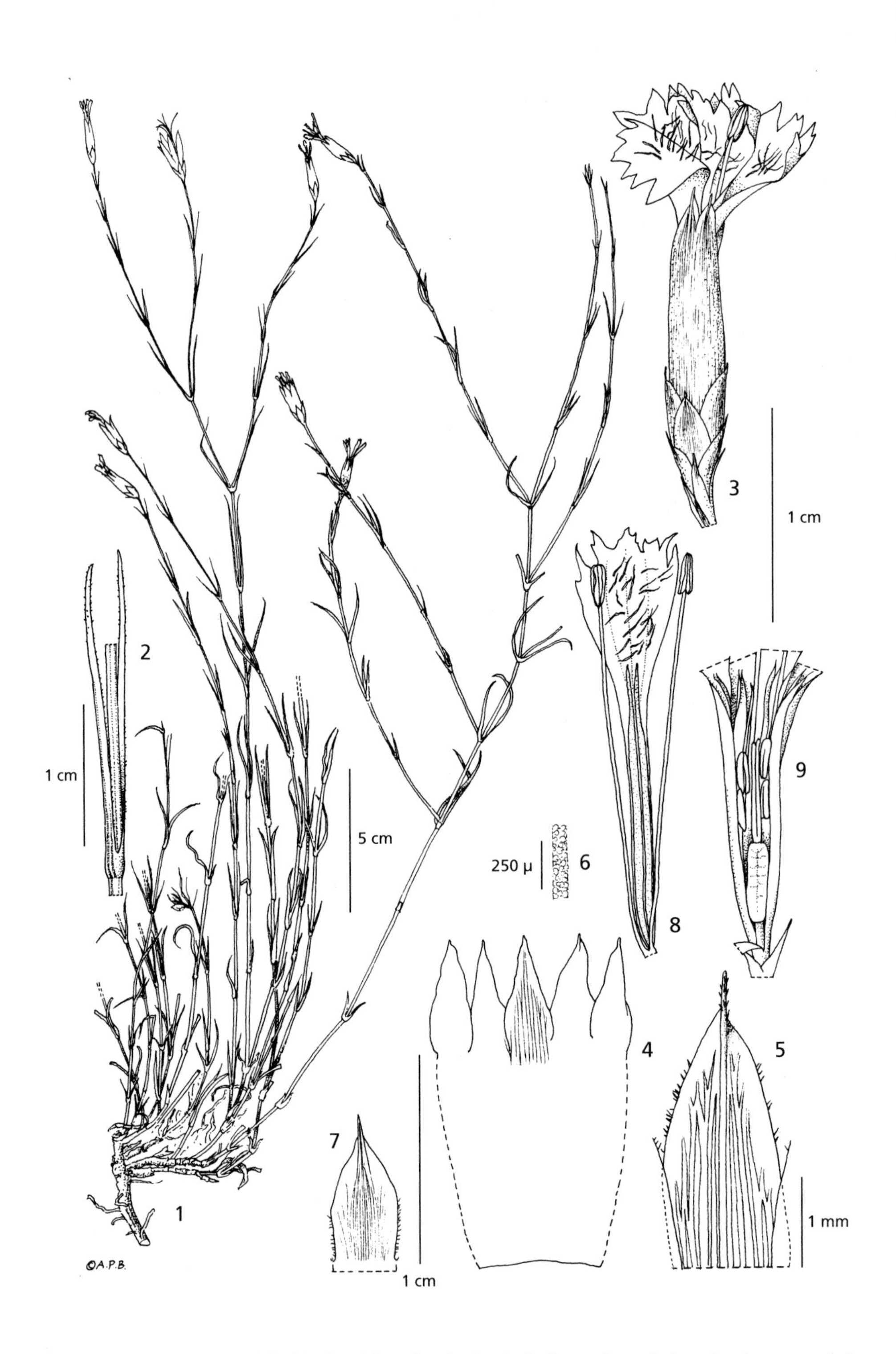

Fig. 23. **Dianthus strictus**. 1, habit; 2, mid cauline leaf pair; 3, flower, lateral view; 4, calyx opened; 5, calyx lobe; 6, calyx rib detail; 7, inner calyx scale; 8, petals with 2 stamens; 9, young flower, lateral view with calyx and one petal removed. 1,2 from *Guest* 2985; 3–9 from *Haines* W469. © Drawn by A.P. Brown.

3. 2–4 of the central veins of each calyx lobe not excurrent into teeth, other marginal veins reaching base of calyx . . . . . .i. subsp. *strictus* var. *gracilior*
All veins excurrent in region of calyx lobe, or only one marginal vein reaching middle of calyx subsp. . . . . ii. subsp. *strictus* var. *subenervis*

a. subsp. **strictus** var. **strictus**

D. multipunctatus Ser. in DC, Prodr. 1: 362 (1824).
D. quadrilobus Boiss. in Tchich., Asie Min. Bot. 1: 222 (1860).
D. sulcatus Boiss., Fl. Or. 1: 483 (1867).

Distrib. **MAM**: 35 km NE of Zakho, *Omar* & *Dabbagh* 45389!; Sarsang, *Haines* W469!; Hammam Ali, *Hussein* s.n!; **MRO**: Khanzad, *Omar, Kaisi* & *Wedad* 49583; Haji Umran, *Jackson* 15064!; Rowanduz, *Guest* 1544!; **MSU**: Jarmo, *Helbaek* 1829!; Penjwin, *Guest* 12938!; Tusluja, *Omar* & *Karim* 47031!; **FUJ**: Mosul, Asi, *Rawi* 4859!; 40 km from Mosul to Qaiyara, *Dabbagh* & *Jasim* 46813!; Sinjar, *Omar* 37599!; **FNI**: Tal Kaif, *Alizzi* & *Omar* 35263!; Mindan Bridge, *Chapman* 26194!; Ain Sifni, *Guest* 4042!; **FKI**: 10–15 km N of Kirkuk, *Rawi* 21564!

E Aegean Is., Turkey (SE Anatolia), Lebanon, Iraq and Iran.

i. subsp. **strictus** var. **subenervis** (Boiss.) Eig, J. Bot. 75: 191 (1937); Reeve in Fl. Turk. 2: 108 (1967); Rech.f., Fl. Iranica 163: 136 (1988).

D. multipunctatus Ser. var. subenervis Boiss., Fl. Orient. 1: 483 (1867).
D. sulcatus Boiss., Boiss., Fl. Orient. 1: 483 (1867).

Distrib. Alpine zone. **MRO**: Halgurd, *anon.* 13815!; *Rechinger* 11391!; Haji Umran, *Rechinger* 11322.

Turkey (SE Anatolia), Lebanon.

ii. subsp. **strictus** var. **gracilior** (Boiss.) Eig, J. Bot. 75: 191 (1937); Reeve, Notes Roy. Bot. Gard. Edinb. 28: 19 (1967); Reeve in Fl. Turk. 2: 108 (1967); Rech.f., Fl. Iranica 163: 136 (1988).

D. multipunctatus Ser. var. *gracilior* Boiss., Fl. Orient. 1: 483 (1867).
D. striatellus Fenzl, Pugillus 10 (1842).
D. paniculatus Pau, Trab. Mus. Nac. Cienc. Nat., Ser. Bot., Madrid 14: 9 (1918).

Distrib. **MAM**: Zinta Gorge, *Chapman* 26129!; Sarsang, *Omar, Kaisi & Hamad* 45762!; Khaira Mountain, nr. Zakho, *Rawi* 22977!; **MRO**: Halgurd Dagh, *Gillett* 12461!; Rayat, *Guest* 13065!; Rowanduz, *Omar, Kazim & Hamid* 38349!; **MSU**: Sarchinar, *Feddo* 5357!; 11 km N of Penjwin, *Rawi* 22894!; **MJS**: Jabal Sinjar, *Gillett* 10926!; **FUJ**: Balad Sinjar, *Gillett* 11070!; between Mosul & Tal Afar, *Karim* 37477!; Mir Qassim, between Balad Sinjar and Tal Afar, *Field & Lazar* 555!; Karim 37487!; **FNI**: Mahad, *Guest et al.* 2604!; Eski Kellek, *Gillett* 8199!; **FKI**: Kirkuk, *anon.* 3952!; Bawanur, *Hooper* 603!; **FPF**: Kani Masi, *Omar & Dabbagh* 45664!; *Omar & Dabbagh* 45567!

Turkey, Syria and Iran.

b. subsp. **velutinus** (Boiss.) Greuter & Burdet in Wildenowia 12: 187 (1982).

D. multipunctatus Ser. var. *velutina* Boiss. Diagn. Ser. 1, 8: 65 (1849).
D. strictus Banks & Soland. var. *velutinus* (Boiss.) Eig, J. Bot. 75: 191 (1937) syn. nov.

Distrib. **MAM**: Ser Amadiya, Dabbagh, *Kaisi & Hamid* 45977!; Gali Zawita, NE of Zakho, *Rawi* 23647!; **MSU**: 20 km EW of Sulaimaniya, *Rawi* 22769!

Syria, Lebanon, Palestine.

2. **Dianthus cyri** *Fisch. & C.A. Mey.*, Ind. Sem. Horti Petrop. 4: 34 (1837); Rech.f., Fl. Lowland Iraq: 246 (1964); Zohary, Fl. Palaest. 1: 108 (1966); Reeve in Fl. Turk. 2: 104 (1967); Rech.f., Fl. Iranica 163: 138, t. 376 (1988); Chamberlain in Fl. Arab. Penins. & Socotra: 1: 228 (1996); Boulos, Fl. Egypt 1: 52 (1999); Ghazanfar, Fl. Oman 1: 79 (2003).

Annuals. Stems ± profusely branched, 15–40 cm, glabrous; internodes 3–7.5 cm. Leaves linear-lanceolate, 25–50 × 2–4 mm, acute, 3-veined with ciliate margins, erect to erecto-patent; sheaths not more than 0.5 × diameter of stem. Inflorescence an irregular many-flowered panicle; flowers solitary. Epicalyx scales 4, cuspidate, base lacking nerves, yellowish, upper part green and nerved, equalling or exceeding calyx. Calyx verruculose, subcylindrical, narrowed slightly above, 10–13(–15) × 3.5–5.5 mm, yellowish, lacking nerves; teeth triangular, 4–6 mm and nerved, mostly green or green-violet between nerves, with hyaline margins, sometimes uppermost part of calyx with violet tinge. Petals pinkish purple; limb ± 5 mm, narrowly ovate, dentate, sometimes barbellate. Capsule as long as calyx.

HAB. Weed in farms, at roadsides and in waste places; alt. s.l– 60 m; fl. May–Jun.
DISTRIB. **LEA**: Aziziya, *Alizzi & Khatib* 33479!; **LCA**: *Aucher* 518! & 4230!; Baghdad, *Lazar* 490!; Abu Ghraib, *Alizzi* 32464!

Aucher 518! was cited as *D. macrolepis* Boiss. in Diagn. ser. 1,1: 19 (1842) and later as *D. cyri* in Fl. Orientalis (1867); it is referable to *D. cyri*.

Transcaucasia, Turkey (E Anatolia), Iran, Arabian Peninsula and Afghanistan.

3. **Dianthus siphonocalyx** Blakelock, Kew. Bull. 3: 397 (1948). — Type: Iraq, Jabal Rubal, Guest 3654 (holo, K!). Reeve in Fl. Turk. 2: 112 (1967); Rech.f., Fl. Iranica 163: 139, t. 378 (1988).

Perennials, suffruticose-caespitose, many-stemmed, 30–50 cm, sometimes branched in upper part, puberulent in lower half, glabrous above; internodes 4–6 cm. Leaves linear to linear-subulate, 25–50 × 0.5–1.5 mm, acute and ± erect, at base with a very narrow hyaline margin, margin sometimes scabrid, 3-veined; cauline leaves shorter than internodes; sheaths of cauline leaves 1–2 × diameter of stem. Flowers solitary. Epicalyx scales 6–8(–12), shortly cuspidate, 3–6 mm, covering 1/3 to ? of length of calyx, tips striate with a narrow ciliate margin (distinct in young flowers). Calyx 13–16 mm, striate; teeth lanceolate, 2–3 mm, also with hyaline, shortly ciliate margins. Petals 16–20 mm; limb irregularly obovate to truncate, dentate, 2.5–4 × 1.5–2.5 mm, yellowish-green to pinkish, ebarbellate; claw shortly exserted from calyx. Capsule exceeding calyx by 1–2 mm. Fig. 24: 1–8.

HAB. Rocky limestone slopes in *Quercus* forest, on steep limestone crags; alt. 600–2000 m; fl. Jun.–Jul.
DISTRIB. Montane region. **MAM**: Jabal Rubal, S of Atrush, *Guest* 3654! (type); Jabal Khantur, N E of Zakho, *Rawi, Tikriti & Nuri* 28965!; Zawita, *Guest* 4471!; **MRO**: Darband, *Nuri & Hamid* 41159!; Jabal Kurak, *Rawi* 23798!; Shaqlawa, *Omar & Sahina* 38255!; **MSU**: Jarmo, *Helbaek* 1995!; Asmir Dagh, *Karim* 39332!

Similar to the Turkish *D. multicaulis* Boiss. & Huet and *D. liboschitzianus* Ser with its slender stalks and many stems, very small petals and 6–8 small epicalyx scales.

Turkey, W Iran.

4. **Dianthus judaicus** *Boiss.*, Diagn. ser. 1, 8: 66 (1849) & Fl. Orient. 1: 485 (1867); Rech.f., Fl. Lowland Iraq: 245 (1964); Blakelock in Kew Bull. 2: 188 (1957); Zohary, Fl. Palaest. 1: 110 (1966); Chamberlain in Fl. Arab. Penins. & Socotra: 1: 229 (1996); Boulos, Fl. Egypt 1: 53 (1999).

D. auraniticus Post., J. Linn. Soc. Bot. 24: 422 (1888).

D. monadelphus Vent. subsp. *judaicus* (Boiss.) Greuter & Burdet, Willdenowia 12: 186 (1982).

Glaucous, perennials, forming small tufts, non-flowering shoots inconspicuous. Stems many, 15–40 cm, sometimes branched in upper part, laxly puberulent in lower part, glabrous above; internodes 2.5–5.5 cm. Leaves linear, to 50(–90) × 1.2–1.7(–2.5) mm, acute, margin ciliate-scabrid, membranous at base, usually plicate, patent or recurved, 5–7-veined; sheaths of cauline leaves little less than diameter of stem. Flowers solitary. Epicalyx scales 4, cuspidate, covering 1/4 – 1/3 of length of calyx; tips green, striate, regularly patent to erecto-patent with narrow hyaline margin. Calyx 22–32 mm, conical first, than cylindrical, striate; teeth lanceolate, 7–9 mm, white margined. Petals 40–45 mm; limb ± 10 mm, entire or slightly crenate or dentate, ebarbellate, whitish to cream, sometimes dark cream. Capsule shorter than calyx.

HAB. Dry steppe and subdesert, on limestone crags and in sandy, gravelly soil; alt. 130–700 m; fl. Mar.–Jul.
DISTRIB. **MJS**: Montafah, *Field & Lazar* s.n.!; **DLJ**: Abu Ghraib, *Barkley & Palmatier* 2262!; Rawa, *Gillett & Rawi* 7053!; **DWD**: 35 km SW of K3, Omar, *Kaisi, Hamad & Hamid* 44690!; Wadi al-Ajrumiya, *Rawi & Nuri* 27144!; 4 km SW of Rutba, *Rawi* 21030!

Distinguished by its epicalyx scales spreading in upper part and having green striate tips.

Egypt, Lebanon, Palestine, Syria.

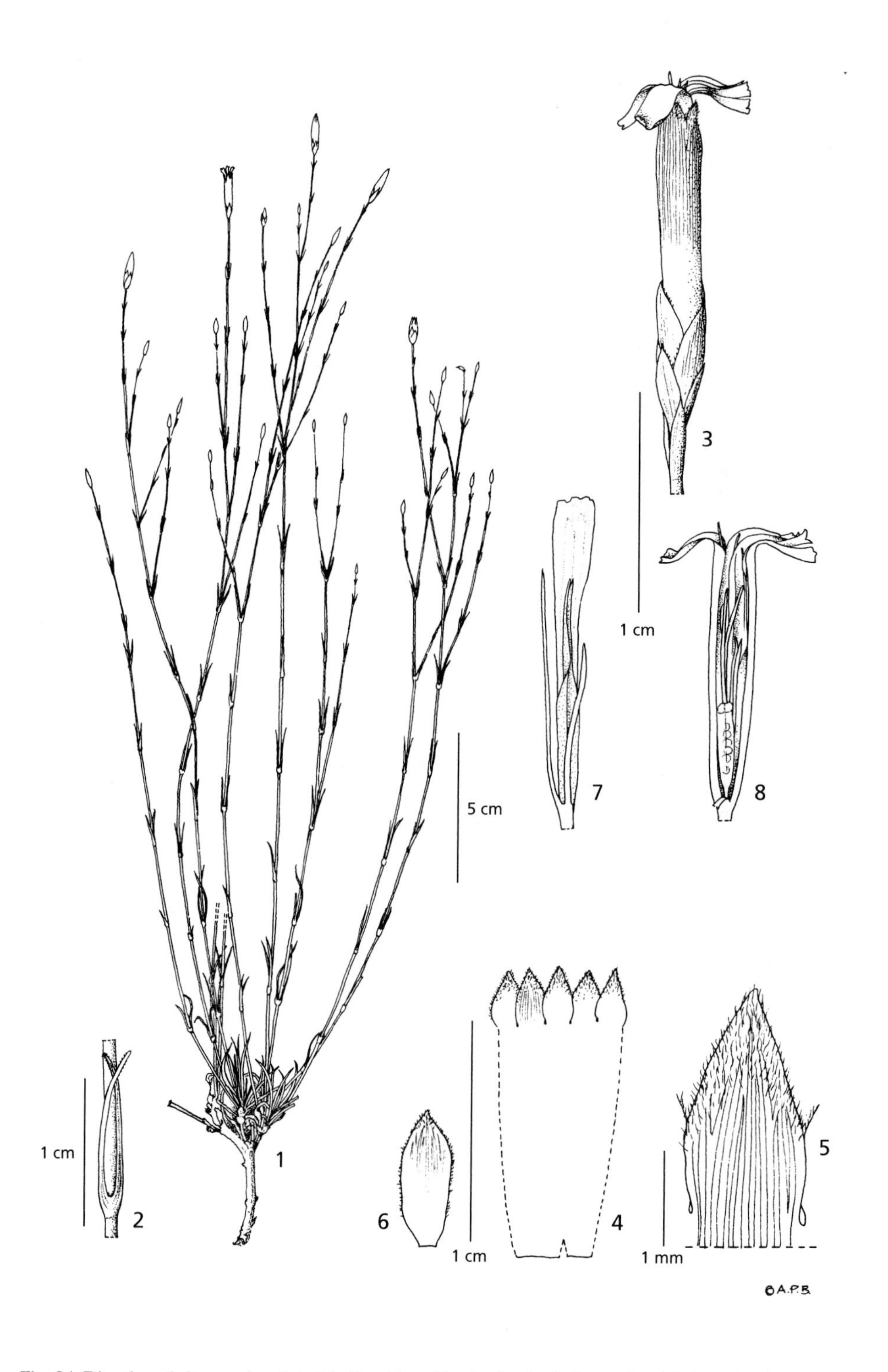

Fig. 24. **Dianthus siphonocalyx**. 1, habit; 2, mid cauline leaf pair; 3, flower, lateral view; 4, calyx opened; 5, calyx lobe; 6, inner calyx lobe; 8, petals with 2 stamens; 9, flower, lateral view with calyx and one petal removed. All from *K.H. Rechinger* 11574. © Drawn by A.P. Brown.

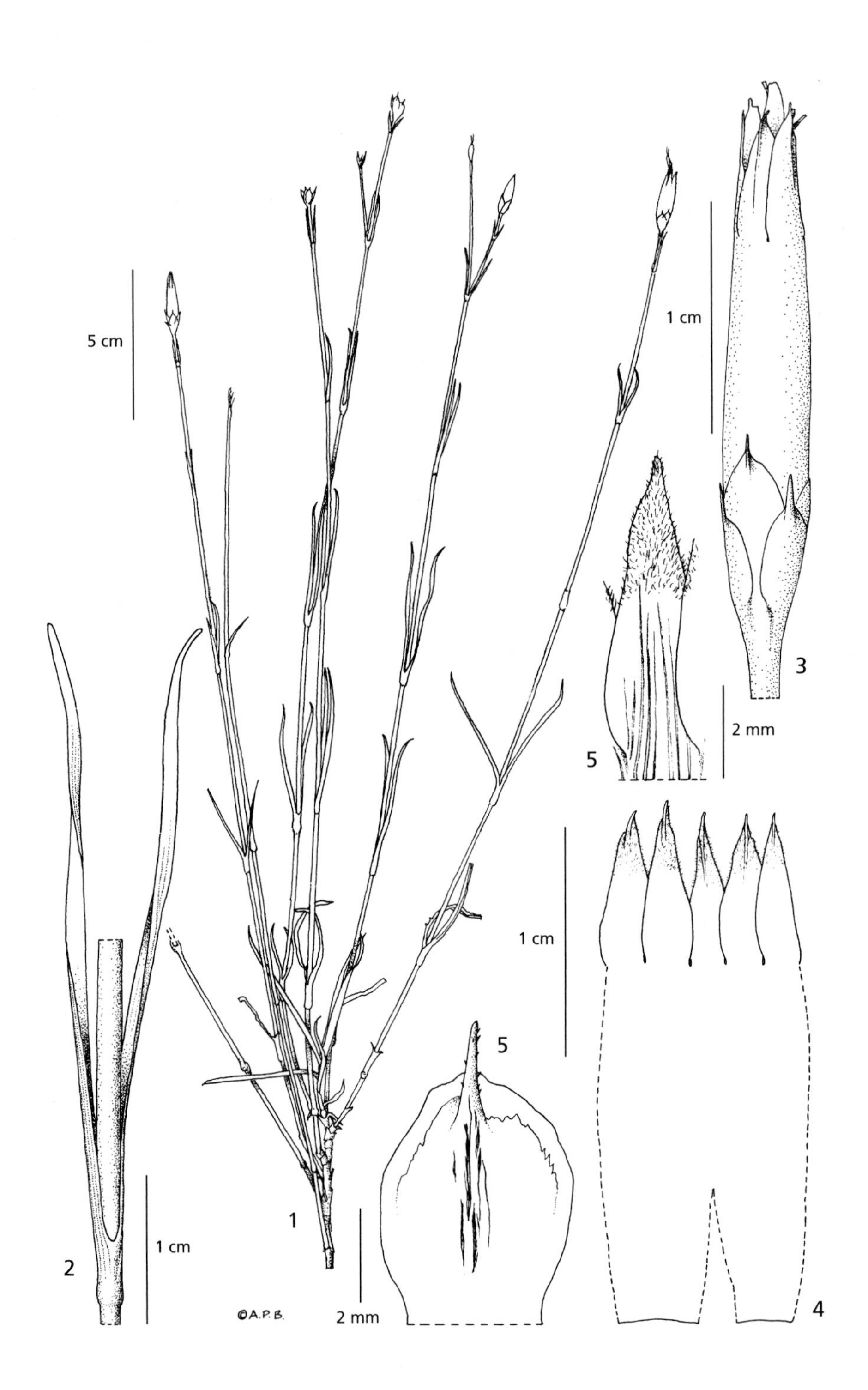

Fig. 25. **Dianthus longivaginatus**. 1, habit; 2, cauline leaves; 3, flower; 4, calyx, opened; 5, calyx lobe; 6, epicalyc lobe. All from *Rechinger* 42769 (type). © Drawn by A.P. Brown.

5. **Dianthus longivaginatus** *Rech.f.*, Pl. Syst. Evol. 142: 244 (1983); Rech.f., Fl. Iranica 163: 153, t. 391 (1988).

Suffruticose perennials without sterile shoots. Stems several, to 40 cm, glabrous, sometimes branched in upper part; internodes 30–50 mm. Leaves lanceolate-linear, 15–32 × 0.7–2 mm, acute, pale green, 3-veined; cauline leaves appressed or semi-appressed or erecto-patent; sheaths of middle cauline leaves 4 mm (3 × diameter of stem). Flowers usually solitary. Epicalyx scales 6(–8), covering to ⅓ of calyx length, ± caudate with ± broadly hyaline margin, outside with green-violet or violet irregular tinge; awn of innermost pair very short (seldom exceeding 0.5 mm). Calyx conical, 21–24 × 3.0–3.5 mm, pale green to purplish, striate, glabrous; teeth 5–6 mm, lanceolate, hyaline margin ± puberulent, occasionally glandular-puberulent. Petal limb 7–9 × 7–8 mm, fimbriate, barbellate; claw little exserted. Fig. 25: 1–6.

Hab. On high mountains in rocky places, on serpentinite; alt. 1500–2600 m; fl. Jul.–Aug.
Distrib. Rare; alpine zone. **mro**: Halgurd Dagh, in valley above Nawanda, *Rechinger* 11373!
This species appears to be related to the *D. orientalis* group. It can be identified by its suffruticose growth form lacking non-flowering shoots and its relatively long sheaths.

NW Iran.

6. **Dianthus orientalis** *Adams* in Weber & Mohr. Beitr. Naturk. 1: 54 (1805); Blakelock in Kew Bull. 2: 189 (1957); Reeve in Fl. Turk. 2: 120 (1967); Ghazanfar & Nasır in Fl. Pak. 175: 114 (1986); Rech.f., Fl. Iranica 163: 155 (1988); Lu Dequan & Turland in Fl. China 6: 105 (2001).

D. fimbriatus Bieb., Fl. Taur.-Cauc. 1: 332 (1808).

Suffruticose perennial, sometimes reduced to one woody creeping caudex. Stems 15–50 cm, glabrous or scabrid to shortly puberulent, usually with few non-flowering shoots, branched mostly in upper part; internodes 25–60 mm. Leaves linear, 15–55 × 0.5–3 mm, pale green, apex long acuminate to shortly acute; upper cauline leaves shorter, erect, erecto-patent or appressed, regularly 3-veined; sheaths of lower cauline leaves 3–4(–5) × diameter of stem. Flowers usually solitary. Epicalyx scales (4–)6–8(–12), covering ¼ to ½ of calyx length, ± caudate, sometimes cuspidate with short or longer awn with 0.3–8 mm; upper margin of lamina of scales without or with broad hyaline margin; tips ± weakly striate. Calyx conical, 17–21(–24) × 2.5–3.5 mm, ± distinctly narrowed towards apex, pale green to purplish, ± striate; teeth 5–8 mm, narrowly lanceolate, acute, shortly puberulent to glabrous. Petal limb pink, fimbriate, barbellate or not; claw exserted.

1. Sterile shoots absent; stems usually longer than 25 cm. . . . . .b. subsp. *macropetalus*
   Sterile shoots present; stems usually shorter than 25 cm 2
2. Awn of innermost scales with broadly hyaline margin, its free tip not more than 0.8 mm. . . . . c. subsp. *nassireddini*
   Awn of innermost scales usually without broad hyaline margin, sometimes with narrow hyaline margins, its free tip at least 1 mm. . . a. subsp. *orientalis*

a. subsp. **orientalis**

Suffruticose, many-stemmed with sterile shoots; stems seldom exceeding 25 cm; epicalyx scales ovate, (4–)6–8(–10); awn of innermost scales without broad hyaline margin, sometimes with narrow hyaline margins near below; free tip of awn at least 1 mm. Fig. 26: 1–9.

Hab. On cliffs, rocky slopes and screes; alt. 700–2000 m; fl. Jun.–Aug.
Distrib. **mam**: Gara Dagh, *Rawi* 9277!; Sarsang, *Haines* W487!; **mro**: Seri Hasan Beg, *Guest* 2903!; Qandil Range, *Rawi & Serhang* 26702!; **msu**: Zalam Mountain, *Rawi, Hosham & Nuri* 29314!

Transcaucasia, Turkey, Iran.

b. subsp. **macropetalus** (Boiss.) Rech.f., Pl. Syst. Evol. 151: 289 (1986); Rech.f., Fl. Iranica 163: 159, t.396 (1988).

D. fimbriatus Bieb. var. *macropetalus* Boiss., Boiss., Fl. Orient. Suppl. 77 (1888).
D. orientalis Adam. var. *macropetalus* (Boiss.) Bornm. in Beih. Bot. Centralbl. 19, 2: 213 (1906); Blakelock in Kew Bull. 2: 189 (1957).

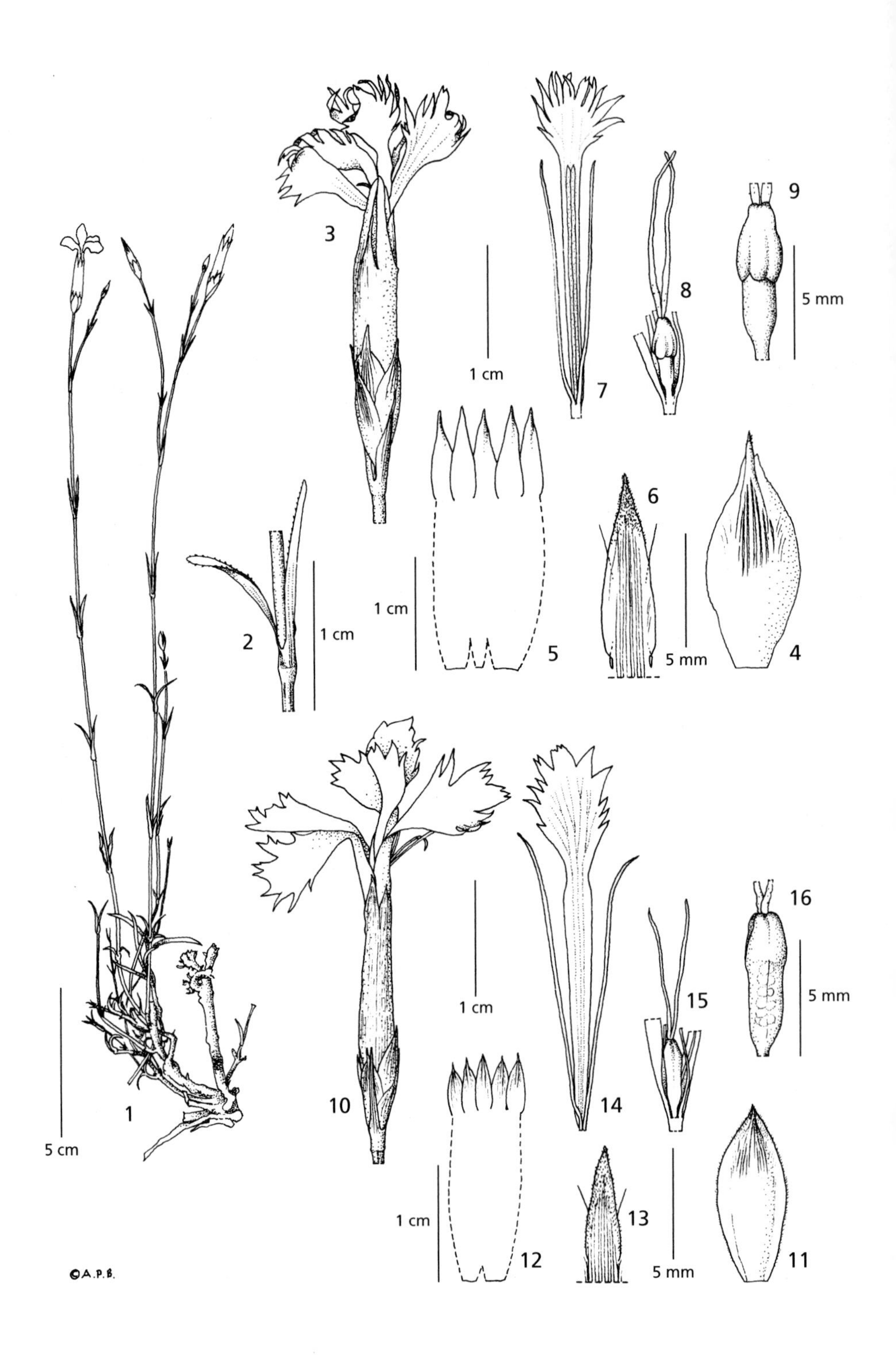

Fig. 26. **Dianthus orientalis** subsp. **orientalis.** 1, habit; 2, mid cauline leaf pair; 3, flower, lateral view; 4, inner epicalyx scale; 5, calyx, opened; 6, calyx lobe detail; 7, petal with 2 stamens; 8, ovary; 9, ovary detail. **Dianthus orientalis** subsp. **macropetalus**. 10, flower, lateral view; 11, inner epicalyx scale; 12, calyx opened; 13, calyx lobe detail; 14, petal; 15, ovary; 16, ovary detail. 1–2 from *Guest* 2903; 3–9 from *Cowie* s.n., 12 May 1915; 10–16 from *Gillett* 9516. © Drawn by A.P. Brown.

Suffruticose, sterile shoots absent; stems usually more than 25 cm; epicalyx scales ovate or ovate-oblong, (4–)6(–8); awn of innermost scales usually without broad hyaline margin, its free tip at least 1 mm. Fig. 26: 10–16.

HAB. Rocky slopes, steps; alt. 900–2100 m; fl. Jun.–Aug.
DISTRIB. **MAM**: Ser Amadiya, *Haines* W2059!; **MRO**: Sula Khal, *Rawi & Serhang* 24683!; Haji Umran, *Rawi* 24266!; Halgurd Dagh, *Gillett* 9516!; **MSU**: Penjwin, *Haines* 2037!; *Kaisi & Hamad* 43521!

Leaf sheaths are usually shorter than 3 mm in *D. orientalis*, occasionaly, a few sheaths may be longer than 3 mm in some specimens belonging to subsp. macropetalus.

W Iran.

c. subsp. **nassireddini** (Stapf) Rech.f., Pl. Syst. Evol. 151: 292 (1986); Rech.f., Fl. Iranica 163: 163, t. 401(1988).

D. nassireddini Stapf, Denkschr. Akad. Wiss. Wien, Math.-Nat. Kl. 51: 279 (1886).
D. fimbriatus Bieb. var. *brachyodontus* Boiss. & Huet in Boiss., Diagn. Pl. Or. Nov. ser. 2, 5: 53 (1856).
D. orientalis Adam. var. *brachyodontus* (Boiss. & Huet Bornm.) Bornm., in Beih. Bot. Centralbl. 19, 2: 212 (1906); Blakelock in Kew Bull. 2: 189 (1957).

Suffruticose, many-stemmed with sterile shoots; stems seldom exceeding 25 cm; epicalyx scales ovate, (4–)6–8(–10); awn of innermost scales with hyaline margin; free tip of awn not more than 0.8 mm. Fig. 27: 1–9.

HAB. Sandy soil, on cliffs, rocky slopes and screes; alt. 1700–3000 m; fl. May–Aug.
DISTRIB. Montane and alpine zones. **MRO**: NE of Haji Umran, *Kass & Nuri* 27811!; Qandil Range, *Rawi & Serhang* 24135!; Rowanduz, *Bornmüller* 962!; *Guest et al.* 2595!

Turkey, Armenia, NW and S Iran.

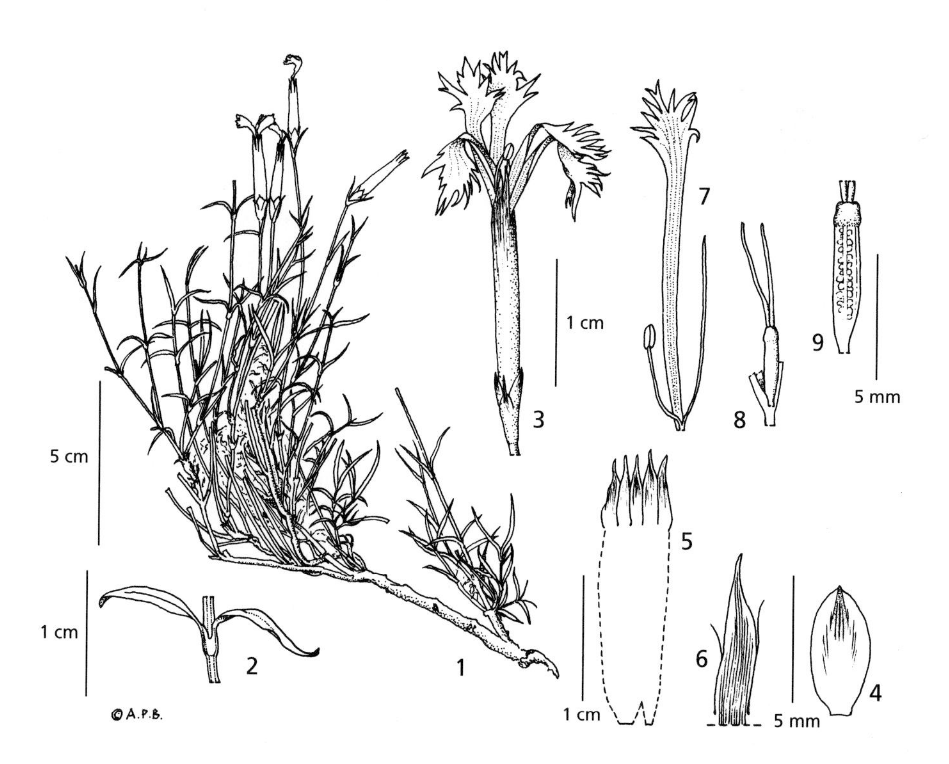

Fig. 27. **Dianthus orientalis** subsp. **nassireddini**. 1, habit; 2, mid cauline leaf pair; 3, flower, lateral view; 4, inner epicalyx scale; 5, calyx, opened; 6, calyx lobe detail; 7, petal with 2 stamens; 8, ovary; 9, ovary detail. 1, 2 from *Guest et al.* 2595; 3–9 from *Bornmüller* 962. © Drawn by A.P. Brown.

7. **Dianthus pendulus** *Boiss. & Blanche* in Boiss., Diagn. Pl. Or. Nov. ser. 2, 6: 28 (1859); Blakelock in Kew Bull. 2: 189 (1957); Rech.f., Fl. Iranica 163: 167, t. 405 (1988).

Suffruticose perennials. Stems several, ascending or pendent (from vertical habit), 20–35 cm, glabrous, sometimes weakly pubescent to papillose below, branched in upper part; internodes 3–7 cm. Leaves grass-like, to 60(–90) × 1.5–2.0(–3.0) mm, linear, long-attenuate, green, glabrous, 3–5-veined; upper cauline leaves erecto-patent to erect; sheaths of lowest cauline leaves 3–5 × diameter of stem. Inflorescence a lax few-flowered panicle with 2–4(–7) solitary flowers. Epicalyx scales (12–)16–30(–40), 5–10 mm, densely imbricate; outer shortly cuspidate; inner ± cuspidate, weakly striate, scarious. Calyx conical, (18–)24–33 mm, exceeding epicalyx scales by 12–18 mm, striate, greenish with a partly violet tinge; teeth lanceolate, 4–5 mm. Petal limb 8–10 mm, fimbriate to $^{1}/_{3}$ – $^{1}/_{2}$ of their length, weakly barbellate. Capsule unknown.

HAB. On vertical rocks and cliffs, especially beside gorges; alt. c. 900 m; fl. Jul.–Nov.
DISTRIB. Occasional in upper forest zone. **MAM**: Sarsang, *Haines* W1243!; Amadiya, *Tikriti* 16394!; 25 km NE of Zakho, *Rawi, Nuri & Tikriti* 29038!; **MRO**: Rowanduz Gorge, *Rechinger* 11335!; Gali Ali Beg, *Rawi*, 24251A!; Kaniwa Rash, *Rawi & Serhang* 23765!; **MSU**: Dokan, *Haines* W2128!; *Haines* W7171! & *Haines* W7130!
This species can be distinguished by its numerous, densely imbricate epicalyx scales.

Lebanon, N Iran.

8. **Dianthus basianicus** Boiss. et Hausskn., Boiss., Fl. Orient. Suppl.: 77 (1888). — Type: In arenosis Derbent-I Bazian, *C. Haussknecht* (holo. G; iso. W); Zohary, Fl. Iraq & Its Phytogeogr. Subdivis.: 54 (1946); Blakelock in Kew Bull. 2: 188 (1957); Rech.f., Fl. Iranica 163: 169, t. 407 (1988).

D. crinitus Sm. var. *tomentellus* Boiss., Fl. Orient. 1: 469 (1867).

Perennial with ± stout woody rootstock, with sterile shoots. Stems few to several simple or branched, 10–60 cm, pubescent or glabrous; internodes 3–4 cm. Leaves lanceolate, long-acuminate, 15–40 × 1–2 mm, coriaceous; upper cauline leaves ± patent; leaf sheaths ± equalling diameter of stem. Flowers solitary. Epicalyx scales (4–)6–8, of variable length, $^{1}/_{3}$ – ? of calyx length, cuspidate, rarely caudate with a white or yellowish membranous margin, upper part distinctly striate, greenish, erecto-patent or recurved. Calyx conical at first than cylindirical, 28–34 × 4–6 mm, distinctly striate, shortly pubescent; teeth lanceolate, mucronate, 8–10 mm with hyaline ciliate margin. Petals white 40–50 mm; lamina long fimbriate, obovate in outline, 8–11 mm, ebarbellate; claw exserted to 5 mm.

HAB. On rocky slopes and steep hills, frequent in the *Astragalus* formation, but also in vineyards; alt. ± 1300 m; fl. May–Jun.
DISTRIB. Rare. **MSU**: Jarmo, *Helbaek* 1820!; *Haines* W245!

W Iran.

9. **Dianthus macranthus** Boiss., Diagn. Pl. Orient. Nov. Ser. 1, 1: 23 (1843); Rech.f. in Fl. Iranica 163: 180 (1988).

Perennial with ± stout woody rootstock. Stems several, simple or branched, 30–40 cm, glabrous to puberulous. Internodes 3–7 cm. Leaves linear-subulate, 18–30 × 1.5 mm, shorter than internodes. Flowers solitary. Epicalyx scales (8–)14(–16), ovate, imbricate, striate, not exceeding ½ of calyx. Calyx tubulous, (28–) 35–40 × 4–6 mm, shortly pubescent or glabrous, striate; teeth 7–9 long, triangular-lanceolate. Petals up to 54 mm, white; lamina obovate-spatulate, fimbriate, 9–12 mm, ebarbellate; claw exserted to 4 mm.

HAB. Rocky ground, slopes; alt. ± 900 m; fl. Jun.–Aug.
DISTRIB. Very rare and known only from a single record from Iraq. **MSU**: Darband-i Bazian, *Haussknecht* 97!

NE & W Iran and SW Afghanistan.

10. **Dianthus libanotis** *Labill.*, Pl. Syr. Dec. 1: 14 (1791); Reeve in Fl. Turk. 2: 118 (1967); Rech.f., Fl. Iranica 163: 185, t. 425 (1988); Boulos, Fl. Egypt 1: 53 (1999).

D. atomarius Boiss., Diagn. Pl. Or. Nov. Ser. 1, 8: 71 (1849).

Perennials. Stems 12–60 cm, usually robust, glabrous, branched or not. Internodes 20–70 mm. Leaves glabrous, lanceolate, 22–60 × 2–7 mm, 3–50-veined, usually recurved, with acute rigid apex; sheaths of middle cauline leaves ± 3 mm. Flowers solitary or 2–5 in clusters. Epicalyx scales (4–)6–8, lanceolate, glabrous, covering up to ½ of calyx length, occasionaly longer, pale, with green, spiny patent or slightly curved apex. Calyx conical to cylindrical, glabrous, striate, 35–42 mm; teeth (4–)9–11 mm, lanceolate, mucronate, glabrous or ciliate on scarious margins. Petal limb white, sometimes with reddish spots, cuneate at base, fimbriate more than ½ of its length, barbellate, 9–13 mm.

Hab. Rocky ground, slopes, road sides; alt. 1300–2850 m; fl. Jun.–Aug.
Distrib. Montane zone, rare. **MRO**: Chiyā-i Mandau, *Guest* 2705!; Halgurd Dagh, *Gillett* 9503!; Baski Hawaran, *Rawi & Serhang* 23967!; SE of Serva, near Ser Kurawa, *Gillett* 9700!

Armenia, Turkey, Syria.

11. **Dianthus zonatus** Fenzl, Pug. 11 (1842).

var. **hypochlorus** (Boiss. & Heldr.) Reeve in Notes R.B.G. Edinburgh 28: 21 (1967); Reeve in Fl. Turk. 2: 128 (1967).

Perennials, sterile shoots few or absent. Stems ascending to erect, 4–30 cm, glabrous, sometimes shortly hairy. Leaves up to 40 mm, linear-acuminate, glabrous or short hairy or with short hairs on margins; internodes 1.2–4.2 cm long. Inflorescence freely branched, pedicels 3–30 mm. Epicalyx scales 4–6(–8), usually shorter than ½ of calyx length, ovate, membranous with scarious margins with aristate tips. Calyx subcylindirical, 12–18 mm, glabrous, pale green usually tinged with purple beneath teeth; teeth 2–3.5 mm, apex mucronate with scarious margins. Petal 18–25 mm; limb pink, ± 5 mm, barbullate.

Hab. Fields; alt. 200–500 m; fl. Jul.–Aug.
Distrib. Very rare. **FPF**: Khanaqin, *B.P. Uvarov* s.n!

Turkey, W Syria.

12. **Dianthus hymenolepis** Boiss., Diagn. Pl. Or. Nov. ser. 1, 8: 64 (1849). — Type: in Mesopotamia loco non indicato. *Th. Kotschy*, Pl. Assyr. Kurd. Mos. 161. (K!); Reeve in Fl. Turk. 2: 124 (1967); Rech.f., Fl. Iranica 163: 186, t. 426 (1988).

Perennial with ± stout woody rootstock, laxly suffrutescent; sterile shoots few or absent. Stems several ± straight, pruniose-puberulent, 20–45 cm, if branched then only in upper part; internodes 3–7 cm. Leaves linear-lanceolate, 30–65 × 1.2–2.5 mm, acute, 3–veined; upper cauline leaves ± erect, sheaths of cauline leaves 1–1.5 × diameter of stem. Flowers 3–5(–7) in shortly pedicellate (0–2 mm) capitate simple corymbs. Epicalyx scales 4–6, coriaceous, ± equalling calyx length; outer more cuspidate; inner more caudate, with 3–5 mm awn, weakly veined; tip greenish, yellowish or dark violet; upper margin of lamina sometimes weakly undulate and also dark violet. Calyx ± tubular, (10–)12–15 × 3–4 mm, ± distinctly striate, very short puberulent; teeth 2.5–4 mm, obtuse with a short mucro, yellowish-green, 2 outer calyx teeth usually dark violet. Petals carmine; limb rounded to obovate, dentate, distinctly barbellate, sometimes with dark violet spots. Capsule a little shorter than calyx.

Hab. On rocky slopes in Quercus forest, also in open rocky grazed slopes; alt. 1000–1700 m; fl. Jun.–Aug.
Distrib. 'Mesop. Kurd. Mosul', Kotschy 161! (type); **MAM**: Jabal Khantur, *Hooper* 10822!; Ser Amadiya, Haines W2075!; Zawita, Haines *et al.* 2144!

This species is recognizable by its flowers in a corymb, epicalyx scales about as long as calyx, the outer more cuspidate, and the inner more caudate, very often with dark violet margins or tips.

Turkey (SE Anatolia).

13. **Dianthus persicus** *Haussknn.*, Mitt. Geogr. Ges. Thür. (Jena) 9: 16 (1890); Rech.f., Fl. Iranica 163: 186, t. 427 (1988).

Glabrous perennial with ± stout woody rootstock, laxly suffrutescent. Stems ± straight, 40–60 cm long; internodes (3–)5–8(–10) cm. Lower cauline leaves linear or narrowly lanceolate, 35–70 × 2–8 mm, middle and upper cauline leaves often more lanceolate, 3–7-veined, broader cauline leaves sometimes net-veined; sheaths of lower cauline leaves at most 2 × diameter of stem. Inflorescence a dense corymb of 4–10 flowers. Epicalyx scales (4–)6, long-cuspidate, ± equalling to calyx. Calyx 18–20 mm, with distinct broad hyaline margins, ± striate. Petals up to 30 mm, pink to pinkish; limb 5–8 mm, obovate, dentate, shortly barbellate. Capsule shorter than calyx. Fig. 28: 1–7.

HAB. Steppe slopes, waste land, open habitats in *Quercus* forest; alt. 1100–1800 m; fl. Jun.
DISTRIB. **MAM**: Sarsang, *Haines* W517!; *Chapman* 26399!; **MSU**: Penjwin, *Rawi* 12206!
This species can be recognised by its flowers in a corymb, middle cauline leaves distinctly broader than lower; and bracteoles all caudate with broad hyaline margins. The main difference between *D. hymenolepis* and *D. persicus* is the longer flowers in *D. persicus*. *D. persicus* is related to the Turkish *D. robustus*, but there are some differences in inflorescence, calyx length and epicalyx scales.

W Iran.

SPECIES DOUBTFULLY RECORDED

Dianthus anatolicus *Boiss.*, Diagn. Pl. Or. Nov. ser. 1, 1: 22 (1843).

D. parviflorus Boiss., Diagn. Pl. Or. Nov. ser. 1,1: 22 (1843).
D. kotschyanus Boiss., Diagn. Pl. Or. Nov. ser. 1, 8: 68 (1849).

Tufted perennials, 10–30 cm with a branched woody stock. Stems many, erect to ascending, swollen at nodes, sparsely pubescent to glabrous, hairs minute. Leaves 7–30(–40) mm, linear, usually congested basally, reducing in size upwards, glabrous with ± ciliate margin, apex acute, midrib thick. Flowers solitary, very rarely 2. Bracteoles 4–6, 6–8 mm, ovate to narrowly ovate, apiculate with a membranous margin. Calyx 11–12(–15) mm, glabrous, contracted above, teeth lanceolate, acute or mucronulate, with a ciliate margin. Petals c. 15 mm, white to rose pink, limb 3–4 mm, subentire to dentate at apex, ebarbulate.

This species is known Greece, Turkey and Pakistan. It was recorded from Amadiya, Rowanduz and Sulaimaniya by Zohary, but there are no specimens at K or E from Iraq.

Dianthus floribundus *Boiss.* in Tchihat., Asie Min. Bot. 1: 221 (1860)

D. noeanus Boiss., Diagn. Ser. 2(5): 52 (1856), non Boiss., Diagn. Ser. 2 (1): 67 (1853).

Perennial, 20–35 cm. Leaves 11–40 × 0.5–1.5 mm, linear, narrowed from the base to an acuminate apex. Stems freely branched, pedicels more than 2 cm. Bracteoles (4–)6, less than ½ calyx length, narrowly acuminate to apiculate, scarious margin distinct. Calyx 14–18 × 2.5–3.5 mm, cylindirical or widest in the lower half; teeth 3–5 mm, acuminate. Petal limb 4–7 mm, white or pinkish white, ebarbulate, finely dentate.

A Turkish species, recorded from Jarmo, Kirkuk Liwa by Helbaek, but no specimens have been seen from Iraq.

Dianthus masmenaeus *Boiss.*, Diagn. Ser. 2 (5): 51 (1856).

var. **glabrescens** *Boiss.*, Fl.Orient. 1: 502: (1867); Reeve, Fl. Turk. 2: 123 (1967).

D. asperulus Boiss. & Huet in Boiss., Diagn. Ser. 2 (5): 51 (1856).
D. transcaucasicus Schischk. in Ber. Tomsker Staatsuniv. 80: 452 (1928).

Perennials up to 34 cm high. Stems pubescent to subglabrous. Leaves linear-oblong, 20–60 × 1.5–4 mm, shorter than internodes, glabrous to hairy with ± scabrid margin, shortly acuminate, margins ± thick. Flowers 3–7 together, pedicels not more than 3 mm. Bracteoles ovate with long setaceous points, usually shorter than calyx. Calyx glabrous, 10–15 × 2–3.5 mm, cylindrical: teeth 3.7 mm, acuminate. Petals pink, limb ± 4 mm, pink, barbullate, dentate.

This taxon was recorded from by Brant & Strangways in 1840 from Kurdistan. Collecting number and full locality had not been indicated on the label and it is likely to be from Turkey.

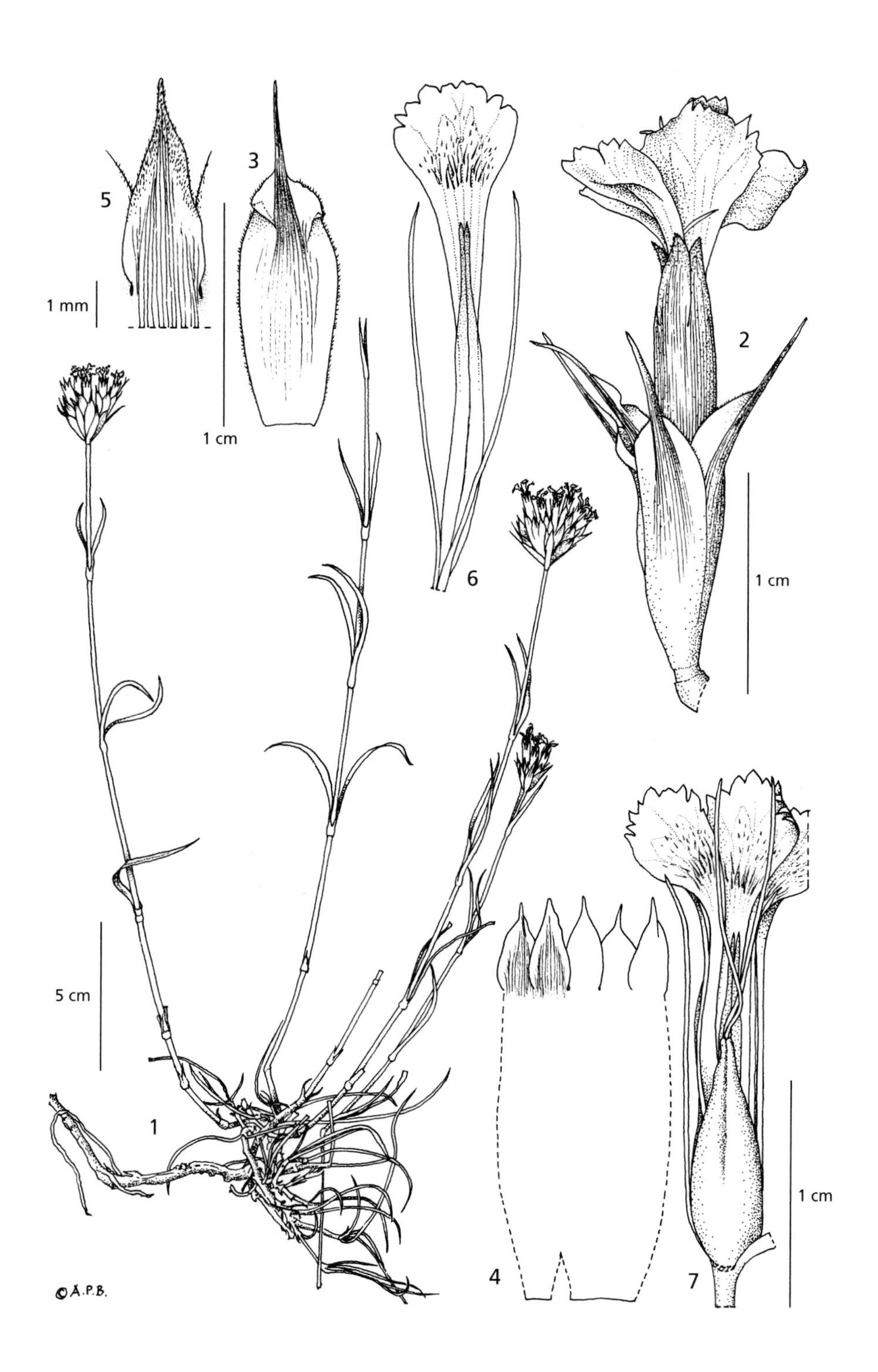

Fig. 28. **Dianthus persicus**. 1, habit; 2, flower, lateral view; 3, inner epicalyx scale; 4, calyx, opened; 5, calyx lobe detail; 6, petal with 2 stamens; 7, flower with petal removed. All from *K.H. Rechinger* 42960. © Drawn by A.P. Brown.

CULTIVATED SPECIES

Dianthus barbatus L., Sp. Pl. 409 (1753).

Short-lived perennial, sometimes annual. Stem prostrate, 20–40 cm, with sterile and flowering shoots, thickened at nodes. Leaves broad- to oblong-lanceolate, 35–60 × 5–18 mm, glabrous, sometimes ciliate on margins. Epicalyx scales as long as or longer than calyx tube. Flowers clustered in many-flowered corymbs. Calyx 9–13 mm. Petals limb reddish, pink to purplish, barbellate, dentate.

In cultivated fields. LCA: Baghdad, *Fawzi* 767!

Dianthus chinensis L. Sp. Pl. 1: 411 (1753).

Annual, or short-lived perennial. Stem ± erect, to 40 cm, canaliculate, glabrous to coarsely ciliate, branched above. Leaves linear-lanceolate to broadly lanceolate, 34–70 × 2–9 mm, acute, short-stalked, finely ciliate on margins. Flowers several. Epicalyx scales 4–6, long-cuspidate, ± as long as calyx. Calyx 12–14 mm, smaller in male flowers. Petal limb 7–10 mm, bright to dull red, seldom white, dentate, usually spotted, weakly barbellate, greenish on outside.

In cultivated fields. LCA: Baghdad, *Sahina* 176!

Dianthus caryophyllus *L.* Sp. Pl. 410 (1753).

Glaucous and glabrous perennial, with densely leafy sterile shoots. Stem to 80 cm, branched from woody base, erect or ascending. Leaves linear-lanceolate, 35–80 × 2–8 mm, acute, scabrid on margin only. Flowers solitary or in clusters with up to 5(–7) flowers, ± 3.5 cm diameter. Epicalyx scales 4–6, green, cuspidate, ¼–⅓ length of calyx, two inner ones shortly cuspidate. Calyx (14–) 22–28 × 8–12 mm, striate; teeth 7–9 mm, acuminate. Petal limb 10–15 mm, purple, rarely pink or white, dentate, ebarbullate.

In cultivated fields; alt. ± 100 m; fl. May–Aug. LCA: Abu Ghraib, *Sahina* 158!; *Mokhtar* 34872! & 35718!; Baghdad, *A.Latif* 744!; *Raja* 795!; Za'faraniya, *Omar* & *Sahina* 34807!

22. PETRORHAGIA (Ser.) Link

Handbuch 2: 235 (1829)

Annual or perennial herbs. Stem simple or branched. Leaves opposite, narrow, linear, oblong or lanceolate. Inflorescence panicle or capitate, or solitary. Bracts present or absent. Calyx obconical, cylindirical or campanulate, 5–ribbed, with 1–5(–7) veins, and hyaline interwals between nerves. Petals 5. Stamens 10. Styles 2. Capsule dehiscing by 4 teeth.

About 30 species found in Mediterranean region to C Asia, with a concentration in the Mediterranean region, especially Turkey and Greece; four species in Iraq.

Ref. P.W. Ball & V.H. Heywood (1964) A. Revision of the genus *Petrorhagia*. Bulletin of The British Museum (Natural History), Botany, London 3, 4: 121–172.

1. Perennial herbs with woody root 2
   Annual or biennial herbs 3
2. Bracts shorter than pedicels; flowers usually solitary .... 4. *P. wheeler-hainesii*
   Bracts longer than pedicels; flowers not solitary .... 1. *P. macra*
3. Stem glandular-pubescent in middle parts; costae of calyx 1-veined .... 2. *P. alpina*
   Stem glabrous; costae of calyx 3-veined .... 3. *P. cretica*

1. **Petrorhagia macra** (*Boiss. & Hausskn.*) *Ball & Heywood,* Bull. Brit. Mus. (Nat. Hist.) Bot. 3,4: 158 (1964); Rech.f., Fl. Iranica 163: 190 (1988).

Tunica macra Boiss. & Hausskn. ex. Boiss., Fl. Or. Suppl. 82 (1888).
T. gracilis Williams, Journ. Bot. 28: 196 (1890).

Perennials with short woody rhizome. Stems several, erect or slightly ascending, 25–40 cm, glabrous, little branched or not. Lower leaves linear-oblong, up to 7 mm, falling early; upper leaves subulate, minute, glabrous. Bracts triangular-lanceolate, carinate, acute, longer than pedicels. Inflorescence usually terminal, sometimes axillary with 1–4 flowers; Pedicels to 3 mm, ± glandular. Calyx ± 6 mm, cylindrical, attenuate at base, glandular-puberulent;

Fig. 29. **Petrorhagia cretica**. 1, habit × 1; 2, flower × 4; 3, petal × 3; 4, calyx × 4. Reproduced with permission from Flora Palaestina Vol. 1 Plates, f 141. Drawn by Ruth Koppel. ©The Israel Academy of Sciences and Humanities.

costae 5 with 3 veins; teeth oblong, obtuse with membranous margin. Petals ± 8 mm, linear, entire. Capsule oblong-cylindrical, equal or slightly shorter than calyx.

HAB. Cliffs; alt. ± 700 m ; fl. Jul.
DISTRIB. Very rare. **MSU**: mt. Hauraman (Avroman), *Haines* W2085!

W Iran.

2. **Petrorhagia alpina** (*Habl.) Ball* & *Heywood*, Bull. Brit. Mus. (Nat. Hist). Bot. 3,4: 145 (1964); Ghazanfar & Nasır in Fl. Pak. 175: 110 (1986); Coode & Cullen in Fl. Turk. 2: 133 (1967); Rech.f., Fl. Iranica 163: 191, t. 370 (1988).

Gypsophila alpina Habl., Neue Nord. Beitr. 4: 57 (1783).
G. stricta Bunge in Ledeb., Icon. Pl. Fl. Ross. 1: 5 (1829).
Tunica stricta (Ledeb.) Fisch. & C A. Mey., Ind. Sem. Horti Petrop. 4: 50 (1837).

Annual or biennial herbs. Stem glabrous, erect, branched, 10–40 cm. Basal leaves 20–30 cm long, 1-nerved, oblanceolate, base attenuate with scabrid margin; cauline leaves linear-lanceolate, ± adpressed to stem. Bracts not enclosing calyx. Inflorescence panicle. Pedicel to 30 mm. Calyx 2–4 mm, campanulate; costae 5, 1-nerved. Petals 3–6 mm, entire. Capsule oblong-ovoid, exserted from calyx.

HAB. Forests, slopes; alt. 900–3000 m ; fl. May–Jun.
DISTRIB. Very rare. **MRO**: Halgurd Dagh, *Hadač* 2076.

Caucasus, Turkey, Iran, Pakistan, Afghanistan and C Asia.

3. **Petrorhagia cretica** (*L.*) *Ball* & *Heywood*, Bull. Brit. Mus. (Nat. Hist). Bot. 3 (4): 142 (1964); Zohary, Fl. Palaest. 1: 105 (1966); Coode & Cullen in Fl. Turk. 2: 133 (1967); Rech.f., Fl. Iranica 163: 192, t. 371 (1988);); Chamberlain in Fl. Arab. Penins. & Socotra: 1: 227 (1996); Georgiou in Fl. Hellenica 1: 337 (1997).

Saponaria cretica L., Sp. Pl., ed. 2, 1: 584 (1762).
Tunica cretica (L.) Fisch. & Mey., Ind. Sem. Hort. Bot. Petrop. 4: 49 (1837) quoad syn.
T. pachygona Fisch. & Mey., Ind. Sem. Hort. Bot. Petrop. 4: 50 (1837).
T. brachypetala Jaub. & Spach, III. Pl. Or. 1: 11, tab. 5 (1842).

Erect annual herbs. Stem up to 30 cm, branched above, glabrous below, glandular-pubescent in middle parts. Basal leaves oblong or oblanceolate, 6–18 mm, 1-veined; cauline leaves linear-lanceolate 12–28 mm, 3-veined with white hyaline margins below. Inflorescence a lax panicle, 3–20-flowered on per stem. Pedicels longer than calyx, glabrous or glandular near base. Bracts not enclosing calyx. Calyx 8–10 mm, ± equalling petals, glabrous, ± obconical; costae 5, 3-veined; commissures papery and crinkled. Petals entire, white. Capsule oblong-ovoid. Fig. 29: 1–4.

HAB. On hillsides, steppes, dry silt slopes, open *Quercus* scrub ; alt. 400–1200 m; fl. May–Jun.
DISTRIB. **MAM**: Sawara Tuka, *Robertson* RC5Y!; Zawita, *Robertson* RC57!; **MSU**: Tawila, *Rawi* 22127!; Jarmo, *Helbaek* 1286!; *Haines* W369!; **FUJ**: Balad Sinjar, *Field* & *Lazar* 646!

Albania, Greece, Turkey, Syria, Palestine, Saudi Arabia, Iran.

4. **Petrorhagia wheeler-hainesii** *Rech.f.*, Bot. Jahrb. 107: 52 (1985). — Type: Iraq, Kopi Qaradagh, *R. Wheeler-Haines* 1504 (holo, E); Rech.f., Fl. Iranica 163: 193, t. 372 (1988).

Perennials. Stems 25–30 cm, usually erect, glandular-puberulous below, glabrous above, branched. Lower leaves linear-oblong, 18–25 × ± 2 mm, glabrous, rarely short hairy near slightly expanded base, attenuate, ± acute at apex; median one linear-filiform; upper leaves subulate. Flowers solitary at end of branches or cyme with 2 flowers. Bracts shorter than pedicels in mature flowers. Pedicels to 12 mm, glandular. Calyx 7–10 mm, cylindrical, attenuate at base, glandular-puberulent, costae 5 with 3 veins; teeth ± obtuse, 1.5 mm with hyaline margins. Petals oblong-linear, rotundate-truncate or emarginated with purpurescent nerves.

HAB. Rocky crevices, limestone slopes; alt. ± 1500 m; fl. May–Jun.
DISTRIB. Rare in mountains of NE Iraq. **MSU**: Kopi Qaradagh, *Haines* W1504! (type) & *Haines* W1134!; Darband-i Khan, *Haines* W1700!

Endemic.

23. **VELEZIA** L.

Sp. Pl. ed. 1: 332 (1753)

Annuals with rigid, dichotomously branched stems. Leaves linear-filiform. Flowers solitary or cymose. Epicalyx absent. Calyx 5–7-toothed, tubular (in our species), sometimes campanulate 5–15-nerved, without membranous intervals. Petals 5, lobed with long claw; coronal scales absent. Stamens usually 10 (sometimes 5). Styles 2. Capsule cylindirical, dehiscining by 4 teeth. Seeds acute with a facial hilum.

Six species distributed from the Mediterranean to Afghanistan; one species in Iraq.

Velezia rigida *L.*, Sp. Pl. ed 1:332 (1753); Boiss., Fl. Orient. 1: 478 (1867); Blakelock in Kew Bull. 2: 218 (1957); Coode in Fl. Turk. 2: 137 (1967); Ghazanfar & Nasir in Fl. Pak. 175: 120 (1986); Rech.f., Fl. Iranica 163: 194 (1988); Strid, Fl. Hellenica 1: 373 (1997).

Stems 8–40 cm, erect to ascending, divaricately branched, glandular-pubescent, sometimes glabrous. Leaves 6–12 × 0.75–1 mm, linear, sessile, ± expanded at base with hyaline ciliate margin, acute at apex. Flowers solitary in the axils of leaves or up to 4 in axillary fascicles. Pedicels up to 3 mm, thick, glabrous to glandular-pubescent. Calyx 12–15 mm, glabrous to glandular-pubescent, 15-nerved; teeth 1 mm. Petals pink, bifid, exceeding calyx; claw as long as the calyx tube. Capsule cylindrical, shorter than calyx. Fig. 30: 1–2.

Hab. Rocky mountain sides, sand banks nr river, open fields; alt. 300–1590 m; fl. May–Jun.
Distrib. **mro**: Saladdin, *Gillett* 11263!; Karokh, *Rawi, Nuri* & *Kass* 27267!; **msu**: Kajan Mountain, nr Penjwin, *Rawi* 22723!; Jarmo, *Helbaek* 1767! & *Haines* W230A!; 2 km E of Sulaimaniya, *Gillett & Rawi* 11697!; **mjs**: Jabal Sinjar, Kaisi, *Khayat* & *Karim* 50852!; *Sharief* & *Hamad* 50198!; **fni**: nr. Eski Kellek, *Gillett* 8196!; **fpf**: Buksaya (Bagsaiya), *Rawi* & *Haddad* 25579!

Mediterranean Region, Caucasus, Turkey, Iran, Afghanistan and Pakistan.

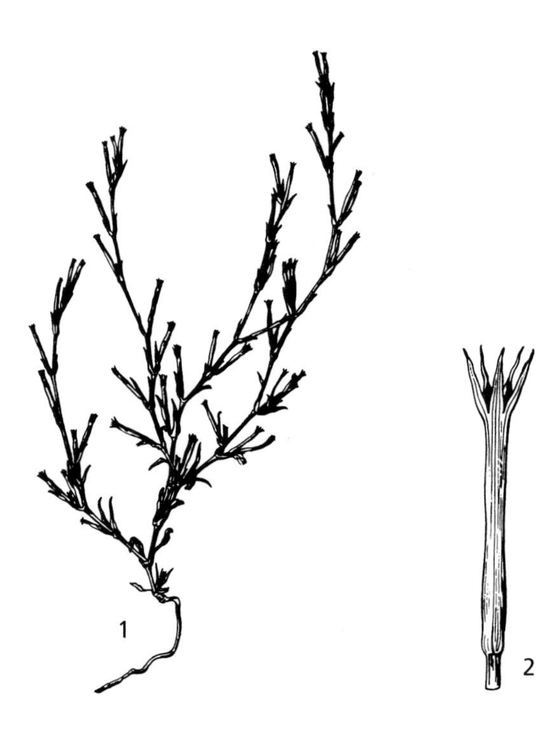

Fig. 30. **Velezia rigida**. 1, habit × ½; 2, flower × 3. Reproduced with permission from Flora of Pakistan 175: f. 18, 1986. Drawn by S. Hameed.

24. SAPONARIA L.

Sp. Pl. 408 (1753); Gen. Plant. ed. 5: 191 (1754)

Annual, biennial or perennial herbs, glabrous or hairy with glandular or eglandular hairs. Leaves simple, linear-lanceolate, spatulate or ovate-elliptic, obtuse or acute. Inflorescence paniculate or cymose. Calyx ebracteolate, cylindrical, many-nerved, without or with very narrow commissures. Petals entire, bifid or lobed; coronal scales present or absent. Stamens 10. Styles 2. Capsule dehiscing apically by 4 valves or teeth. Seeds reniform, tuberculate or flat.

A genus of about 40 species distributed in temperate Eurasia, mainly in the Mediterranean and Irano-Turanian regions; three species in Iraq.

1. Chasmophytic perennials .... 3. *S suffruticosa*
   Annuals, not chasmophytic 2
2. Petals divided into 3 linear-triangular lobes .... 2. *S tridentata*
   Petals bifid to retuse .... 1. *S viscosa*

1. **Saponaria viscosa** *C A. Mey.*, Verz. Pfl. Cauc. 212 (1831); Zohary, Fl. Iraq & Its Phytogeogr. Subdivis.: 54 (1946); Blakelock in Kew Bull. 2: 204 (1957); Hedge in Fl. Turk. 2: 142 (1967); Rech.f. in Fl. Iranica 163: 201 (1988).

Annual herbs, usually divaricately branched, 7–25 cm, covered with long capitate glandular and short eglandular hairs. Leaves 10–35 × 2–7 mm; linear-lanceolate; lower ones attenuate, obtuse; upper ones shortly petiolate or sessile. Corymbs many-flowered. Pedicels 3–7 mm, shorter than calyx. Calyx 8–10 mm, narrowly oblong-cylindrical, glandular-villose; teeth ± 2 mm, lanceolate, acute. Petals reddish or pink, linear, longer than calyx, bifid or retuse; coronal scales absent. Capsule ovoid-oblong, slightly longer than calyx. Fig. 31: 1–9.

HAB. Rocky slopes, on serpentine; alt. 900–1400 m; fl. May–Jun.
DISTRIB. Rare. **MRO**: Serderian, between Arbil and Rowanduz, *Nábělek* 4192; **MSU**: Sulaimaniya, Penjwin, *Rechinger* 12294.

South Caucasus, Turkey, Iran, Azerbaijan.

2. **Saponaria tridentata** *Boiss.*, Diagn. ser. 1(1): 17 (1843); Blakelock in Kew Bull. 2: 203 (1957); Hedge in Fl. Turk. 2: 143 (1967).

Annual divaricately branched herb, 10–20 cm, with a dense indumentum of long glandular hairs covering the entire plant except for the lower leaves. Leaves oblong or linear, shortly petiolate. Inflorescence loose, many-flowered. Pedicels erect-spreading, longer, rarely shorter, than calyces. Calyx 9–11 mm, narrowly cylindrical with very short acute teeth, thickly covered with long, capitate glandular hairs. Petals pink to reddish pink, ± 13 mm, with an obovate lamina deeply divided into 3 (or 5) linear-triangular lobes. Capsule oblong, slightly longer than calyx.

HAB. Rocky slopes, fallow fields; alt. 1± 000 m; fl. May–Jun.
DISTRIB. One doubtful record from the border area between Turkey and Iraq. **MAM**: ?Amadiya region, *Nábělek* 4185.

No specimens at K; cited record from Blakelock (l.c.).

Turkey.

3. **Saponaria suffruticosa** *Nábělek* in Publ. Fac. Sc. Univ. Masaryk, Brno No. 35, 41 (1923). — Type: in Kurdistaniae Turcicae district Berwari in fussuris rupium calcar. montis Choarra-Sia supra pagum 'Ain Nune inter Hasitha et Amadia, alt. ca. 1500 m, 16. VI. 1910, *Nábělek* 4181 (SAV). Zohary, Fl. Iraq & Its Phytogeogr. Subdivis.: 54 (1946); Blakelock in Kew Bull. 2: 203 (1957); Dönmez, Hacettepe J. Biol. & Chem. 37, 3: 181–187 (2009).

Chasmophytic perennial, forming mats in rock fissures, woody at base, stem much branched at base, 8–15 cm, with few sterile and many fertile branches, densely to sparsely white hairy with and glandular patent hairs. Leaves of fertile and sterile branches different in size; leaves broadly obovate, 5–12 × 3–8 mm, abruptly constricted into a short petiole, obtuse or slightly acute at apex, slightly ciliate at margins, lamina with white finely glandular and eglandular hispid hairy, one-nerved. Inflorescence with many flowers at apex with densely glandular and finely hispid hairs; bracts small, linear-lanceolate, acute, 6–9 × 1–2 mm; pedicels 1–1.5 mm. Flowers of two kind; fertile flowers with well developed capsules with mature seeds; sterile flowers with stamens and ovary without developed seeds. Calyx 8–10 × 3–4 mm in fertile flowers, 8–10 × 1–2 mm in sterile flower, cylindric or oblong-cylindric, with eglandular patent and glandular hairs, obscurely 13–15-nerved, brownish

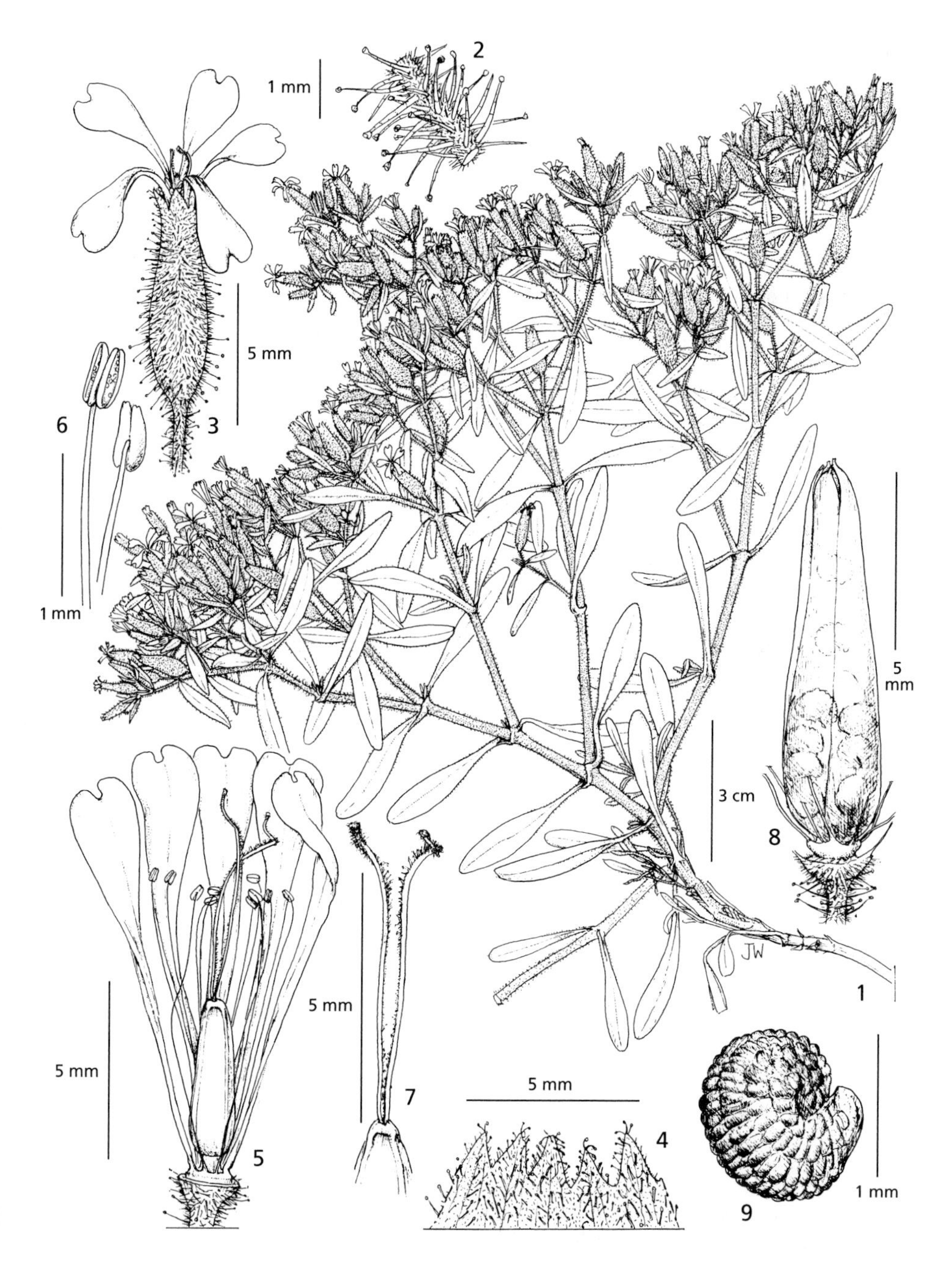

Fig. 31. **Saponaria viscosa**. 1, habit; 2, stem detail; 3, flower, side view; 4, calyx lobe detail; 5, flower with calyx removed; 6, anthers; 7, style detail; 8, capsule with calyx removed; 9, seed. All from *Bowles Scholarship Bot. Exped.* 1221. © Drawn by Juliet Beentje.

to green, teeth 2–2.5 × 1–1.2 mm. Petals reddish to pale pink or whitish, 10–12 mm, claw gradually enlarging upwards into limb top of the calyx tube, coronal scales absent, limb 2 × 2.5 mm in wide. Stamens 10(–11). Capsule 6–8 × 3–3.5 mm in fertile flowers with 2 styles, oblong, opening with 4 teeth with 6–8 seeds; carpophore 2 mm, concealed in calyx; capsule of sterile flowers with rudimentary seeds in calyx tube. Seeds peltate, tuberculate.

HAB. Limestone crevices; alt. 500–1700 m; fl. May–Jun.
DISTRIB. One doubtful record from the border area between Turkey and Iraq. **MAM**: ?Amadiya region, *Nábělek* 4181.

The species was described from between "Hašitha" and Amadiya by Nábělek in 1910. The modern (Kurdish) name of this locality is Asutka, and it lies just N of the border inside Turkey. Dönmez commented that the species occurs in Iraq (l.c.) but was unable to pinpoint the locality "Choarra Sia" which has not been traced. No Iraqi specimens at K; description from Dönmez (l.c.).

Turkey (SE Anatolia).

25. **GYPSOPHILA** L.

Sp. Pl. 406 (1553)

Annual, biennial or perennial herbs, rarely subshrubs, glaucous, glabrous, eglandular or frequently glandular hairy. Leaves usually linear to lanceolate, sometimes ovate or elliptic. Flowers small in dichasial cymes, panicles or heads. Bracts green or scarious. Bracteoles absent. Calyx campanulate, turbinate, rarely tubular, 5–toothed, without commissural nerves, usually with druses. Petals 5, linear to cuneate, without coronal appendages, inserted on a cup-shaped disc. Stamens 10. Androphore very short. Styles 2. Capsule unilocular, globular or oblong, 4-valved. Seeds auriculate, compressed both sides, usually tuberculate.

Gypsophila is a predominantly Eurasian genus. It is native to Europe, Asia and North Africa with a centre of diversity in the Black Sea region, Anatolia, Caucasus, N Iraq and N Iran. 14 native species occur in Iraq. *G. elegans* Bieb. is grown in gardens as a cultivar.

Refs: F.N Williams (1889) A revision of genus *Gypsophila*. J. Bot. 27: 321–329.
G. Stroh (1937) Die Gattung *Gypsophila*. Beih. Bot. Centr. 59: 455–477 (1937).
S. Zmarzty (1993) A Reconsideration of *Gypsophila* Series *Deserticolae* Barkoudah (Caryophyllaceae) including *G. capillaris* (Forssk.) C Chr. Kew Bull. 48(4): 683–697 (1993).

Key to Species

1. Annuals. 2
   Perennials. 5
2. Stems patent hispid, hairy in middle parts; pedicels becoming deflexed; calyx longer than 4 mm . . . . 13. *G. pilosa*
   Stems glabrous or hairy, not patent hispid; pedicels erect or spreading; calyx shorter than 4 mm 3
3. Calyx glandular hairy . . . . 9. *G. linearifolia*
   Calyx glabrous 4
4. Bracts mostly linear and foliaceous . . . . 10. *G. capillaris*
   Bracts mostly triangular and scarious. . . . . 8. *G. heteropoda* subsp. *heteropoda*
5. Stem leaves cordate-amplexicaul . . . . 5. *G. ruscifolia*
   Stem leaves not cordate-amplexicaul 6
6. Fibrous collar present at stem base . . . . 1. *G. caricifolia*
   Fibrous collar absent at stem base 7
7. Flowers sessile or subsessile in dense globose clusters . . . . 2. *G. sphaerocephala* var. *sphaerocephala*
   Flowers pedicellate, not in dense globose clusters 8
8. Low shrubs with rigid dead stems at base . . . . 6. *G. nabelekii*
   Perennial herbs, without rigid dead stems at base 9
9. Calyx tubular, glandular . . . . 14. *G. boissieriana*
   Calyx campanulate, glabrous 10

10. Calyx teeth ciliate along margins . . . . 4. *G. libanotica*
    Calyx teeth not ciliate along margins. 11
11. Stems completely glabrous or glandular-pubescent below, glabrous above 12
    Stems glabrous below, glandular-hairy above, at least in part 13
12. Stems completely glabrous . . . . 10. *G. capillaris*
    Stems glandular-pubescent below, glabrous above . . . . 3. *G. perfoliata* var. *anatolica*
13. Stem more or less viscous . . . . 12. *G. persica*
    Stems not viscous. 14
14. Basal leaves lanceolate to linear-lanceolate, coriaceous, acute to acuminate, 1–3 × 0.3–0.7 cm. . . . . 7. *G. pallida*
    Basal leaves spatulate, not coriaceous, obtuse, 2–6 × 1–2.5 cm 1
    . . . . 1. *G. polyclada* var. *polyclada*

1. **Gypsophila caricifolia** *Boiss.*, Diagn. Pl. Orient. Nov. Ser. 1,1: 14 (1843); Barkoudah in Wentia 9: 66, pl. 2, f. 33–39 (1962); Rech.f., Fl. Iranica 163: 215, t. 262 (1988).

Silene caricifolia (Boiss.) Bornm., Verh. Zool.-Bot. Ges. Wien 60: 82 (1910).

Perennials with woody caudex, collar fibrous and rosette present. Stems 2–40 cm, erect, glabrous, simple or branched in upper part; internodes up to 8 cm long. Basal leaves linear, 3–7(–12) × 2–5 mm, 3- multinerved, glabrous with narrow hyaline margins, expanded near base, with long acuminate apex; cauline leaves smaller. Inflorescence compact cymules forming ± capitate, terminal and/or axillary, 9–20 mm in diameter. Bracts up to 3.5 mm, glabrous, scarious, acuminate longer or shorter than pedicels. Pedicels glabrous, equal or shorter than calyx or absent. Calyx campanulate, 4–5 mm, glabrous; teeth 1–2 mm, ovate, apiculate. Petals pink, clearly longer than calyx, cuneate, retuse or sinuate at apex. Stamens longer than petals. Ovules about 12. Capsule as long as calyx, usually 2 seeded; seeds with small flat tubercles.

HAB. Stream sides on mountains; alt. 2000–2500 m; fl. Aug.
DISTRIB. Very local in NE Iraq. **MRO**: Warshanka-Magar Range, *Rawi* & *Serhang* 24321!

Iran.

2. **Gypsophila sphaerocephala** *Fenzl* ex *Tchihat.*, Asie Min. Bot. 1: 205 (1860). — Type: Iraq: in monte Gara Kurdistaniae, 1843, *Kotschy* 318 (holo. G; iso W, K!). Blakelock in Kew Bull. 2: 195 (1957); Barkoudah in Wentia 9: 87, pl. 5, f. 42–48 (1962); Huber-Morath in Fl. Turk. 2: 154 (1967); Rech.f., Fl. Iranica 163: 219, t. 264 (1988).

var. **sphaerocephala**

Perennials, glabrous with woody rhizome. Stems several, to 70 cm, rigid, woody at base, unbranched or little branched, withish to glaucous, sometimes glandular on upper nodes or near capitula; internodes 2–5 cm long. Leaves linear 10–70 × 1–2 mm, flat or triquetrous, narrowly acuminate. Inflorescence dense capitate, terminal and axillary, 7–10 mm diameter. Bracts up to 3 mm, longer than pedicels, oblong to ovate, acuminate, basal ones caudate, glabrous or scattered glandular hairy. Calyx 3.5–5 mm, turbinate, usually glabrous; teeth ovate to acuminate, shorter than tube. Petals pink to white, 4–7 mm, linear to obtuse. Stamens as long as petals or longer. Capsule equalling calyx or slightly shorter. Seeds with acute tubercles. Fig. 34: 5–7.

HAB. On dry rocky hills; alt. 500–2000 m; fl. Jul.
DISTRIB. Very local in N Iraq and known only from the type specimen. **MAM**: Monte Gara Kurdistaniae, *Kotshy* 318! (type).

G. pilulifera Boiss. & Heldr. is close to *G. sphaerocephala* but, it differs from *G. sphaerocephala* in having shallowly emarginate or dentate petals, apiculate calyx teeth and heads less than 1 cm in diameter. The leaves are also narrower in *G. pilulifera.* One specimen from Sarsang (Haines W494!)has emarginate petals. It was identified as *G. pilulifera* by Barkoudah, but, its calyx teeth seem like *G. sphaerocephala* and its general habit is closer to *G. sphaerocephala* var. *sphaerocephala.*

Rhodes (Greece), Turkey.

3. **Gypsophila perfoliata** *L.*, Sp. Pl., 408 (1753).

G. trichotoma Wenderoth, Linnaea 11: 92 (1837).

var. **anatolica** (*Boiss. & Heldr.*) Barkoudah in Wentia 9: 104 (1962); Rech.f., Fl. Iranica 163: 224 (1988).

G. anatolica Boiss. & Heldr. in Boiss., Diagn. Pl. Or. Nov. Ser. 1,8: 57 (849); Blakelock in Kew Bull. 2: 192 (1957).

G. trichotoma Wenderoth var. *anatolica* (Boiss. & Heldr.) Bornm., Verh. Zool. Bot. Gesel. Wien 60: 81 (1910).

Perennials. Stems 25–60 cm, ascending at base, much branched above, glandular-pubescent below, glabrous above. Leaves oblong or linear-lanceolate, 2–8 × 1–3.5 cm, acute to obtuse, glandular-puberulent to glabrous, 3–7-veined. Inflorescence lax spreading paniculate, glabrous. Bracts to 2.5 mm, clearly shorter than pedicels, scarious, triangular, acuminate, glabrous. Pedicels to 15 mm, glabrous. Calyx campanulate, 2–2.5 mm, glabrous; teeth ovate, obtuse, shorter or equal to tube. Petals white to pink, twice to three times as long as calyx, oblong, obtuse to emarginate, claws constricted below lamina. Stamens shorter than petals. Capsule slightly longer than calyx with 4–8 seeds; seeds with small flat tubercles.

HAB. Sandy meadows, steppe, slopes; fl. Jul.–Aug.
DISTRIB. Very local in NW Iraq. **MJS**: Jabal Sinjar, *Handel-Mazzetti* 1636 (not seen).

No specimen at K; cited record from Blakelock.

Caucasus, Armenia, Turkey, Syria, Iran.

4. **Gypsophila libanotica** *Boiss.*, Diagn. Pl. Orient. Nov. ser. 1,1: 12 (1842); Barkoudah in Wentia 9: 110, pl. 10, f. 18–24 (1962); Huber-Morath in Fl. Turk. 2: 160 (1967).

G. pallidifolia Bark., Wentia 9: 115, Plate 10, f. 55–60 (1962).

Perennials with branched woody stock. Stems many, 25–55 cm, ± rigid, branched, ascending to erect, glabrous or glandular-hairy above. Leaves linear-lanceolate to lanceolate, 7–44 × 1–4 mm, glaucous, acute to acuminate with prominent nerves. Inflorescence panicle with many flowers. Bracts ± 1 mm, distinctly shorter than pedicels, lanceolate, acuminate, scarious. Pedicels to 3 mm, glabrous. Calyx campanulate, 2–3 mm, glabrous; teeth ovate-rotundate, equal to longer than tube with ciliate hyaline margins, calcium oxalate druses prominent. Petals 4–6 mm, oblong, truncate to emarginated. Stamens shorter than petals. Capsule shorter than calyx. Seeds with acute tubercles. Fig. 32: 1–5.

HAB. Rocky mountains; alt. ± 1400 m; fl. Jun.
DISTRIB. Very local in NE Iraq. **MSU**: Mt. Avroman, *Rawi et al.* 19770! & 19791!

Turkey, Lebanon, Syria.

5. **Gypsophila ruscifolia** *Boiss.*, Diagn. Pl. Or. Nov. ser. 1,1: 12 (1842); Barkoudah in Wentia 9: 111, pl. 10, f. 25–32 (1962); Huber-Morath in Fl. Turk. 2: 162 (1967); Rech.f., Fl. Iranica 163: 224, t. 273 (1988).

G. diaphylla Azn. in Mag. Bot. Lap. 16: 10 (1918).

Glabrous perennials with woody stock. Stems woody at base, 35–100 cm, numerous, yellowish, rigid, branched in upper half, glabrous, sometimes glandular above nodes. Leaves coriaceous, light green, sessile, with prominent nerve-net, acut to acuminate; basal leaves lanceolate 15–25 × 4–10 mm; stem leaves cordate-amplexicaul, up to 20–25 mm wide. Inflorescence dichasial, paniculate, many-flowered. Bracts ± 1 mm, shorter than pedicels, linear, acute. Pedicels 3–8 mm, sometimes glandular-hairy at base. Calyx campanulate, 2–2.5 mm; teeth ovate, obtuse to apiculate, equal to tube or slightly longer, calcium oxalate druses prominent. Petals white to pink, linear-oblong, obtuse, longer than calyx. Capsule longer than calyx. Seeds with flat tubercles. Fig. 32: 6–9.

HAB. On dry stony hills and slopes, on mountains in Quercus forests; alt. 700–2000 m; fl. Jul–Aug.
DISTRIB. Common in mountain region. **MAM**: Badi, *Guest* 4422!; Ser Amadiya, *Dabbagh, Kaisi & Hamid* 45982!; Daimka, *Omar & Dabbagh* 45393!; 23 km NE of Zakho, *Rawi* 23197!; **MRO**: Rowanduz, *Bornmüller* 955!; Khalana, *Rawi* 13828!; **MJS**: Jabal Sinjar, *Kaisi et al.* 50931!

Turkey, Lebanon, Palestine, Syria, NW Iran.

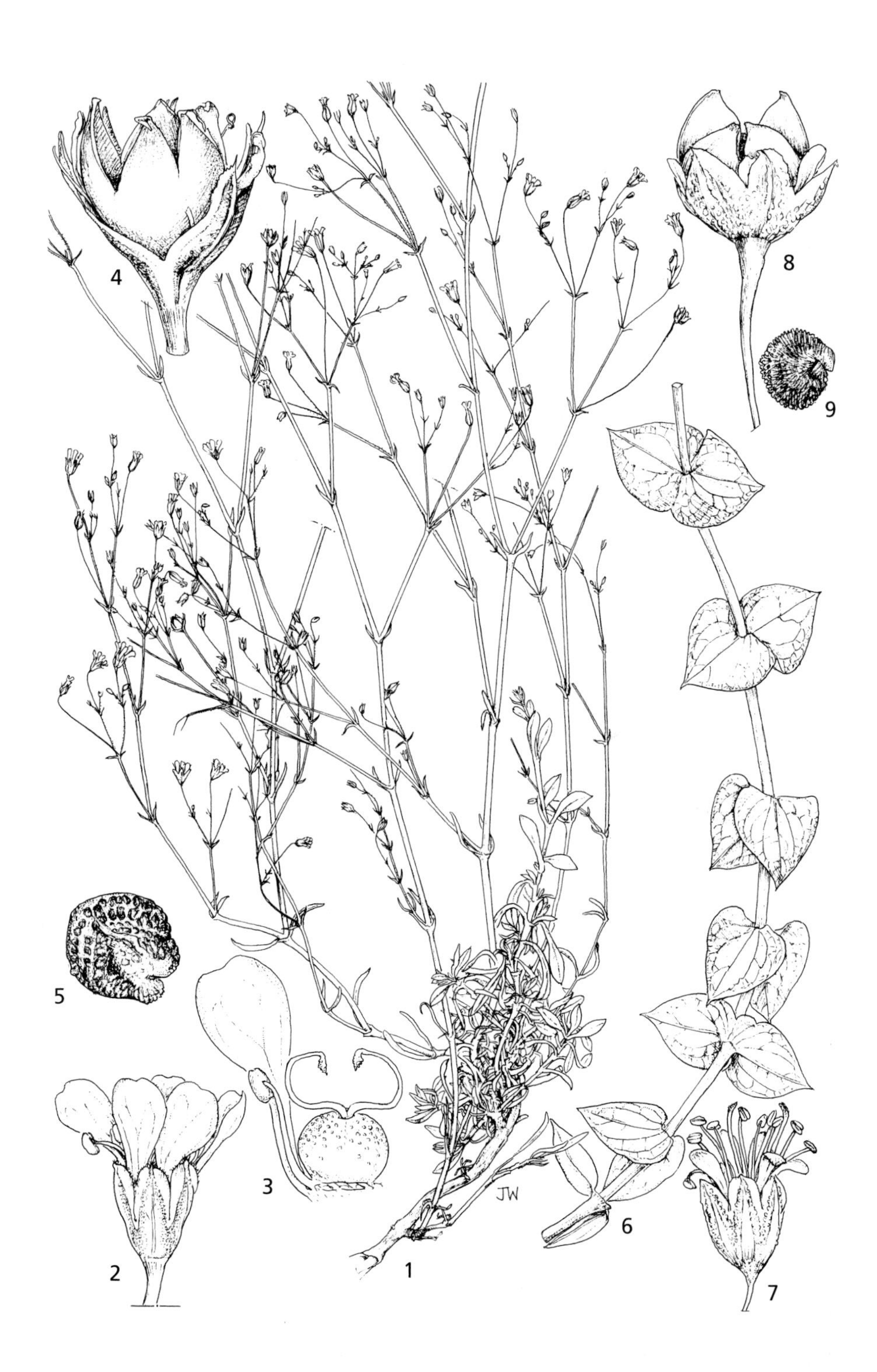

Fig. 32. **Gypsophila libatonica**. 1, habit × ⅔; 2, flower × 8; 3, flower dissected × 12; 4, capsule with calyx × 8; 5, seed × 12. **Gypsophila ruscifolia**. 6, vegetative shoot × ⅔; 7, flower × 12; 8, capsule × 12; 9, seed × 12. 1–4 from *Daris* 20171; 5 from *Rawi et al.* 19791; 6 from *Kasim et al.* 41082; 7 from *Rawi* 23372; 8, 9 from *Omar* 37728. © Drawn by Juliet Beentje.

6. **Gypsophila nabelekii** *Schischk.*, Candollea 3: 475 (1928); Blakelock in Kew Bull. 2: 194 (1957); Barkoudah in Wentia 9: 108, pl. 10, f. 1–9 (1962); Huber-Morath in Fl. Turk. 2: 159 (1967); Rech.f., Fl. Iranica 163: 225, t. 274 (1988).

G. lignosa Náb., Publ. Fac. Sci. Univ. Masaryk Brno 35: 42 (1923) nec Hemsl. & Lace (1891).
G. lipskyi Schischk Candollea 3: 475 (1928).
G. nabelekii Schischk. var. *lipskyi* (Schischk) Bark., Wentia 9: 109 (1962), syn. nov.

Subshrubs with a woody stock, branched. Stems numerous, 5–17 cm, glandular-pubescent below or glabrous or papillose; dead stems persistent, erect and stiff. Leaves linear, glandular-pubescent or glabrous-papillose; lower leaves subimbricate, 10–20 × 1–2 mm; cauline leaves smaller with 2–5 pairs. Inflorescence dichasial, with 3–9 flowers. Bracts up to 2 mm, shorter than pedicels, usually scarious, triangular, acuminate, glabrous. Pedicels 1–4 mm, equal to calyx or longer or shorter, glabrous. Calyx 3–3.5 mm, glabrous; teeth ovate, equal or shorter than tube, obtuse with scarious margin. Petals pale purple, clearly longer than calyx, emarginate, to scarcely bilobed, base attenuate. Stamens shorter than petals. Capsule as long as calyx. Seeds with flat tubercles.

HAB. In rocky cliff, rocky slopes; alt. 2700–3500 m; fl. Jul.–Aug.
DISTRIB. Rare, alpine zone. **MRO**: Qandil Range, *Rawi & Serhang* 18257!; Halgurd Dagh, *Rechinger* 11460! *Ludlow-Hewitt & Guest* 2873!

Transcaucasica, Turkey, Iran.

7. **Gypsophila pallida** Stapf, Denkschr. Akad. Wiss. Wien, Math.-Nat. Kl. 51: 281 (1886); Barkoudah in Wentia 9: 112, pl. 10, f. 41–48 (1962); Huber-Morath in Fl. Turk. 2: 162 (1967); Rech.f., Fl. Iranica 163: 225, t. 275 (1988).

G. haussknechtii Boiss., Fl. Or. Suppl. 86 (1888).
G. pallida Stapf var. *haussknechtii* (Boiss.) Bark., Wentia 9: 114 (1962), syn. nov.

Perennials with woody stock. Stems numerous, 30–60 cm, ascending to erect, branched in upper part glabrous below, glandular-hairy above. Leaves lanceolate to linear-lanceolate, coriaceous, acute to acuminate, 10–32 × 3–7 mm, with 1–3 prominent nerves. Inflorescence dichasial, paniculate with many-flowered. Bracts up to 2 mm, shorter than pedicels, scarious, lanceolate-acuminate. Pedicels 3–8 mm, glabrous, sometimes glandular belove. Calyx campanulate, 2–2.5 mm, glabrous; teeth ovate, obtuse, apiculate, equal to tube or shorter or longer. Petals 3–5 mm, linear, obtuse to retuse, with broad claw. Stamens longer than petals. Capsule longer than calyx. Seeds with obtuse tubercles.

HAB. Sandy rocky soil, in Quercus forests; alt. 800–2000 m; fl. Jun.–Aug.
DISTRIB. Upper forest zone. **MRO**: Rowanduz, *Bornm.* 954!; **MSU**: Chemchemal, *Gillett & Rawi* 11609!; *Haines* W358!; Jarmo, *Haines* W266!; **FUJ**: between Sinjar and Tall Afar, *Haussknecht* 181! (type of. *G. haussknechtii*).

Turkey, Syrian Desert, Iran.

8. **Gypsophila heteropoda** *Freyn & Sint.* in Bull. Herb. Boiss. ser. 2,3: 865 (1903); Blakelock in Kew Bull. 2: 194 (1957); Barkoudah in Wentia 9: 126, pl. 12, f. 35–43 (1962); Rech.f., Fl. Lowand Iraq: 242 (1964); Huber-Morath in Fl. Turk. 2: 164 (1967); Rech.f., Fl. Iranica 163: 229 (1988).

G. nanella Grossh. & Schischk. in Pl. Or. Exs. fasc. 1:8 (1924).

subsp. **heteropoda**

Annual herbs. Stems 4–25 cm, slender, erect, usually branched above, sometimes unbranched, glabrous or viscous with sessile glands above. Leaves linear to linear-lanceolate, 5–30 × 1.3.5 mm, sessile, glabrous, obtuse to acute. Inflorescence lax dichasial. Bracts 0.5–1 (–5) mm, clearly shorter than pedicel, triangular, scarious, acute. Pedicels capillary, 2–15 mm, glabrous. Calyx wide-campanulate, glabrous, 2–2.5 mm; teeth ovate, obtuse to apiculate, longer than tube. Petals white, 2.5–5 mm, oblanceolate to cuneate, truncate, emarginate or sinuate at apex. Stamens usually shorter than petals. Capsule hardly exceeding calyx. Seeds obtuse to acute tubercles.

HAB. Sandy slopes, gypsaceous subdeserts; alt. 50–250 m; fl. Feb.–Mar.

DISTRIB. **DWD**: 5 km W of Kerbela, *Gillett & Rawi* 6370!; **DSD**: 5 km of Salman, *Gillett & Rawi* 6189!

Transcaucasia, Turkey, Iran, Afghanistan and Kazakhstan.

9. **Gypsophila linearifolia** (*Fisch.* & *Mey.*) *Boiss.*, Fl. Or. 1:550 (1867); Barkoudah in Wentia 9: 133, pl. 13, f. 38–45 (1962); Rech.f., Fl. Lowland Iraq: 242 (1964); Huber-Morath in Fl. Turk. 2: 165 (1967); Rech.f., Fl. Iranica 163: 231, t. 284 (1988).

Dichoglottis linearifolia Fisch. & Mey., Ind. Sem. Horti Petrop. 1: 26 (1835).
Gypsophila szowitsii Fisch. & Mey. var. *glandulosa* in Ledeb., Fl. Ross. 1: 289 (1842).
G. trichopoda Boiss., Diagn. Pl. Or. Nov. ser. 1, 1: 10 (1843); Blakelock in Kew Bull. 2: 195 (1957).

Annual herbs. Stems 7–23 cm, thin, dichotomously branched throughout, glandular-puberulent. Leaves linear, 10–60 × 1–2.5 mm, puberulent, ± papillose, obtuse. Inflorescence very lax dichasial. Bracts (1–)3–8(–12) mm, glandular-pubescent linear, mostly foliaceous , uppermost sometimes scarious and clearly shorter than pedicel. Pedicels capillary, 5–18 mm, glabrous. Calyx campanulate, 1.5– 2 mm, glandular-puberulent; teeth ovate, obtuse with scarious margin, longer than tube in fruit. Petals white, 2–3.5 mm, linear, bilobed to emarginated. Capsule as long as calyx. Seeds with acute tubercles. Fig. 33: 1–4.

HAB. Gypsum hills, semi-deserts; alt. 50–250 m; fl. May–Jun.
DISTRIB. **FUJ**: Qaiyara, *Bayliss* 104!; **FKI**: Kirkuk, Altin Kupri, *Bornm.* 958 & 959; Salahiya, *Nábelek* 4113.

S Russia, Turkey, Iran and Kazakhstan.

10. **Gypsophila capillaris** (Forssk.) C Chr., Dansk. Bot. Arkiv. 4(3): 19 (1922); Blakelock in Kew Bull. 2: 193 (1957); Barkoudah in Wentia 9: 139, pl. 14, f. 40–47 (1962); Zmarzty in Kew Bull. 48(4): 693 (1993); Chamberlain in Fl. Arab. Penins. & Socotra: 1: 225 (1996).

Rokejeka capillaris Forssk., Fl. Aegypt. Arab. 90 (1775).
Gypsophilla rokejeka Delile, Fl. Egypt. 87, t. 29 (1842).
G. antari Post & Beauverd ex Dinsmore, Pl. Post. & Dinsm. fasc. 1: 4 (1932); Barkoudah in Wentia 9: 140, pl. 15, f. 1–8 (1962); Rech.f., Fl. Lowland Iraq: 242 (1964); Huber-Morath in Fl. Turk. 2: 167 (1967); Rech.f., Fl. Iranica 163: 233, t. 287 (1988).

Annual or perennial glabrous herbs. Stems single or few, 10–55 cm, branched throughout. Leaves bright to grey-green, glabrous, papillose beneath; basal leaves 10–50 × 3–8(–11) mm, lanceolate, obovate to spatulate, base attenuate, apex obtuse to acute; cauline ones linear to lanceolate, decreasing in size upwards. Inflorescence lax dichasial. Bracts much shorter than pedicels; lower bracts linear, 1–6 mm, leafly; upper minute, scarious. Pedicels capillary, 8–20 mm. Calyx campanulate, 2–3.5 mm, glabrous; teeth ovate to elliptic, acute to obtuse with scarious margin, usually longer than tube. Petals white, obovate, base cuneate, apex obtuse to rounded, longer than calyx. Stamens shorter than petals. Ovary with 4–16 ovules. Capsule usually dehiscent, 1-seeded; perianth caducous or persistent in fruit. Seed testa black, smooth or prominently tuberculate.

Perianth ± persistent in fruit; seeds 4–8 in per capsule, smooth . . . . . . .a. subsp. *capillaris*
Perianth caducous in fruit; seeds 1–4 in per capsule, tuberculate. . . . . . .b. subsp. *confusa*

subsp. **capillaris**

HAB.Sandy gravelly soils, limestone hills and plains, steppes; alt. 80–500 m; fl. Mar.–May .
DISTRIB.: Common in lowland areas. **MJS**: Jabal Sinjar, *Kaisi, Khayat & Karim* 50990!; **FUJ**: 39 km S E of Tal Afar, *Abbas-al-Ani & Bharucha* 8816!; 6 km W of Tal Afar, *Gillett* 11169!; **FKI**: 2 km N of Kirkuk, *Rawi* 21554!; **FPF**: Jabal Hamrin, *Gillett & Rawi* 7323!; 25 km S E of Mandali, *Rawi* 20713!; **DLJ**: Tikrit, *Omar, Khayat & Kaisi* 52355!; **DGA**: Adhaim, *Robertson R.B.* 25!; **DWD**: Rutba, Traibel, *Alizzi & Husain* 34087!; 260 km N W of Ramadi, *Rawi* 20938!; **DSD**: 33 km W N W of Ansab, *Guest, Rawi & Rechinger* 19016; Khadhar al-Ma'i, *Rawi, Khatib & Tikriti* 29124!; 114 km W of Karbala, *Rawi* 30940!; nr. Um Qasir, *Khatib & Alizzi* 33454!; Busaiya, *Fawzi, Hazim & Hamid* 38946!; **LEA**: Fakka, *Rawi* 25864!; **LCA**: Baghdad, *Haines* W167!

Egypt, Iraq, Kuwait and NE Saudi Arabia.

subsp. **confusa** Zmarzty, Kew Bull. 48(4): 694 (1993). — Type: Iraq, in cretac. inter Erbil et Altun Koprii, *Haussknecht* 186 (holo. JE).

G. arabica Bark. in Wentia 9: 139, pl. 14, f. 48–55 (1962); Rech.f., Fl. Iranica 163: 233 (1988).
G. obconica Bark. in Wentia 9: 140, pl. 15, f. 9–16 (1962); Rech.f., Fl. Lowland Iraq: 243 (1964); Rech.f., Fl. Iranica 163: 234, t. 288 (1988).

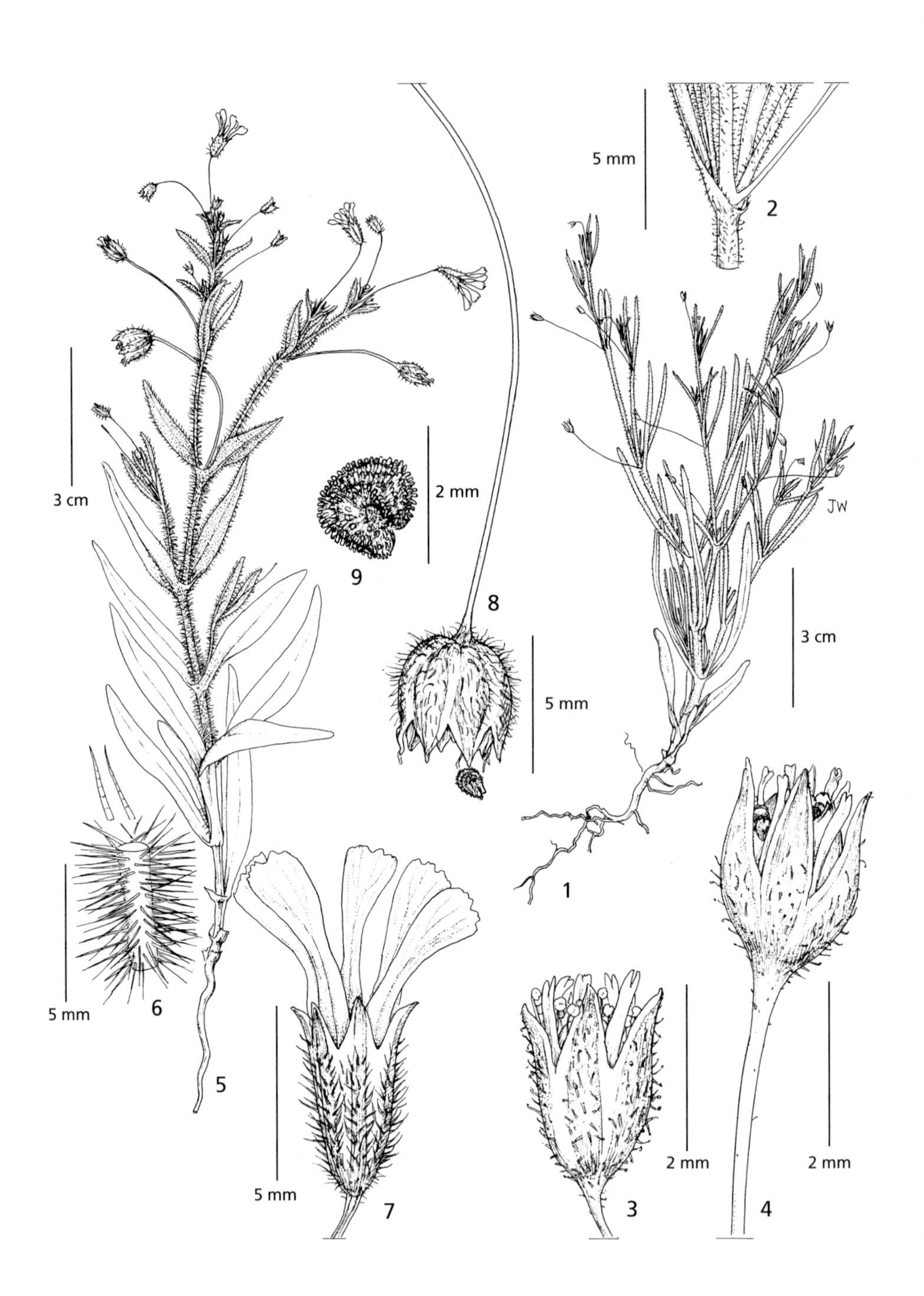

Fig. 33. **Gypsophila linearifolia**. 1, habit × ²⁄₃; 2, stem with leaves × 4; 3, flower × 12; 4, capsule with calyx × 12. **Gypsophila pilosa**. 5, habit × ²⁄₃; 6, stem indumentum × 4; 7, flower × 5; 8, capsule × 3; 9, seed × 10. 1–4 from *Bayliss* 104; 5×7 from *Omar et al.* 52383; 8, 9 from *F. A. Barkley & Haddad* 5488. © Drawn by Juliet Beentje.

Hab. Sandy gravelly soils, limestone hills and plains, steppes; alt. 90-500 m; fl. Jun.–July.
Distrib. Occasional. **fki**: Baba Gurgur, *Guest* 4013!; Kor Mor, nr. Tuz, *Guest* 0311!; 25 km N of Kirkuk, *Katib & Tikriti* 29723!; **fpf**: 4 km N of Sa'diya, *Kaisi* 42892!; **dga**: Ghurfa Plain, *Guest* 4003!; **dwd**: 200 km E of Rutba, *Omar & Kaisi* 43930!

Two specimens from Adhaim (*Robertson* S/1031!) and Tal Afar (*Kaisi et. al.* 50840!) are 4–7-seeded and have a tuberculate testa. They were indicated as intermediate by Zmarzty (l.c.).

Sinai, Jordan, Israel, Syria, Lebanon, W Iran.

11. **Gypsophila polyclada** *Fenzl* ex *Boiss.*, Fl. Or. 1: 542 (1867); Blakelock in Kew Bull. 2: 194 (1957); Barkoudah in Wentia 9: 142, pl. 15, f. 32–39 (1962); Rech.f., Fl. Iranica 163: 235, t. 290 (1988).

G. pulchra Stapf, Denkschr. Akad. Wiss. Wien, Math.-Nat. Kl. 51: 13 (1886).
G. koeiei Rech.f., Dansk Bot. Arkiv 15,4: 20 (1955).

var. **polyclada**

Perennial with woody root. Stems single or few, 20–60 cm, erect, branched nearly throughout, usually glandular hairy above nodes. Leaves glaucous; basal leaves spatulate, obtuse to obtuse, 2–6 × 1–2.5 cm; cauline leaves lanceolate, 2–7 × 5–20 mm, acute to acuminate. Inflorescence lax dichasial. Bracts minute, triangular, acuminate, scarious. Pedicels capillary, 10–20 mm, glabrous or glandular below. Calyx campanulate, 2.5–3 mm, glabrous; teeth ovate, acute to obtuse with scarious margins, equal to tube or shorter. Petals oblanceolate, rounded at apex, 5–7 mm, twice or three times as long as calyx. Stamens shorter than petals. Capsule globose, up to 3 mm, longer than the persistent calyx.

Hab. Rocky foothills, slopes, fallow fields, in Quercus forests; alt. 700–1800 m; fl. Jun.–Jul.
Distrib. Common in E and N of Iraq: **mam**: Gunde Sergele 9 km E of Amadiya, *Dabbagh & Hamid* 45906!; 10 km S of Sarsang, *Kaisi & Hamad* 54811!; **mro**: Mergadreija, nr Haji Umran, *Rawi* 9130!; Saran Village, nr Karokh, *Kass & Nuri* 27272!; **msu**: Kopi Qaradagh, *Haines* W1101!; Sulaimaniya, *Rechinger* 10540!; Mela Kowa, *Rawi* 22438!; Penjwin, *Rawi* 22614!

Iran.

12. **Gypsophila persica** *Bark.*, Wentia 9: 141, Plate 15, f. 25–31 (1962); Rech.f., Fl. Iranica 163: 238, t. 293 (1988).

Perennial herbs. Stem several, 40–60 cm, 3–4 mm in diameter near base, branched above, glaucous below, glandular pubescent above, more or less viscous. Lower leaves linear to spatulate, 15–40 × 4–7 mm, glabrous, acute; cauline leaves linear to lanceolate, 20–50 × 5–15 mm, glabrous, acute. Inflorescence lax dichasial. Bracts up to 2.5 mm, much shorter than pedicels, triangular, acuminate, scarious. Pedicels 10–18 mm, glabrous. Calyx campanulate, 2 mm, glabrous; teeth ovate, apiculate or acutish with scarious margins, equal to tube or shorter. Petals white, oblanceolate, 5–7 mm, apex rounded, twice or three times as long as the calyx. Stamens shorter than petals. Capsule globose, ± 4 mm, clearly longer than the persistent calyx.

Hab. In *Quercus* forest; alt. 1400–2000 m; fl. Apr.–May.
Distrib. Very local. **msu**: Kopi Qaradagh, *Poore* 446! *Gillett* 7932!

NW Iran.

13. **Gypsophila pilosa** *Hudson*, Philos. Trans. 56: 252 (1767); Barkoudah in Wentia 9: 151, pl. 17, f. 1–9 (1962); Rech.f., Fl. Lowland Iraq: 243 (1964); Huber-Morath in Fl. Turk. 2: 170 (1967); Ghazanfar & Nasır in Fl. Pak. 175: 92 (1986); Rech.f., Fl. Iranica 163: 240, t. 295 (1988); Chamberlain in Fl. Arab. Penins. & Socotra 1: 224 (1996).

Silene porringens Gouan ex L., Syst. Nat. ed. 12,3: 230 (1767).
Saponaria porringens (Gouan) L., Mantissa alt. 239 (1771).
Hagenia filiformis Moench, Meth. 61 (1794).
Gypsophila porrigens (L.) Boiss., Fl. Or. 1: 557 (1867); Blakelock in Kew Bull. 2: 195 (1957).
Pseudosaponaria pilosa (Huds.) Ikon., Nov. Syst. Pl. Vasc. Leningrad 15: 145 (1979).

Annual herbs. Stems 18–80 cm, erect, branched above, glabrous at base, patent hispid hairy in middle parts, glabrous or hispid in inflorescence region, branched above, 30–90

× 8–20 mm, lanceolate, acuminate, 3–5-veined, glabrous or glandular hairy. Inflorescence lax, dichasial. Bracts much shorter than pedicels; lower bracts up to 4(–12) mm, linear-lanceolate, foliaceous, hairy; upper bracts smaller, hairy; uppermost minute, glabrous. Pedicels capillary, 10–35 mm, becoming deflexed, glabrous. Calyx campanulate-tubular, 5–6 mm, glandular-hispid; teeth triangular, obtuse to acute with scarious margins, shorter than tube. Petals white to pale pink, 8–10 mm, emarginate to shallowly bilobed. Stamens shorter than calyx and petals. Capsule broadly ovoid, as long as or slightly exserted from than calyx. Seeds with acute tubercles. Fig. 33: 5–9.

HAB. Ruins, sandy clay soils, fields; alt. 90–500 m; fl. Apr.–Jun.
DISTRIB. Widespread but scattered in dry steppe areas. **MSU**: Jarmo, Helbaek 1166!; **FUJ**: Mosul, Gillett 11176!; 28 km from Mosul to Kasik, Dabbagh 46730!; Hatra (Hadhr), Bharucha & Brahim 8650!; **FNI**: Nimrud, Helbaek 946!; **FKI**: 40 km N of Kirkuk, Barkley *et al.* 1351!; **FPF**: Sa'adiya, Kaisi 42879!; **DWD**: 6 km above Ana, Gillett & Rawi 6977!; Al-Bert Village, between Nabiya & Qaim, Omar *et al.* 44394!; **DSD**: Salman, Karim *et al.* 40129!; **LCA**: between Baghdad & Falluja, *Wedad* 47543!

Transcaucasia, Turkey, Jordan, Palestine, Syria, Iran, Arabian Peninsula, Afghanistan, Pakistan and Turcomania.

14. **Gypsophila boissieriana** *Hausskn.* & *Bornm.*, Beih. Bot. Centrbl. 28, 2: 137 (1911). — Type: Iraq: Arbil, Mt. Sakri Sakran, 2100 m, *Bornmüller* 957! (holo, B†; iso. BR, JE, K!, P). Barkoudah in Wentia 9: 152, pl. 17, f. 17–22; Rech.f., Fl. Iranica 163: 242, t. 298 (1988).

Saponaria boissieriana (Hausskn. & Bornm.) Preobr. ex. Popov, Trudy Turkest. Gosud. Univ. 4: 24 (1922).
Gypsophila platyphylla auct. non Boiss., Blakelock in Kew Bull. 2: 194 (1957).

Perennials with woody root. Stems few, 30–60 cm, erect, whitish, glaucous and glaucous to glandular-pubescent in inflorescence region, branched throughout. Leaves glabrous; basal ones spatulate to ovate, 7–11 × 2–3 cm, 5–7-nerved, attenuate at base, acute to obtuse at apex; lower cauline leaves similar to basal ones; upper leaves broadly lanceolate, smaller than basal ones. Inflorescence dichotomous and trichotomous. Bracts up to 6 mm, lanceolate, acuminate, scarious, at least upper ones glandular-hairy. Pedicels 7–11(–20) mm, glandular hairy. Calyx tubiform, 6–9 mm, glandular hairy; teeth triangular, acuminate, with scarious margins, shorter than tube. Petals white to pink, linear-cuneate, truncate, twice as long as calyx. Stamens longer than calyx, shorter than petals. Ovules 12–16. Fig. 34: 1–4.

HAB.Nr. stream, Quercus forests; alt. 1400–2100 m; fl.Jun.–Aug.
DISTRIB. Occasional in mountainous areas. **MRO**: Mt. Sakri Sakran, *Bornmüller* 957 (type); Qandil Range, N of Pishtashan Village, *Rawi & Serhang* 18301!; **MSU**: Kopi Qaradagh, *Haines* W1168; Penjwin, *Rechinger* 10539!

This species is similar to *G. platyphylla* Boiss. which is known from Kermanshah, M. Avorman in W Iran, very close to the Iraq-Iran border. No collections of *G. platyphylla* are known from Iraq, but it is likey to be present in Iraq. *G. platyphylla* is distinguished by its glandular-pubescent, more or less viscous stems, 4–5 mm long turbinate-tubiform calyx, petals with a shallow contraction between the limb and the claw, stamens shorter than calyx, and ovary with 8 ovules.

W Iran.

SPECIES IMPERFECTLY KNOWN OR DOUBTFULLY RECORDED FROM IRAQ

Gypsophila venusta *Fenzl*, Pugill. Pl. Nov. Syria et Taur 9 (1842); Blakelock in Kew. Bull. 2: 195 (1957); Barkoudah in Wentia 9: 156, pl. 18, f. 10–17 (1962); Huber-Morath in Fl. Turk. 2: 171 (1967).

G. wiedemanni Boiss., Fl. Or. 1: 541 (1867).

Perennials. Stems few, ascending, whitish, glabrous, branched above, 50–100 cm; internodes 2–4 cm long. Leaves lanceolate, 2–5 × 5–15 mm, 3–5-veined. Inflorescence lax paniculate-dichasial. Bracts lanceolate to triangular, acuminate with scarious border. Pedicels to 2.5 mm, capillary. Calyx tubiform-turbinate, 3.5–5 mm with large scarious intervals; teeth semicircular, green band tri-veined. Petals 1½–2 × as long as calyx, cuneate to linear-cuneate, truncate to shallowly emarginate, white with reddish veins. Capsule obovoid-globose, mostly dehiscing irregularly, rarely four-valved, as long as calyx.

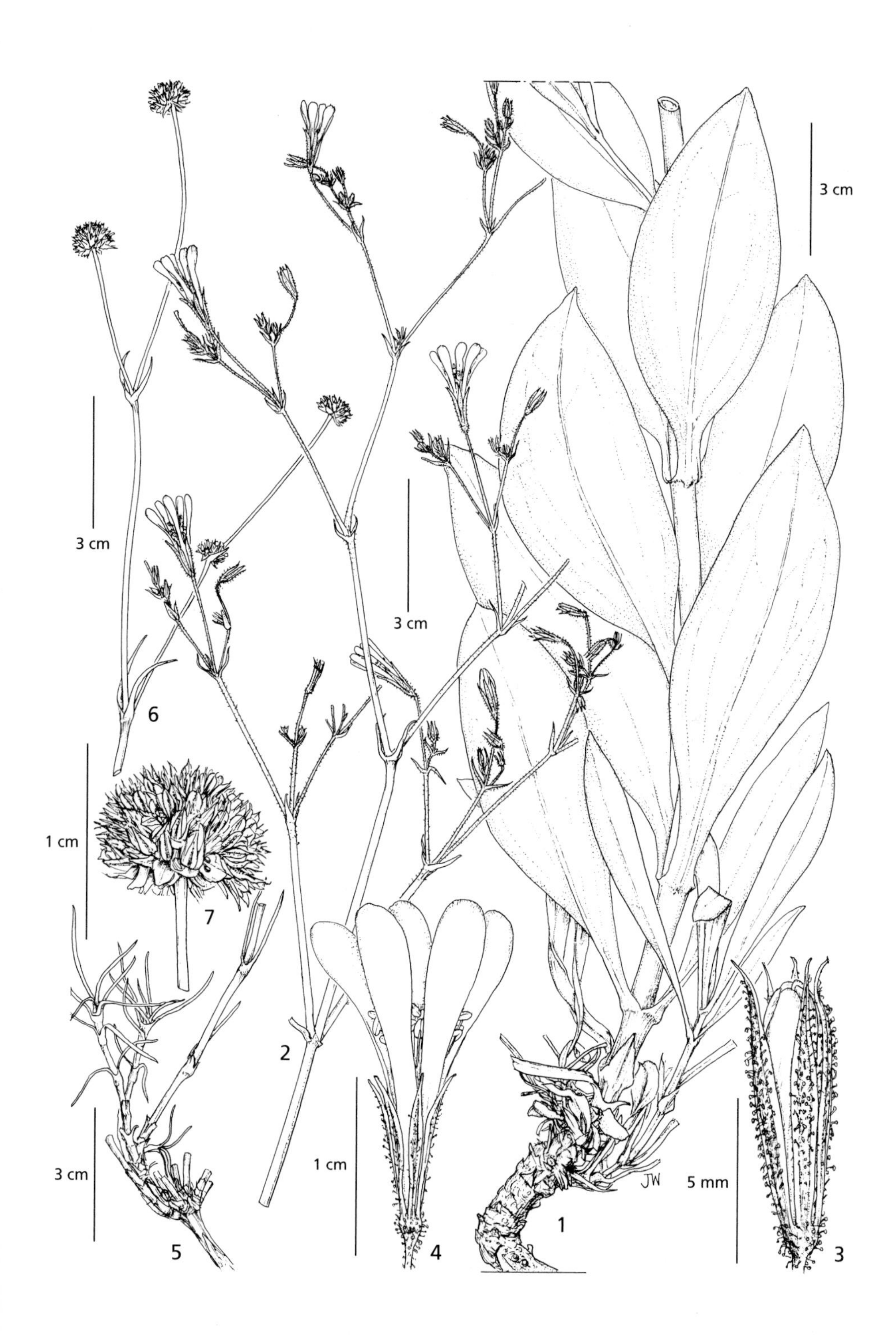

Fig. 34. **Gypsophila boissieriana**. 1, habit × ²/₃; 2, upper part of inflorescence × ²/₃; 3, flower bud × 5 4, flower × 2 ²/₃. **Gypsophila sphaerocephala** var. **sphaerocephala**. 5, habit × ²/₃; 6, part of inflorescence × ²/₃; 7; flower head × 3. 1–4 from *Haines* W1168; 5–7 from *Haines* W494. © Drawn by Juliet Beentje.

This species was recorded by Blakelock from ?Pir Omar Gudrun, June 1867, *Haussknecht* 182 (BM!). Its presence in Iraq needs confirmation.

Gypsophila lurorum *Rechinger* f., Bot. Jahrb. 75: 354 (1951); Barkoudah in Wentia 9: 144, pl. 15, f. 40–45 (1962); Rech.f., Fl. Iranica 163: 239, t. 294(1988).

Perennials. Stems 60–80 cm, erect, branched trichotomously nearly throughout, lower branches short, upper ones nearly as long as the main axis, glandular hairy, viscid throughout, yellowish green. Leaves spatulate, semipetiolate, acute, slightly connate at the base; lower ones glabrous, upper ones glandular hairy on lower surface, obscurely 3–5-nerved. Inflorescence with many flowers, lax dichasial. Bracts lanceolate, scarious-bordered, glandular hairy on the lower surface. Pedicels to 2.5 cm long, capillary, glandular-hairy. Calyx campanulate 3 mm; teeth ovate, mucronulate, green bands as broad as the scarious intervals, equal to tube or shorter. Petals twice or three times as long as the calyx, linear to spatulate, emarginate to shallowly bilobed. Ovary ovoid. Styles divergent; ovules 8.

HAB. In Quercus forest; alt. 730–1500 m; fl. May–Jun.
DISTRIB. **MSU**: Kopi Qaradagh, *Haines* W1104!

Gypsophila lurorum is recorded as endemic to Iran; Haines's specimen determined as *G. lurorum* by himself, lacks flowers and has glabrous pedicels. More collections are needed to ascertain the presence of this species in Iraq.

W Iran.

26. **ANKYROPETALUM** Fenzl

Bot. Zeit. 1: 393 (1843)

Perennial rigid plants, unbranched or intricately branched with woody caudex. Leaves linear, mostly caducous. Inflorescence lax dichasial. Calyx tubiform, 5-ribbed with hyaline intervals. Petals 5, with linear claw and deeply (3–)5-lobed limb. Stamens 10, exserted. Ovary ovoid, sessile with two long styles. Capsule ovoid to oblong, dehiscing from base by longitudal slits, 1–3-seeded; seeds reniform-globose, covered with granular wartlets.

Four species distributed from eastern Mediterranean to W Iran and Armenia; one species in Iraq.

Ankyropetalum gypsophiloides *Fenzl*, Bot. Zeit. 1: 393 (1843); Barkoudah in Wentia 9: 173, pl. 18, f. 18–28 (1962); Rech.f., Fl. Lowland Iraq: 244 (1964); Huber-Morath in Fl. Turk. 2: 148 (1967); Rech.f., Fl. Iranica 163: 247 (1988).

Ankyropetalum gypsophiloides Fenzl var. *glandulosum* Bornm., Beih. Bot. Centrbl. 28, 2: 136 (1911).
A. gypsophiloides Fenzl var. *viscosum* Bark.,Wentia 9: 174 (1962); Rech.f., Fl. Iranica 163: 248(1988), syn. nov.
A. gypsophiloides Fenzl var. *coelesyriacum* (Boiss.) Bark. Wentia 9: 174 (1962), syn. nov.
Gypsophila gypsophiloides (Fenzl) Blakelock in Kew Bull. 2: 193 (1957).
G. subaphylla Rech.f., Bot. Jahrb. 75: 355 (1951).

Perennial with thick woody caudex. Stems 40–75 cm, several, erect, branched throughout, glabrous to glandular-hairy below, glandular-hairy above, sometimes viscous with sessile glands. Leaves linear, 18–40 × 1–2 mm, caducous. Bracts triangular, minute. Pedicels capillary, to 20 mm, glabrous to glandular-hairy. Calyx tubiform, 4–5 mm, glandular-hairy; teeth ovate, obtuse, much shorter than tube. Petal limb (3–)5-lobed, central lobe entire, deltoid. Capsule elliptic with 2 seeds, shorter than the persistent calyx. Fig. 35: 1–9.

HAB. Dry hills, stony limestone slopes; alt. 500–700 m; fl.Jun.
DISTRIB. Widespread in the dry steppe zone. **MAM**: Atrus, *Guest* 3620!; **MRO**: Rowanduz, *Bornmüller* 953!, Kani Rash, *Rawi* & *Serhang* 23761!; nr. Rayat, *Guest* 15930!; **MSU**: Darband-i Bazian, *Rawi* 22850!; mt. Hauraman (Avroman), *Nuri, Chak, Rawi* & *Alizzi* 19331!; Tawila, *Rawi* 22339!; Sulaimaniya, *Field* & *Lazar* s.n!; **FUJ**: 85 km NW of Mosul, *Rawi* 23138!; **FAR**: Brizia, *Gillett* 9637!; Arbil, *Rechinger* 11268!; Mahmour Dagh, *Gillett* 44248!; **FKI**: Altun Kopri, *Guest* 4022!; Kani Dolman, *Guest* 4350!; **FPF**: Jabal Hamrin, *Haines* W1462!; nr Tursak, *Haines* W994!

Barkoudah (l.c.) separated this species into three varieties: var. *gypsophiloides*, var. *viscosum* Bark. and var. *coelesyriacum* (Boiss.) Bark. using indumentum of the stem as the separating charcater. This character is variable on the speciemens seen and does not justify varietal status.

Lebanon, Syria, Palestine, Sinai, N Iraq and Iran.

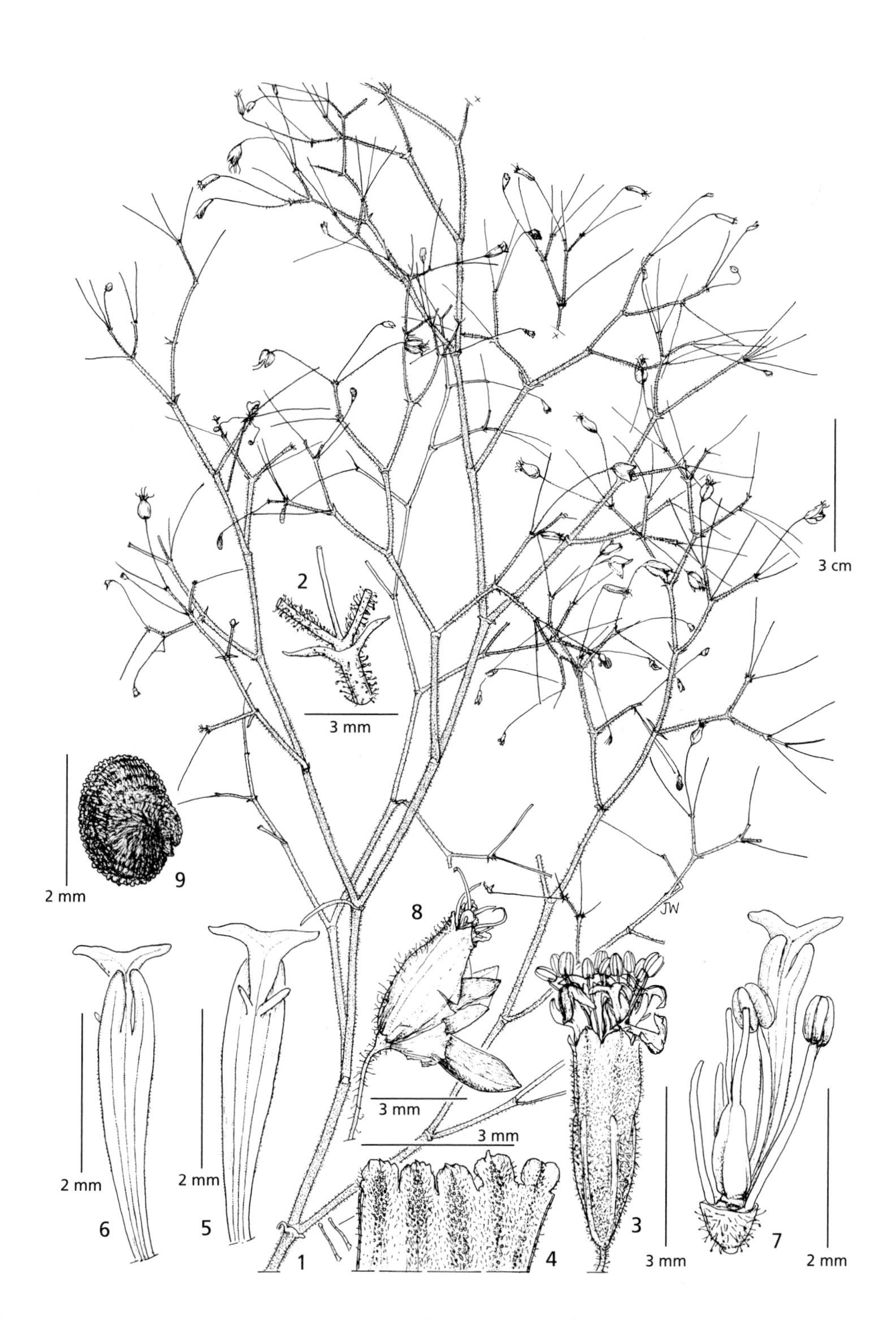

Fig. 35. **Ankyropetalum gypsophiloides**. 1, habit × ⅔; 2, stem hairs × 5; 3, flower × 8; 4, calyx lobe detail× 8; 5, petal, abaxial × 12; 6, petal, adaxial × 12; 7; flower dissected × 12; 8, casule with calyx × 5; 9, seed × 10. 1–2, 8–9 from *Rawi & Serhang* 23761; 3–7 from *Guest* 3620. © Drawn by Juliet Beentje.

27. **ACANTHOPHYLLUM** C.A.Meyer

Verz. Pfl. Cauc. 210 (1831)

Small shrub with acerose spiny leaves. Flowers white or pink, sessile, clustered in sessile or short peduncled heads. Leaves linear to linear-lanceolate, spiny. Bracts and bracteoles usually spiny. Calyx tubular, sometimes turbinate, 5 toothed, 5–15-nerved with membranous hyaline intervals between nerves. Petals 5, with entire or retuse limb. Stamens 10; anthers exserted. Styles 2; ovary unilocular, 4–12-ovuled. Capsule 1-seeded; seeds reniform.

Acanthophyllum is an Irano-Turanian genus with about 60 species worldwide; six species in Iraq.

M. Golenkin (1893) Verzeichnis der Arten der Gattung *Acanthophyllum* CA.Mey. Acta Horti Petrop. 13 (6): 77–88.

A. Gilli (1964) Die Gattungsabgrenzung von *Gypsophila* und *Acanthophyllum*, Österreichische Botanische Zeitschrift 111: 285–290.

Key to Species

1. Bracteoles with broad hyaline margin 2
   Bracteoles without broad hyaline margin 3
2. Clusters of flowers globose; bracts obovate, 6–8 mm; calyx 6–8 mm . . . . 1. *A. bracteatum*
   Clusters of flowers elongated; bracts sublanceolate-obovate, 10–12 mm; calyx 10–11 mm . . . . 2. *A. khuzistanicum*
3. Plant densely pulvinate; internodes shorter than 5 mm, calyx usually pinkish above . . . . 6. *A. caespitosum*
   Plant not densely pulvinate; internodes longer than 6 mm, calyx never pinkish 4
4. Calyx tube glabrous or scabrous . . . . 4. *A. verticillatum*
   Calyx tube velvety or tomentose 5
5. Clusters of flowers at least 12 mm diam. bracteoles 6–9 mm, clearly exceeding the calyx . . . . 5. *A. acerosum*
   Clusters of flowers not much 10 mm diam; bracteoles 2–4 mm, usually equalling to calyx or shorter . . . . 3. *A. kurdicum*

1. **Acanthophyllum bracteatum** *Boiss.*, Pl. Diagn. Orient. Nov. Ser. 1, 1: 43 (1843); Rech. f., Fl. Lowland Iraq: 244 (1964); Leonard, Contrib. Fl. et Veg. Deserts d'Iran 6: 12 (1986); Schiman-Czeika in Fl. Iranica 163: 270, t. 322 (1988).

Small shrub, intricately branched from base. Stems 10–22 cm, glabrous or shortly hirsute, sometimes with glandular hairy. Leaves 14–50 × 1–2 mm, much longer than internodes, rigid, subulate, convex beneath, glabrous or shortly hirsute, horizontally to subhorizontally spreading, ending in glabrous hard spine; floral leaves longer than bracts. Clusters of flowers terminal and axillary, ± globose, 9–11 × 10–16 mm diameter flowers sessile. Bracts 6–7 × 3–4.5 mm with hyaline margins, shortly hairy, sometimes glandular hairs also present; bracteoles 6–7 × 3–5 mm, glandular hairy, with broad hyaline part with wavy margin, base attenuate, apiculate to cuspidate at apex. Calyx 5.5–8 mm, glandular hairy, sometimes eglandular hairs also present; teeth with broad hyaline margins, obtuse to minutely apiculate. Petals white, slightly longer than calyx; lamina linear-spatulate, entire. Stamens clearly exserted. Fig. 36: 1–7.

HAB. On hillsides, steppes, sandy-clay soils; alt. 100–700 m; fl. Apr.–Jun.
DISTRIB. Scattered in desert and dry steppe areas. **DWD**: 160–190 km WN of Ramādi to Rutba, Rawi 31355!; 260 km NW of Ramādi, *Rawi* 20959!;10 km SW of Rutba, *Rawi* 21071!; **LEA**: 5 km SE of Shihabi Police Station, *Rawi & Haddad* 25620!; 75 km N of Amara, *Hazim & Nuri* 30644!

Iran, Pakistan.

2. **Acanthophyllum khuzistanicum** *Rech.f.*, Flora Iranica 163: 273, t. 324 (1988).

Sub-shrub, 10–20(–30) cm tall, much-branched at the base, stems scabrid or smooth; internodes ± 25 mm. Inflorescence branches often in pairs. Leaves (20–)25–35(–45) × 1.5 mm, triquetrous from base, becoming subulate, rigidly spreading-patent; spring leaves

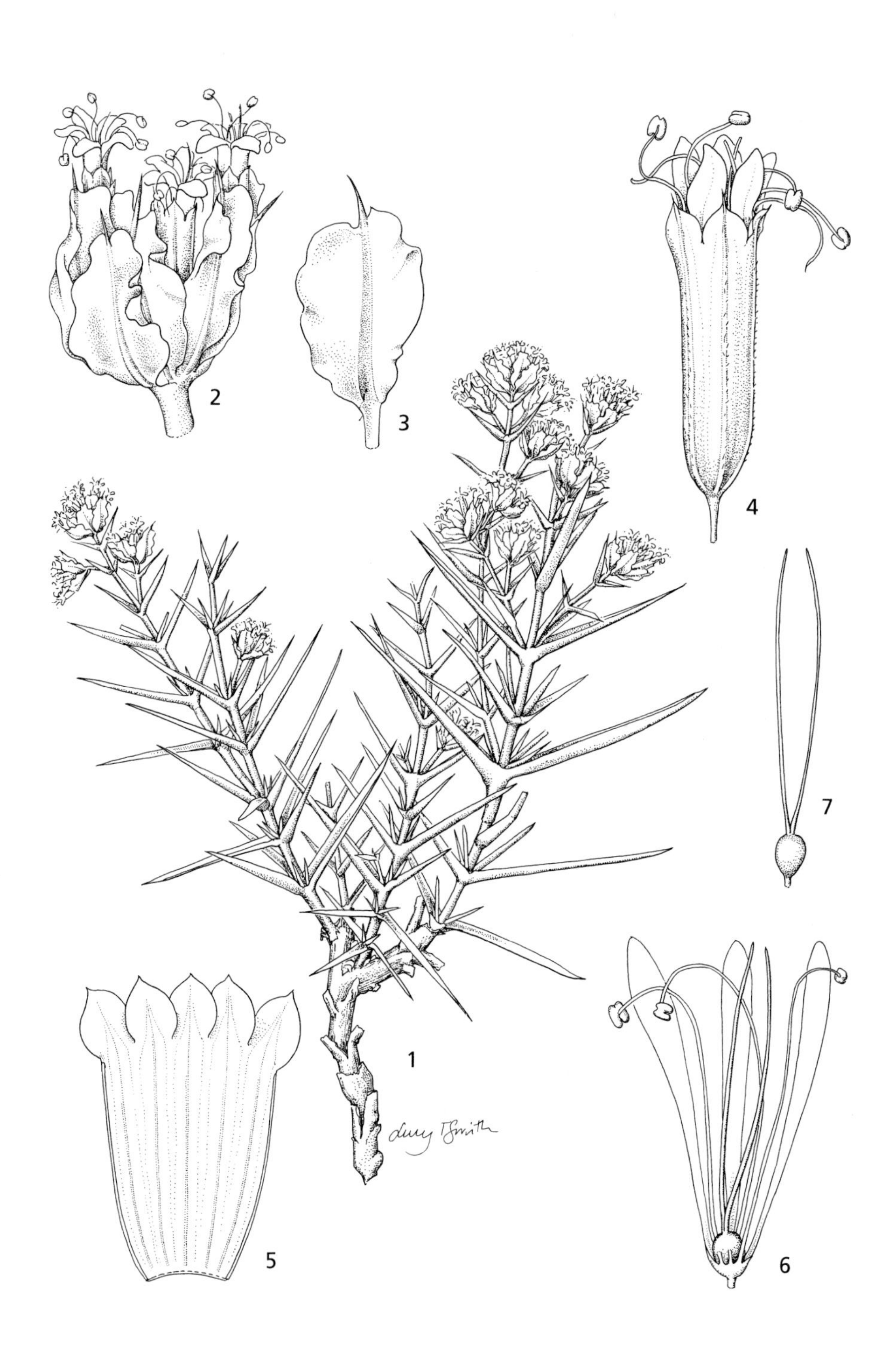

Fig. 36. **Acanthophyllum bracteatum**. 1, habit × 2/3; 2, inflorescence × 4; 3, bracteole × 4; 4, flower × 6; 5, calyx opened × 6; 6, flower with front petals removed × 6; 7, ovary with styles × 9. All from *Rawi* 20959. © Drawn by Lucy T. Smith.

clothing base of branches. Glomerules lax, ovoid, partial cymes 1–3-flowered; pedicels to 2 mm. Bracts 10–12 × 4(–5) mm, sublanceolate-obovate, abruptly narrowed at base, with 1–2 mm mucro at tip and with 1–2 mm broad hyaline margins. Bracteoles similar but smaller. Calyx 10–11 mm, narrowly cylindrical, very short-pilose, with 5 prominent nerves; calyx teeth distinctly unequal, 1–3 mm, ± triangular, mucronate. Petals ± equalling calyx; blade to 0.75 mm wide. Filaments exceeding the petals. Ovary 4-ovulate.

HAB. Sandstone cliff; alt. ± 200 m.
DISTRIB. Rare. **LEA**: Kut-al Imara, 30 km SE of Badrah, *Rechinger* 9162.

Iran.

3. **Acanthophyllum kurdicum** *Boiss.* & *Hausskn.* in Boiss., Fl. Or. Suppl. 90 (1888); Schiman-Czeika in Fl. Iranica 163: 277, t. 327 (1988).

Sub-shrub, 15–25 cm. Stems numerous, ascending, pubescent, glandular and eglandular hairs present. Leaves 10–18 mm, longer than internodes, linear lanceolate, rigid, pubescent to glabrous-papillose, ending in a spine. Flowers clustered on main stem and axillary branches, sessile. Bracts 3–4 mm with ciliate scarious margins; bracteoles up to 5 mm, not clearly exceeding the calyx, usually equalling to calyx or slightly shorter, ± subulate, mucronulate with ciliate scarious margins. Calyx 2.2–3 mm, tubular-obconic, short pilose or velutinous, usually glandular; teeth ± 1 mm, unequal, ovate with narrow hyaline margin, apex mucronulate. Petal 4–5 mm white; lamina oblong-spatulate, slightly dentate. Filaments clearly exceeding the petals.

HAB. In valley, rocky places; alt. 1000–1450 m; fl. Aug.–Sep.
DISTRIB. Occasional in montane areas. **MRO**: Inter Rayat et Hajji Umran, *Rechinger* 11279!; Haji Umran, *Rawi* 24963!; Chiya-i Mandali, *Guest* 2653!; Sidaka, *Guest* & *Husham* 15862!; Kani Rash, *Rawi* & *Serhang* 26688!; **MSU**: Pira Magrun, *Haines* 1868!

Iran.

4. **Acanthophyllum verticillatum** (*Willd.*) *Hand. Mazz.* in Ann. Nat. Hofmus. Wien 26: 152 (1912); Blakelock in Kew Bull. 2: 180 (1957); Huber-Morath in Fl. Turk. 2: 176 (1967); Schiman-Czeika in Fl. Iranica 163: 281, t. 330 (1988).

Arenaria verticillata Willd., Sp. Pl. ed. 4,2: 275 (1799).
Acanthophyllum mucronatum Fenzl, Ann. Wiener Mus. Naturgesch. 1: 37 (1836), non. CA. Mey. (1831).
Acanthophyllum tournefortii Fenzl, Ann. Wiener Mus. Naturgesch. 2: 310 (1840).

Sub-shrub, 15–40 cm. Stems numerous, scabrid to pilose, sometimes glabrous to papillose, ascending to erect. Leaves 8–22 mm, shorter or longer than internodes, glabrous, ± papillose, needle-like, ending in a spine. Flowers congested in terminal and axillary glomerules, sessile. Bracts linear, 4–6 mm, glabrous or ciliate at margins; bracteoles up to 6 mm, glabrous or scabrous with ciliate margins, not exceeding the petals. Calyx tubular, 4–6.5 mm, glabrous or scabrous; teeth unequal, lanceolate, mucronulate with hyaline margin. Petals white, linear or linear-lanceolate, longer than calyx; lamina, emarginate, usually longer than calyx.

HAB. Stony slopes; alt. ± 1500 m; fl. Jul.–Sep.
DISTRIB. Rare. **MRO**: Sakri-Sakran, *Bornmüller* 952.

No specimens at K; record from Flora Iranica.

Turkey, Iran.

5. **Acanthophyllum acerosum** *Sosn.*, Monit. Jard. Bot. Tiflis 11: 7 (1915) non Barkoudah in Wentia 9: 180 (1962); Huber-Morath in Fl. Turk. 2: 176 (1967); Schiman-Czeika in Fl. Iranica 163: 282, t. 332, 484, f. 1 (1988).

A. confertiflorum Rech. f. Anz. Math.-Nat. Kl. Österr. Akad. Wiss. 98: 10 (1961).

Subshrub, 10–18 cm. Stem numerous, lax pulvinate or not, papillose-pubescent, ascending. Leaves needle-like, 10–14 mm, longer or shorter than internodes, pubescent, ending with a spine. Flowers sessile and usually clustered on terminal branches and in axils of uppermost leaves with many flowers; glomerules (incl. bracts) 15–20 mm diam. Bracts 6–9 mm, clearly longer than calyx, linear-lanceolate, spiny, pilose, with membranous ciliate margins at base;

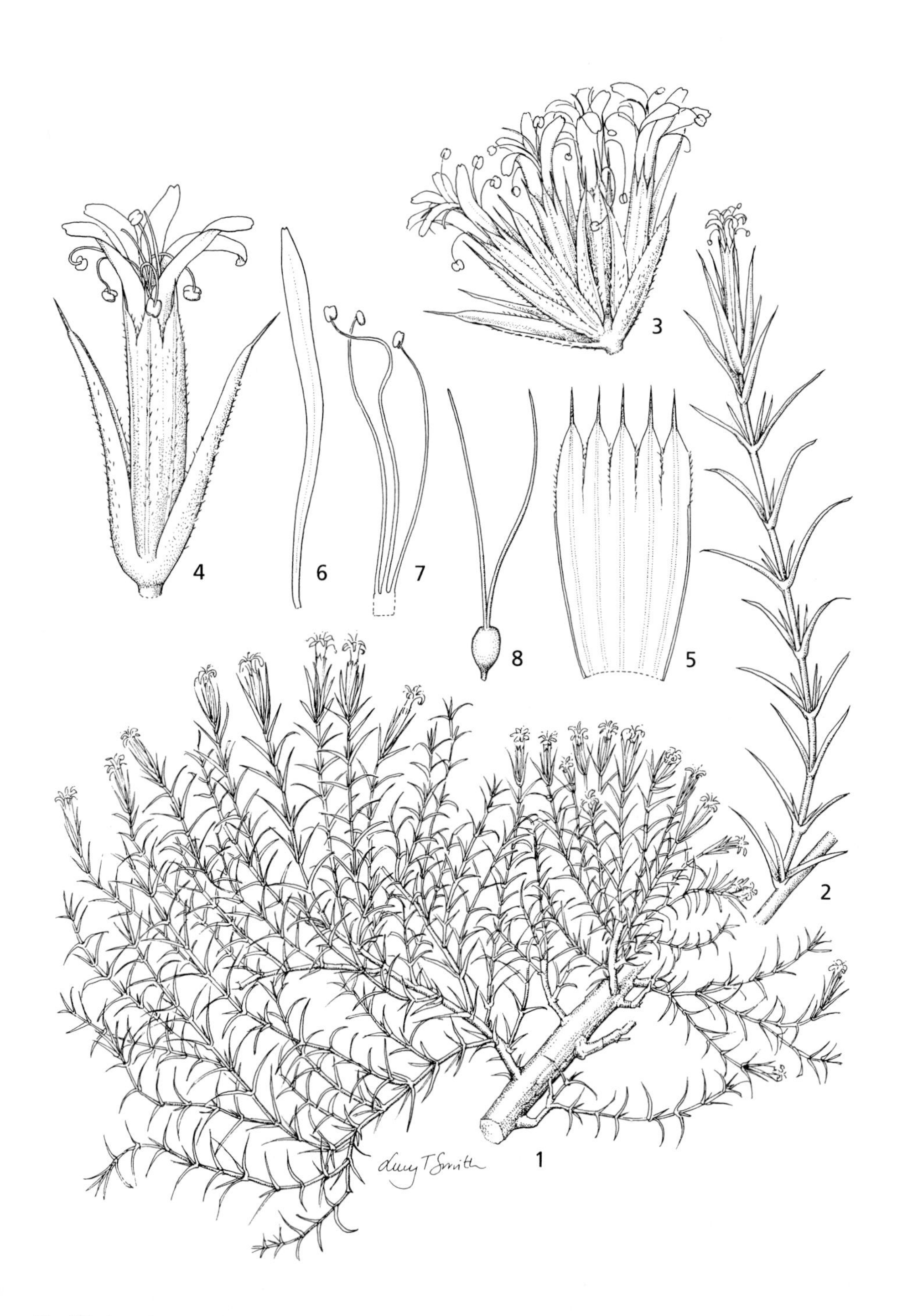

Fig. 37. **Acanthophyllum caespitosum**. 1, habit × 1; 2, flowering branch × 2; 3, inflorescence × 3; 4, flower with bracteoles × 5; 5, calyx opened × 5; 6, petal × 5; 7, stamens × 5; 6, ovary with styles × 5. 1, 2 from *Rawi* et al. 29591; 3–8 from Gillett 11742. © Drawn by Lucy T. Smith.

bracteoles similar to bracts. Calyx tubular, 4–6 mm, densely pubescent; teeth inequal, 1–3 mm, lanceolate, prickly. Petals white; lamina narrowly obovate, ± retuse, exceeding the calyx.

HAB. In valley, mountain steppes; alt. 1100–2200 m; fl. Jul.–Sep.
DISTRIB. Rare. **MRO**: Mt. Qandil, *Rechinger* 11098!; Rayat to Haji Umran, *Rechinger* 11278! (type of *A. confertiflorum*); Mergadreija, nr. Haji Umran, *Rawi* 9155!

Turkey, Iran.

6. **Acanthophyllum caespitosum** *Boiss.*, Diagn. Pl. Orient. Nov. Ser. 1, 1: 42 (1843); Blakelock in Kew Bull. 2: 180 (1957); Schiman-Czeika in Fl. Iranica 163: 318 , t. 360, 477 (1988).

Densely pulvinate perennials. Stems to 7 cm, decumbent to ascending, hirsute, also glandular hairs present in inflorescence region; internodes shorter than 4 mm. Leaves linear-triangular, 4–7 mm, with smaller leaves in their axils, hirsute, much longer than internodes, ending in a long spine; lower leaves ± recurved. Inflorescence glomerules terminal, 8–13 mm diameter with (3–)5–15 sessile flowers. Bracts 9–12 mm, linear-triangular, subulate, hairy; bracteoles similar to bracts, 9–10 mm, not exceeding the petals. Calyx tubular, 8–11 mm with 10 costa, mostly glandular and sparsely eglandular hairy, usually pinkish above; teeth 2–3 mm, unequal, usually pinkish, ending in a yellowish spine. Petal longer than calyx; lamina usually pinkish, emarginate, bilobed. Fig. 37: 1–7.

HAB. Hillsides, steppes; alt. 950–1500 m; fl. Jun.–Jul.
DISTRIB. Very local. **MSU**: 10 km W of Tawila, *Rawi* 22167!; Biyara, *Rawi et al.* 29591!; Biyara, Jabal Avroman, *Gillett* 11742!, *Rechinger* 10180!

Iran.

28. **VACCARIA** Wolf

Gen. III (1776); Gen. Sp. 234 (1781)

Glabrous annual herbs. Leaves sessile. Flowers in spreading dichasial panicles. Calyx without commissural veins, but with 5 green wings. Petals without coronal scales. Stamen 10. Styles 2. Capsule papery, dehiscing by 4 teeth.

One or 4 species in C and E Europe, Mediterranean, Temperate Asia; represented with *V. hispanica* in Iraq.

Vaccaria hispanica (*Miller*) *Rauschert*, Wiss. Zeitschr. Univ. Halle Math.-Nat. 14: 496 (1965); Rech.f., Fl. Iranica 163: 338 (1988); Chamberlain in Miller & Cope, Fl. Arab. Penins. & Socotra 1: 222 (1996); Greuter, Fl. Hellenica 1: 333 (1997); Lu Dequan, Lidén & Oxelman in Fl. China 6: 102 (2001).

Saponaria vaccaria L., Sp. Pl.: 409 (1753).
Vaccaria pyramidata Medik., Philos. Bot. 1: 96 (1789).
V. grandiflora (Fisch. ex DC) Jaub. & Spach, III. Fl. Or. 3: tab. 231 (1848).
V. oxyodonta Boiss., Diagn. Pl. Or. Nov. Ser. 2,1: 68 (1854).
V. hispanica var. *oxydonta* (Boiss.) Léonard, Bull. Jard. Nat. Belg. 55: 298 (1985).

Stems 18–50 cm, divaricately branched above. Leaves 20–90 × 6–22 mm, the lowermost petiolate; upper ones cordate-lanceolate; ± amplexicaul, sessile. Pedicels 10–40 mm. Calyx 9–14 mm with 5 conspicuous wings, obclavate at first, pyramidal in fruit; teeth, broadly triangular or triangular-acuminate. Petals 13–22 mm; limb cuneate-cordate, pink (rarely white); coronal scales absent. Stamens 10. Styles 2. Capsule ovoid, many-seeded. Fig. 38: 1–6.

HAB. Hills, mountain slopes, cultivated lands, sandy gravelly soils, shady places near forest, roadsides; alt. 60–2200 m; fl. Apr.–Jun.
DISTRIB. Widespread. **MAM**: 8 km S of Dohuk, *Botany Staff* 43375!; **MRO**: Pishtashan, *Rawi & Serhang* 26552!; Magar mountain, Haji Umran, *Rawi & Serhang* 24310!; Rayat, *Guest* 13027!; **MSU**: Dokhan, *Robertson* 73!; Tawila, *Rawi* 21836!; Penjwin, *Rawi* 22599!; Sulaimaniya, *Omar & Karim* 38096!; **MJS**: 10 km from Sinjar to Tal Afar, *Wedad & Khayat* 53406!; **FUJ**: Between Mosul & Hamam Ali, *Chakravanty et al.* 32131!; 28 km from Mosul to Kasik, *Dabbagh* 46726!; **FNI**: Eski Kellek, *Botany Staff* 43280!; **FAR**: 7 km from Altun Kupri to Erbil, *Botany Staff* 43229!; **FKI**: Kirkuk, *P. Grigg* 67!; 6 km E of Kirkuk to Sulaimaniya, Botany *Staff* 42998!; **FPF**: Dakka, *Sa'ad* 1747!; **DGA**: Simarra, *Rawi* 22881; **DSD**: Ma'niya, *Alizzi & Omar* 35543!; 60 km N W of Ramadi, *Alizzi* 35143!; **LCA**: Baghdad, *Lazar* 3901!; Sudur, *Omar et al.* 44290!

Temperate to subtropical regions.

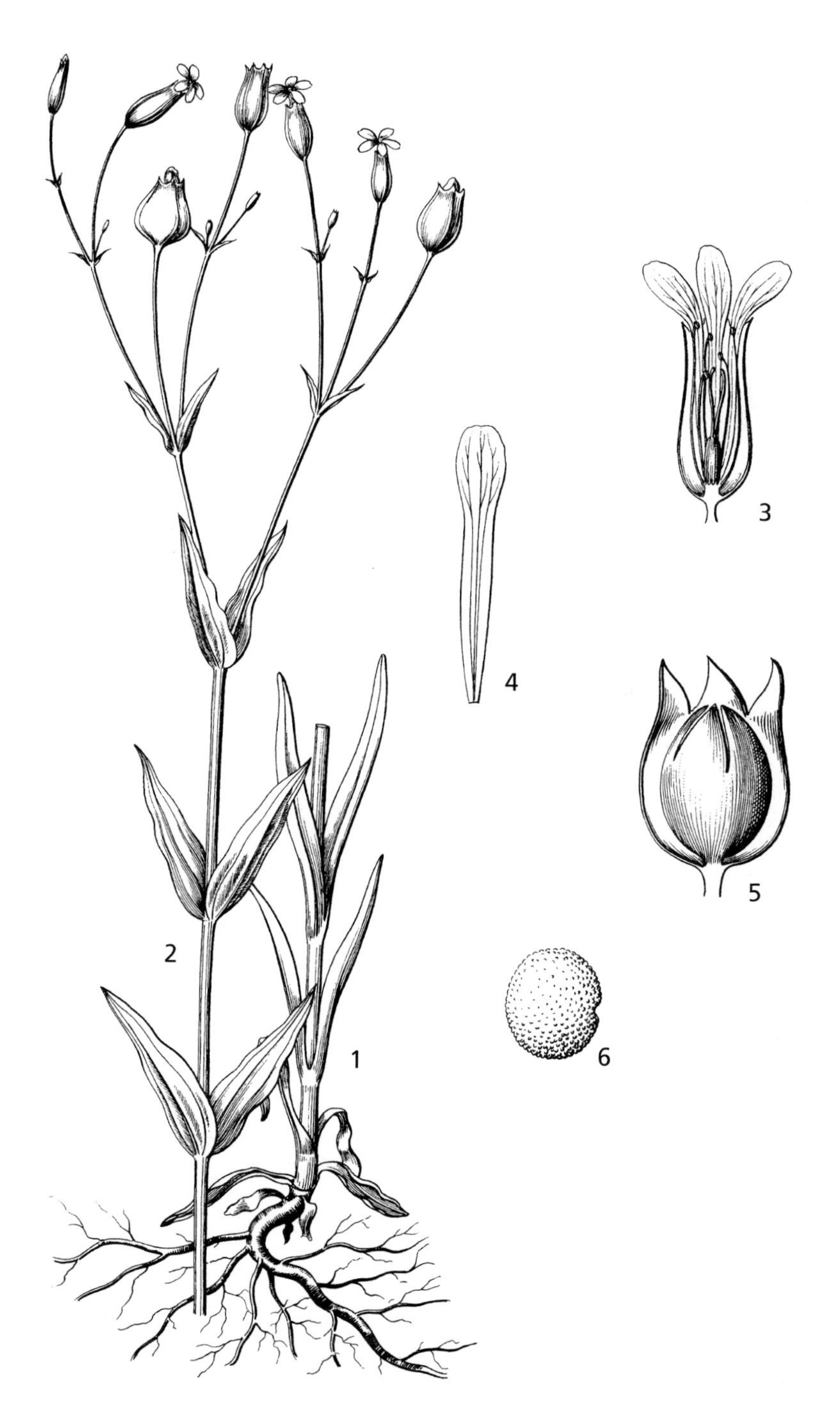

Fig. 38. **Vaccaria hispanica**. 1, Habit lower part × 1/2; 2, flowering shoot × 1/2; 3, flower with front sepals and petals removed × 2; 4, petal × 2; 5, fruit with calyx detail; 6, seed detail. Reproduced with permission from Flora of China Vol. 6, fig. 96 (2001). Drawn by Qian Cunyuan.

26. **SILENE** L.

Sp. Pl.: 416 (1753); Gen. Pl.: ed. 5, 195 (1854)

Annual, biennial and perennial herbs. Stems erect to ascending to decumbant with usually a woody rootstock in perennials. Leaves radical and cauline, opposite, lamina various, usually entire, sessile to cuneate at base, exstipulate. Flowers single or in dicasial or paniculate cymes or in racemes; pedicellate. Bracts and bracteoles as leaves. Sepals 5, united to form a cylindrical to conical calyx, sometimes infalted, 5–toothed above, 5–(10–30)-nerved, glabrous, glandular to pubescent. Petals white to variously pink, 5, free, each with a claw and limb; claw sometimes expanded apically to form auricles (Section *Auriculatae*); limb bilobed to almost entire or sometimes fimbriate. Stamens 10, in two whorls. Styles 3–5. Ovary, stamens and petals borne on an anthophore. Fruit capsular, dehiscing by twice the number of styles. Seeds usually reniform or globose, smooth or variously sculptured, back often with a groove, sometimes winged or papillose.

A large genus with over 500 species mainly temperate in distirbution; represented in Iraq by 37 species.

Characters of the calyx, lobation of the petals and the presence or absence of auricles on the petal-claw are important in the identification of species.

Ref.: Rohrbach, P. (1868) Monographie der Gattung Silene. Engelmann, Leipzig.
Ghazanfar, SA. (1983) New Taxa, combinations and notes on typification and nomenclature in *Silene*. Notes RBG Edinb. 41: 97–107.
Oxelman, B., Lidén, M., Rabeler, R. K. & Popp, M. (2001) A revised generic classification of the tribe *Sileneae* (Caryophyllaceae). Nord. J. Bot. 20: 515–518.

1. Perennials usually with a woody stock and often giving annual branches. 2
   Annuals (occasionally biennial or short-lived perennial) but without a woody stock 24
2. Calyx inflated, prominently so in fruit. 3
   Calyx not inflated 8
3. Petals divided into linear segments . . . . 12. *S. schizopetala*
   Petals shallowly bilobed but not divided into linear segments. 4
4. Flowers 3–5 congested in cymules; inflorescence a raceme . . . . 11. *S. ampullata*
   Flowers solitary or 2 or 3–7 in dichasia 5
5. Calyx 20 mm or more 6
   Calyx less than 20 mm. 7
6. Leaves 12–50 × 3–10 mm, linear-lanceolate, elliptic or oblanceolate; flowers 3(–7); calyx 20–28 mm . . . . 14. *S. rhynchocarpa*
   Leaves 15–27 × 1–1.5 mm, triquetrous or linear; flowers 1(–2); calyx 28–30 mm . . . . 22. *S caryophylloides* subsp. *subulata*
7. Plant gynodioecious; leaves narrowly linear; anthophore pubescent, 7–8 mm in bisexual flowers . . . . 18. *S. pungens*
   Plant not gynodioecious; leaves lanceolate to narrow-lanceolate; anthophore glabrous, 3–5 mm . . . . 21. *S. odontopetala*
8. Calyx glabrous 9
   Calyx pubescent. 12
9. Calyx more than 20 mm 10
   Calyx less than 20 mm. 11
10. Flowers on lateral branches; calyx not viscid (26–35 mm) . . . . 3. *S. chlorifolia*
    Flowers in terminal cymes; calyx viscid, ± 25 mm . . . . 4. *S. avromana*
11. Leaves 11–15 mm wide; calyx 10–11 mm; petal-limb 13–14 mm . . . . 1. *S. longipetala*
    Leaves ± 2 mm wide; calyx 7–9 mm; petal-limb 3–4 mm. . . . . 2. *S.* cf. *saxatilis*
12. Calyx 10 mm or less. 13
    Calyx more than 10 mm 15
13. Basal leaves obovate, cauline leaves ovate-cordate to orbicular; calyx pilose . . . . 16. *S. monantha*
    Basal and cauline leaves linear; calyx glandular-hairy. 14
14. Plant gynodioecious; leaves somewhat rigid and recurved; calyx 9–10 mm, if 8 mm then flower ♀ . . . . 6. *S. spergulifolia*
    Plant not gynodioecious; leaves herbaceous; calyx 6–8 mm . . . . 5. *S. stenobotrys*

15. Flowers in panicles, lateral branches ending in 1–3 flowers 16
    Flowers 1–2 or 3 in a dichasium 18
16. Stems with dense white eglandular hairs; bracts embracing the calyx . . 10. *S. eriocalycina*
    Stems and bracts not as above 17
17. Capsule as long as the anthophore; anthophore hairy . . . . 7. *S. pruinosa*
    Capsule twice the length of the anthophore; anthophore minutely puberulent . . . . 8. *S. montbretiana*
18. Cauline leaves 3–5–nerved, ovate-cordate to broadly lanceolate . . . 15. *S. commelinifolia*
    Cauline leaves 1–nerved, linear to linear-lanceolate 19
19. Calyx less than 25 mm 20
    Calyx 25 mm or more 22
20. All leaves 3–nerved, glandular-pubescent . . . . 17. *S. erimicana*
    All leaves 1–nerved, ± glandular-pubescent 21
21. Flowers solitary; anthophore glabrous . . . . 20. *S. retinervis*
    Flowers 3 in a dichasium; anthophore pubescent . . . . 9. *S. oreophila*
22. Sems densely glandular and viscid above; leaves 3–nerved, pubescent ± glandular; flowers 1–2 or 3 in a dichasium; anthophore twice length of capsule . . . . 19. *S. brevicaulis*
    Stems glandular-pubescent, not viscid above; leaves 1–nerved, with whitish grey hairs; flowers 3–(5–7) in dichasia; anthophore ± equalling capsule . . . . 13. *S. microphylla*
24. Flowers in compact, umbellate heads . . . . 23. *S. compacta*
    Flowers in mono- or dichasial cymes, panicles or solitary 25
25. Calyx 15–30-nerved 26
    Calyx 10–nerved 27
26. Calyx 15–20-nerved, hairs rigid with swollen bases; fruiting calyx not inflated at base . . . . 36. *S. coniflora*
    Calyx 25–30-nerved, hairs without swollen bases; fruiting calyx distinctly inflated at base . . . . 37. *S. conoidea*
27. Bracts membranous . . . . 27. *S. aegyptiaca*
    Bracts herbaceous 28
28. Seeds with conspicuous flat or undulate wings 29
    Seeds without wings 30
29. Calyx campanulate-clavate, 6.5–9.5(–11.5) mm, teeth triangular-acuminate . . . . 24. *S .apetala*
    Calyx cylindrical, 11–16 mm, teeth ligulate, obtuse . . . . 31. *S. colorata* subsp. *oliveriana*
30. Green nerves of calyx slightly winged and covered with multicellular hairs . . . . 25. *S. dichotoma* subsp. *racemosa*
    Nerves of calyx not winged 31
31. Calyx distinctly striate, membranous between nerves 32
    Calyx not striate 37
32. Inflorescence paniculate with wide-angled, symmetrical, dichasial branches; calyx teeth to 2 mm . . . . 1. *S. longipetala*
    Inflorescence and calyx teeth not as above 33
33. Plant to 60 cm tall, branched from at or above the middle; calyx teeth to 5.5 mm 34
    Plant to 25 cm tall, branched from base; calyx teeth 2–3 mm 35
34. Claw of petals distinctly exserted from calyx; anthophore 4–8 mm . . . . 35. *S. linearis*
    Claw of petals not exserted from calyx; anthophore 3–5 mm . . . . 34. *S. chaetodonta*
35. Calyx nerves covered with multicellular hairs; calyx 14–19 mm . . . . . . 26. *S. lagenocalyx*
    Calyx nerves lacking multicellular hairs; calyx to 12 mm 36
36. Leaves scabridulous-pubescent; calyx strictly cylindrical . . . . 33. *S. arenosa*
    Leaves glandular-pubescent; calyx weakly campanulate . . . . 30. *S. arabica*
37. Upper part of calyx green to pale green, lower part whitish-green or membranous and ± striate; calyx 16–23 mm; fruiting capsule reflexed . . . . 32. *S. villosa*
    Whole of calyx pale green 38
38. Calyx 9–12 mm, finely puberulent; teeth ± semicircular, to 1.5 mm . . . 29. *S. diversifolia*
    Calyx 12–19.5 mm, glandular-pubescent; teeth triangular, 2–2.5 mm . . . 28. *S. atocioides*

Sections are based on morphological characters and generally follow Coode and Cullen in Flora of Turkey (1967) except for *S. pungens* and *S. odontopetala* which are placed in Section Auriculatae.

1. Calyx 15–60-nerved . . . . Sect. Conoimorpha
   Calyx 10–nerved 2
2. Dichasium condensed into a capitate or umbellale cyme. . . . . Sect. Compactae
   Dichasium not condensed 3
3. Petal-limb divided into linear segments. . . . . Sec. Fimbriatae
   Petal-limb not divided into linear segments 4
4. Flowers in few-flowered cymules forming panicles, dichasial and dichotomous cymes or solitary 5
   Flowers in racemes, congested racems or monochasial cymes. 10
5. Petal-claw conspicuously auriculate . . . . Sect. Auriculatae
   Petal-claw without auricles 6
6. Plants with rigid stems; inflorescense irregularly dichotomous . . . . Sect. Rigidulae
   Plants with herbaceous or ± woody stems; inflorescence not dichotomous. 7
7. Calyx hairy 8
   Calyx glabrous 9
8. Inflorescence a compound dichasium; calyx glandular- or eglandular-pubescent, often inflated in fruit, but not contracted above in fruit . . . . Sect. Atocion
   Flowers in dichotomous cymes; calyx conspicuously hairy or papillose, contracted in fruit. . . . . Sect. Lasiocalycinae
9. Petal-claw and filaments hairy . . . . Sect. Lasiostemones
   Petal-claw and filaments glabrous. . . . . Sect. Sclerocalycinae
10. Calyx inflated in fruit . . . . Sect. Ampullatae
    Calyx not inflated in fruit 11
11. Flowers in congested racemes or panicles; plants often gynodioecious. . . . . Sect. Spergulifoliae
    Flowers in unilateral racemes (monochasium); plants not gynodioecious 12
12. Calyx contracted in fruit. . . . . Sect. Lasiocalycinae
    Calyx not contracted in fruit. . . . . Sect. Bipartitae

SECT. LASIOSTEMONES (Boiss.) Schischk.

1. **Silene longipetala** *Vent.*, Jard. Cels. 83, t. 83 (1800); Sibth. & Sm., Fl. Gr. 5: t. 419 (1825); Boiss., Fl. Or. 1: 636 (1867); Rohrb., Monogr., 211 (1868); Post, Fl. Syr. Palest. & Sinai 1: 185 (1932); Rawi in Dep. Agr. lraq Tech. Bull. 14: 49 (1964); Rech.f., Fl. Lowland lraq 240 (1964); Chater in Fl. Europ. 1: 165 (1966); Mouterde, Nouv. Fl. Liban et Syr. 1: 502 (1966); Zohary, Fl. Palaest. 1: 84 (1966); Tackh., Stud. Fl. Egypt ed. 2, 85 (1974); Coode & Cullen in Fl. Turk. 2: 194 (1967); Melzh. in Fl. Iranica 163: 383 (1988); Fragman, Levy-Yamamori & Christodoulou, Fl. Eastern Med: 254 (2001); Boulos, Fl. Egypt Checklist Rev.: 37 (2009).

Perennial (rarely biennial), 35–60 cm, with a small woody stock. Stems erect, with short retrorse hairs, glabrous and often viscid above. Leaves rosulate, 30–55 × 11–15 mm, lanceolate to ovate-lanceolate, puberulent, apex acute, base attenuate in the lower leaves, sessile in the cauline leaves. Bracts and bracteoles similar to cauline leaves, but reducing in size upwards. Inflorescence a lax panicle, the lateral branches usually ending in 7–flowered dichasia. Calyx 10–12 mm, cylindric and tapering below, clavate and slightly constricted below the capsule in fruit, glabrous, distinctly striate, membranous between nerves; teeth triangular, with a ciliate margin. Petals white, pinkish or lurid; limb 13–14 mm, deeply bilobed (to $^2/_3$) into linear lobes; claw 7–8 mm, hairy, auriculate; coronal scales oblong. Filaments hairy; anthophore 2–4 mm, pubescent. Capsule 6–9 mm, conical, included in the calyx or exserted for 1 mm. Seeds ± reniform. Fig. 39: 1–5.

HAB. Fields and open and rocky places; alt. 500–2600 m; fl. Mar.–Jun., fr. Jun.–Jul.
DISTRIB. **MSU**: Sulaimaniya, Gweija Dagh [Qara Dagh], *Rawi* 8881A; **MAM**: Sarsang, *Haines* 976; **FUJ**: Mosul, Matina, *Rawi* 8716;

Egypt, Cyprus, Greece, Turkey, Jordan, Lebanon, Palestine, Syria, SW Iran.

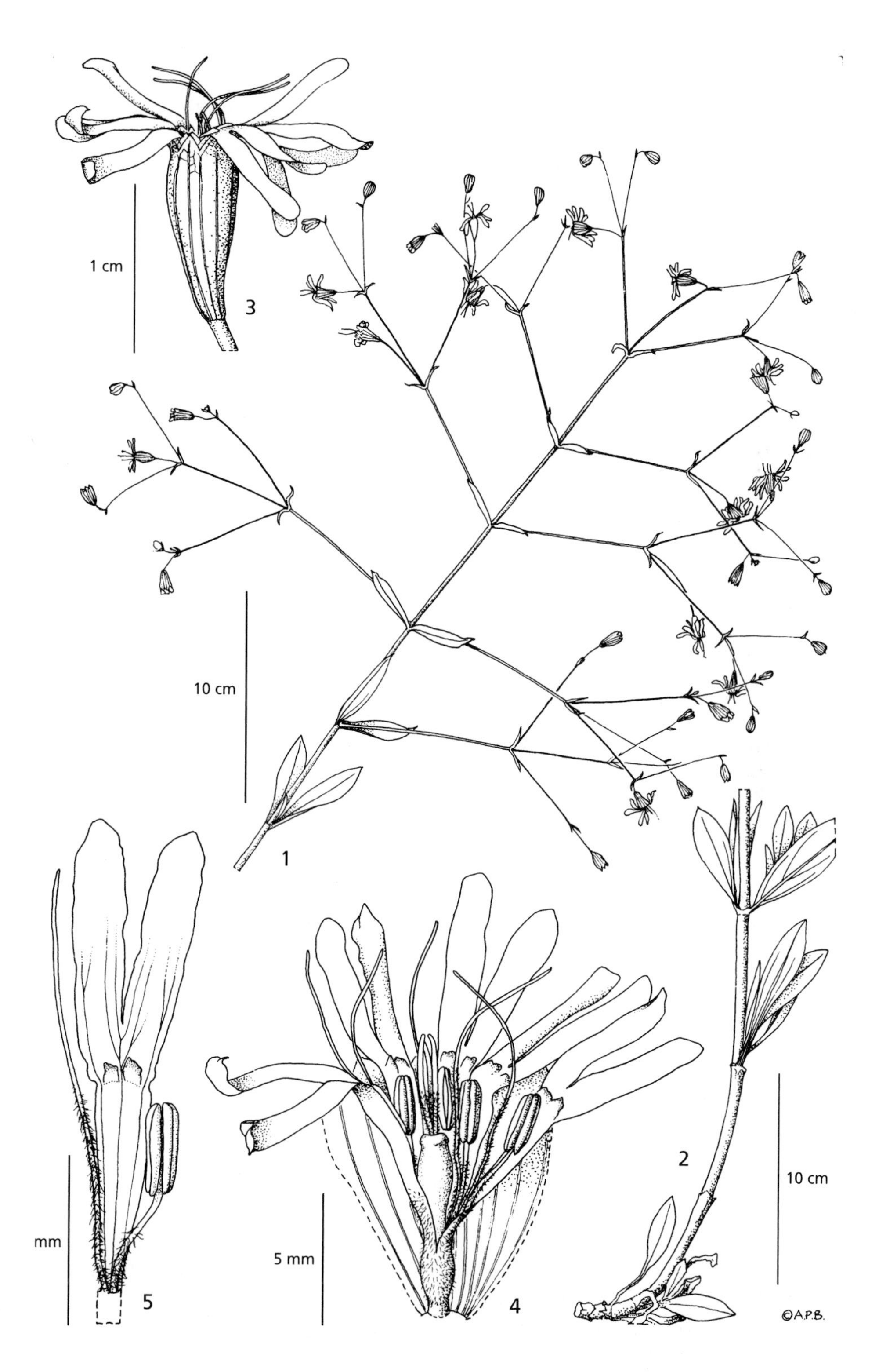

Fig. 39. **Silene longipetala**. 1, habit 2, lower paortion of stem; 3, flower; 4, flower opened; 5, petal with stamen. 1, 3–5 from *Maitland* 96; 2 from *Davis* 4267. © Drawn by A.P. Brown.

2. **Silene saxatilis** *Sims* in Bot. Mag. 18: t. 689 (1803); Boiss., Fl. Orient. 1: 635 (1867); Coode & Cullen in Fl. Turk. 2: 195 (1967).

Perennial, ± ceaspitose, with a branched woody stock. Stems ascending to decumbent, up to 20 cm, glabrous. Leaves 10–25 × 2 mm, obovate to spatulate , the cauline leaves reduced in size, surfaces glabrous, margin ciliate. Flowers in racemes (see below). Calyx 7–9 mm, cylindric-clavate, clavate in fruit, glabrous; teeth ovate. Petals white; limb c. 4 mm, 3/4 divided; claw c. 5.5 mm, auriculate, hairy; coronal scales oblong. Filaments hairy; anthophore 3–4 mm, glabrous. Capsule ± 8 mm, conical, partially exserted from the calyx.

HAB. Rocks; alt. ± 3000 m; fl. ?Jun.
DISTRIB. Kurdistan, *Brant & Strangways* 1840!

S. saxatilis has not been recorded from Iraq and it is not certain whether this plant was collected from Iraq. The plant described above fits closely to the description of *S. saxatilis.* The inflorescence appears to be a raceme but since part of the specimen is damaged it is difficult to be certain

Iran, Armenia, Turkey (E and SE Anatolia).

SECT. SCLEROCALYCINAE (Boiss.) Chowdh.

3. **Silene chlorifolia** *Sm.*, Ic. ined. 1: 14, t. 13 (1789); Bot. Mag. 21: t. 807 (1805); Boiss., Boiss., Fl. Orient. 1: 640 (1867); Rohrb.; Monogr., 177 (1868); Rawi in Dep. Agr. Iraq Tech. Bull. 14: 48 (1964); Zohary, Fl. Palest. ed. 2, 186 (1932); Mouterde, Nouv. Fl. Liban et Syr. 1: 503 (1966); Coode & Cullen in Fl. Turk. 2: 199 (1967); Gabrielian & Fragman-Sapir, Fl. Transcauc. & adj. areas: 140 (2008).

Perennial, up to 80 cm, with a woody stock. Stems leafy, many, erect, glabrous, viscid above. Basal leaves few, withering away. Cauline leaves 15–36 × 12–35 mm, reducing in size above, orbicular, ovate-cordate to ovate-lanceolate, obovate or lanceolate; apex sharply acute to mucronate, base amplexicaul or attenuate, surfaces glabrous, margin ± ciliate. Leaves and stems glaucescent. Bracts similar to cauline leaves but very reduced in size. Inflorescence a spreading panicle, with the lateral branches bearing 1–3 flowers. Calyx 26–35 mm, cylindric, tapering below; in fruit clavate and constricted below the capsule, coriaceous, glabrous, purple-red when young, yellow-brown at maturity. Calyx teeth triangular, ciliate. Petals white or greenish; limb 10–11 mm, ½ divided; claw 22–23 mm, auricles obscure; coronal scales oblong. Anthophore 14–15 mm, glabrous. Capsule ± 17 × 8 mm, conical, partially exserted from calyx. Seed ± 2.4 mm, smooth, grooved at back.

Distribution (of species): Throughout the Caucasus, Greece, Armenia, Turkey, Lebanon, Palestine, Syria, N Iran.
S chlorifolia is extremely variable in the shape and size of its cauline leaves. Another species, *S swertiifolia* Boiss. is a closely related species with similar morphological characters and the same distribution range as *S. chlorifolia. S. swertiifolia* is differentiated from *S. chlorifolia* basically on its cauline leaves which are ovate to obovate in the former and orbicular to ovate-orbicular in the latter. Both species exhibit intermediate forms as well where the two occur together; based on leaf-shape alone it is superfluous to recognise *S. chlorifolia* and *S. swertiifolia* as two distinct species, and therefore *S swertiifolia* is best treated at a subspecific rank awaiting molecular and cytological studies.

Cauline leaves orbicular to orbicular-ovate; base amplexicaul. . . . . . . a. subsp. **chlorifolia**
Cauline leaves obovate, ovate to ovate-lanceolate or lanceolate;
base sessile or attenuate . . . . b. subsp. **swertiifolia**

a. subsp. **chlorifolia.** Fig. 40: 1.

S chlorifolia f. *macrocalyx* Bornm., Beih. Bot. Centr. 28 (2): 144 (1911). — Type: Kurdistan, Riwandes [Rowandaz] in Mt. Sakri Sakran, 24 Jun. 1893, *Bornmüller* 984 (lecto. K!)

HAB. Slopes and open stony places, steep mountain slopes, on rocks and in *Quercus* forest; alt. 1000–1600 m; fl. May–Jun., fr. Jun.–Aug.
DISTRIB. Locally common and found frequently. **MAM**: Dohoka village, 12 km NE Sarsang, *Janan* 46152!; Bigdaoda village, 6 km S of Kamilosi, *Omar & Dabbagh* 45686!; Bikhair, nr. Zakho, *Rawi* 23047! **MRO**: Karoukh [Karokh], *Kass & Nuri* 27353!; valley between Gunda Shor & Darband, *Gillett* 12401!; Gali Ali Beg, nr. last bridge to Rowandaz, *Kass, Nuri & Serhang* 27209!; Shaklawa, Jabal Saffan, *Omar et al.* 38256!; Hab as Sultan [Haibat Sultan], *Rawi, Nuri & Kass* 28258!; **MSU**: Penjwin, *Rawi* 22629! **FUJ**: Gara Mt., *Dabbagh & Jasim* 46940A!; ibid, *Joseph & Kasi* 19526!; Iraq, *sine loc. Wheeler Haines* W 2036!

Iran, Lebanon.

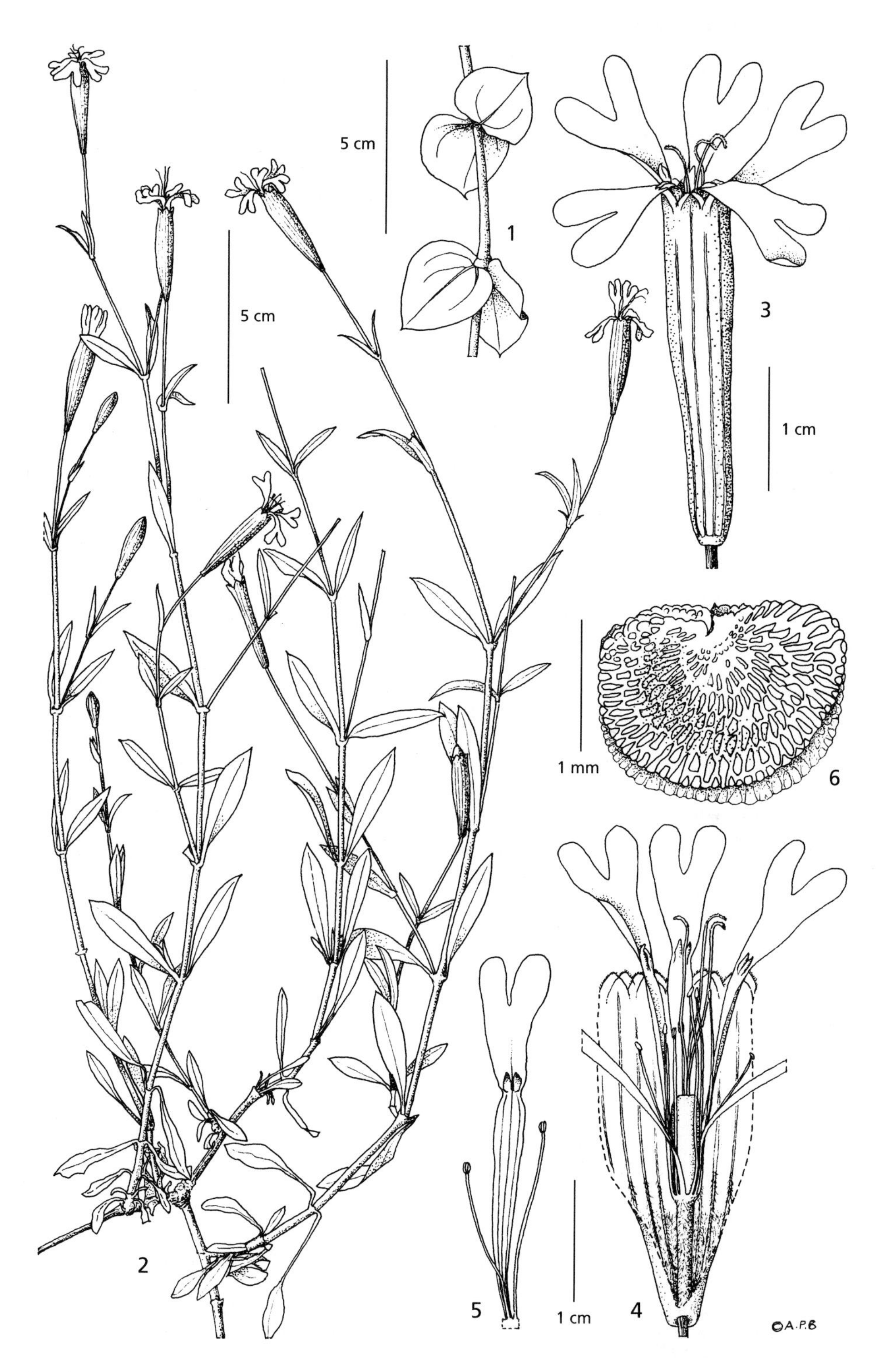

Fig. 40. **Silene chlorifolia** subsp. **chlorifolia**. 1, habit showing leaves. **Silene chlorifolia** subsp. **swertiifolia**. 2, habit; 3, flower lateral view; 4, flower opened; 5, petal with stamens; 6, seed. 1 from *Haines* W2036; 2 from *Barkley & Haddad* 5869; 3–5 from *Rechinger* 10365; 6 from *Rawi* 21928. © Drawn by A.P. Brown.

b. subsp. **swertiifolia** Ghaz. **stat. nov.** Fig. 40: 2–6.

S swertiifolia Boiss., Diagn. Pl. Or. Nov. ser. 1, 1: 32 (1843); Boiss., Fl. Orient. 1: 640 (1867); Zohary, Fl. Palaest. 1:k (1966); Mouterde, Nouv. Fl. Liban et Syr. 1: 503 (1966).
S swertiifolia ß *stenophylla* Boiss., Boiss., Fl. Orient. 1: 641 (1867).
S chlorifolia var. *swertiifolia* Rohrb., Monogr. Silene: 177 (1868); Rawi in Dep. Agr. Iraq Tech. Bull. 14: 48 (1964).

HAB. Slopes and open stony places; alt. 1300–2000 m; fl. May–Jun., fr. Jun.–Aug.
DISTRIB. MAM: Sulamaniya, Mt. Avroman, *Rechinger* 10365!; Tawila, *Rawi* 21928!; 48 km south of Sulimaniya, *Barkley & Haddad* 5869!

Iran, Lebanon, Turkey, Jordan, Palestine, Syria.

4. **Silene avromana** Boiss. & Hausskn. in Boiss., Suppl. 105 (1888). — Type: Iraq, Mt. Avroman et Schahu, VIl.1867, *Haussknecht,* (iso. G., G-BOISS!); Rawi in Dep. Agr. Iraq Tech. Bull. 14: 48 (1964); Melzh. in Fl. Iranica 163: 409 (1988).

Caespitose perennial. Stems 25–35 cm, erect. Leaves 10–20(–40) × 3–10 mm, oblong to linear, pubescent or glabrous, apex mucronate, base attenuate. Flowers 3–5 in terminal cymes. Calyx 25–30 mm, cylindrical, glabrous, viscid, clavate in fruit; teeth triangular-lanceolate. Petals brownish; limb 8–10 mm, ½ divided; claw ± 12 mm, exauriculate; coronal scales ovate. Anthophore ± 12 mm, glabrous. Capsule oblong, about as long as the anthophore.

HAB. alt. 2000–3000 m; fl. Jul.
DISTRIB. MSU: Mt. Avroman [Hawraman], Mt. Shahu, *Haussknecht* 204 (type).

The species has been described from Mount Avroman [Hawraman] on the border between Iraq and Iran. It is also reported by Rawi, l.c., to be found in the Sulaimaniya district of Iraq.

SECT. SPERGULIFOLIAE (Boiss.) Chowdh.

5. **Silene stenobotrys** Boiss. & Hausskn. in Boiss., FI. Orient. 1: 611 (1867); Rohrb., Monogr. Silene: 195 (1868); Zohary, Fl. Palaest. ed. 2, 182 (1932); Post, Fl. Syr. Palest. & Sinai 1: 182 (1932); Rawi in Dep. Agr. Iraq Tech. Bull. 14: 49 (1964); Rech.f., Fl. Lowland Iraq: 240 (1964); Mouterde, Nouv. FI. Liban et Syr. 1: 499 (1966); Coode & Cullen in Fl.Turk. 2: 204 (1967); Melzh. in Fl. Iranica 163: 419 (1988).

S. paniculata Ehrenb. ex Rohrb. , Monogr. Silene 195 (1868).

Perennial, up to 60 cm, with a slender woody stock. Stems erect, simple, with short retrorse white hairs, glandular above. Leaves 10–20 × 0.5 mm, linear, with eglandular hairs; fascicles present in the axils of sterile shoots; cauline leaves distantly placed. Inflorescence paniculate, with lateral cymules of usually 3 flowers; cymules congested, forming pseudoverticels. Calyx 6–8 mm, cylindrical, cylindrical-clavate in fruit, with long dense glandular and eglandular hairs; teeth acuminate. Petals white; limb 2–3 mm, deeply bilobed into linear lobes; claw 11–12 mm, pilose. Filaments pilose. Anthophore 1–2 mm, pilose. Capsule 5–6 mm, conical, included in calyx.

HAB. Mountain slopes; alt. 800–1200 m; fl. Jun. fr. Jul.–Aug.
DISTRIB. In the mountains and the upper Jazira district of Iraq. MAM: Sarsang, 1200 m, *Rechinger* 11650; Jabal Khantur, 1200 m, *Rechinger* 10751 & ibid., 1480 m, *Rechinger* 10850. FUJ: Mosul, 8 km S of Zakho, *Rechinger* 10669!; Bakhair nr. Zakho, *Field & Lazar* 777!

S. stenobotrys can be identified by its pilose filaments and petal-claws. Eglandular hairs on the calyx are sometimes once forked.

NW Iran, Syrian Desert, Turkey.

6. **Silene spergulifolia** (*Desf.*) *Bieb.*, Fl. Taur.-Cauc. 3: 305 (1819); Boiss., Fl. Or. 1: 612 (1867) ; Rohrb., Monogr., 206 (1868); Zohary, Fl. Palaest. ed. 2, 1: 182 (1932); Post, Fl. Syr. Palest. & Sinai 1: 182 (1932); Rawi in Dep. Agr. Iraq Tech. Bull. 14: 49 (1964); Coode & Cullen in Fl.Turk. 2: 206 (1967); Melzh. in Fl. Iranica 163: 413 (1988); Gabrielian & Fragman-Sapir, Fl. Transcauc. & adj. areas: 142 (2008).

Cucubalus spergulifolia Desf., Choix PI. 73 (1808).
Silene armeniaca Rohrb. in App. Ind. Hort. Berol. 5(1867).
S. spergulifolia var. *elongata* Boiss., Fl. Orient. 1: 612 (1867).
S. spergulifolia var. *arbuscula* Boiss., Fl. Orient. 1: 612 (1867).

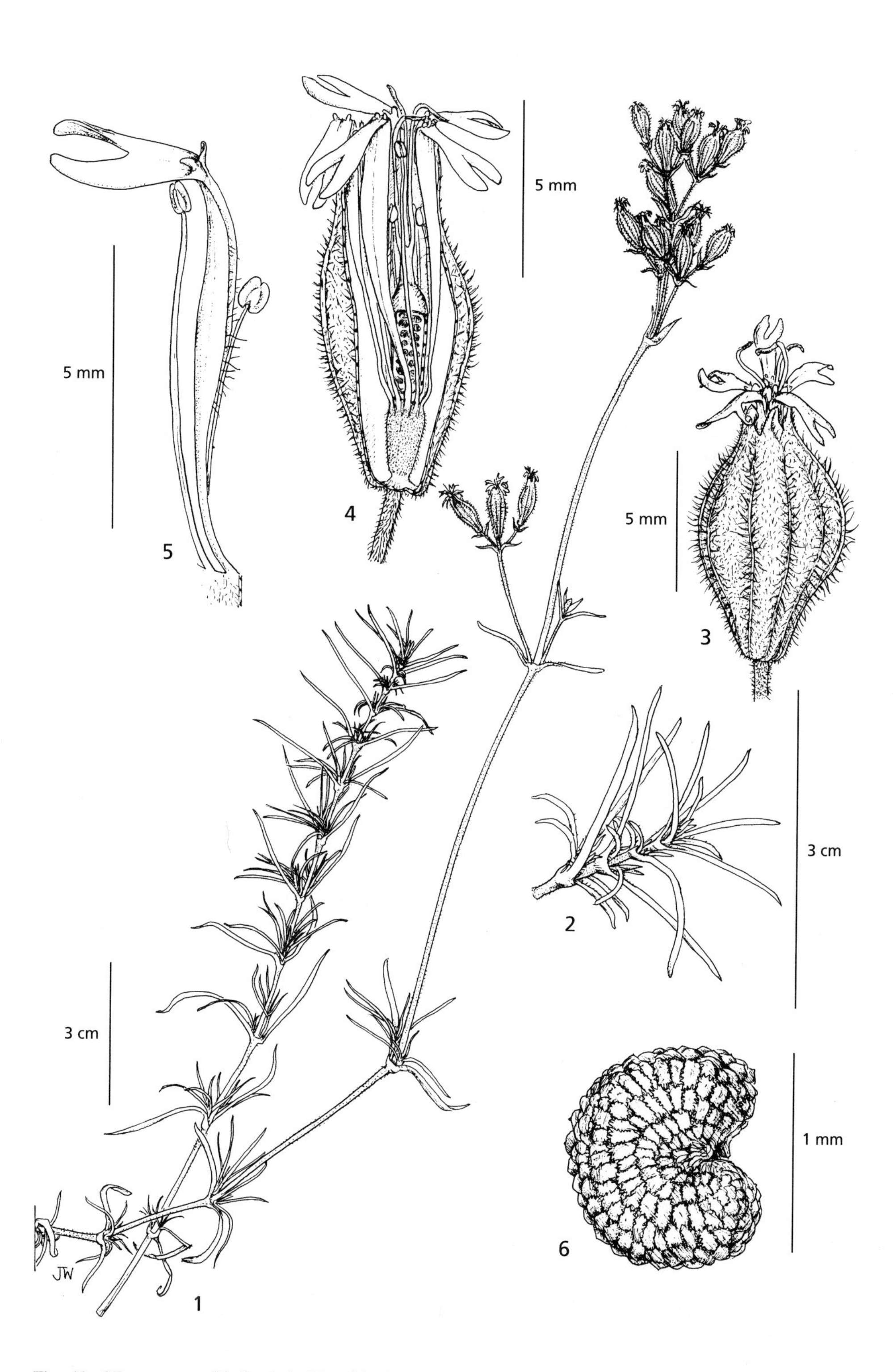

Fig. 41. **Silene spergulifolia**. 1, habit × ⅔; 2, vegetative shoot × 1½; 3, flower × 4; flower opened × 5; 5, petal and stamens × 8; 6, seed × 28. 1–2, 6 from *Rawi* 23350; 3–5 from *Rawi* 23547. © Drawn by Juliet Beentje.

Gynodioecious perennial, with a branched woody stock. Stems up to 140 cm, slender, erect and ascending, with white retrorse hairs, glandular above. Leaves 8–25 × 0.5–1.0 mm, linear, somewhat rigid and recurved, with sterile shoots in axils. All leaves puberulent, the older ± glabrous. Inflorescence a lax or congested panicle, with 3–7 flowers in dichasia. Bracts similar to leaves, ± glandular. Calyx of hermaphrodite flowers 9–10 mm; of female flowers ± 8 mm, fusiform, slightly constricted below the capsule in fruit, glandular-pubescent; teeth lanceolate. Petals white; limb 3–4 mm; claw 6–7 mm; auricles obscure or absent; coronal scales small. Anthophore in hermaphrodite flowers 4 mm, in female flowers 2 mm, puberulent. Capsule in hermaphrodite flowers 7–8 mm, in female flowers ± 6 mm, conical, included in calyx. Fig. 41: 1–6.

HAB. Stony and rocky mountain slopes, in *Quercus* forest; alt. 900–2000 m; fl. May–Jul.
DISTRIB. **MAM**: Mosul, Mt. Zawita, 1600 m, *Rechinger* 10913!; Sarsang, 1200 m, *Haines* 975; Zawita, *Guest* 4667!, Rawi 23618, 23547; Khantur NE of Zakho, *Rawi* 23261; ibid, *Rechinger* 10751; **FAR**: Arbil, Zeita, *Agnew & al.* 5887!

Common throughout the Caucasus (Armenia, Georgia, Azerbaijan), Turkey, ?Syria, Iran.

7. **Silene pruinosa** *Boiss.*, Diagn. Pl. Or. Nov. ser. 1, 1: 23 (1843); Boiss., Fl. Orient. 1: 612 (1867); Rawi in Dep. Agr. Iraq Tech. Bull. 14: 48 (1964).

S. supina Bieb. subsp. *pruinosa* (Boiss.) Chowdh. in Notes RBG Edinb. 22: 266 (1957); Coode & Cullen in Fl. Turk. 2: 206 (1967).

Perennial, stems erect or ascending, up to 50 cm. Leaves densely puberulent; sterile shoots in leaf axils absent. Inflorescence a lax panicle, flowers in dichasia. Calyx 12–16 mm, pubescent, with ± glandular hairs. Petals greenish yellow; limb deeply bifid, auricles absent or obscure. Anthophore hairy. Capsule about as long as anthophore.

HAB. Not recorded.
DISTRIB. **DSD**: Jebel Sanam, 1935, *Anthony*!

S. pruinosa is very similar to *S. spergulifolia* (Desf.) Bieb., but differs in not having sterile shoots in axils of leaves. A single record from Iraq which I have not been able to locate. Its presence in Iraq is doubtful.

Armenia,Turkey, Iran, Syria.

8. **Silene montbretiana** *Boiss.*, Diagn. Pl. Or. Nov. ser. 1, 1: 26 (1843); Zohary, Fl. Palaest. ed. 2, 1:183 (1932); Post, Fl. Syria, Palest. & Sinai 1: 183 (1932); Chowdh. in Notes RBG Edinb. 22: 266 (1957); Coode & Cullen in Fl. Turk. 2: 208 (1967).

S. ispirensis Boiss. et Huet in Boiss., Diagn. Pl. Or. Nov. ser. 2, 5: 55 (1856).
S. hohenackeri Boiss., Diagn. Pl. Or. Nov. ser. 2, 1: 75 (1859).
S. aucheriana (non Boiss.) Boiss. sensu Boiss., Fl. Orient. 1: 617 (1867), non Diagn. Pl. Or. Nov. ser. 1, 1: 27 (1843); Rawi in Dep. Agr. Iraq Tech. Bull. 14: 48 (1964).
S. bornmulleriana Freyn in Österr. Bot. Zeitschr. 40: 363 (1890).

Gynodioecious perennial, with a branched slender woody stock. Stems up to 40 cm, erect or ascending, retrorsely pubescent. Basal leaves rosulate. Leaves 20–50 × 3–6 mm, narrowly elliptic to linear-lanceolate or lanceolate, basal leaves with an attenuate base, cauline leaves sessile, puberulent. Bracts similar to cauline leaves. Inflorescence paniculate, flowers 1–3. Calyx 12–16 mm, cylindric, clavate in fruit, glandular-puberulent; teeth triangular. Petals white; limb 7–8 mm, $1/2$ divided; claw 7–8 mm, with obscure auricles; coronal scales oblong. Anthophore 3–5 mm, minutely puberulent. Capsule 9–10 mm, oblong-conical, included in the calyx.

HAB. Hillsides and on serpentine rocks; alt. 1300–2000 m; fl. Jun.–Jul.
DISTRIB. **MRO**: Helgurd [Algurd Dagh], *Rawi* 13731!; Pushtashan, 15 km NE of Raina, *Rawi & Serhang* 24207!; **MAM**: Khantur, *Rawi* 8789!; Matina, *Rawi* 8684!

The female flowers of the species are smaller in size. *S. montbretiana* is very similar to *S. arguta* Fenzl, found in Armenia, N & W Iran and E Turkey, but can be distinguished by its 1-nerved cauline leaves (in *S. arguta* the cauline leaves are lanceolate to falcate and are 3–5-nerved). Melzheimer in Fl. Iranica separates the Iranian *S. aucheriana* Boiss. from the Turkish *S. montbretiana* on the basis of leaf characters and maintains both as good species.

S. montbretiana var. *microphylla* Boiss., Suppl. 98 (1888) is found in N Syria; it has smaller basal leaves (4–7 mm) and narrower cauline leaves. The calyx is also densely pubescent.

Vernacular GA SANGOAL (*Rawi* 13731).

Turkey, Azerbaijan, Armenia, N Syria.

9. **Silene oreophila** *Boiss.*, Fl. Orient. 1: 617 (1867); Rohrb., Monogr. Silene: 136 (1868); Rawi in Dep. Agr. Iraq Tech. Bull. 14: 49 (1964); Coode & Cullen in Fl. Turk. 2: 209 (1967).

S oreophila Boiss. var. *latifolia* Chowdh. in Notes RBG Edinb. 22: 267 (1957).

Perennial, 30–40 cm, with a slender branched woody stock. Stems erect to ascending, usually simple, shortly pubescent below, glandular above. Basal and cauline leaves similar, 30–70 × 3–8 mm, oblong to linear-lanceolate, basal leaves with an attenuate base, cauline leaves sessile, apex acute to acuminate, cauline leaves faintly 3-nerved; all leaves pubescent. Flowers usually 3 in a dichasium, rarely solitary. Calyx 18–24 mm, cylindric, slightly inflated, clavate and constricted below the capsule in fruit, glandular-pubescent; teeth triangular. Petals white, limb 6–7 mm, $^{1}/_{3}$ divided; claw 10–11 mm, auriculate; coronal scales oblong. Anthophore 10–12 mm, oblong-conical, glabrous, included in the calyx.

HAB. Mountain slopes; alt. ± 1000 m; fl. Apr.
DISTRIB. **MSU**: Penjwin, *Rawi* 8810!

S. oreophila resembles *S. montbretiana* Boiss. and *S. rhynchocarpa* Boiss. in habit and vegetative characters, but can be distinguished from the former by its longer calyx and anthophore, and from the latter by its less inflated calyx and calyx-nerves that lack the reddish purple reticulations found in *S. rhynchocarpa.*

Vernacular: GLUNCA SPIA or CRNAROKA (*Rawi* 8810).

Turkey (C Anatolia).

10. **Silene eriocalycina** *Boiss.*, Diagn. Pl. Or. Nov. ser. 1, 1: 28 (1843). — Type: Mesopotamia, *Aucher-Eloy* 461, G-BOISS!; Boiss., Fl. Orient. 1: 615 (1867); Rohrb., Monogr. Silene: 190 (1868); Bornm. in Beih. Bot. Centr. 28 (2): 141 (1911); Rawi in Dep. Agr. Iraq Tech. Bull. 14: 48 (1964); Melzh. in Fl. Iranica 163: 426 (1988).

Perennial, 20–50 cm tall, with a slender woody stock. Stems usually simple, slender, erect and ascending, leafy, with dense white retrorse or patent hairs. Basal and cauline leaves similar, 28–65 × 4–3 mm, lanceolate to elliptic. Basal leaves with an attenuate base, cauline leaves sessile, apex acute, densely pubescent. Bracts 5–20 × 1–5 mm, linear. Calyx (14–)18(–20) mm, cylindric-clavate, clavate in fruit, with long dense eglandular hairs; teeth lanceolate. Petals reddish (?); limb ± 8 mm, $^{2}/_{3}$ divided; claw 11–12 mm, auricles small; coronal scales small. Anthophore 8–9 mm, puberulent. Capsule 10–15 mm, conical, included in calyx.

HAB. Mountain slopes; common; alt. 1300–2600 m; fl. Jun., fr. Jul.
DISTRIB. **MRO**: Haji Omran, *Kass & Nuri* 27869 & 27778!; Sakran, 15 km SE Ghunam, *Dabbagh & Hamad* 46256!; Mergan, nr. Bardanas, *Rawi & Serhang* 24345!; Algurd Dagh, *Guest* 2940, 2932!; Chia Mandali, *Guest & Ludow-Hewit* 2769!; Karoukh mt., *Kass & Nuri* 27527; ibid., 97342; Seiwaka village, *Kass & Nuri* 27569!; Gali Warta, *Rawi, Nuri & Kass* 28726!; Kani Khanjar Khan, NE of Helgord Mt., *Rawi & Serhang* 24701; Khantur Mt., NE of Zakho, *Rawi* 23414!; **MSU**: Penjwin, *Rechinger* 10510!; Tawila, *Rawi* 21954!; **FAR**: Sakri Sakran, *Bornmüller* 980!; Molla Khort, *Rawi et al.* 29465!

S. eriocalycina is diagnostic in its dense white eglandular pubescence and linear bracts which curl upwards and embrace the calyx. The species is most likely gynodioecious. Specimen no. 29465 from Molla Khort mountains has only female flowers with a calyx length of 14–15 mm. Collections from Sarsang, *Chapman* 26385, *Wheeler Haines* 549, Rowandiz, *Field & Lazar* 894 and Shaqlawa, *Wheeler Haines* 611 are larger plants with large leaves (to 65 mm) and large calyces (to 20 mm). This may be recognised at a subspecific rank to *S. eriocalycina* .

Turkey, Armenia, Iran, Syria.

SECT. AMPULLATAE Boiss.

11. **Silene ampullata** *Boiss.*, Diagn. Pl. Or. Nov. ser. 1, 1: 26 (1843); Boiss., Fl. Orient. 1: 606 (1867); Rohrb., Monogr. Silene: 82 (1868); Bornm., Beih. Bot. Centr. 28 (2): 140 (1911); Rawi in Dep. Agr. Iraq Tech. Bull. 14: 48(1964); Coode & Cullen in Fl. Turk. 2: 210 (1967); Melzh. in Fl. Iranica 163: 420 (1988).

Gastrocalyx ampullatus (Boiss.) Schischk. in Bull. Mus. Cauc. 12: 200 (1919); Schischkiniella ampullata (Boiss.) Steenis, Blumea 15: 145 (1967).

Perennial, 15–30 cm, with a thick woody stock. Stems many, erect to ascending, shortly pubescent. Sterile shoots ± glandular, leafy. Leaves 15–30 × 0.5–1 mm, linear, base sessile, apex acute, glandular-pubescent. Flowering shoots with a few distantly placed leaves. Inflorescence a congested raceme, with 3–5–flowered cymules. Calyx 8–18 mm, cylindric, slightly inflated, in fruit membranous and inflated, greenish yellow, contracted at the apex, glandular-pubescent; teeth triangular. Petals yellowish green; limb 7 mm, $^{1}/_{2}$ divided; claw 5–6 mm, auriculate; coronal scales oblong. Anthophore 2.5–3 mm, shortly pubescent. Capsule 5–6 mm, subglobose, included in the calyx, 1(–2)-seeded.

HAB. Commonly found on limestone on mountain sides, frequently in *Quercus* forest; alt. 1200–2000 m; fl. Apr.–May, fr. Jun–Jul.
DISTRIB. **MAM**: Sarsang, *Rechinger* 11650! (W); Zakho, *Field & Lazar* 777!; Suwara Tuka, *Rechinger* 11569! (W); Jabal Khantur, *Rechinger* 10751! (W). **MRO:** Sefin Dagh above Shaqlawa, *Gillett* 8146 (BAG). **MSU**: Penjwin, *Khalil Feddo* 3447 (BAG); **FUJ**: Mosul, 8 km S of Zakho, *Rechinger* 10669! (W).

S. ampullata is similar in habit and vegetative characters to *S spergulifolia* Boiss. but can be distinguished by its calyx which is prominently inflated, especially in fruit.

Armenia, NW Iran, Syria, Turkey (E Anatolia).

SECT. FIMBRIATAE Boiss.

12. **Silene schizopetala** *Bornm.* in Bull. Herb. Boiss.: 117 (1899) & in Beih. Bot. Centr. 28 (2): 144 (1911); Rawi in Dep. Agr. Iraq Tech. Bull. 14: 49 (1964); Melzh. in Fl. Iranica 163: 381 (1988). — Type. Iraq, Arbil, Kuh Safin, Shaqlawa, *Bornmüller* 951 (B†)

Perennial or biennial, with a short woody stock. Stems thick, 30 cm or more, pilose-glandular, viscid above. Leaves 30–70 × 20–50 mm, ovate, basal leaves with an attenuate base, cauline leaves sessile; apex acute to acuminate; all leaves glandular-pilose, 3–5–nerved. Bracts ovate-lanceolate. Flowers 3–5, in dichasial cymes. Pedicels slender, glandular-pilose. Calyx 17–25 mm, campanulate, inflated, green, membranous, inflated in fruit and slightly contracted at the apex; teeth ovate lanceolate. Petals white; limb 6–8 mm, divided into linear lobes; claw 12–13 mm, exauriculate; coronal scales absent. Anthophore 2–2.5 mm, glabrous. Capsule 10–12 mm, subglobose, included in the calyx. Fig. 42: 1–6.

HAB. Dry slopes in *Quercus* forest; alt. 1200–1350 m; fl. May–Jun.
DISTRIB. **MAM**: Bakirma, *Rawi* 8575!; **MSU**: Kopi Karadagh, *Wheeler-Haines* 1078!

Endemic.

SECT. AURICULATAE Boiss.

13. **Silene microphylla** *Boiss.*, Diagn. ser 1, 1: 33 (1843); Boiss., Fl. Orient. 1: 634 (1867); Rohrb., Monogr. Silene: 135 (1868); Chowdh., Notes RBG Edinb. 22: 243 (1957); Rawi in Dep. Agr. Iraq Tech. Bull. 14: 49 (1964); Ghazanfar in Notes RBG Edinb. 41(1):106 (1983); Melzh. in Fl. Iranica 163: 449 (1988).

Perennial, 15–30 cm with a branched woody stock. Stems ascending, glandular-pubescent. Basal leaves 16–16 × 2–5 mm, oblanceolate to narrowly lanceolate, base attenuate; apex acute; cauline leaves similar to basal leaves but without an attenuate base. All leaves with whitish grey hairs mixed with glandular hairs. Bracts 5–6 × 2 mm, ovate-lanceolate. Flowers 3–(5–7) in dichasial cymes. Calyx 27–28 mm, cylindric-clavate, clavate and slightly constricted below the capsule in fruit; densely glandular. Petal-limb 7–8 mm, $^{2}/_{3}$ divided; claw 14–15 mm, auriculate; auricles obtuse at apex; coronal scales oblong. Anthophore 11–12 mm, glabrous. Capsule 11–13 mm, oblong-ovoid, included in the calyx.

HAB. Mountains and rocky areas; alt. 1300–2900 m; fl. & fr. May–Jun.
DISTRIB. **MRO**: Mt. Qandil, NE Rania, *Rawi & Serhang* 26805!

Iran, Turkmenistan, Afghanistan.

14. **Silene rhynchocarpa** *Boiss.*, Diagn. Pl. Or. Nov. ser. 1, 1: 33 (1843); Boiss., Fl. Orient. 1: 618 (1867); Rohrb., Monogr. Silene: 126 (1868); Chowdh., Notes RBG Edinb. 22: 242 (1957); Rawi in Dep. Agr. Iraq Tech. Bull. 14: 49 (1964); Coode & Cullen in Fl. Turk. 2: 217 (1967); Melzh. in Fl. Iranica 163: 438 (1988).

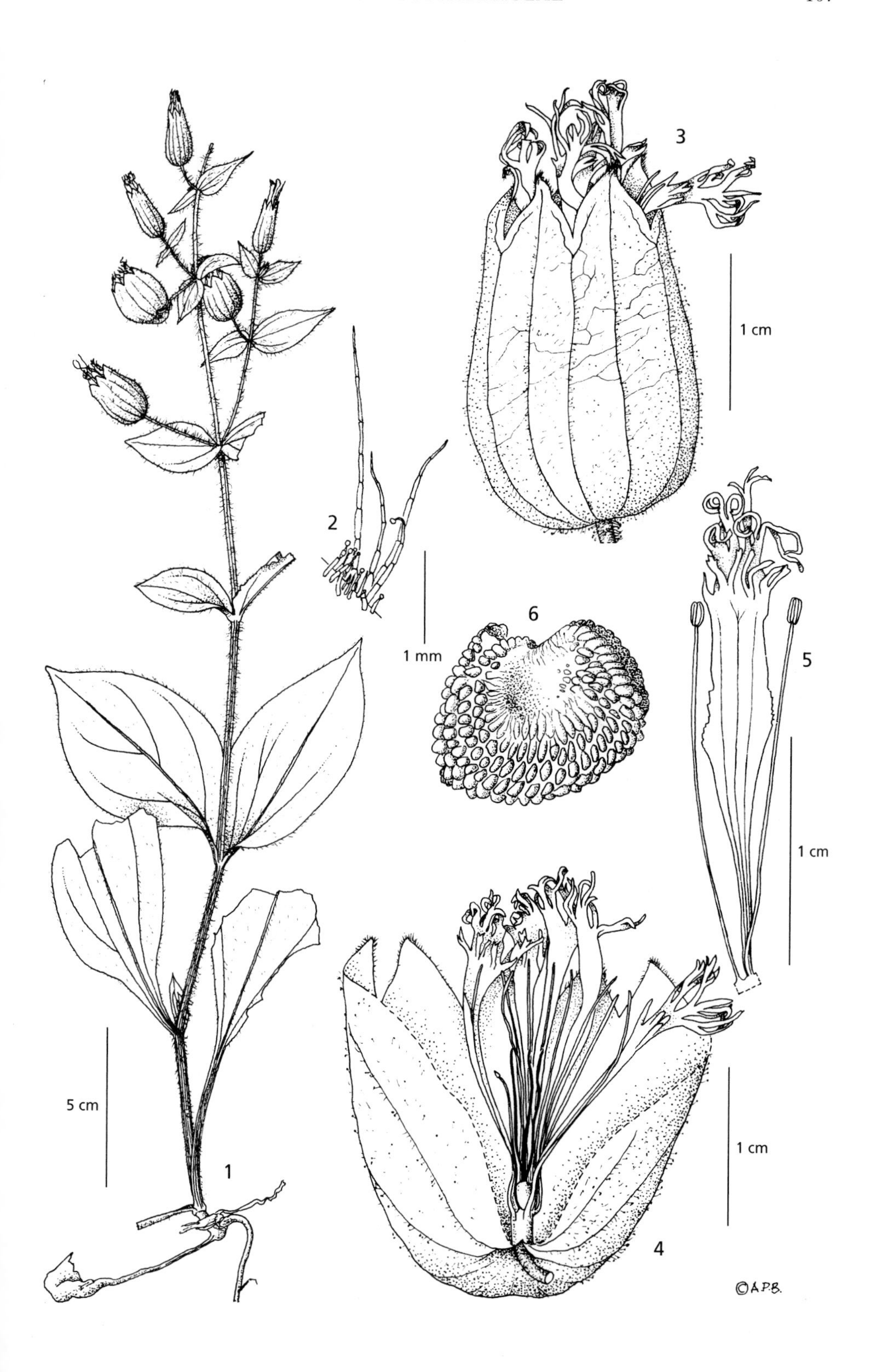

Fig. 42. **Silene schizopetala**. 1, habit 2, young stem indumentum; 3, flower; 4, flower opened; 5, petal with stamens; 6, seed. 1–2 from *Rawi* 8575; 3–6 from *Haines* W1078. © Drawn by A.P. Brown.

S. lycia Boiss., Diagn. Pl. Or. Nov. ser. 1, 8: 90 (1849).
S. rhynchocarpa var. *lycia* Boiss., Fl. Orient. 1: 618 (1867).

Perennial, 10–30 cm, with a branched woody stock. Stems erect or ascending, with greyish retrorse hairs, glandular above. Leaves 12–50 × 3–10 mm, linear-lanceolate, elliptic or oblanceolate, puberulent to pilose; the basal leaves with a narrow attenuate base; apex acute. Leaves usually crowded at the base, few and distant on the flowering stems. Bracts and bracteoles similar to cauline leaves but reduced in size. Flowers 3(–7) in dichasia, rarely solitary. Calyx 20–28(–30) mm, cylindric to cylindric-clavate, slightly inflated, clavate, inflated and slightly constricted below the capsule in fruit, glandular-pubescent; teeth ovate-lanceolate. Petals white or pink; limb 10–11 mm, $^1/_3$ divided; claw 13–14 mm, auriculate; auricles acute at apex; coronal scales oblong. Anthophore 11–14(–15) mm, puberulent or glabrous. Capsule ± 14 mm, ovoid, included in the calyx. Fig. 43: 1–6.

HAB. Cliffs and mountain slopes; alt. 2400–3800 m; fl. & fr.: Jun.–Aug.
DISTRIB. Rare in the mountain zone of NE Iraq. **MRO**: Mt. Qandil by Iranian frontier, *Rechinger* 11798 (W!); Mt. Helgord [Algurd Dagh] on W side of summit, *Rechinger* 1149 (W!); Malikh Mt., Qandil range, *Rawi & Serhang* 24036, 24480; NE Qandil, mountain side, *Rawi & Serhang* 24405!; Mergan, nr Bardanas, *Rawi & Serhang* 24354!; Helgord [Algurd Dagh], *Guest & Husham* 15830!; ibid., *Guest* 2841!; ibid., *Gillett* 9583, 12348A!; ibid., *Rawi & Serhang* 24833A!; Chia Mandali, *Guest* 2698!

Turkey, W Iran.

15. **Silene commelinifolia** *Boiss.*, Diagn. Pl. Or. Nov. ser. 1, 1: 33 (1843); Boiss., Fl. Orient. 1: 624 (1867); Rohrb., Monogr. Silene: 128 (1868); Rawi in Dep. Agr. Iraq Tech. Bull. 14: 48 (1964).

S. commelinifolia var. *isophylla* Bornm. in Beih. Bot. Centr. 19: 216 (1905).
S. commelinifolia var. *heterophylla* Fenzl in Kotschy (in sched.?).
S. commelinifolia var. *ruscifolia* Huber-Morath & Reese in Feddes Rep. 52: 44 (1943).
S. ruscifolia (Huber-Morath & Reese) Huber-Morath in Notes RBG Edinb. 28: 4 (1967); Coode & Cullen in Fl. Turk. 2: 218 (1967).

Pulvinate perennial, forming clumps, with a branched woody stock. Stems 12–40 cm, branched, postrate to ascending, puberulent, sparsely glandular below, densely glandular above. Leaves 18–87 × 15–80 mm. Basal leaves lanceolate, spatulate or ovate-lanceolate; base attenuate; apex acute; cauline leaves lanceolate to ovate-cordate, opposite and decussate, base sessile, apex acute to mucronate, 3–5-nerved. All leaves glandular. Flowers in 3–5(–7)-flowered dichasia, rarely solitary. Bracts and bracteoles similar to the cauline leaves. Calyx (15–)18–24(–32) mm, cylindric-clavate, slightly inflated, clavate in fruit, glandular-pubescent; teeth ovate-lanceolate. Petals white; limb 8–9 mm, $^1/_2$ divided; claw 11–12 mm, auriculate; coronal scales triangular. Anthophore 8 mm, puberulent. Capsule 9–13 mm, oblong-ovoid, included in the calyx.

HAB. Common on rocky places and mountain slopes, growing gregariously; alt. 1300–3800 m; fl. & fr.: Jun.–Aug.
DISTRIB. Rare in the mountain zone of NE Iraq. **MRO**: Rowanduz, N slope of Mt. Potin, *Agnew, Hadač & Harris* 6129!; Sarcal, *Hadač* 2243, 2154!; Helgord (Algurd Dagh), *Hadač* 2088!

A widespread species, very variable in its basal and cauline leaves and bracts. Some collections from N Iraq, W Azerbaijan, Iran and Turkey have lanceolate bracts, on the basis of which var. *ruscifolia* (*S. ruscifolia*) has been separated. This variation is not present in specimens from a distinct geographical area as other collections from Iran and Iraq that I have examined have bracts that are ovate to ovate-lanceolate or ovate-cordate or lanceolate. Based on the shape of the bracts throughout the range of the species, any taxa below the the rank of species seems unjustified.

Caucasus, Iran, Turkey.

16. **Silene monantha** *Boiss. & Hausskn.* ex *Boiss.*, Fl. Orient. Suppl.: 99 (1888) — Type: Hab. in fissuris montis Sindjar, *Haussknecht* s.n.; Rawi in Dep. Agr. Iraq Tech. Bull. 14: 49 (1964).

Caespitose perennial with a thick short woody stock, covered with remains of old leaf bases. Stems 9–18 cm, erect to ascending, simple, grey-pilose. Leaves 9–22 × 22–10 mm, grey-green. Basal leaves obovate, with a long attenuate base, but soon withering; cauline leaves ovate-cordate to rotund, base sessile, clasping; apex acute, densely pilose. Flowers 1–2. Calyx 9–10 mm, cylindric, in fruit clavate and constricted below the capsule, pilose; teeth triangular.

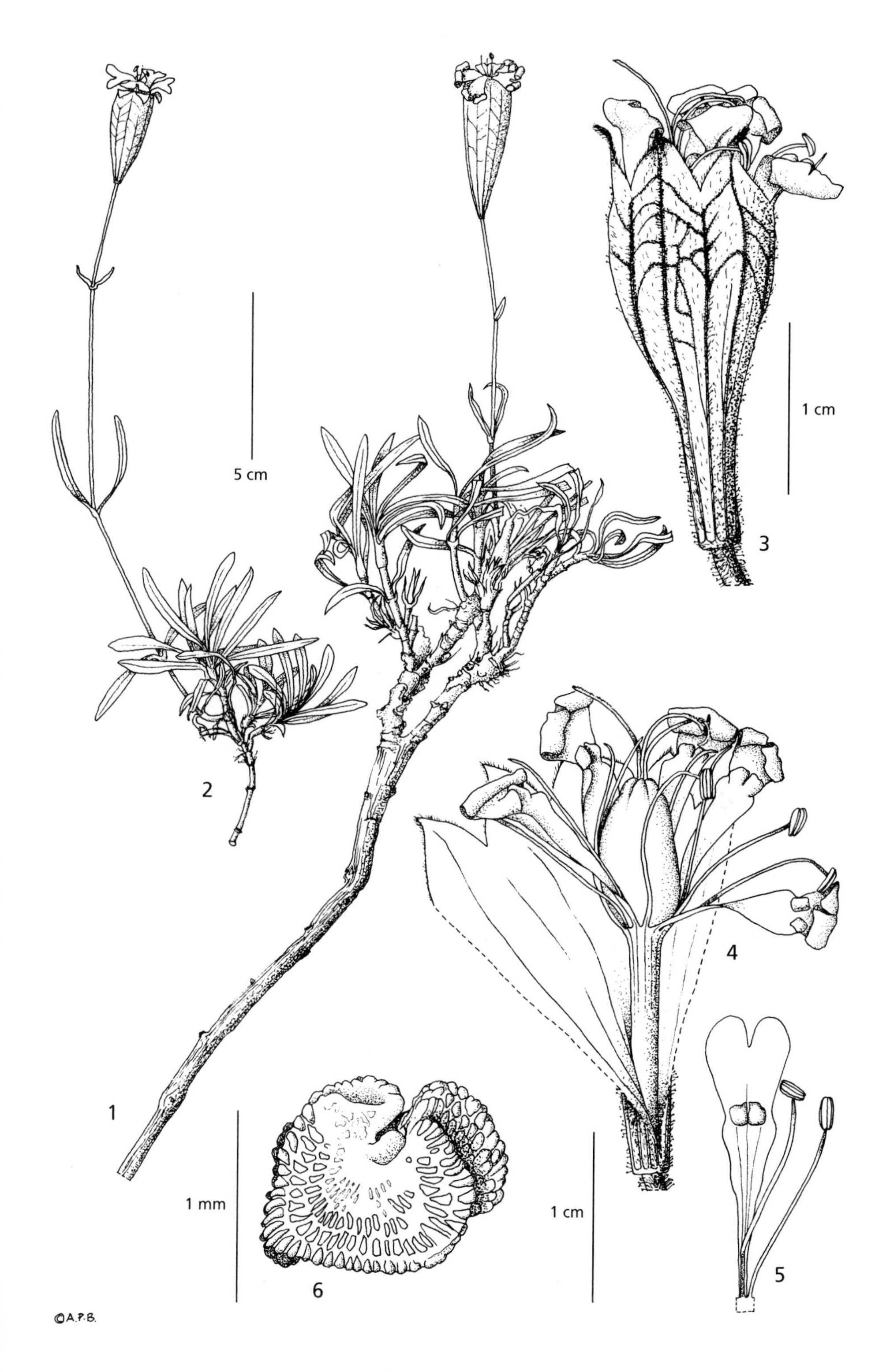

Fig. 43. **Silene rhynchocarpa**. 1, 2, habit; 3, flower; 4, flower opened; 5, petal with stamens; 6, seed. 1 from *Guest* 2841; 2 from *Rawi & Serhang* 24480; 3–6 from *Guest* 2698. © Drawn by A.P. Brown.

Petals greenish; limb 5–4 mm, 1/2 divided; claw 7–8 mm, auricles obscure. Anthophore ± 2 mm, glabrous. Capsule 8–9 mm, sub-globose, nodding, included in the calyx.

HAB. In crevices on mountain; alt. not specified; fl. May.
DISTRIB. **MJS**: Mt. Sinjar, *Haussknecht* s.n.

Probably endemic.

17. **Silene erimicana** *Stapf* in Denkschr. Akad. Wiss. Wien, Math.-Nat. Kl. 51: 17 (1886); Coode & Cullen in Fl. Turk. 2: 219 (1967).

S. commelinifolia Boiss. var. *erimicana* (Stapf.) Blakelock, Kew Bull. 12(2): 211 (1957).

Perennial, with a branched woody base. Stems many, erect, covered with adpressed eglandular hairs below, hairs glandular above, viscid. Leaves 20–45 × ± 3 mm, linear, base attenuate, apex acute. Cauline leaves distantly placed, 7–12 × 1–2 mm, linear. All leaves glandular-pubeseent, 3–nerved. Bracts ± 3 mm, linear-lanceolate. Flowers 1–2, or 3 in a dichasium. Calyx 17–18 mm, clavate-cylindric, clavate in fruit, glandular-pubescent; teeth triangular. Petal-limb 5–6 mm, 1/2 divided; claw 10–11 mm, auriculate, auricles obtuse. Anthophore 8–9 mm, puberulous. Capsule 9–10 mm, ovoid, included in calyx. Fig. 44: 1–5.

HAB. Mountains, mountains slopes, near streams; alt. 2500–2600 m; fl. Jul.–Aug., fr. Aug.
DISTRIB. **MRO**: Mergan, nr. Bardanas, *Rawi & Serhang* 24352!

NW and W Iran, NE Turkey.

18. **Silene pungens** *Boiss.*, Diagn. Pl. Or. Nov. ser. 1, 1: 32 (1843); Boiss., Fl. Orient 1: 625 (1867); Rohrb., Monogr. Silene: 77 (1868); Rawi in Dep. Agr. Iraq Tech. Bull. 14: 48 (1964); Coode & Cullen in Fl. Turk. 2: 220 (1967); Ghazanfar in Notes RBG Edinb. 41(1): 106 (1983); Melzh. in Fl. Iranica 163: 373 (1988).

Gynodioecious perennial, with a short woody stock. Stems 15–25 cm, many, simple or branched, erect or ascending, retrorsely pubescent, hairs glandular above. Leaves 15–27 × ± 1 mm, pubescent, narrowly linear, slightly arcuate, distantly placed on the flowering stems, base sessile, apex acute. Bracts 2–3 mm, lanceolate, pubescent, with ± glandular hairs. Flowers solitary or 3 in a dichasium. Calyx in hermaphrodite flowers 13–18 mm, in female flowers 11–12 mm, inflated, cylindric, inflated in fruit, glandular-pubescent; teeth ovate, obtuse. Petal limb 8–9 mm, 1/2 divided; claw 8–10 mm, auriculate; coronal scales oblong, denticulate at the top. Anthophore 7–8 mm in hermaphrodite flowers, 4–5 mm in female flowers, puberulous. Capsule 10–11 mm, included in calyx.

HAB. Mountain slopes; alt. 2000–3800 m; fl. & fr. Jul.–Aug.
DISTRIB. Rare in montane zone of NE Iraq. **MRO**: Helgord (Algurd Dagh), *Rechinger* 11436 (G!); Qandil, *Rechinger* 11157 (W!); ibid., *Rawi & Serhang* 20207!, 24573!; Jabal Khantur, *Rechinger* 12092 (W!).

S. pungens is easily distinguished by its patent, slightly arcuate leaves.

NW and W Iran, N and W Turkey.

19. **Silene brevicaulis** *Boiss.*, Diagn. Pl. Or. Nov. ser. 1, 1: 34 (1843); Boiss., Fl. Orient. 1: 623 (1867); Coode & Cullen in Fl. Turk. 2: 218 (1967); Ghazanfar in Notes RBG Edinb. 41(1): 106 (1983).

S. brevicaulis var. *latifolia* Boiss., Fl. Or. 1: 623 (1967).

Caespitose perennial, with a branched woody stock. Stems 7–15(–20) cm, erect to ascending, leafy, simple, densely pubescent, ± glandular, densely glandular and viscid above. Leaves 11–60 × 2–10 mm, sometimes 3-nerved, lanceolate to linear-lanceolate, or oblanceolate; base attenuate of the basal leaves, sessile in the cauline leaves; apex acute, pubescent, ± glandular. Bracts and bracteoles ovate-lanceolate. Flowers 1–2 or 3 in a dichasium; pedicels densely glandular. Calyx 25–29 mm, cylindric-clavate, clavate and constricted below the capsule in fruit, densely glandular-pubescent; teeth ovate-lanceolate. Petals white, limb 6–8 mm, 1/3 divided; claw 9–10 mm, auriculate; coronal scales lanceolate. Anthophore 14–18 mm, glabrous. Capsule 6–8 mm, included in calyx.

HAB. Mountains and calcareous rocks and in rock crevices; alt. 1600–2000 m; fl. & fr. Jul.–Aug.

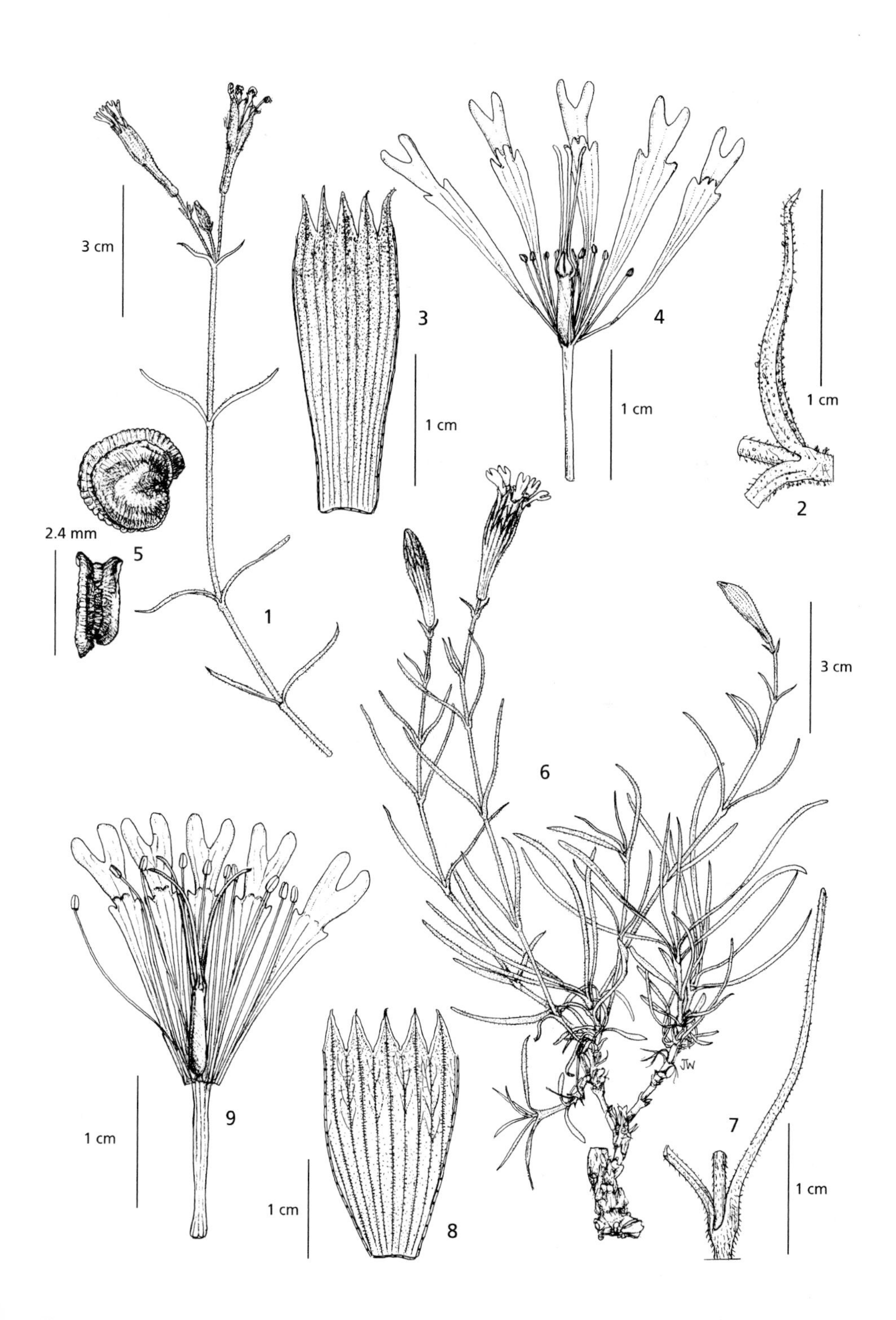

Fig. 44. **Silene erimicana**. 1, habit × ⅔; 2, leaf × 3; 3, calyx opened × 1½; flower opened with calyx removed × 2; 5, seeds × 16. **Silene caryophylloides** subsp. **subulata**. 6, habit × ⅔; 2, leaf × 2; 3, calyx opened × 1½; flower opened with calyx removed × 2. 1–5 from *Th. Pichler* s.n.; 6–9 from *Guest* 2792. © Drawn by Juliet Beentje.

DISTRIB. **MSU**: Avrioman mt., [Hauraman], *Rawi, Chakravarty, Nuri & Alazizi* 19747!; Qopi Karadagh, *Haines* W1083!; ?Kurdistan, *Haussknecht* 1865!

S. brevicaulis has not been recorded from Iraq before. It is found in SE Turkey and on Iran (W)/Iraq border. The species is variable in the size and pubescence of its leaves.

Turkey (SE Anatolia), Syria.

20. **Silene retinervis** *Ghaz.* in Kew Bull. 37(3) 458 (1982). — Type: Iraq: top of Qandil range, between Perrish and Bardanas (Iraqi-Iranian border), ± 3000 m, 29 Aug. 1957, *Rawi & Serhang* 24304 (holo, K!, iso, BAG).

Caespitose perennial, with a branched woody stock. Stems slender, 10–15 cm, decumbent or ascending, simple, usually glabrous below, with white eglandular and ± glandular hairs above. Basal leaves 12–15 × 0.5–2 mm, obovate to linear-lanceolate, base sessile, apex acute. Cauline leaves 1–2 pairs, 7–10 × 0.5 mm, linear-lanceolate, base sessile, apex acute, densely pubescent, ± glandular. Flowers solitary; pedicels densely glandular. Calyx 19–21 mm, cylindric-clavate, densely glandular; teeth ovate-lanceolate, ciliate. Petal-limb 6–7 mm, $^1/_3$ divided; claw 13–14 mm, partially exserted from the calyx, auriculate; auricles prominent, acute; coronal scales ± 1 mm. Anthophore 10–11 mm, pubescent. Capsule not known.

HAB. Rocky mountians; alt. ± 3000 m; fl. & fr. Jul.-(?)Aug.
DISTRIB. Rare in the mountain zone of NE Iraq. **MRO**: Top of Qandil range, between Perrish and Bardanas (Iraq-Iran border), *Rawi & Serhang* 24304 (type).

S. retinervis resembles *S. microphylla* Boiss. in facies but can be distinguished by its solitary flowers and a pubescent anthophore.

Endemic to N Iraq.

21. **Silene odontopetala** *Fenzl*, Pugill. 9 (1842); Boiss., Fl. Orient. 1: 625 (1867); Rohrb., Monogr. Silene: 78 (1868); Zohary, Fl. Palaest. ed. 2, 1: 183 (1932); Rawi in Dep. Agr. Iraq Tech. Bull. 14: 49 (1964); Tackholm, Stud. Fl. Egypt ed. 2, 85 (1974); Mouterde, Nouv. Fl. Liban et Syr. 1: 500 (1966); Melzh. in Fl. Iranica 163: 374 (1988).

S. odontopetala var. *cerastifolia* Boiss., Fl. Orient. 1: 625 (1867).
S. odontopetala var. *glabrifolia* Blakelock in Kew Bull. 1939: 88 (1939).
S. odontopetala var. *congesta* Boiss., Fl. Or. 1: 626 (1867).
S. odontopetala subsp. *congesta* (Boiss.) Melzh. in Fl. Iranica 163: 376 (1988).

Perennial, up to 40 cm, with a branched woody stock. Stems simple, glabrous to hairy, with ± glandular hairs (usually present above). Basal and cauline leaves similar, but the cauline leaves reducing in size upwards. Leaves 10–80 × 7–1 mm, oblanceolate, lanceolate to ovate-lanceolate or narrowly lanceolate; apex acute, base attenuate in the basal leaves, sessile in cauline leaves; all leaves hairy with ± glandular hairs on surfaces, sometimes glabrous. Bracts similar to cauline leaves. Flowers 1–(2–3). Calyx campanulate, (12–)14–16 mm, becoming inflated in fruit, puberulous with ± glandular hairs. Petals white to pinkish; limb 5–6 mm, $^1/_2$ divided; claw 14–15 mm, auriculate, auricles obtuse at apex; coronal scales ± 1.5 mm. Anthophore 3–5 mm, glabrous. Style hairy. Capsule ± 10 × 7 mm, conical, included in calyx.

HAB. Rocky mountain sides and cliffs; alt.1500–4000 m; fl. & fr. Jun.–Jul., Aug.
DISTRIB. **MAM**: Zawita N Zakho, 2100 m, *Rechinger* 10940 Gali Zawita NE Zakho, Rawi, 23623!; **MRO**: Mt. Sakri Sakran nr. Rowandaz, *Bornmüller* 982!, 933!; Algurd Dagh (Helgurd), *Rechinger* 11874!; ibid, *Rawi & Serhang* 24826!; Mt. Qandil *Rechinger* 11115!; Chiya-i-Mandau, *Guest* 2803!; Sersang, *Wheeler-Haines* 483!; **MSU**: Avroman [Hauraman], *Rawi et al.* 19746!; **FNI**: Chiya Gara, *Rawi* 9286!

S. odontopetala is variable in the shape, size and indumentum of its leaves and stems. Post (1932) has recognised five varieties based on such variations. Blakelock (Kew Bull. 88 (1939)), recognised var. *glabrifolia* on the basis of its leaves and stems. I have examined the type specimen of the variety at K, and seen that the specimens on the type sheet do not all have glabrous leaves and stems, therefore this variety does not justify subspecific recognition. Var. *cerastiifolia* Boissier, l.c. is differentiated in having lanceolate leaves with a long cuneate base and acuminate calyx-teeth, and var. *congesta* (Boiss.) Melzh. separated on the basis of its congested inflorescence. These characters are variable and best recognised as variations based on the habitat where they are found and included within the range of the species.

Egypt (Sinai), Lebanon, Palestine, Syria, Iran, Turkey.

22. **Silene caryophylloides** (*Poir.*) *Otth* in DC, Prodr. 1: 369 (1842); Boiss., Fl. Orient. 1: 619 (1867); Rohrb., Monogr. 127 ((1868); Zohary, Fl. Palaest. ed. 2, 183 (1932); Post, Fl. Syr. Palest. & Sinai 1: 183 (1932); Rawi in Dep. Agr. Iraq Tech. Bull. 14: 48 (1964); Mouterde, Nouv. Fl. Liban. et Syr. 1: 499 (1966); Coode & Cullen in Fl. Turk. 2: 221 (1967).

Caespitose perennial, densely pulvinate, with a low ± branched woody stock. Stems 4–25 cm, erect or ascending, with eglandular and ± glandular hairs. Leaves 15–27 × 1–1.5 mm, triquetrous or linear, straight or arcuate, base sessile, apex acute, indurate, 1–3-nerved. Basal leaves rosulate; cauline leaves 2–4 pairs, linear to linear-lanceolate. All leaves pubescent or puberulent or pilose or sometimes setaceous, with ± glandular hairs. Bracts similar to cauline leaves. Flowers 1(–2). Calyx 28–35 mm, cylindric to cylindric-clavate, clavate and often inflated in fruit, pubescent, frequently with glandular hairs, but varying. Petals white, pale green or pinkish, limb 1/2 divided; claw auriculate. Anthophore 10–20 mm, glabrous. Capsule included or partially excluded from the calyx.

S. caryophylloides is variable in length of calyx and pubescence of calyx and leaves. Three subspecies are recognised based on the length and shape of calyx, pubescence of vegetative parts and the capsule to anthophore ratio. Only subsp. *subulata* is found in Iraq, the other two in Turkey.

subsp. **subulata** (*Boiss.*) *Coode & Cullen* in Notes R B G Edinb. 28: 6 (1967); Coode & Cullen in Fl. Turk. 2: 221 (1967); Ghazanfar in Notes R. B. G. Edinb. 41(1): 107 (1983)

S. subulata Boiss., Diagn. Pl. Or. Nov. ser. 1, 1: 33 (1843); Boiss., Fl. Orient. 1: 619 (1867); Rawi in Dep. Agr. Iraq Tech. Bull. 14: 51 (1964).
S. caryophylloides var. *nardifolia* Boiss. ex Rohrb., Monogr. Silene: 127 (1868).
S. nardifolia Boiss. et Huet, nom. nud. (in sched.).

Leaves with glandular hairs and setaceous margins. Calyx 28–30 mm, cylindric-clavate, clavate in fruit, not inflated, densely glandular. Anthophore 15–16 mm. Capsule 11–14 mm. Fig. 44: 6–9.

HAB. Rocks and crevices on mountains; alt. ± 3000 m; fl. Jun.–Jul.
DISTRIB. **MRO**. Chiya- i- Mandali [Chiya-i Mandau], *Guest* 2792!

Turkey (C Anatolia), Lebanon.

SECT. COMPACTAE BOISS.

23. **Silene compacta** Fisch., Cat. Jard. Gorenki ed. 2: 60 (1812) [nom. nud.] in Hornem, Hort. Bot. Hafn. 1: 417 (1813); Coode & Cullen in Fl. Turk. 2: 224 (1967); Melzh. in Fl. Iranica 163: 490 (1988); Gabrielian & Fragman-Sapir, Fl. Transcauc. & adj. areas: 140 (2008).

Atocion compactum (Fisch.) Tzvelev in Novosti Sist. Vÿssh. Rast. 33: 97 (2001);
S.vandasii Nábělek Spisy Pír. Fak. Masarykovy Univ. 35: 43 1923. — Type: Hasitha Kurdistaniae, inter Amadia et Gulamerk, *Nábělek* 4255 (BRA).
S. reuteriana Boiss. & C.I.Blanche, Diagn. Pl. Orient. Ser. 2, 5: 54 (1856).

Biennial or short-lived perennial, 20–40(–65) cm tall; stems simple or sparingly branched above, glabrous throughout. Leaves 20–50 × 10–12 mm; basal leaves forming a rosette, sessile, spatulate to spatulate-lanceolate, obtuse, glabrous; cauline leaves ovate, acute to obtuse, amplexicaul, reducing in size upwards. Inflorescence of many flowers forming a dense, compact, umbellate head. Pedicels ± 2 mm. Bracts ovate, ± 1 mm, acuminate, membrabous. Calyx 12–13 mm, cylindrical, constricted below the capsule in fruit; teeth triangular-acuminate; nerves and anastomoses weak, glabrous. Petals bright pink to purplish pink, exserted from calyx; limb 4–6 mm, entire. Coronal scales 2–4 mm, emarginate. Anthophore 8–10 mm, glabrous. Capsule 6–8 mm, oblong-ovoid, included in the calyx. Seeds semi-circular. Fig. 45: 1–5.

HAB. Edges of forest, fallow fields; alt. not recorded on specimen from Kurdistan; in Turkey found from 1000–1300 m; fl. Jul.–Sep.
DISTRIB. Mountain region of Kurdistan, a single collection without precise locality, 1840, *J Brant & W.H.F. Strangeways* s.n.

This species is doubtfully recorded from Iraq, and may not have been collected there.

Greece, Rumania, Bulgaria, S Russia, Armenia, Georgia, Azerbaijan, Turkey, Lebanon, Syria, NW Iran.

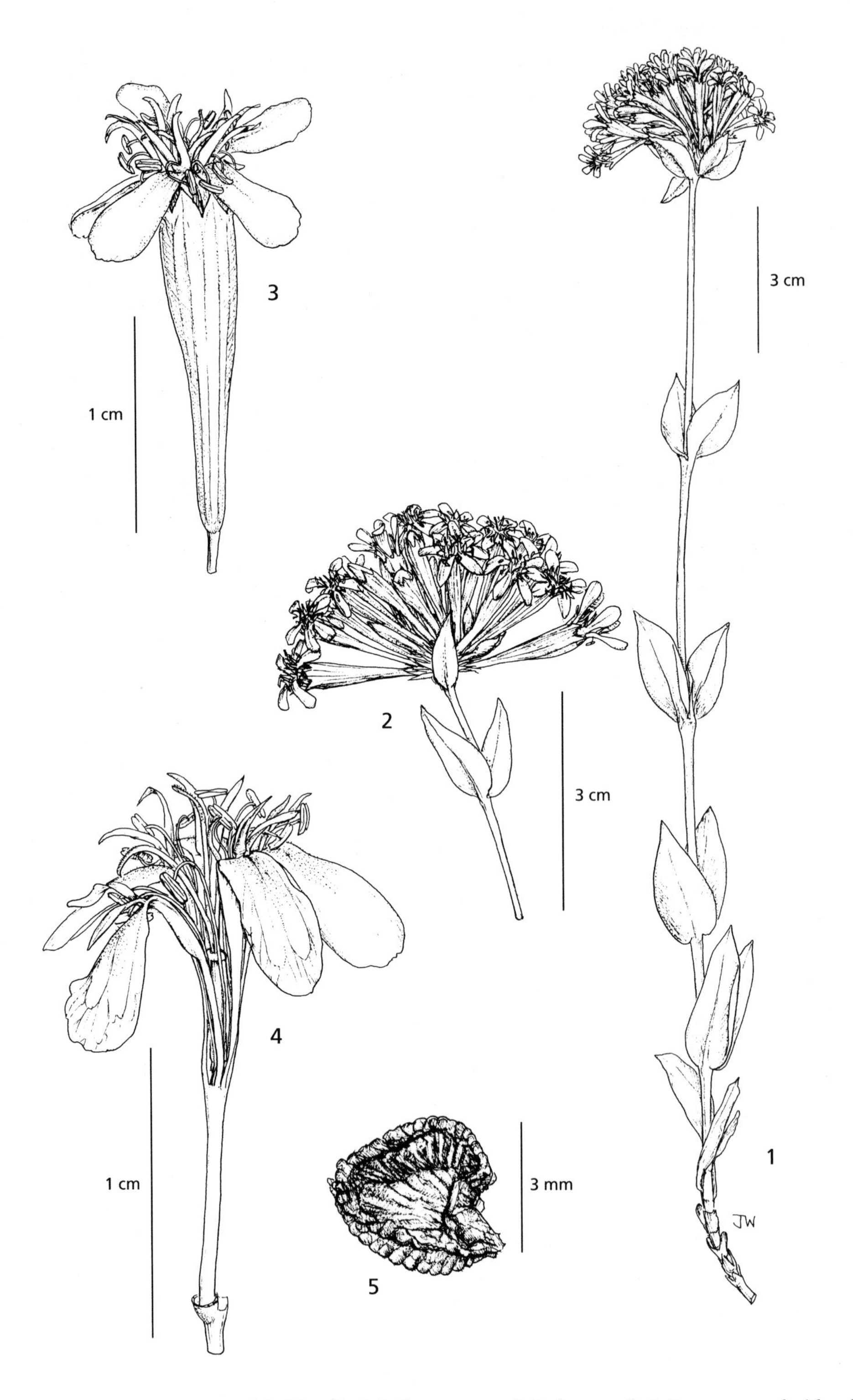

Fig. 45. **Silene compacta**. 1, habit × $^2/_3$; 2, inflorescence × 1; 3, flower × 3; 4, flower opened with calyx removed × 4; 5, seed × 60. 1 from *Brant & Strangeways*, 1840 s.n.; 2–4 from *Hagemann et al.* 1914; 5 from *Haradjian* 389. © Drawn by Juliet Beentje.

SECT. LASIOCALYCINAE Boiss.

24. **Silene apetala** *Willd.*, Sp. Pl. ed. 4, 2: 703 (1799); Boiss., Fl. Orient. 1: 596 (1867); Rohrb., Monogr. Silene: 118 (1868); Schischk. in Fl. USSR 6: 614 (1936); Zohary, Fl. Palaest. 1: 98 (1966); Coode & Cullen in Fl. Turk. 2: 239 (1967); Ghazanfar & Nasir in Fl. Pak. 175: 84 (1986); Melzh. in Fl. Iranica 163: 493 (1988); Chamberlain in Fl. Arab. Penins. & Socotra 1: 210 (1996); Ghazanfar, Fl. Oman 1: 76 (2003).

Annual, 2–25 cm tall, branched above, seldom below, appressed-puberulent. Basal leaves forming a lax rosette, soon withering; lamina ± spatulate to linear-lanceolate, acute, ciliolate. Inflorescence a reduced dichasium, sometimes 1–flowered. Pedicels usually longer than calyx. Bracts narrowly lanceolate. Calyx 6.5–9.5(–11.5) mm, campanulate-clavate, ± striate, not umbilicate; teeth 2–3 mm, triangular-acuminate, with membranous margins, laxly ciliate; anastomoses very weak, nerves pubescent. Petals white or pinkish, short, not or rarely exserted from calyx, bipartite to ± ½, sometimes petals absent. Anthophore 1–4 mm, glabrous. Capsule 3–7.5 mm, globose to ovoid, slightly exserted from calyx. Seeds flat, deeply grooved on back, auriform with undulate margins. Fig. 46: 1–3.

HAB. Cultivated fields, aluvium hummocks, depressions under shade of overhanging rocks, river banks, wadi bottoms, disturbed and waste places, in gravelly and sandy soils; alt. 30–300 m; fl. Feb.–Apr.
DISTRIB. In the foothills and the western and southern deserts. **FKI**: Abu Graib, 24 km E of Ana Ramadi Liwa, *Barkley, Juma & Ibrahim Mohammed* 613!; **FPF**: Jabal al-Muwaila, E of J. Hamrin, *Guest, Rawi & Rechinger* 17547!; **DSD**: 5 km SE of Zubair, *Guest, Rawi & Rechinger* 16816!; 10 km E of Salman, *Guest, Rawi & Rechinger* 18806!; nr. Ansab, ± 145 km E of Salman, *Guest, Rawi & Rechinger* 18970!; Zubair, *Haines* W822!; Khanaqin, *Guest* 1868!; nr. Zubair, *Gillet & Rawi* 6061!; bottom of Hawran valley, *Chakravarty, Rawi & Alizzi* 31614!; **DWD**: 20 km E of Anato Al Qaim, *Omar, Kaisi & Hamid* 44450!

Mediterranean area, Turkey, Syria, Jordan, Palestine, Saudi Arabia, Iran, Afghanistan, Pakistan.

25. **Silene dichotoma** *Ehrh.*, Beitr. Naturk. 7: 144 (1792); Boiss., Fl. Orient. 1: 588 (1867); Rohrb., Monogr. Silene: 94 (1868); Schischk. in Fl. USSR 6: 610 (1936); Zohary, Fl. Palaest. 1: 92 (1966); Coode & Cullen in Fl. Turk. 2: 234 (1966); Melzh. in Fl. Iranica163: 490 (1988).

subsp. **racemosa** (*Otth*) *Graebn. & Graebn. f.* in Aschers. & Graebn., Syn. Mitteleur. Fl. 5(2): 93 (1920).
S. trinervis Banks & Sol. in Russell, Nat. Hist. Aleppo ed. 2, 2: 252 (1794).
S. racemosa Otth in DC, Prodr. 1: 384 (1824).
S sibthorpiana Reichenb., Fl. Germ. Excurs.: 815 (1832);
S. dichotoma subsp. *sibthorpiana* (Reichenb.) Rech.f. in Denkschr. Akad. Wiss. Wien, Math.-Nat. Kl. 105 (1): 166 (1943).

Annual, or rarely biennial, erect or ascending, 13–45(–70) cm tall, ± branched, puberulent to crisp-pubescent. Basal leaves very variable in shape, subspatulate to lanceolate with petiole, soon withering; cauline leaves in 3–7 pairs, ovate-lanceolate or oblong-lanceolate or lanceolate, 30–50 × 12–25 mm, acute, 3– or seldom 5–nerved, with short petiole, or subsessile, leaves of auxillary fascicles narrower, usually crisp-pubescent beneath, crispate hairs ± scattered above. Inflorescence a dichasium below passing into monochasia above, flowers sometimes subsessile or pedicels 1–5 mm, usually spreading at anthesis, ascending in fruit. Bracts linear-lanceolate, to 6 mm, scarious throughout or only along margins, 3–nerved. Calyx ovoid-cylindrical, ovoid-oblong in fruit, 8–13 mm, umbilicate, nerves slightly winged, green, covered in multicellular hairs, teeth ± lanceolate, acute, 1–2 mm, green. Petals white or rarely reddish, limb 4–8 mm, bilobed to ± ½. Anthophore 1–2.5 mm, glabrous. Capsule ovoid-oblong, 6–10 × 4–5.5 mm, included in calyx. Seeds reniform.

Subsp. *dichotoma* (not recorded from Iraq) has a somewhat longer calyx and the indumentum of leaves and calyx is somewhat denser than in subsp. *racemosa*. These features are not correlated with geographical separation, as can be seen for example in the distribution map in Coode & Cullen in Fl. Turk. where both subspecies are sympatric.

HAB. Fields, steppe, banks, forests; alt. ± 1200 m; fl. Apr.–Jun.
DISTRIB. **MRO**. Shaqlawa, *Haines* W605!

Turkey, Cyprus, Syria, Palestine, Iran.

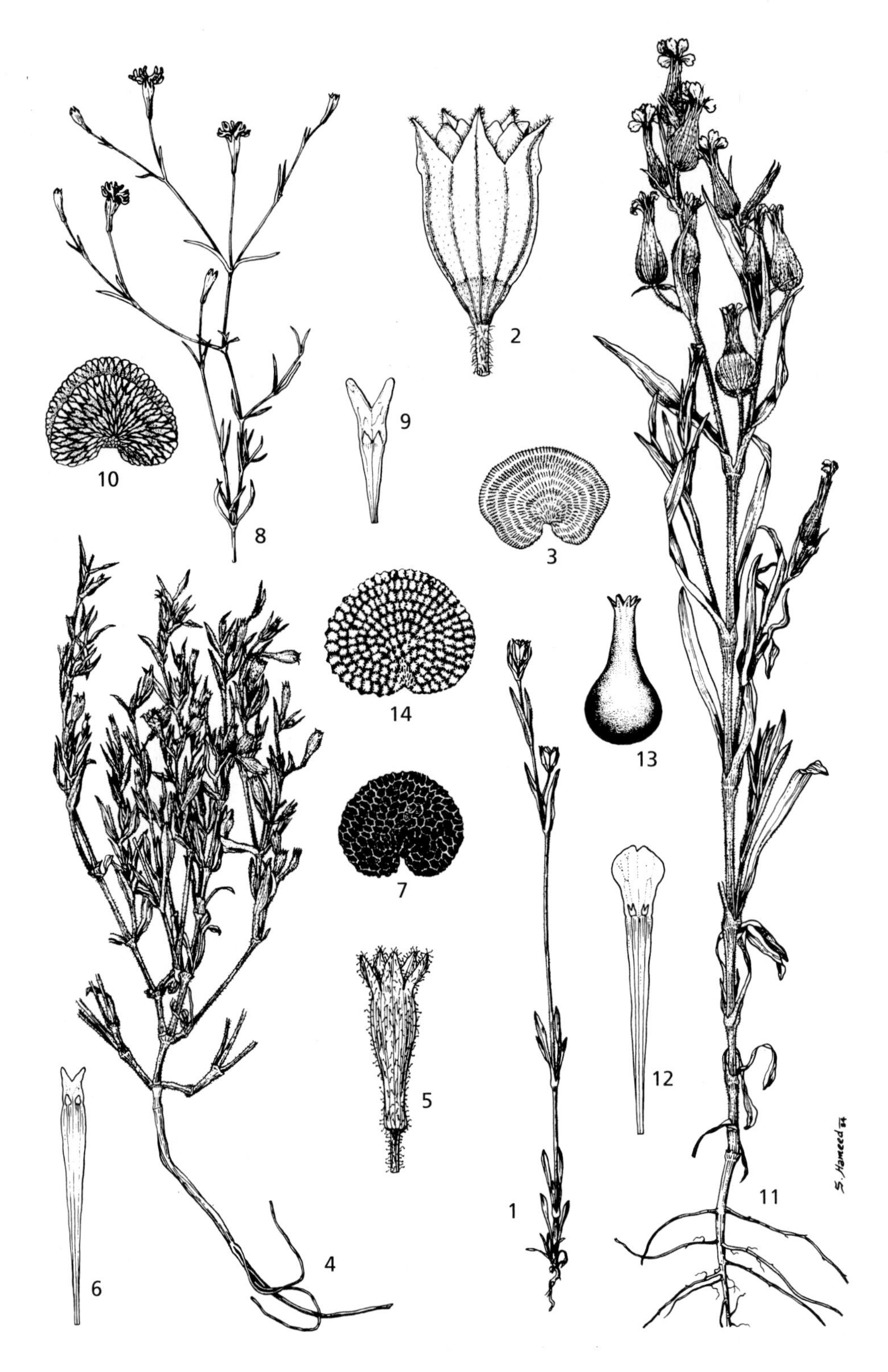

Fig. 46. **Silene apetala**. 1, habit × ½; 2, calyx and capsule × 4; 3, seed × 20. **Silene villosa**. 4, habit × ½; 5, calyx × 2; 6, petal × 3; 7, seed × 20. **Silene arenosa**. 8, habit × ½; 9, petal × 2; 10, seed × 20. **Silene conoidea**. 11, habit × ½; 12, petal × 1½; 13, capsule × 1½; 14, seed × 20. Reproduced with permission from Flora of Pakistan 175: f. 13, 1986. Drawn by S Hameed.

26. **Silene lagenocalyx** *Fenzl ex Boiss.*, Fl. Orient. 1: 587 (1867); Rohrb., Monogr. Silene: 93 (1868); Melzh. in Fl. Iranica 163: 494 (1988).

Annual or rarely biennial, ascending or erect, 10–25(–40) cm tall, often branched at base, loosely crisp-pubescent. Basal leaves spatulate to obovate or lanceolate, with petiole of variable length, soon withering; cauline leaves with 4–6 pairs, ovate-lanceolate or oblong-lanceolate or lanceolate, 20–45 × (3–)6–13 mm, acute, 1–3–nerved, scabrous to pubescent; leaves of axillary fascicles narrower. Inflorescence a dichasium below, passing into monochasia above, flowers usually with a short pedicel 1–4 mm, sometimes subsessile, ± villous. Bracts linear-lanceolate, to 17 mm, 1–nerved. Calyx cylindrical, clavate in fruit, 15–19 m, umbilicate, only green on nerves and teeth, nerves ± densely covered with multicellular hairs, teeth lanceolate, acute, ± 2 mm, margin membranous, weakly ciliate. Petals white or pinkish, limb 5–8 mm, bilobed to ± ½. Anthophore 4–5.5 mm, glabrous. Capsule ± ovoid, 6–10 × 4–5 mm, included in calyx. Seeds reniform.

HAB. Fields, steppe, banks, forests; alt. 600 m; fl. Apr.–Jun.
DISTRIB. Very rare, recorded only once from Iraq. **MSU**. Darbandikhan, 21 Apr. 1961, *Haines* s.n.

Iran.

SECT. ATOCION Otth

27. **Silene aegyptiaca** (*L.*) *L. f.*, Suppl.: 241 (1782); Boiss., Fl. Orient. 1: 600 (1867); Rohrb., Monogr. Silene: 156 (1868); Zohary, Fl. Palaest. 1: 89 (1966); Coode & Cullen in Fl. Turk. 2: 229 (1967); Melzh. in Fl. Iranica 163: 499 (1988).

Cucubalus aegyptiacus L., Sp. Pl.: 415 (1753).
Silene atocion Jacq., Hort. Bot. Vindob. 2: 19 (1777).
S. orchidea L.f. , Suppl.: 241 (1782).
S. atocion Jacq. ß *umbrosa*. — Type: Kurdistan, in sched.

Annual, 10–30(–40) cm tall, usually much-branched from base, crisply pubescent below, glandular-viscid above. Leaves variable in size, basal spatulate to oblong-spatulate , petiolate, acute or obtuse; cauline leaves in 4–6 pairs, oblong-ovate to oblong-elliptic, 15–35 × 5–9 mm, subsessile to sessile, subacute, subglabrous to puberulent. Inflorescence a ± congested dichasium with erect flowers. Bracts distinctly membranous, broadly ovate, 2–4 × 1.5–2.7 mm. Pedicels seldom longer than 6 mm. Calyx ± cylindrical, somewhat inflated, in fruit ± clavate, (12–)14–18(–19.5) mm, umbilicate, frequently purplish, membranous except for green nerves with glandular hairs and with fine multicellular eglandular hairs between nerves; teeth triangular, obtuse, 2–2.5 mm, pubescent at apex. Petals pink, limb 5–7 mm, retuse to emarginate, on each side with a lateral acute tooth. Anthophore 6–8(–9) mm, glabrous. Capsule ovoid to roundish, 5–8.5 × 3–4 mm, included in calyx. Seeds ± roundish. Fig. 47:

HAB. Moist cultivated ground, waste places, roadsides; alt. 1200–1400 m; fl. Sep.
DISTRIB. **MRO**: Kuh-e-Sefin, Schaklawa [Shaqlawa], near Arbil, *Bornmüller* 975; *O.Polunin* 5025; Baradost, *Thesiger* 809.

Turkey, Palestine, Jordan, Lebanon, Iran.

28. **Silene atocioides** *Boiss.*, Diagn. Pl. Or. Nov. ser. 1, 5: 83 (1844); Boiss., Fl. Orient. 1: 600 (1867); Rohrb., Monogr. Silene: 156 (1868); Melzh. in Fl. Iranica 163: 499 (1988).

Annual; stems 15–25 cm, ± branched from base, greenish or purplish in lower part, crisp-pubescent below, glandular above. Basal leaves oblong-spatulate to orbicular, petiolate; cauline leaves in 4–6 pairs, 15–32 × 5–9 mm, oblong-ovate to elliptic, subacute; leaves pubescent, 1-nerved with a crisp-pubescent petiole. Inflorescence a ± dense dichasium, many-flowered. Bracts herbaceous, 4–8(–12) mm, linear-lanceolate. Calyx ± cylindrical, becoming inflated after anthesis to ± clavate, 14–18 mm, umbilicate, often suffused with purple; teeth 2–2.5 mm, obtuse, pubescent at tip, margins ciliate. Petals pink; claw 5–7 mm, emarginate; limb 6–7 mm. Filaments glabrous. Anthophore 6–9 mm, glabrous. Capsule 5–8 × 3–4 mm, ovoid to subglobose, included in calyx. Seeds ± orbicular.

HAB. Moist cultivated ground, waste places, roadsides, amongst rocks; alt. 1300–1400 m; fl. Jan.–May.
DISTRIB. **MAM**: Amadia, *Guest* 1272 (BAG); Zakho pass, *Guest* 2251 (BAG); Zawita gorge, *Enberger et al.*,

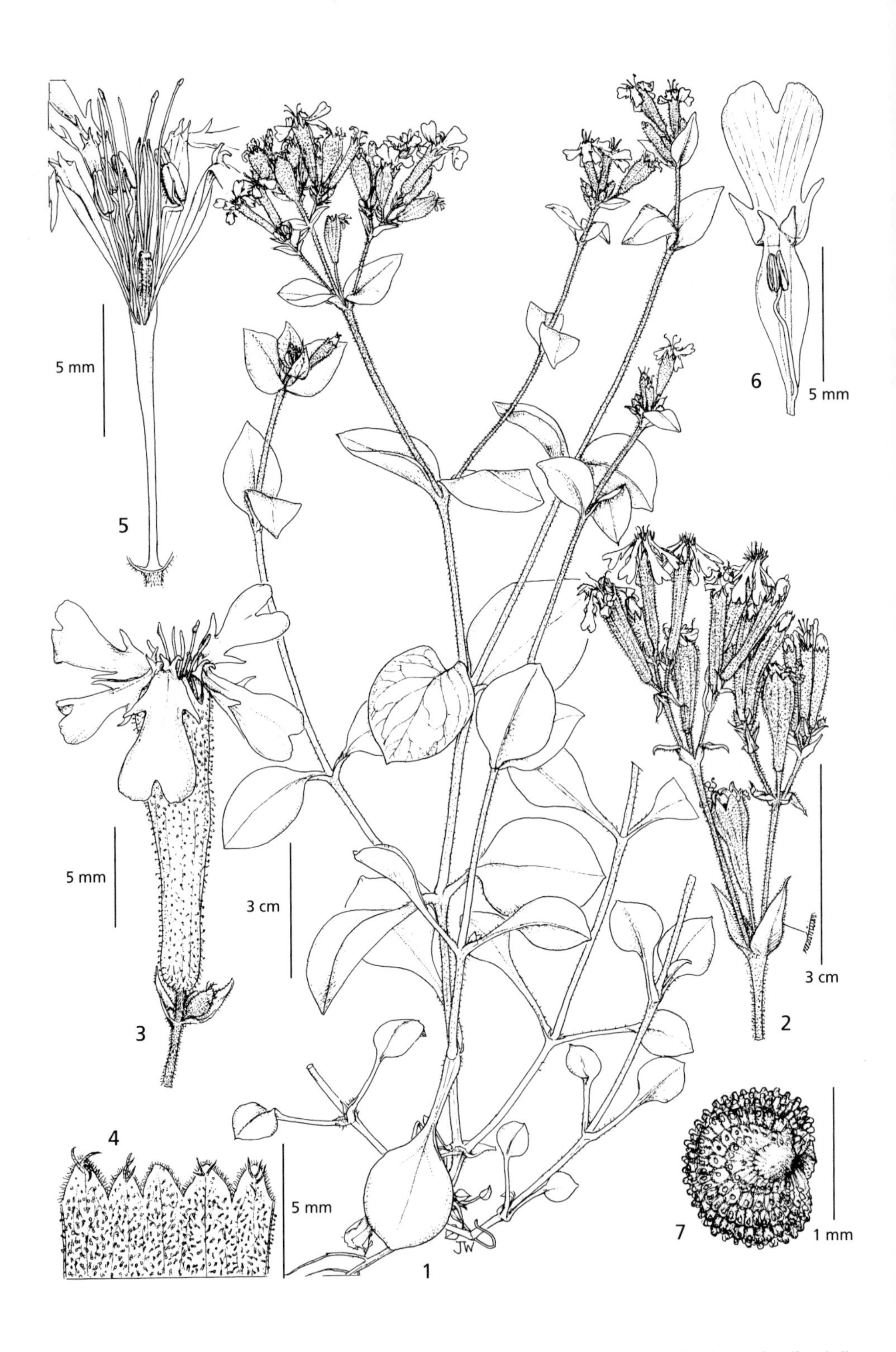

Fig. 47. **Silene aegyptiaca**. 1, habit × ⅔; 2, inflorescence × 1; 3, flower × 3; 4, calyx lobes detail × 4; 5, flower with calyx removed × 4; 6, petal × 4; 7, seed × 20. 1–6 from *ARI* 2098, 7 from *Davis* 2732. © Drawn by Juliet Beentje.

15361 (BAG). **MRO**. Arbil, *Agnew et al.* 5997!; Algurd, *Hadac* 2190; Sersang, *Rechinger* 11912!; 80 km SW Haj Omran, *Shahwani* 25274 (BAG).

Turkey, Cyprus, Lebanon, Syria, C Iran.

29. **Silene diversifolia** Otth, Prodr. 1: 378 (1824); Brummitt, Taxon 42: 487 (1993).

Silene rubella L., Sp. Pl.: 419 (1753); Boiss., Fl. Orient. 1: 598 (1867); Rohrb., Monogr. Silene: 155 (1868); Zohary, Fl. Palaest. 1: 90 (1966); Melzh. in Fl. Iranica163: 498 (1988); *S babylonica* Boiss., Diagn. Pl. Or. Nov. ser. 2, 1: 71 (1854); *S rubella* var. *laxa* Boiss., Fl. Orient. 1: 598 (1867).

Annual, erect, branched or not, 8–30(–40) cm tall, glaucescent, subglabrous or sparsely pubescent below, more densely pubescent above. Basal leaves spatulate to oblong-spatulate , soon withering; cauline leaves in 4–7 pairs, oblong-spatulate or ovate-oblong or lanceolate, sessile and shortly contracted at base, 10–50 × 4–15 mm, obtuse, somewhat undulate. Inflorescence a many-flowered dichasium, seldom 1-flowered; branches sometimes ± congested. Pedicels of variable length. Bracts lanceolate, to 10 mm, shortly pubescent, subacute. Calyx ± turbinate, sub-clavate in fruit, 9–12 mm, umbilicate, weakly striate, whitish or purplish-tinged, papery, very finely puberulent; teeth short, ± semi-circular, not more than 1.5 mm. Petals pink, limb to 3 mm, emarginate. Anthophore 3–5 mm, pubescent. Capsule ovoid, 6–8 × 3–3.5 mm, not or slightly exserted from calyx. Seeds reniform.

HAB. Fallow and irrigated fields; alt. 30–150 m; fl. Feb.–May.
DISTRIB. **LCA**. Baghdad, *Wheeler-Haines* 128!

Mediterranean area except Balkans; N Africa (Algeria, Egypt, Libya, Morocca, Tunisia); Palestine, Jordan, Lebanon, Syria, NW and N Iran.

SECT. BIPARTITAE Boiss.

30. **Silene arabica** *Boiss., Fl. Orient.* 1: 593 (1867); Zohary, Fl. Palaest. 1: 94 (1966); Melzh. in Fl. Iranica 163: 495 (1988); Chamberlain in Fl. Arab. Penins. & Socotra 1: 216 (1996).

S affines (non Godr.) Boiss., Diagn. Pl. Or. Nov. ser. 2, 1: 72 (1854); *S. arabica* var. *moabitica* Zoh., Fl. Palaest. 1: 342 (1966).

Annual, 6–15(–20) cm tall, much branched from base, glandular- and eglandular-pubescent below, glandular-pubescent above. Basal leaves few, lanceolate, soon withering; cauline leaves in 4–6 pairs, linear-lanceolate or narrowly long-ovate, not tapered at base, 10–28 × 1.5–3(–5) mm, acute or obtuse, 1-nerved, at base often 3- or sometimes even 5-nerved, indumentum as for stems. Inflorescence a short monochasium, 1– to many-flowered. Pedicels shorter than calyx, spreading in fruit and capsule erect. Bracts oblong-linear, to 10 mm. Calyx weakly campanulate, 9–12 mm, umbilicate, glandular-pubescent as are the bracts, striate, but often only in lower half; teeth ligulate, 2–2.5 mm. Petals white to reddish, bifid to ¾, claw after flowering exserted from calyx by 2–3 mm. Anthophore 3–6.5 mm, shortly puberulent. Capsule oblong, 5.5–9 × 3–3.5 mm, not or slightly exserted from calyx. Seeds ± reniform.

HAB. Sandy deserts, waste places, fallow fields; alt. 20–50 m; fl. Mar.–May.
DISTRIB. **DSD**. Ethell, nr. Zubair, 18 Mar. 1957, *Haines* s.n.!; Basra, Jabal Sanam, *Rechinger* 8552!; **LCA**. Faluja desert, 14 May 1955, *Haines* s.n.!; Baghdad, *Rechinger* 8579 (E!); c. 50 km from Baghdad, 3 May 1961, *Agnew, Hadač & Haines* s.n.!

Sometimes misidentified for *S colorata* Poir. subsp. *oliveriana* (Otth) Rohrb. or *S villosa* Forssk., but can be distinguished by the pink flowers and winged seeds in *S colorata* subsp. *oliveriana*, and by the longer and glabrous anthophore in *S villosa*.

Palestine, Sinai, Arabia, Iran, ?Afghanistan, ?Pakistan.

31. **Silene colorata** *Poiret*, Voy. Barbarie 2: 163 (1798); Rohrb., Monogr. Silene: 114 (1868); Zohary, Fl. Palaest. 1: 97 (1966); Coode & Cullen in Fl. Turk. 2: 239 (1967); Meikle, Fl. Cyprus 1: 247 (1977); Greuter, Med-Checklist 1: 253 (1984); Chamberlain in Fl. Arab. Penins. & Socotra 1: 217 (1996).

subsp. **oliveriana** (*Otth*) *Rohrb.*, Monogr. Silene: 116 (1868).

S. oliveriana Otth in DC, Prodr. 1: 373 (1824); *S. colorata* Poir. var. *oliveriana* (Otth) Muschler, Man. Fl. Egypt 1: 338 (1912).

Ascending or erect annual, 10–30 cm tall, branched or unbranched, appressed-pubescent, eglandular. Basal leaves narrowly obovate, soon withering; lower cauline leaves lanceolate or sub-spatulate, 2–5 × 0.3–1.2 cm, pubescent, petiole margin with crisp hairs; upper cauline leaves similar but narrower, often lanceolate to linear-lanceolate, ± sessile, with indumentum becoming sparser upwards. Inflorescence a raceme-like monochasium. Bracts narrow, lanceolate or linear, foliaceous. Pedicels slender, spreading at anthesis, erect in fruit. Calyx cylindrical, clavate in fruit, 11–15(–16) mm, umbilicate, papery, only its nerves and teeth green or sometimes reddish green, nerves shortly pubescent, teeth ligulate, 2–3 mm with broad membranous ciliolate margin. Petals pink, limb 6–9 mm, divided to ½. Anthophore 5.5–8 mm, minutely pubescent. Capsule broadly ovoid, 6–9 × 4.5–5.5 mm, not or only slightly exserted from calyx. Seeds broadly reniform with two conspicuous flat or undulate wings.

HAB. Sandy or stony field, steppes and deserts; alt. ± 200 m; fl. Feb.–Apr.
DISTRIB. **DWD**: about 10 km E of Salman, *Guest, Rawi & Rechinger* 18777 (BAG); 20 km S of Nukhaib, *Alizzi & Omar* 35525 (BAG). **DSD**. 30 km W by N Busaiya, *Guest, Rawi & Long* 14170 (BAG); southern desert, Diwaniya, *Rechinger* 9289 (E!).

?Egypt, Turkey, Palestine, Jordan, Syria, Iran; Mediterranean area.

32. **Silene villosa** *Forssk.*, Fl. Aegypt.-Arab.: 88 (1775); Boiss., Fl. Orient. 1: 592 (1867); Rohrb., Monogr. Silene: 110 (1868); Zohary, Fl. Palaest. 1: 93 (1966); Ghazanfar & Y. Nasir in Fl. Pak. 175: 83 (1986); Leonard, Contrib. Fl. et Veg. Deserts d'Iran 6: 17 (1986); Melzh. in Fl. Iranica 163: 496 (1988); Chamberlain in Fl. Arab. Penin. & Socotra 1: 217 (1996).

Annual, 8–25 cm tall, ± branched at base, viscid-glandular on stems and leaves. Basal leaves few, lanceolate, soon withering; cauline leaves with 3–6 pairs, broadly linear, lanceolate to elongate-ovate, 15–45 × 2–8(–10.5) mm, obtuse, 1-nerved at base or sometimes 3-nerved. Inflorescence a short, ± regular dichasium, few- to many-flowered. Pedicels shorter than calyx, reflexed in fruit. Bracts pale green, oblong-linear or lanceolate, up to 10 mm. Calyx subcylindrical at anthesis, ± clavate in fruit, 16–23 mm, umbilicate, weakly striate, lower part soon turning yellow, upper part pale green, glandular-pubescent; teeth ligulate, 1.7–2.6 mm. Petals white to reddish; limb 6–10 mm, bilobed to ¾, exserted from calyx by up to 6 mm after flowering. Anthophore 9–12(–13.5) mm, glabrous. Capsule oblong-cylindrical, 7–10(–11.5) × 3–4 mm, not exserted from calyx. Seeds ± reniform. Fig. 46: 4–7.

HAB. Sandy deserts, on loose sand and low sand dunes, sandy waste places; alt. 50–150 m; fl. Mar.–May.
DISTRIB. Found in the southern desert of Iraq. **DSD**: Al Zarqa, 12 km SE of Samara, *Alkas* 17672; south desert, 35 km E by Busaiya, *Guest & Rawi* 14186; J. Samara, 23 Aug. 1919, *Watson & Sharples* s.n.; nr. Chilwa, 110 km SW of Basra, *Guest, Rawi & Rechinger* 17208; J. Sanam, 30 km S of Zubair, *Guest, Rawi & Rechinger* 16946; southern desert, Al Urmaigh, ± 30 km S of Tall Lahn, *Guest& Ibrahim Mahmoud* 15305.

Egypt, Sinai, Arabian Peninsula, Palestine, Jordan, Iran, Pakistan.

SECT. RIGIDULAE Boiss.

33. **Silene arenosa** *C Koch*, Linnaea 15: 711 (1841); Boiss., Fl. Orient. 1: 603 (1867); Rohrb., Monogr. Silene: 161 (1868); Schischk. Fl. USSR 6: 682 (1936); Ghazanfar & Y. Nasir in Fl. Pak. 175: 82 (1986); Melzh. in Fl. Iranica 163: 503 (1988); Chamberlain in Fl. Arab. Penin. & Socotra 1: 220 (1996).

S. leyseroides Boiss., Diagn. Pl. Or. Nov. ser. 1, 1: 41 (1843).
S. salsa Boiss, Diagn. Pl. Or. Nov. ser. 1, 8: 77 (1849).

Annual, 5–15(–20) cm tall, ascending or erect, ± strongly branched at base, subglabrous below, glabrous above with short viscid sections at upper internodes. Basal leaves linear-subulate, soon withering; cauline leaves in 3–5 pairs, linear-lanceolate, 10–25(–35) × 0.5–2.5 mm, rostrate, slightly canaliculate, ± scabridulous-puberulent. Inflorescence ± richly dichotomously branched, regularly many-flowered, with pedicels of various lengths. Bracts triangular, 2–4 mm, with narrow hyaline ciliate margin. Calyx oblong-clavate, 9–15 mm, umbilicate, striate, green only on nerves and teeth, anastomoses inconspicuous but present, with some solitary hairs or glandular hairs, sometimes pubescent, teeth acuminate, 2–3 mm, margin membranous, villous. Petals white to pinkish; limb 4.5–6 mm, bilobed to ⅔.

Stamens ± distinctly exserted from calyx. Anthophore 4–7 mm, slightly puberulent. Capsule oblong, 5–9 × 1.7–2.5 mm, included in calyx. Seeds reniform. Fig. 46: 8–10.

Hab. Open places in sandy deserts, sandy seashores and sandy areas of farm land; alt. 40–780 m; fl. Apr.–Jun.
Distrib. Found in the alluvial plains of the central plains and in the southern desert. **msu**: 10 km from Sulimaniya to Saiyid Sadiq, *Omar, Kaisi & Wedad* 49418!; ?Mesopotamia, 1919, *Capt. Watson* s.n.!; **lea**: Fakka, 70 km E of Amara, *Rawi* 25815!; **lca**: 70 km NW of Fallujah, *Chakraverty & Rawi* 30363!; Baghdad, Faluja desert, *Haines* W164!; **dsd**: S of Zubair, *Rawi* 25863!

Sometimes misidentified as *S. linearis* DC. and *S. chaetodonta* Boiss. but distinct from both in its spreading habit and generally smaller flowers; there are also records of natural hybridisation between *S. chaetodonta* and *S. arenosa.*

Kuwait, Jordan, Iran, Afghanistan, Pakistan (Baluchistan).

34. **Silene chaetodonta** *Boiss.*, Diagn. Pl. Or. Nov. ser. 1, 1: 39 (1843); Boiss., Fl. Orient. 1: 605 (1867); Rohrb., Monogr. Silene: 162 (1868); Schischk. in Fl. USSR 6: 682 (1936); Coode & Cullen in Fl. Turk. 2: 228 (1967); Melzh. in Fl. Iranica163: 501 (1988).

S debilis Stapf, Denkschr. Akad. Wiss. Wien, Math.-Nat. Kl. 51: 282 (1896).

Annual, erect, 10–40(–50) cm tall, branched at middle, shortly pubescent, upper internodes wtih viscid sections. Basal leaves linear-lanceolate, soon withering; cauline leaves with 5–7 pairs, 15–35(–48) × 0.5–5 mm, sometimes with axillary fascicles of narrower and shorter leaves; all leaves pubescent, upper also with some glandular hairs. Inflorescence a ± regular dichasium with rigid branches, few- many-flowered. Pedicels to 50 mm. Bracts linear, 8–12 mm, ciliate. Calyx cylindrical-clavate, 11–17 mm, umbilicate, striate, anastomoses indistinct, nerves prominent at flowering time, with very short glandular hairs and in region of teeth only englandular-pubescent; teeth acuminate, 2–5.5 mm, margins membranous, ciliate to weakly villous. Petals pinkish; limb 4–6 mm, bilobed to about half. Stamens little or not exserted from calyx. Anthophore 3–5 mm, shortly pubescent. Capsule oblong, 7–10.5 × 3.5 mm, included in calyx. Seeds semi-circular to reniform.

Hab. Steep stony slopes, mountain slopes; alt. ± 1000 m; fl. May–Jul.
Distrib. ?Kurdistan, Mardin, Kasimi, *P. Sintensis* 1128 (E!).

The long calyx teeth and short petals, after flowering even shorter than calyx teeth, are good characteristics for this species. Natural hybrids are possible between *S. chaetodonta* and *S. arenosa*; they have the habit and calyx teeth of *S. chaetodonta* and the exserted stamens of *S. arenosa.* Two specimens from **dwd**: Al Niser, Rutba, *Alizzi & Omar* 36192! and **dwd**: 230 km NW of Ramadi (Ramadi-Rutba highway), *Rawi* 20933! have calyces about 10 mm, smaller than typical material. Further investigations are needed to establish whether this is a distinct population.

Cyprus, Palestine, Lebanon, Syria, Iran, Turkmenistan, Afghanistan, Pakistan.

35. **Silene linearis** *Decne.*, Ann. Sci. Nat. (Bot.) ser. 2, 3: 276 (1835); Boiss., Fl. Orient. 1: 602 (1867); Rohrb., Monogr. Silene: 162 (1868); Post, Fl. Syr. Palest. & Sinai 1: 180 (1932); Zohary, Fl. Palaest. 1: 88 (1966); Chamberlain in Fl. Arab. Penins. & Socotra 1: 221 (1996).

Annual, 20–50 cm tall, erect with stem branching from above, puberulent in lower part, glabrous and with viscid sections of internodes above. Basal leaves soon withering; cauline leaves with 5–8 pairs, lanceolate to linear-subulate, 7–35 × 0.5–2 mm, scabridulous-pubescent. Inflorescence dichotomously branched. Pedicels longer or shorter than calyx. Bracts triangular, 2–4 mm, with narrow hyaline ciliate margins. Calyx oblong-clavate, 10–17 mm, umbilicate, striate, nerves and teeth green, anastomoses weak, nerves with some glandular, short hairs; teeth acuminate, 2–4(–5) mm, margin membranous, villous. Petals usually pink or pinkish; limb 5–8 mm, bilobed to up to half. Anthophore 4–8 mm, puberulent to pubescent. Capsule oblong, 5–11 × 2–3.5 mm, included in calyx. Seeds reniform. Illustration

Hab. Sandy, stony and gravelly hillsides, dry silt ridges, in *Quercus* forest; alt. 150–1300 m; fl. May–Jul.
Distrib. Frequent annual in the Mountain Region: **mam**: Sersang, *Wheeler-Haines* W562!; Dohuk, *Wheeler-Haines* W 457!; Bikhair, nr. Zakho, *Rawi* 22963!; **mro**: Pishtashan, 15 km NE of Raina, lower slope of Qandil Range, *Rawi & Serhang* 23841!; **msu**: Mela Kowa, *Rawi* 22455!; Jarmo, *Helbaek* 1768!; **fuj**: 5 km NW of Mosul, hillside, *Rawi* 23132!; **fpf**: Koma Sank Police stn., nr. Mandali, *Rawi* 20623!; Chilat Police stn., *Rawi & Wedad* 25688.

S. arenosa C.Koch and *S. linearis* Decne are often confused, but they are not synonymous (contrary to Greuter in Med-Checklist 1: 264 (1984)). *S arenosa* is a species of lower altitudes than *S. linearis*, with a smaller calyx and the Iranian *S austro-iranica* Rech.f. (nor represented in Iraq) is a species of higher altitudes than either *S. arenosa* or *S. linearis.*

Eastern part of N Africa, Egypt (Sinai), Arabian Peninsula (Yemen, Oman, UAE, Saudi Arabia), Palestine, Jordan.

SECT. CONOIMORPHA Otth

36. **Silene coniflora** *Nees ex Otth* in DC, Prodr. 1: 371 (1824); Boiss., Fl. Orient. 1: 578 (1867); Rohrb., Monogr. Silene: 89 (1868); Post, Fl. Syr. Palest. & Sinai 1: 171 (1932); Schischk. in Fl. USSR 6: 691 (1936); Zohary, Fl. Palaest. 1: 99 (1966); Coode & Cullen in Fl. Turk. 2: 241 (1967); Ghazanfar & Y. Nasir in Fl. Pak. 175: 85 (1986); Melzh. in Fl. Iranica163: 486 (1988); Chamberlain in Fl. Arab. Penin. & Socotra 1: 213 (1996); Ahmad in Fedd. Repert 124: 2 (2013).

Erect annual, 3.5–18(–25) cm tall, often branched from base, densely hairy with short, fine eglandular hairs below, above also with glandular hairs. Basal leaves linear to lanceolate, 12–30 × 2.5–6 mm, acuminate; cauline leaves in few or many pairs, ± lanceolate, rarely linear, 15–45(–60) × 1–8 mm; all leaves ± densely hairy on both surfaces, margins of petiole ciliate, sometimes also margins of leaves. Inflorescence a ± regular dichasium, 2–30-flowered. Pedicels 4–15 mm. Bracts linear, calyx ovate to elongate-ovate, 9–15(–17) mm, 15–20-nerved, with many rigid glandular hairs on green nerves, hairs with a swollen base, calyx constricted at apex, with 3–6 mm subulate teeth. Petals white or pink; limb 4–6 mm, entire, emarginate or bilobed to ¼. Filaments ciliate at base. Anthophore 0.1 mm. Capsule sessile, ovoid-oblong, 8–13 × 4.5–6.5 mm, included in calyx. Seeds elliptical or reniform.

HAB. Fields, roadside, mountain slopes, amongst *Quercus* and *Acer* trees; alt. 1700–1900 m; fl. & fr. Jun.–July.
DISTRIB. **MSU**: Ballkha Mt., *Ahmad* 11-965, 11-1013, 12-1028, 12-1046 (SUFA). **FUJ**: South of Sinjar, nr. Pa'aj, *Guest* 13380 (BAG). **DWD**: 18 km S of Rutba, *Rawi* 14636 (BAG); 30 km S Shnana, *Rawi* 15723 (BAG).

SW Asia, eastwards to Pakistan.

37. **Silene conoidea** *L.*, Sp. Pl. ed. 1, 418 (1753); Boiss., Fl. Orient. 1: 580 (1867); Rohrb., Monogr. Silene: 92 (1868); Post, Fl. Syr. Palest. & Sinai 1: 172 (1932); Schischk. in Fl. USSR 6: 690 (1936); Zohary, Fl. Palaest. 1: 99 (1966); Coode & Cullen in Fl. Turk. 2: 241 (1967); Ghazanfar & Y. Nasir in Fl. Pak. 175: 84 (1986); Leonard, Contrib. Fl. et Veg. Deserts d'Iran 6: 17 (1986); Melzh. in Fl. Iranica163: 121 (1988); Chamberlain in Fl. Arab. Penin. & Socotra 1: 213 (1996); Zhou Lihua, Lidén & Oxelman in Fl. China 6: 66 (2001); Gabrielian & Fragman-Sapir, Fl. Transcauc. & adj. areas: 142 (2008).

Pleconax conoidea (L.) Sourkova, Österr. Bot. Zeitschr. 119: 579 (1972).

Erect annual, 10–45(–50) cm, branched at apex or at base, with short, dense retrorse indumentum below, densely glandular above, seldom lacking glandular hairs. Basal leaves lanceolate, acute, 15–42 × 2.5–10 mm; cauline leaves with few to many pairs, oblong or oblanceolate to lanceolate, 35–85 × 3–13(–20) mm; all leaves puberulent. Inflorescence a dichasium, 1–30-flowered. Pedicels 6–35 mm. Bracts long-attenuate from ovate base, 8–14(–22) mm, 3–5-nerved, with glandular and eglandular hairs, margins ciliate. Calyx tubular-conical, 16–32 mm, constricted at apex, with 30 nerves, nerves green wtih short glandular hairs, fruiting calyx ± strongly inflated, 10–15 mm broad at base; teeth 6–8 mm, subulate. Petals pink, reddish or lilac; limb 12–16 mm, bilobed to about ¼, or emarginate. Anthophore 0–3(–4) mm, pubescent. Capsule ovate with a long narrow neck, 12–18(–21) × 5–8(–10) mm, included in calyx. Seeds reniform to semi-orbicular. Fig. 46: 11–14.

HAB. Fields, waste and desert places, stony slopes, Quercus forest, in sand, gravel and loamy soils; alt. 300–2000 m; fl. Mar.–Aug.
DISTRIB. Common in the mountains and plains. **MAM**: 10 km to Dohuk from Mosul, *Kaisi & Wedad* 51955! **MRO**: Sanam village, Malkani, *Nuri & Kass* 27288! Haji Umran, *Rawi, Kass & Nuri* 27866! **MSU**: Kamarspa on road between Halabja & Taweela, *Rawi* 22261! Kopi Qara Dagh, M.ED. Moore 615! **MJS**: Jabal Sinjar, *Kaisi & Wedad* 52169; ibid., *Omar, Khayat & Kaisi* 52439! **FUJ**: Mosul, Guest 1492!; N of

Qaiyara, *E & F. Barkley*! **DWD/DLJ**: 17 km SE of Haditha, *Kaisi & Hamad* 48960! 17 km SE of Alba Hayst, sandy island, *Kaisi & Hamad* 53121! Abu Hayyat, Gharraf village SE of Haditha, *Al Khayat & Hamad* 51915! Khumran, 30 km W of *Tikrit, Alizzi & Hussain* 33745! **DSD**: Shaiba [?Shabicha], 23 Aug. 1919, *Watson & Staples* s.n!

SE Mediterranean, SW Asia to Pakistan.

A doubtful record of *S. muscipula* L. (Anthony in Notes Roy. Bot. Gard. Edinb. 18: 283, 1932) has not been authenticated; it could perhaps have been resulted from misidentification of *S. lagenocalyx* Fenzl ex Boiss.

26. **AGROSTEMMA** L.

Sp. Pl.: 435 (1753)

Annual herbs. Stems erect. Leaves sessile, connate basally and forming a sheath, narrowly lanceolate to linear, margin entire, exstipulate. Inflorescences solitary or lax dichasial, flowers axillary; pedicellate. Bracts leaflike. Sepals 5, united to form a cylindrical to ovoid calyx, teeth 5, calyx teeth foliaceous, exceeding the petals, calyx tube strongly 10-ribbed. Petals 5, red to purplish, free, shorter than calyx teeth. Stamens10. Styles 5, hairy. Ovary 1-locular. Fruit capsular ovoid, dehiscing by 5 ascending teeth.

A genus of two species native to the Mediterranean but naturalised in most parts of the world; one species in Iraq.

Agrostemma githago L., Sp. Pl.: 435 (1753); Cullen in Fl. Turk. 2: 244 (1967); Lu Dequan, Linden & Oxelman in Fl. China 6: 100 (2001).

Githago segetum Desf., Tabl. 159 (1805).

Annual, 25–40cm. Stem erect, with dense long white pilose eglandular hairs. Leaves narrowly lanceolate to linear, 20–120 mm × 2–8 mm, apex acute-acuminate, densely pubescent-pilose with eglandular hairs. Bracts similar to cauline leaves but smaller in size. Pedicel 2–8.5 cm. Flowers 2–(3) in lax dichasial cymes. Calyx 35–43 mm, cylindrical to ovoid; tube 14–18 mm long, inflated in fruit, with pubescent-pilose eglandular hairs outside, teeth 21–27 mm long, apex acute. Petal 25–28 mm long; limbs purple-mauve, 15–14 × ± 11 mm, apex emarginate, shorter than calyx teeth; claw white, 11–13 mm long, narrowly cuneate. Stamens and styles exserted.

HAB. Mountain slopes, under *Populus* trees, near water; alt. 1350–2200 m.; fl. Jun.–Aug.
DISTRIB. Rare in mountain-forest region of Iraq: **MRO**: Magar range, Haj Omran, 25 Aug. 1957, *Rawi* 24313 (BAG!); Haj Omran, 21 Aug. 1976, *Al-Dabbagh & Hamad* 46279 (BAG!); **MSU**: Tawela, 18 June 1957, *Rawi* 22324 (BAG!).

Europe to China.

91. AIZOACEAE

Pflanzenfam., ed. 2, 16C: 180–233, p.p. (1934)
(incl. Ficoidaceae)

C. Jeffrey[9]

Mostly herbs, often succulent. Leaves simple, alternate, opposite or whorled, sometimes with membranous stipules. Inflorescence cymose, axillary or terminal. Flowers regular, usually hermaphrodite. Calyx of usually 5 united sepals. Petals absent. Staminodes sometimes present, then often petaloid. Stamens 5-many, epigynous or episepalous. Ovary superior or inferior, of 2–5-many united carpels, or of 1 carpel; loculi as many as carpels; ovules 1-many per loculus, basal, axile, apical or sometimes parietal. Fruits usually capsular, loculicidal or circumscissile; seeds subreniform, embryo usually curved.

A family related to the Portulacaceae; mainly tropical, with its greatest development and diversity in South Africa.

Klak, C., Khunou, A., Reeves, G. & Hedderson, T. (2003). A phylogenetic hypothesis for the Aizoaceae (Caryophyllales) based on four plastid DNA regions. Amer. J. Bot. 90(10) 1433.

1. Petaloid staminodes absent; stamens 7–15 in groups of 1–3 alternating with the calyx lobes; ovary superior. . . . . 1. *Aizoanthemum*
   Petaloid staminodes many; stamens numerous, not in fascicles; ovary inferior or semi-inferior. 2
2. Flowers sessile or subsessile, inconspicuous; ovary semi-inferior . . . . 2. *Mesembryanthemum*
   Flowers long-stalked, conspicuous; ovary inferior. . . . . 3. *Dorotheanthus*

1. AIZOANTHEMUM

Dinter ex Friedrich in Mitt. Bott. Staatssaml. München 2: 343 (1957)

Herbs or sometimes subshrubs, often succulent; leaves alternate or opposite, exstipulate. Flowers hermaphrodite, solitary or clustered, often appearing in forkings of branches; sepals 4–5, united below; staminodes absent; stamens many, inserted on calyx tube, often in small fascicles alternating with sepals. Ovary superior, of 4–10 united carpels, 4–10-locular; ovules 2-many per loculus, axillary; fruit a loculicidal capsule, valves as many as or twice as many as carpels.

Native to Angola and Namibia with about 5 species; one species in Iraq.

1. **Aizoanthemum hispanicum** (*L.*) *H.E.K.Hartmann*, Illustr. Handb. Succ. Pl.: Aizoaceae A-E (ed. E.K.Hartmann) 29 (2002 publ. 2001).

Aizoon hispanicum L., Sp. Pl. 1: 488 (1753); DC., Hist. Pl. Grasses, 30 (1800); Boiss., Fl. Orient. 2: 765 (1872); Zohary, Fl. Pal. ed. 2, 1: 327 (1932); Blakelock in Kew Bull. 3: 431 (1948); Zohary in Dep. Agr. Iraq Bull. 31: 50 (1950); Rech.f., Fl. Lowland Iraq: 217 (1964); Zohary, Fl. Palaest. 1: 74 (1966); Miller in Fl. Arab. Penins. & Socotra 1: 166 (1996); Boulos, Fl. Egypt 1: 46 (1999).

Small spreading annual succulent herb; branches 5–15 cm or more, ascending to erect, branched dichotomously. Leaves opposite, subopposite or alternate, sessile, oblong-lanceolate, obtuse, 1-nerved, subsucculent, glabrous, minutely papillate, 13–60 × 2–12 mm. Flowers sessile or shortly stalked in axils of branches, solitary; calyx lobes 5, triangular-lanceolate, equal, green outside, white inside 6–7 mm in flower, somewhat longer in fruit; staminodes absent; stamens 7–12 in fascicles of 1–3 alternating with calyx lobes; ovary 4–5-locular, 4–5-angled.Capsule 4–5-valved, valves each with a central sharp keel on inner face, spreading horizontally when moist and exposing the seeds in a central mass; seeds many, subreniform, concentrically ridged. Fig. 48: 1–6.

HAB. Sandy or gravelly desert, slightly saline canal banks, also in waste places, abandoned fields, ancient dry mud ruins; alt. up to 550 m; fl. & fr. Feb.–Apr.

---

[9] Komarov Botanical Institute, St Petersburg, Russia

Fig. 48. **Aizoanthemum hispanicum**. 1, habit × ⅔; 2, flower showing calyx × 2; 3, flower opened × 2; 4, Capsule with spreading valves showing seeds × 2; 5, capsule, view from below × 2; 6, seeds × 10. All from *Rawi* 31022. Drawn by Derrick Erasmus.

DISTRIB. Common in desert plateau and Lower Mesopotamian regions of Iraq: ?DLJ: between Ana and Dair (Dair-as-Zur, in Syria), *Bornmüller* s.n.; DWD: Wadi al-Ghadaf, *Rawi* 14739!; 4850!; 80 km NW of Nukhaib, *Rawi* 23717!; ± 5 km W of Ramadi, *Guest & Rawi* 13692!; DSD: nr Zubair, *Gillett & Rawi* 6044!; nr Jabal Sanam, *Guest, Rawi & Rechinger* 17040!; DWD/LCA: Hindiya Barrage, *Rawi & H. Walter* 16693!; LCA: Abu Ghraib, Baghdad, *Anis Susi* 13590!; 3 km SE of Baghdad, *O. Polunin* 5000!; LCA: Babylon, *Nábělek* 300, 301; LEA: 15 km from Baghdad on old Kut road, *C. Boswell & Haines* W130!; LBA: halfway between Qurna and Basra, *Rawi* 15035!

Egypt, Arabian Peninsula, Palestine, Iran to Pakistan; S Mediterranean region through N Africa to Canary Islands, Spain and Balearic Islands.

2. MESEMBRYANTHEMUM L. nom. conserv.

Sp. Pl.: 480 (1753)

Gasoul Adans., Fam. 2: 243 (1763)

Mostly annual succulent papulose herbs; leaves opposite, upper sometimes also alternate. Flowers cymose or solitary, terminal, short-stalked. Calyx lobes 5, unequal, united below into a tube; staminodes many, linear, petaloid, united below; stamens many, arising from staminodal tube. Ovary semi-inferior, of usually 5 carpels, usually 5-locular; ovules many, axile. Fruit usually a 5-valved capsule; valves membranous-winged, with 2 central closely contiguous narrow erect unwinged expanding keels, opening when moist; cell wings absent, the seeds thus exposed in the open capsule.

About 70 species distributed in South Africa, the Mediterranean region, Arabia; S Australia; Californial one species in Iraq.

1. **Mesembryanthemum nodiflorum** L., Sp. Pl. ed. 1, 481 (1753); DC., Hist. Pl. Grases 88 (1802); Boiss., Fl. Orient. 2: 764 (1872); Zohary, Fl. Pal. ed. 2, 1: 486 (1932); Blakelock in Kew Bull. 3: 431 (1948); Zohary in Dep. Agr. Iraq Bull. 31: 50 (1950).

Gasoul nodiflorum (L.) Rothm. in Notizbl. Bot. Gart. Berlin 15: 143 (1941).
Cryophytum nodiflorum (L.) L. Boulus in S. Afr. Gard. 17: 327 (1927).

Small diffuse annual succulent herbs; branches glabrous, green, 1–10 cm or more. Leaves opposite, or upper often alternate, sessile, succulent, papulose, semi-terete, 6–20 mm. Flowers solitary, subsessile, inconspicuous, white. Calyx lobes unequal, 3 spreading, c. 4 mm, 2 erect, much shorter; staminodes many, petaloid, linear, white, scarcely exceeding calyx lobes, united below; stamens 12–15, uniseriate. Ovary conical in upper part, of 5 united carpels. Capsule rounded above, c. 4 mm diameter, opening when moist by 5 valves. Seeds sub-angular-reniform, slightly rough. Fig. 49: 1–5.

HAB. Saline desert soils (sand, gravel, silt); alt. up to 300 m; fl. & fr. Mar.–May.
DISTRIB. FPF: 70 km N of Amara, *Hazim & Nuri* 30624 (BAG); Central and Southern Iraq: DWD: S of Bahr al-Malh, *Haines* W707!; DSD: nr Najaf, *Rawi* 19975! 19975A!; 35 km W of Samawa, *Rawi* 14877!; Waqsa well, *Guest, Rawi & Rechinger* 19212!; 5 km SE by S of Zubair, id. 16809!; 28 km SE by S of Zubair, id. 16873!; LCA: Babylon, *Hand.-Mazzetti* 891, *Nábělek* 296, *Rogers* 0299!, 0434!; LBA: Basra, *Dowson* 1555!; Ma'qil, *Guest, Rawi & Rechinger* 16743!

Baluchistan, S Iran, Arabian Peninsula, Palestine, Jordan, Syria, Cyprus, Aegean Is., and from Egypt W through N Africa to the Canary Is. Introduced into South Africa, Australia and California.

3. DORTHEANTHUS Schwantes

in Möllers Deutsch Gärtner-Zeit. 42: 283 (1927).

Small annual succulent herbs; leaves opposite, papulose, ± sheathing at base. Flowers solitary, long-pedunculate. Calyx lobes 5, unequal; staminodes many, sub-biseriate; stamens many. Ovary inferior, of usually 5 united carpels, usually 5-locular, flat or depressed above; ovules many, parietal; stigmas persistent. Fruit a 5-valved loculicidal capsule; valves apically 2-awned, each awn representing an indurated half-stigma; expanding keels broadly trigonous, lateral, widely separated, wholly adnate to valves, each produced at apex into a free membranous awn; cell wings present, covering the seeds in the open capsule.

One species in Iraq.

1. **Dortheanthus gramineus** (Haw.) Schwantes in Möllers Deutsch Gärtner-Zeit. 42: 283 (1927).

Mesembryanthemum gramineum Haw., Obs. Mesemb. 470 (1795).
[*Lampranthus violaceus* (non (DC.) Schwantes) Blakelock in Kew Bull. 12: 461 (1957).

Small annual succulent herb to c. 10 cm; leaves opposite, succulent, linear, terete, papulose, 20–50 mm. Flowers solitary, with elongate peduncles 15–90 mm, very conspicuous, purplish, 30–40 mm diameter. Calyx lobes 5, unequal, 3 larger c. 15 mm long, 2 shorter; staminodes petaloid, linear, conspicuous, purplish, paler towards base; stamens many, deep purple. Capsule c. 9 mm diameter, flattened above, opening when moist by 5 recurving apically 2-awned valves. Seeds subangular-reniform, smooth.

HAB. and alt. unrecorded; fl. May.
DISTRIB. Introduced into Iraq and apparently established at Basra: **LBA**: Basra, "grows prolifically here, never previously found in this part of Iraq", *L.G. Wylde* H.600/49!

Native of the Cape Penisula of South Africa.

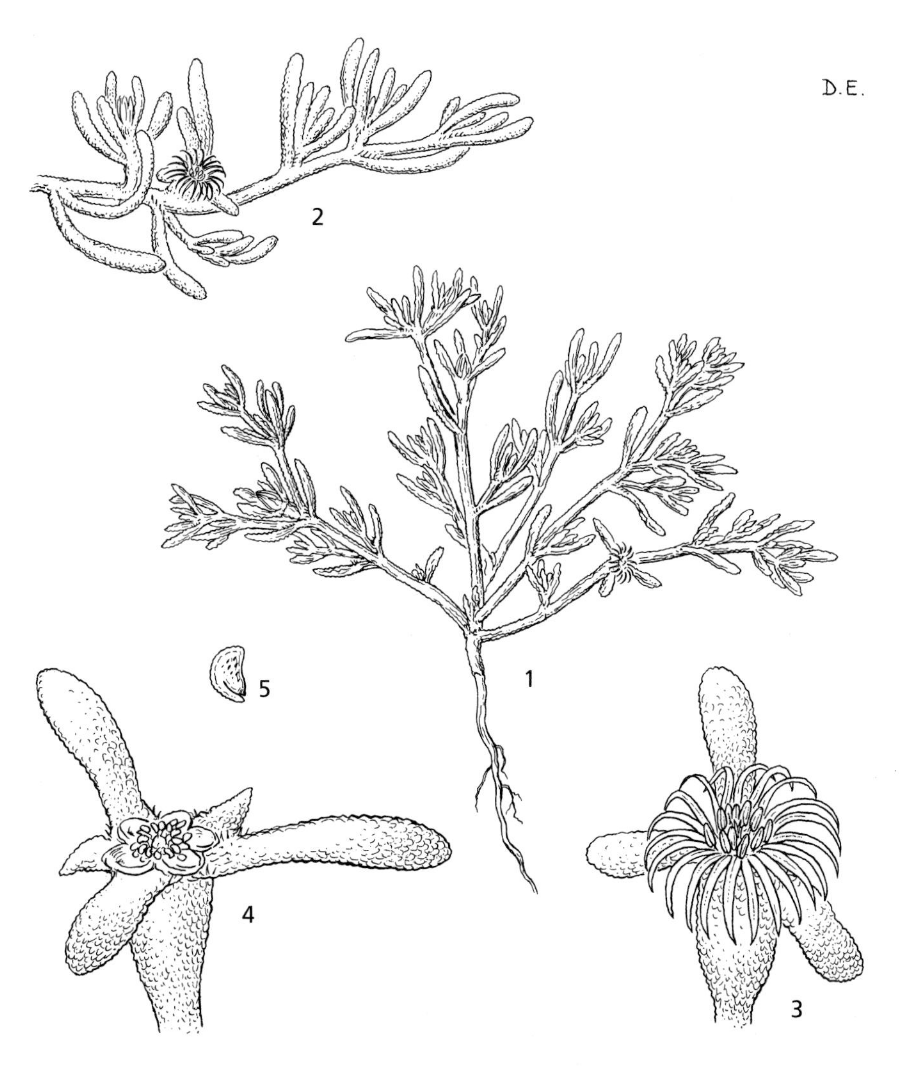

Fig. 49. **Mesembryanthemum nodiflorum**. 1, habit × ⅔; 2, flowering branch × ⅔; 3, flower × 3; 4, capsule (in wet state) × 3; 5, seeds × 6. All from *Rawi* 14877. Drawn by Derrick Erasmus.

92. PORTULACACEAE Engl.

Pflanzenfam. ed. 2, 160: 234 (1934)

C. Jeffrey[10]

Mostly herbs. Leaves opposite or alternate, simple, usually fleshy; stipules scarious or setaceous. Flowers cymose, racemose or solitary, regular, hermaphrodite. Sepals usually 2, free or united. Petals usually 4–6, free or united below. Stamens 4–6, opposite the petals, or more to many, sometimes borne on corolla. Ovary superior or rarely semi-inferior, of 2 or more united carpels, 1-locular; ovules 1-many, basal, fruit usually a circumscissile or valvular capsule.

A family of some 19 genera and 350 species allied to the Aizoaceae, cosmopolitan but with its greatest development in the New World; one genus in Iraq.

PORTULACA L.

Sp. Pl. 445 (1753); Gen. Pl. ed. 5: 204 (1754)

Herbs, ± succulent. Leaves simple, opposite or alternate, fleshy; stipules usually setaceous, represented by tufts of hairs. Flowers sessile, solitary or cymose, surrounded by a false whorl of 2 or more subapical leaves, hermaphrodite, regular. Sepals 2, united below. Petals 4–6, free or ± united, fugaceous. Stamens 4-many, inserted on corolla tube. Ovary semi-inferior, unilocular; ovules many, free central; stigma 1, styles 2 or more. Fruit a circumscissile capsule. Seeds 1-many, usually reniform.

Some 20 species in the tropics and subtropics; only 2 species (one native, the other introduced and sometimes cultivated) in Iraq.

Portulaca (from *Portilaca*, the old Latin name of the plant used by Pliny); Purslane, small half-hardy annuals with fleshy leaves, of which two species only concern us here. The Purslane Tree (*Portulacaria afra*), a small shrubby species in another genus of the family, is sometimes grown in gardens in Egypt but has not been reported from Iraq.

Petals 4–8 mm; stipular hairs inconspicuous, caducous; native . . . . 1. *P. oleracea*
Petals ± 20 mm; stipular hairs long, prominent, persistent; cultivated . . . 2. *P. grandiflora*

1. **Portulaca oleracea** L., Sp. Pl. ed. 1: 445 (1753); Boiss., Fl. Or. 1: 757 (1867); Hand.-Mazz. in Ann. Naturh. Mus. Wien 26: 145 (1912); Nábělek in Publ. Fac. Sci. Univ. Masaryk 35: 50 (1923); Zohary, Fl. Pal. ed. 2, 1: 220 (1932); Guest in Dep. Agr. Iraq Bull. 27: 76 (1933); Anthony in Notes Roy. Bot. Gard. Edinb. 18: 283 (1935); Kusneva in Fl. SSSR 6: 386 (1936); Blakelock in Kew Bull. 3: 403 (1948); Dickson, Wild Fls. Kuwait: 76 (1955);Rechinger, Fl. Lowland Iraq: 219 (1964); Rawi in Dep. Agr. Iraq Tech. Bull. 14: 53 (1964); Rawi & Chakravarty, ibid. 15: 77 (1964); Mouterde, Nouv. Fl. Syr. 1: 450 (1966); Zohary, Fl. Palaest. 1: 78 (1966); Coode in Fl. Turk. 2: 13 (1967); Takholm, Stud. Fl. Egypt ed. 2: 76 (1974); Husain & Kasim, Cult. Pl. Iraq: 97 (1975); Rechinger, Fl. Iranica 117: 2 (1976); Meikle, Fl. Cyprus 1: 287 (1977); Ghafoor in Fl. Pak. 4: (1972); Miller, Fl. Arab. Penins. & Socotra 1: 170 (1996).

Annual glabrous fleshy ± spreading herb, branches up to 30 cm or more. Leaves alternate, tending to be crowded towards ends of branches, sessile or subsessile, obovate-spatulate, tapered at base, rounded or subtruncate at apex, to 30 × 12 mm. Flowers 1–5, terminal, surrounded by a false whorl of apical leaves; sepals 2–4 mm; petals yellow, 4–8 mm, united below. Stamens 7–12. Ovary ovoid; stigmas 3–6. Capsule dehiscent about the middle. Seeds granulate. Fig. 50: 1–8.

HAB. Lower mountain slopes, among rocks, on stony clay, by a mountain stream – also on the plain as a weed in gardens, orchards etc.; alt. up to 1500 m; fl. & fr. (Apr.) Jun.–Jul.(Aug.).
DISTRIB. Occasional ain the forest zone, adventive in the desert regon on the irrigated alluvial plain. **MAM**: Sharanish, *Rawi* 23695!; **MRO**: Ser-i Hasan Beg, ± 13 km NE of Rowanduz, *Guest* 30411!; Umar Awa (Umarawa), ± 20 km E by S of Rowanduz and 35 km on road from Haji Umran, *Qaisi & K. Hamad* 43539!; **MSU**: Hauraman range (Avroman), *Rawi, Chakravarty, Nuri & Alizzi* 19823!; Penjwin, *Rawi* 22642 (BAG); **FUJ/FNI**: on banks of R.Tigris, nr Mosul, *Nábělek* 304. **FAR**: 19 km S Erbil, *Botany Staff* 48916 (BAG). **LCA**: in a garden nr Aqarquf, *Barkley & Ani* 2530!; Abu Ghraib, *Chakravarty* 32786; Baghdad, *Paranjpye* s.n.! & *Ali 'Abd al-Latif* 39497; Za'faraniya, *Schira & T. Sakin* 34770! **LSM**: Amara, in Anthony, l.c., *Khatib & Alizzi* 32610; "Babylonia", *Noë* s.n.

[10] Komarov Botanical Institute, St Petersberg, Russia

Fig. 50. **Portulaca oleracea**. 1, habit and root × ⅔; 2, inflorescence × 2; 3, flower × 3; 4, flower with petals and sepals removed × 3; 5, stamen detail; 6, capsule × 4; 7, seed × 8. 1 from *Kotte* s.n.; 2–6 from *Morris* 78. Drawn by Derrick Erasmus.

C Mediterranean and SW Europe (to Greece, SW Russia, Crimea), Cyprus, Syria, Lebanon, Palestine, Jordan, Sinai, Egypt, Arabian Peninsula, Turkey, Caucasus, Iran, Turkmenistan, Afghanistan C Asia; N Africa (Morocco to Libya); Macaronesia (Canary Is., Azores). Introduced into many other parts of the world and now almost cosmopolitan.

Common or Garden Purslane (Eng.) — also Pursley, Pussely, Pussly, Pusley (Am. — all corruptions of the Old English name Purslane), Pigweed (Austral.), Pouprier, Potager (Fr.); BARBÎN (Ir.), BAQLA ("greens, vegetable") (Ar., Syr., *Hakim* 21 in Hand.-Mazz., l.c.), BAQLATU 'L-HAMQA ("fool's greens"), Ar., Hava), RIJLA ("foot-weed", Eg., Sud.), ?PARPÎNA (Kurd., Turk.). Often collected or sometimes cultivated and sold in the markets in Iraq as a salad or potherb for stews, and also for pickling. Though of no great food value the plant is avidly grazed by livestock.

The medicinal properties of purslane and its use as a vegetable have been known the world since ancient times. Its use as a potherb and salad has long been known throughout Europe, often used in French salads, the older shoots being pickled or used for the pot. Dickson (1955) states that purslane is cultivated in gardens and eaten as a spinach or in salads at Kuwait. Many writers dwell on the anti-scorbutic properties of the plant.

2. **Portulaca grandiflora** Hook., Bot. Mag. 56: t. 2885 (1829); Guest (sub P. sp.) in Dep. Agr. Iraq Bull. 27: 77 (1933); Blakelock in Kew Bull. 12: 461 (1957); Husain & Kasim, Cult. Pl. Iraq: 97 (1975).

Annual hairy fleshy herb; branches ascending, 15–20 cm or more; leaves alternate, scattered, fleshy, sessile, cylindrical, acute, 4–40 mm. Flowers 3–4, terminal, sessile, surrounded by an involucre of apical leaves; sepals ovate, green. Petals conspicuous, red, orange or yellow, 20–25 mm, united below; stamens many, inserted on corolla; ovary conical; stigmas 6–9; capsule dehiscent above base. Seeds granulate.

HAB. An exotic plant known to have been cultivated as an ornamental for its brightly-coloured flowers in gardens in lower Iraq for the past 50 years or more; alt. up to 50 m; fl. & fr. Jun.–Jul. onwards.
DISTRIB. Two records from the Baghdad environs, but widespead according to the literature: **LCA**: Za'faraniya (cult.), *H. Ahmed* 9843; Baghdad (cult.), *Sahira* C. 561.

Native of S America (Argentina), widely introduced and cultivated as an ornamental for its brightly-coloured flowers in many parts of the world.

According to Guest it is generally known in Iraq by its Turkish name YALDIZ or YALDIZ CHICHAJI or sometimes ALDŌZ. This gay little annual plant is grown as a summer ornamental for its brightly coloured (scarlet, yellow, orange, white, etc.) flowers which open in the early morning sunshine and close in the middle of the day.

93. **POLYGONACEAE**

Juss. (1789), nom. conserv.

J.R. Edmondson[11] and J.R. Akeroyd[12]

Shrubs, climbers and perennial, sometimes annual or biennial, herbs. Leaves usually alternate, simple, petiolate to sessile; stipules fused into a sheath (ochrea) clasping the stem. Inflorescences spicate, fasciculate or paniculate, flowers hermaphrodite or unisexual, actinomorphic. Perianth 3-6-merous, segments fused below, often accrescent in fruit. Petals absent. Stamens 6–9(–16), anthers medifixed. Ovary superior, 1-locular; styles 2–4; ovule 1, basal. Fruit a nut, usually trigonous or lenticular, often enclosed by the persistent perianth.

A large family of 46 genera and 1200 species, ± cosmopolitan, but predominantly in the northern temperate regions.

Brandbyge, J. (1993). Polygonaceae. Pages 531-544. In: Klaus Kubitzki (ed.), Jens G. Rohwer, and Volker Bittrich (vol. eds), *The Families and Genera of Vascular Plants* volume II. Springer-Verlag: Berlin; Heidelberg

Decraene, R., & Akeroyd, J.R. (1988). Generic limits in *Polygonum* and related genera (Polygonaceae) on the basis of floral characters. Bot. Journ. Linn. Soc. 98: 321–371 (1988).

[11] Honorary Research Associate, Royal Botanic Gardens, Kew
[12] Independent Scholar, Hindon, Wilts.

Rechinger, K.H. (1949). Rumices asiaticae (Vorarbeiten zu einer Monographie der Gattung Rumex VII). Candollea 12: 9–152.

Tavakkoli, S., Osaloo S.K. & Maassoumi, A.A. (2010). The phylogeny of *Calligonum* and *Pteropyrum* (Polygonaceae) based on nuclear ribosomal DNA ITS and chloroplast *trn*L-F sequences. Iranian J. Biotech. 8(1): 7–15.

1. Shrubs and subshrubs, usually much-branched, leaves soon deciduous 2
   Annual, biennial or perennial herbs, sometimes with a woody stock 4
2. Stamens inserted at base of perianth; fruit sharply angled but unwinged . . . . 5. **Atraphaxis**
   Stamens inserted at mouth of perianth; fruit with 3 bilobed wings 3
3. Stamens 10–16; fruit 4-winged, sometimes clothed with soft bristles; leaves caducous . . . . 7. **Calligonum**
   Stamens 6–8; fruit 3-winged, not clothed with bristles; leaves linear, persistent . . . . 6. **Pteropyrum**
4. Perianth segments 4; leaves basal, long-petiolate, blade reniform to triangular-cordate . . . . 3. **Oxyria**
   Perianth segments 5–6; leaves lanceolate to linear, hastate or sagittate 5
5. Flowers unisexual (monoecious); outer tepals with a terminal stout spine . . . . 1. **Emex**
   Flowers hermaphrodite; outer tepals with soft marginal spines or glabrous 6
6. Stamens 9; leaves large, broad, palmately veined . . . . 4. **Rheum**
   Stamens 6–8; leaves much longer than broad, pinnately veined 7
7. Twining annual; fruit black (rare introduced weed) . . . . 10. **Fallopia**
   Annuals, biennials or perennials, not twining; fruit usually brown or blackish-brown 8
8. Perianth segments 6, the 3 inner much more enlarged in fruit; stamens 6. . . . 2. **Rumex**
   Perianth segments 5, remaining ± equal in fruit; stamens usually 8 9
9. Leaves triangular, about as wide as long, venation palmate; nut 2× length of perianth . . . . 11. **Fagopyrum**
   Leaves twice or more as long as wide, venation pinnate; nut usually less than 2× length of perianth 10
10. Flowers in axillary or terminal racemes, panicles or spike-like inflorescences; ochreae ± fimbriate, chartaceous . . . . 8. **Persicaria**
    Flowers solitary or in small axillary clusters; ochreae entire but lacerate, hyaline . . . . 9. **Polygonum**

1. EMEX Campd.

Monogr. Rumex 56 (1819)

Monoecious annuals. Leaves alternate; ochrea membranous, becoming lacerate. Flowers unisexual. Female flowers at base of inflorescence of ± distant whorls of flowers arising from axils of small bracts. Perianth segments 6, free in male flowers, connate in female flowers. Stamens 4–6. Fruits triquetrous, the perianth segments fused below, indurate in fruit, forming a polygonal case around the nut, the outer 3 tipped with a patent spine.

Only two species, distributed in coastal Mediterranean, South Africa, Australia, eastern Africa, and parts of SW Asia; one species in Iraq.

Emex spinosa (*L.*) *Campd.*, Monogr. Rumex 58 (1819); Boiss., Fl. Orient. 4: 1005 (1879); Graham in Fl. Trop. East Africa Polygonaceae: 3 (1958); Rech.f. Fl. Lowland Iraq: 172 (1964); Zohary, Fl. Palaest. 1: 67 (1966); Cullen in Fl. Turk. 2: 293 (1967); Rech.f. & Schiman-Czeika in Fl. Iranica 56: 2 (1968); Nyberg & Miller in Fl. Arab. Penins. & Socotra 1: 139 (1996); Boulos, Fl. Egypt 1: 24 (1999); Qaiser in Fl. Pak. 205: 165 (2001).

Glabrous annual, much-branched at base, the stems trailing to suberect, 14–30 cm, with small, greenish dentate ochreae. Basal leaves ovate to hastate, to 12 × 3 cm, petiolate; cauline leaves smaller, sessile. Lowest flowers often near or below ground. Male flowers in axillary and terminal, pedunculate racemes, female flowers in axillary whorls, sessile. Fruits 3–5 mm, ovoid-triquetrous, pale brown or reddish, the 3 outer perianth segments with a stiff reflexed spine. Fig. 51: 1–2.

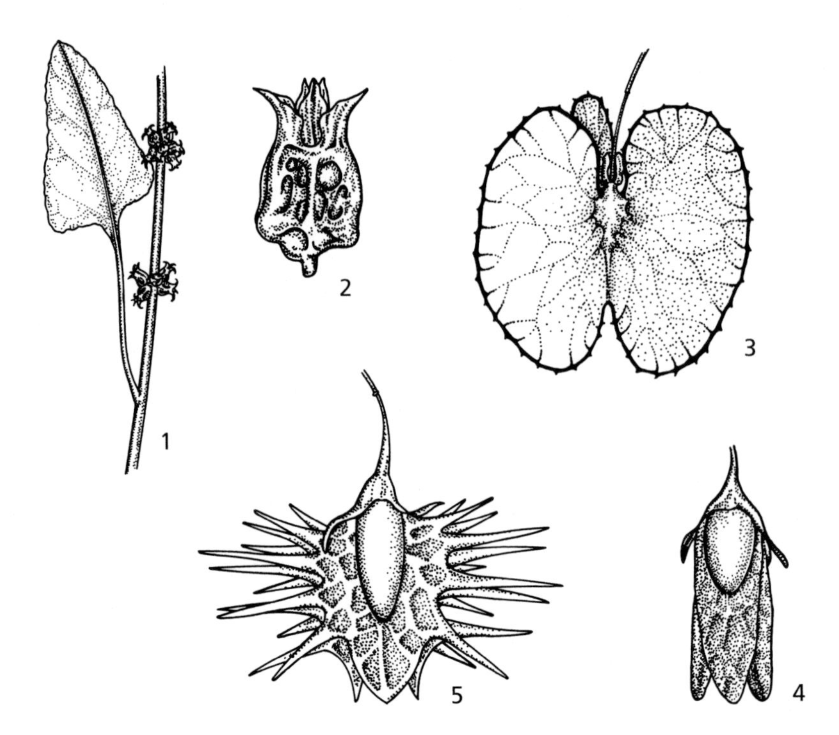

Fig. 51. **Emex spinosa**. 1, habit × 1; 2, fruit × 7. **Rumex cyprius**. 3, fruit × 5. **Rumex conglomeratus**. 4, fruit × 7. **Rumex dentatus**. 5, fruit × 7. Reproduced with permission from Flora of the Arabian Peninsula & Socotra 1: f., 24, 25. 1996.

HAB. Compact sandy soil with small stones and grit, wadi beds, *haswa* desert, sandy desert, moist depressions in silty soil; alt. 5–420 m; fl. Mar.–Apr.

DISTRIB. Mainly in the southern desert region of Iraq. **DWD**: Al Khasfa, between Ana and Haditha, *Kaisi et al.* 56789. **DSD**: Ath-Thulaima nr Jumaima, 145 km SW of Salman, at Saudi frontier, *Guest, Rawi & Rechinger* 19088!; Al-Rubaitha, 20 km N of Salman, *Botany Staff* 42086!; 25 km from Busaiya to Khidr al-Mai, *Karim, Sharif, Hamad & Hamid* 48046!; nr Zubair, *Guest & Rawi* 14330! *Graham* 409!; W of Zubair, *Guest, Eig & Zohary* 5038!; 5 km SE by E of Zubair, *Guest, Rawi & Rechinger* 16804!; 23 km SE by E of Zubair, *Guest, Rawi & Rechinger* 16847!; Shu'aiba, 1919, *Jackson* s.n.!, *Graham* 342!; 8 km E of Shithatha, *Rawi & Gillett* 6468!; Umm Qasr, *Alizzi & Omar* 35013!; *Guest, Rawi & Rechinger* 16888!; Wadi Dwairi, *Rawi* 14955!; Al-Batin, c. 65 km SW of Basra, *Guest, Rawi & Rechinger* 17070!; Al-Baniya, 65 km WSW of Basra, *Guest, Rawi & Rechinger* 17292!; 30 km W by N of Busaiya, *Guest, Rawi & Long* 14162! 45 km S.E. by E. of Busaiya, *Guest & Rawi* 14246! 20 km NW of Umm Qasr, *Khatib & Alizzi* 33426!;12 km W of Ukhaidir, *Rawi* 30856!; 10 km W of Karbala, *Rawi* 30768!; Jadida W of Nukhaib on Saudi border, *Rawi* 31054!; 80 km W of Shabicha, *Chakr., Khatib, Rawi & Tikriti* 30073!; 10 km on road between Zubair and Safwan, *Rawi* 25975!; 5 km S of Jumaima, *Kaisi, Hamad & Hamid* 48195! Ukhaidir, *Kaisi, Khayat, Jenan, Sahira & Wedad* 51202; around Jabal Sanam, *Kaisi, Hamad & Hamid* 48595; **FPF**: Jabal Hamrin, Khanaqin road, *Rawi* 17765. **LCA/FPF**: Jabal Hamrin near the new dam, *Kaisi* 48684. **LBA**: Nahr al-Khora, *Sutherland* 427!; Jabal Sanam, *Haddad* H391!

N Africa, S Europe, Cyprus, Lebanon, Palestine, Kuwait, Bahrain, Saudi Arabia, Oman.

2. **RUMEX** L.

Sp. Pl.: 333 (1753)

Annual, biennial or perennial herbs with tubular ochreae. Leaves simple, alternate, petiolate or subsessile. Flowers bisexual or unisexual (plant monoecious), polygamous or dioecious, in whorled racemose or paniculate inflorescences. Perianth segments 6, the inner 3 persistent and much accrescent in fruit, 1 or all 3 usually with a globose to elongate corky tubercle. Stamens 6. Ovary with 3 styles and a single ovule. Fruit a trigonous glabrous nut, enclosed by the inner perianth segments (valves).

Rumex (from Lat. *rumo*, I suck, alluding to the practice among Romans of sucking sorrel leaves to allay thirst.)

A large genus of some 200 species distributed in temperate regions of the world; nine species in Iraq.

1. Leaves sagittate, ovate-triangular or triangular, tasting sour acidic 2
   Leaves not sagittate or triangular, not tasting sour acidic........................ 4
2. Leaves sagittate; valves not more than 5 mm wide .... 1. *R. tuberosus*
   Leaves not sagittate; valves more than 10 mm wide 3
3. Margins of valves without prominent vein, entire...................2. *R. vesicarius*
   Margin of valves with prominent purplish-red vein, minutely spinulose...3. *R. cyprius*
4. Cauline leaves few, linear; valves 10–15 mm .... 5. *R. angustifolius*
   Cauline leaves several to many, lanceolate to oblong; valves less than 10 mm 5
5. Margins of valves entire or minutely toothed 6
   Margins of valves with distinct teeth or spines.............................. 8
6. Branches arcuate-divaricate, often tangled; tubercles of valves rugose .... 8. *R. pulcher*
   Branches divaricate but not tangled; tubercles of valves smooth................ 7
7. Valves 2–3 mm, ovate-oblong, entire .... 7. *R. conglomeratus*
   Valves usually more than 3 mm, orbicular or orbicular-triangular, margins entire or minutely toothed............................. 8
8. Valves less than 7 mm; leaf-margins usually crisped ....6. *R. crispus*
   Valves 9–12 mm; leaves ± flat ....4. *R. ponticus*
9. Plant little-branched; marginal teeth of valves 3–6 mm .... 9. *R. dentatus*
   Plant often profusely branched; marginal teeth of valves less than 3 mm 5
10. Pedicels thickened, rigid; tubercles rugose .... 8. *R. pulcher*
    Pedicels slender (abruptly thickened below valve); tubercles smooth... 10. *R. halepensis*

1. **Rumex tuberosus** *L.*, Sp. Pl. ed. 2: 481 (1761); Rechinger & Schiman-Czeika in Fl. Iranica 56: 6 (1968); Cullen in Fl. Turk. 2: 285 (1967).

Perennial, several roots forming fusiform tubers, with often solitary, erect stem 15-60 cm. Basal leaves up to 12 × 4 cm, petiolate, sagittate to cordate at the base, acute or obtuse, cauline leaves few, lanceolate to linear, usually acute. Ochreae lacerate. Inflorescence a raceme of spreading branches, often repeatedly branched and dense in fruit. Valves 3-5 × 3-5 mm, cordate-orbicular, papery, pale brown, each with a small tubercle. Outer perianth segments 2.5-3 mm, reflexed in fruit. Fruits c. 2 mm, brown.

Valves 4–5 mm wide..a. subsp. *horizontalis*
Valves 3–4 mm wide... b. subsp. *turcomanicus*

a. subsp. **horizontalis** (*C. Koch*) *Rech. f.*, Candollea 12: 31 (1949); Cullen in Fl. Turk. 2: 286 (1967); Rechinger & Schiman-Czeika, Fl. Iranica 56: 6 (1968).

Rumex horizontalis C.Koch, Linnaea 22: 211 (1849); Boiss., Fl. Orient. 4: 1016 (1879).

Branches of inflorescence themselves branched. Valves 4–5 mm wide.

HAB. Open grassy clearings in forest, rocky mountains between cracks of rocks; alt. 600–2000 m; fl. Jun.
DISTRIB. **MRO**: Handiyan at foot of mt. Baradost, *Thesiger* 683; Jebel Baradost nr Diana, *Field & Lazar* 895! **MSU**: Qopi Qaradagh, *Haines* W435! **MJS**: Jabal Sinjar, E. slope, *Omar, Kaisi & Khayat* 52591.

Greece, Turkey (N and E Anatolia), Armenia, N Iran.

b. subsp. **turcomanicus** (*Rech. f.*) *Rech. f.*, Candollea 12: 31 (1949); Rechinger & Schiman-Czeika, Fl. Iranica 56: 6 (1968).

Rumex tuberosus L. var. *turcomanicus* Rech.f., Feddes Repert. 49: 1 (1940).

Branches of inflorescence repeatedly and more densely branched. Valves 3–4 mm wide.

HAB. Alpine slopes, shady places in *Quercus* forest (N. aspect); alt. 1100–2000 m; fl. Jun.
DISTRIB. **MRO**: Halgurd Dagh, above Nowanda, *Rechinger* 11382!; above Sarsang, *Rechinger* 11907; Sarsang, slope of Gara Dagh, *E. Chapman* 26384; 13 km from Amadia to Sarsang, *Sharief & Hamad* 50358.

Georgia, N Iran, Turkmenistan.

Distribution of species: S Europe from Italy to SE Russia, Crete, Cyprus, Syria, Turkey, Caucasus, N and NW Iran.

2. **Rumex vesicarius** *L.*, Sp. Pl. 336 (1753); Boiss., Fl. Orient. 1017 (1879); Rechinger, Fl. Lowland Iraq: 174 (1964); Rechinger in Fl. Iranica 56: 8 (1968); Zohary, Fl. Palaest. 2: 61 (1972); Leonard, Contrib. Fl. et Veg. Deserts d'Iran 5: 86 (1985); Nyberg & Miller in Fl. Arab. Penins. & Socotra 1: 136 (1996); Boulos, Fl. Egypt 1: 34 (1999); Qaiser in Fl. Pak. 143: 169 (2001); Breckle et al., Vasc. Pl. Afghan. Checkl.: 410 (2013).

Glaucous annual 10–40 cm, stems branched from base, erect or ascending. Leaves sour to taste, 2–5 × 1–3 cm, truncate or subhastate at the base, broadly ovate to ovate-triangular, obtuse, the upper ovate-oblong, subacute; petiole long. Inflorescence a lax raceme. Pedicels slender, articulate, with 1–2 flowers, a primary and a smaller secondary flower. Primary flower: valves 12–20 mm, subequal, suborbicular, papery, pink, with a reticulation of dense fine purplish-red veins that does not reach the margins, 2 with small tubercles. Nut 3.5–5 mm, ovoid, greyish-brown. Nut of secondary flower 2.8–4 mm, dark brown. Fig. 52: 1–10.

HAB. Stony desert, sandy soil under shade of *Astragalus,* rocky volcanic hillside with sandy patches, sandy and gypsacous soil; alt. 150-210 m; fl. Feb.–Apr.
DISTRIB. Widespread in the desert region of Iraq: **DWD**: 50 km N. of Rutba, *Tall al-Nasir, Rawi & Nuri* 27126; **DSD**: Faidhat al-Gelaib, 217 km N of Neutral Zone, *Rawi, Khatib & Tikriti* 29211!; Shabicha, *Rawi & Gillett* 6207; 4 km N.W. of Rawa, *Rawi & Gillett* 7079; 17 km E. by N. of Salman, *Guest, Rawi & Long* 14109; Jabal Sanam (50 km S. by W. of Basra), *Guest, Rawi & Schwan* 14371; 20 km S.W. of Najaf, *Alizzi & Omar* 35388; 5 km from Salman to Samawah, *Kaisi, Hamad & Hamid* 48237; Al-Sawah, *Kaisi, Hamad & Hamid* 48437; Khidhr al-Mai, *Kaisi, Hamad & Hamid* 48488.

North Africa, Arabia, S Greece, Syria, Palestine, S Iran, Afghanistan to Pakistan and India.

The leaves are edible, though acidic.

3. **Rumex cyprius** *Murb.*, Acta Univ. Lund., ser. 2, 2 (14): 21 (1907); Rechinger, Fl. Lowland Iraq: 174 (1964); Rechinger in Fl. Iranica 56: 7 (1968); Zohary, Fl. Palaest. 2: 61 (1972); Nyberg & Miller in Fl. Arab. Penins. & Socotra 1: 136 (1996); Boulos, Fl. Egypt 1: 34 (1999).

Rumex roseus L., Syst. ed. 10, 2: 990 (1759); Boiss., Fl. Orient. 4: 1018 (1879); Campd., Monogr. Rumex 69, 129 (1819), non L., Sp. Pl. 337 (1753).

Slightly fleshy, pale green annual 10–30 cm, stems branched from base, weakly erect. Leaves sour to taste, 2–5 × 1–3 cm, truncate at the base, ovate-triangular, obtuse or acute, the upper lanceolate, subacute; petiole long. Inflorescence a lax raceme. Pedicels slender, articulate, with 1–2(–3) flowers, a primary and a smaller secondary flower. Primary flower: valves 10–22 mm, unequal, suborbicular, papery, pink, with a reticulation of dense fine crimson veins that reaches the margins to form a minutely spinulose marginal vein, 2 with small tubercles. Nut 3–5 mm, ovoid, pale brown. Nut of secondary flower 2.8–4 mm, darker brown. Fig. 51: 3.

HAB. Sandy debris below riverside cliff, barren degraded conglomerate hills, sandy stony hills, sandy gravelly valleys; alt. 80–240 m; fl. Mar.–Apr.
DISTRIB. Mainly in the dry steppe and desert regions of Iraq. **MSU**: Shahrazur, Wadi Ahmar, *Omar, Karim, Hamza & Hamid* 37122. **FKI**: Jabal Hamrin, nr Injana, *Guest* 690; 6 km S. of Shithatha, *Rawi & Gillett* 6341; Ba'quba, Khanaqin road, *Rawi & Gillett* 7342; Tauq bridge, *Haines* W.1382! Jabal Hamrin, Sharaban to Jalaula, *Rechinger* 14240; **FPF**: Khanaqin, *Rechinger* 14125; 10 km E of Mandali, *Rechinger* 9637; 13 km SE Badrah, *Rechinger* 9151, 13947; Hashima, between Badra and Mandali, *Kaisi & Khayat* 50748; Jabal Ruhaila, Shatt at Tib, 70 km N of Amarah, *Rechinger* 8900, 8952; 10 km E. of Zurbatiya, *Kaisi & Hamid* 46541. **DWD**: Ajrumiya, *Alizzi & Husain* 34039. **LEA**: Ba'quba, Khanaqin road, *Rawi & Gillett* 7342; Tib (Al-Sharhani), *Kaisi & Khayat* 50579;. **LCA**/ **FPF**: Jabal Hamrin, nr new dam, *Kaisi* 48792.

HAMTH, HAMUTH (Ir.-Ba'quba & Shithatha, *Rawi & Gillett*); JURSHKA (Kurd.-Ba'quba, *Rawi & Gillett* 7342). The leaves are said to be edible.

North Africa, Cyprus, Syria, Palestine, Arabia, S Iran.

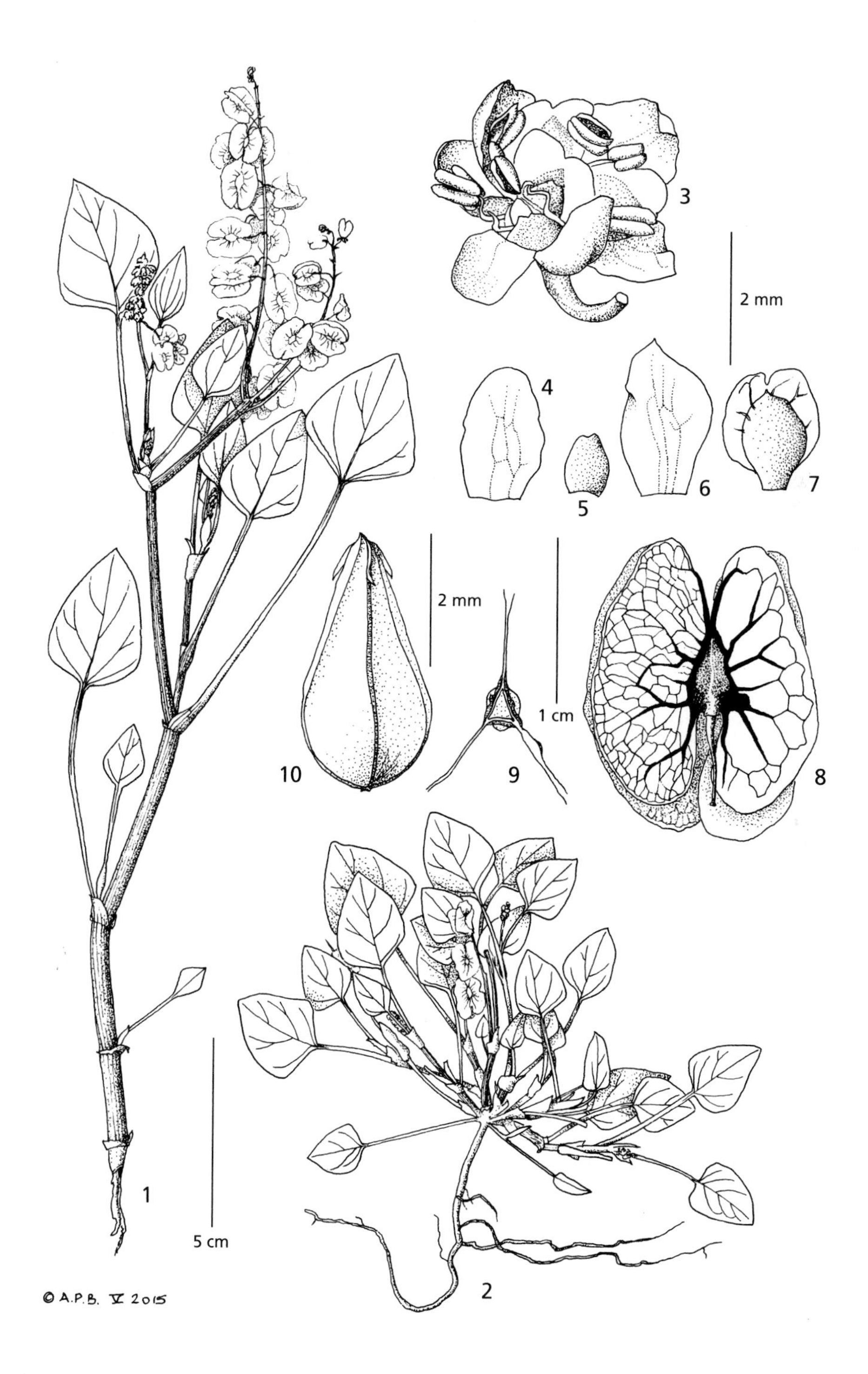

Fig. 52. **Rumex vesicarius**. 1, 2, habit; 3, flower pair; 4, secondary flower outer perianth segment; 5, secondary flower inner perianth segment; 6, primary flower outer perianth segment; 7, primary flower inner perianth segment; 8, fruit with enlarged inner perianth segments; 9, transverse section of fruit with perianth; 10, nutlet. 1, 3–7, 10 from *Kaisi et al.* 48437; 2 from *Nat. Herb.* 42094; 8–9 from *Rawi et al.* 29211. © Drawn by A.P. Brown.

4. **Rumex ponticus** *E.H.L. Krause*, Beih. Bot. Centralbl. xxiv. II. 15 (1908) [emend Rech.f., Feddes Repert 31: 235 (1933)]; Cullen in Fl. Turk. 2: 288 (1967); Rechinger & Schiman-Czeika, Fl. Iranica 56: 12 (1968).

Rumex patientia L. var. *kurdica* Boiss., Fl. Orient. 4: 1009 (1879).

Perennial, stems 80–120 m tall with erect or somewhat spreading branches. Basal leaves oblong-ovate or elliptical-lanceolate, acute, broadly cuneate to truncate; petiole about as long as lamina; cauline leaves similar, sessile to shortly petiolate. Fruiting panicle often dense, with few leafy bracts. Inner tepals enlarged considerably in fruit, 9–11×12–15 mm, cordate-orbicular, strongly reticulate, minutely denticulate, one of them with an ovate-globose tubercle. Nut c. 4.5 × 3.5 mm long, brown. Fruiting pedicels longer than tepals, slender, articulated below middle.

HAB. On limestone & on metamorphic rock,; alt. 1700–3500 m;
DISTRIB. Confined to the highest mountains of NE Iraq. **MRO**: Halgurd Dagh, *Gillett* 12351! *Guest* 2930! N. side of Molla Khort mt., *Rawi, Hakim & Nuri* 29575!; mt. Karoukh, S. slope, *Kass, Nuri & Serhang* 27488! **MSU**: mt. Hawraman, above Darimar, *Gillett* 11869!

E Turkey, W Iran.

5. **Rumex angustifolius** *Campd.*, Monogr. Rumex 63 (1819); Cullen in Fl. Turk. 2: 289 (1967); Rechinger & Schiman-Czeika, Fl. Iranica 56: 14 (1968).

Perennial, with a thick rhizome and erect unbranched stems 20–40 cm. Basal leaves narrowly elliptic or lanceolate, 6–10 × 1–3 cm, tapering at the base, acute, petiolate; cauline leaves few, linear, sessile. Inflorescence simple, spicate, with interrupted whorls of 5–8 flowers. Pedicels rather stout. Valves 10–12 × 8–15 mm, cordate-reniform, entire or with margins irregularly crenulate, each bearing a tubercle. Fruit 6–7 × 4 mm, trigonous, acuminate, brown.

In Iraq, only subsp. *angustifolius* is found. subsp. *macranthus* (Boiss.) Rech.f., with branched stems and broader valves, occurs in E Anatolia, and subsp. *libanoticus* Rech.f., shorter (to 13 cm), with cordate-triangular valves in SE Anatolia and Lebanon.

HAB. Stony slopes, in *Quercus* forest, nr stream; alt. 1300–2000 m; fr. May–Jun.
DISTRIB. Occasional in the upper forest zone of NE Iraq. **MRO**: Between Seiwaka & Dargala villages, E of Jabal Karoukh, *Kass, Nuri & Serhang* 27608!; Gali Warta, 30 km NW of Rania, *Rawi, Nuri & Kass* 28739! **MSU**: Penjwin, *Rawi* 8919, 22623!

Lebanon, Syria, Turkey, Armenia, Georgia, NW and W Iran.

6. **Rumex crispus** *L.*, Sp. Pl. 335 (1753); Rechinger, Fl. Lowland Iraq: 174 (1964); Cullen in Fl. Turk. 2: 289 (1967); Rechinger & Schiman-Czeika, Fl. Iranica 56: 13 (1968); Zohary, Fl. Palaest. 2: 63 (1972); Boulos, Fl. Egypt 1: 32 (1999); Qaiser in Fl. Pak. 143: 150 (2001); Li Anjen, Grabovskaya-Borodina & Mosyakin in Fl. China 5: 337 (2003); Breckle et al., Vasc. Pl. Afghan. Checkl.: 409 (2013).

Biennial to perennial herb, stems erect, 40–100(–150) cm. Basal leaves lanceolate, 20–40 × 4–6(–10) cm, slightly fleshy, acute, margins usually strongly crisped; petiole usually shorter than lamina. Stem leaves up to 25 cm, narrowly lanceolate, sessile or petiole short. Panicle lax or dense, usually with short suberect branches, flowers in crowded whorls of 10–30; pedicels 2–10 mm, filiform. Valves 3–5(–6.5) × 3–4.5(–6) mm, orbicular-triangular, cordate or subcordate at the base, obtuse or rarely subacute, finely reticulate, entire or rarely minutely crenulate, each with a tubercle up to 2.5 mm long (sometimes only 1 valve tuberculate). Fruits 1.5–3 mm, trigonous, brown.

HAB. In plantations and gardens, cultivated hillside, edge of river, *Quercus* forest, roadside, mountain slope; alt. (250–)1060–1830 m.; fr. Jun.–Sep.
DISTRIB. Occasional and local in N Iraq; possibly an introduced weed. **MRO**: Ser-i Hasan Beg, *Guest* 3027!; Kholan, *Rawi* 13703! ; Walash, *Thesiger* 307! Penjwin, *Rawi* 22578; 11 km N. of Penjwin, *Rawi* 22875; Pushtashan, lower slope of Qandil range, *Rawi* 24206; Haji Umran on Iranian frontier, *Rawi & Serhang* 24935. **MSU**: Arbet, *Karim* 39295. **FKI**: Dibis, 48 km N of Kirkuk, *Rawi* 36208!

TIRSHI (Kurd.-Walash, *Thesiger* 307).

Temperate Eurasia S to Yemen, eastwards to Japan, N America; introduced worldwide.

7. **Rumex conglomeratus** *Murray*, Prodr. Stirp. Gott. 52 (1770); Rechinger, Fl. Lowland Iraq: 175 (1964); Rechinger in Fl. Iranica 56: 15 (1968); Zohary, Fl. Palaest. 2: 63 (1972); Nyberg & Miller in Fl. Arab. Penins. & Socotra 1: 138 (1996); Qaiser in Fl. Pak. 143: 153 (2001); Breckle et al., Vasc. Pl. Afghan. Checkl.: 409 (2013).

Perennial herb, with erect branched stems 30–70(–120) cm. Leaves 10–30 × 2–6 cm, oblong to lanceolate, cuneate to truncate at the base, subacute; petiole longer than lamina. Panicle rather lax, usually with long ascending branches and flowers in regular clustered whorls of 10–20, many with shortly petiolate ovate or lanceolate leafy bracts. Pedicels slender. Valves ovate-oblong, 2–3 × 1–1.6 mm, entire, each bearing an oblong tubercle that covers most of the valve. Fruits 1.3–1.8 × 1–1.4 mm, brown or dark reddish-brown. Fig. 51: 4.

HAB. Humid shaded areas, gardens, stream bank, orchard near watercourse, at splash of waterfall; alt. 560–1830 m.
DISTRIB. In the lower forest zone of NE Iraq. **MAM**: Zawita, *Guest* 3745!; Bidaho 15 km W of Kani Masi, *Hamid & Fadhil* 45458!; **MRO**: Galala, *Omar* 37642! *Omar, Sahira, Karim & Hamid* 38440!; Ser-i Rost, *Haley* 96! Gali Ali Beg, *Omar, Kaisi & Wedad* 49546. **MSU**: Sulaimaniya, *Field & Lazar* 965!; 5 km from Sulaimaniya to Sayyid Sadiq, *Omar, Kaisi & Wedad* 49382!; Pira Magrun, Rawi 12130; ?**MSU**: Ahmad Awa, *Omar, Kaisi & Wedad* 49395!; **LEA**: Ba'quba, *Graham* s.n.!; **LCA**: Daltawa (Khalis), *Guest* 2456!

In Europe and elsewhere this species frequently forms hybrids with several other *Rumex* species. For the hybrid *R.* × *baqubensis Rech.f.* (=*R. conglomeratus* × *dentatus* subsp. *mesopotamicus*) see under 9. *R. dentatus.*

TIRSHOK (Kurd.-Zawita, *Guest* 3745), TIRSHKA (Kurd.-Pira Magrun, *Rawi* 12130).

Europe, NW Africa, SW Asia.

8. **Rumex pulcher** *L.*, Sp. Pl. 336 (1753); Rechinger, Fl. Lowland Iraq: 175 (1964); Rechinger & Schiman-Czeika, Fl. Iranica 56: 16 (1968); Zohary, Fl. Palaest. 2: 64 (1972); Boulos, Fl. Egypt 1: 32 (1999).

Biennial (sometimes annual) or short-lived perennial, 20–50 cm tall, much-branched, often tangled. Basal leaves 4–10 × 3.5–5 cm, oblong or sometimes panduriform, truncate or cordate at the base, obtuse. Stem leaves few, oblong or lanceolate, subacute. Panicle with numerous arcuate-divaricate branches and flowers in remote whorls, a few with subsessile leafy bracts. Pedicels thick, rigid; fruiting flowers often persisting on old stems. Valves 4–6 × 3–4 mm, ovate-triangular or suborbicular, truncate or subcordate at the base, with prominent reticulation of veins; usually all bearing unequal, oblong rugose tubercles. Nut 2.5 × 1.5 mm, trigonous, dark brown.

Margins of valves entire or with 1–2 teeth . . . . a. subsp. *anodontus*
Margins of valves with up to 8 teeth . . . . b. subsp. *woodsii*

a. subsp. **anodontus** (*Hausskn.*) *Rech.f.*, Beih. Bot. Centralbl. 49(2): 34 (1932); Rechinger & Schiman-Czeika, Fl. Iranica 56: 17 (1968).

Valves narrowly ovate-triangular, with 1 or 2 meshes of venation on each side of the tubercle, margins entire or with 1–2 teeth up to 0.5 mm long near the base.

HAB. Moist places, by stream, on sand dunes, alt. 40–550 m.
DISTRIB. Widespread in the alluvial plains and foothills of northern Iraq. **MRO**: Rowanduz gorge, *Guest* 3018!; Kuh-i Sefin nr Shaqlawa, *Bornmüller* s.n.; between Dukan and Mirza Rostam, *Rechinger* 11934. **FPF**: between Khanaqin and Baghdad, *Rechinger* 2152. **FNI**: Gerwona nr Ain Sifni: *Field & Lazar* 729! **FUJ**: Tal Afar to Balad Sinjar, *Field & Lazar* 520!; 86 km N of Mosul, *Rechinger* 10631! **LEA**: Shahraban (Muqdadiya), *Rechinger* 129! 9739! 33624! **LCA**: near Baghdad, *Field & Lazar* 519!; Baghdad, *Schaefli* 125! *Haussknecht* s.n.!; Mahmudiya, *Guest* 925!; nr Dukhala, 30 km N of Baghdad, *anon.* 1979!; al-Majarrah 50 km SW of Falluja, *Omar, Kaisi, Hamad & Hamid* 44515!

SW Asia, N Africa; introduced elsewhere.

b. subsp. **woodsii** (*De Not.*) *Arcangeli*, Comp. Fl. Ital. 585 (1882) (subsp. *divaricatus* (*L.*) *Murb.*)

Valves suborbicular or ovate-triangular, with usually 3 meshes of venation on each side of the tubercle and up to 8 rather irregular teeth 0.5–1 mm long on each margin.

HAB. Rocky slopes, sandy stony valley, sand dunes, 700–1200 m; fl. May–Jun.
DISTRIB. Occasional in N. Iraq. **MAM**: 5 km S. of Zakho, *Rawi* 23105. **MRO**: Gali Ali Beg, *Anders.* 1416; **MSU**: 20 km NW of Sulaimaniya, Dukan highway, *Rawi* 21757.

S Europe, Lebanon, Turkey and Iran.

Distribution of species: S and W Europe, N Africa, SW Asia; introduced in N America and elsewhere.

9. **Rumex dentatus** *L.*, Mantissa Alt. 226 (1771).

Annual or biennial 20–70 cm tall, the stem erect, unbranched or with a few erect branches. Basal leaves small, 2–3 times as long as wide, lanceolate or (rarely) panduriform, truncate or subcordate at the base, weakly hairy or glabrous, long-petiolate. Panicle long, with clustered whorls, remote with a few linear-lanceolate bracts which have cuneate bases; pedicels longer than the valves (sometimes almost twice as long), jointed well below the middle, rather slender. Valves 4–6 × 2–3 mm, triangular, subacute, all or only one with a tubercle; margins with 1–5 straight spine-like teeth 3–6 mm long. Fruit 2–2.5 mm. Fig. 51: 5.

subsp. **mesopotamicus** *Rech.f.*, Beih. Bot. Centralbl. 49 (2): 15 (1932). — Type: Baghdad, *Haussknecht* s.n. (W destroyed); Rech.f. & Schiman-Czeika, Fl. Iranica 56: 22 (1968); Boulos, Fl. Egypt 1: 32 (1999).

All valves bearing tubercles; nerves of the valves stout, with two reticulation-masses on each side of the tubercle.

HAB. Weed of cultivated fields of *Oryza, Triticum, Hordeum* and *Medicago*, often in wet or subsaline places and on waste ground; alt. 3–110 m; fl. ?Apr.–Jun.
DISTRIB. Widespread on the Mesopotamian plains. **FNI**: Mandali, Ain Kebrid, *Handel-Mazzetti* 1183; Nineveh, *Bornmüller* s.n. **FPF**: Mandali, *Guest* 886!; **FPF**: Between Khanaqin and Baghdad, *Rechinger* 2153. **LEA**: nr Hafriya, 60 km S of Baghdad, *Rawi* 18391!; Ruz plain, *Sutherland* 431b! 432!; Tigris plain, *Low* 337! **LCA**: Baghdad, *Haussknecht* s.n. (type), *Colvill* s.n.! *Guest* 3238!; **LCA**: Baghdad to Zafraniya, *Rechinger & Naib* 21!; Rustam farm nr Baghdad, *Lazar* 292! *Rogers* 0261!; Aziziya, *Guest* 261 p.p.!; Hilla, *Graham* 239!; 15 km SE of Hai, *Thamer & Abdel Jabar* 50065!; Baghdad to Zafraniya, *Rechinger & Naib* 21!; 46 km SE of Musaiyib, *Thamer & Wedad* 47270!; Shatra farm, *Omar & Karim* 36766!; Abu Ghraib, *Hasan* 34130! *Alizzi, Thaib & Janan* 32591!; 64 km NW of Nasiriya, *Khatib & Alizzi* 32777! **LSM**: Musaida nr Amara, *Field & Lazar* 56!; **LSM**: Hor Hawiza, 10 km E of Qurna, *Thamer* 46699!; Qal'at Salih, *Botany Staff* 42469!; Mesopotamia, *Watson* s.n.! 'Tigris plain', *Low* 337; Lower Diyala, *Robertson* 226!

Egypt, Cyprus, Arabia, Syria, Palestine, W and S Iran.

A polymorphic species, with seven subspecies described. There are two old records of subsp. *reticulatus* (Bess.) Rech. f., which differs in having only one valve bearing a tubercle, with three reticulation-masses on each side of the tubercle: (Mosul, *Noë* s.n.; nr Baghdad, *Olivier* s.n.). As no modern material has been seen, and material from these collectors is sometimes inaccurately localised, these must be regarded as of doubtful occurrence.

Rumex × baqubensis Rech. f., Fl. Iran 56: 22 (1968) (*angustifolius* × *dentatus*) was found growing with the parents in a *Phoenix dactylifera* plantation: **LEA**: Baquba, *Rechinger* 9751 (W holo; B! K! iso). It is intermediate between the parents. Other hybrids in *Rumex* are likely to occur.

Distribution of the species: East Europe to Russia, SW and E Asia.

10. **Rumex chalepensis** *Miller*, Gard. Dict. ed. 8, no. 11 (1768).

R. denticulatus C.Koch, Linnaea 22: 208 (1849), non Campd.; *R. syriacus* Meisn. ap. DC, Prodr. 14: 53 (1856)
R. dictyocarpus Boiss. & Buhse, Mém. Nouv. Soc. Nat. Mosc. 12: 192 (1860).

Perennial, 40–100 cm tall, much-branched, with pale brown stems and branches. Basal leaves up to 20 × 8 cm, oblong or ovate-oblong, cordate at the base, usually obtuse. Stem leaves few, ovate-oblong or -lanceolate, obtuse or acute. Panicle with ascending or divaricate branches and flowers in remote whorls, the lower whorls with leafy bracts. Pedicels slender but abruptly thickened just below valves. Valves 5-7 × 5-6 mm, ovate-triangular, cordate at the base, subobtuse, with reticulation of prominent raised veins; all bearing subequal

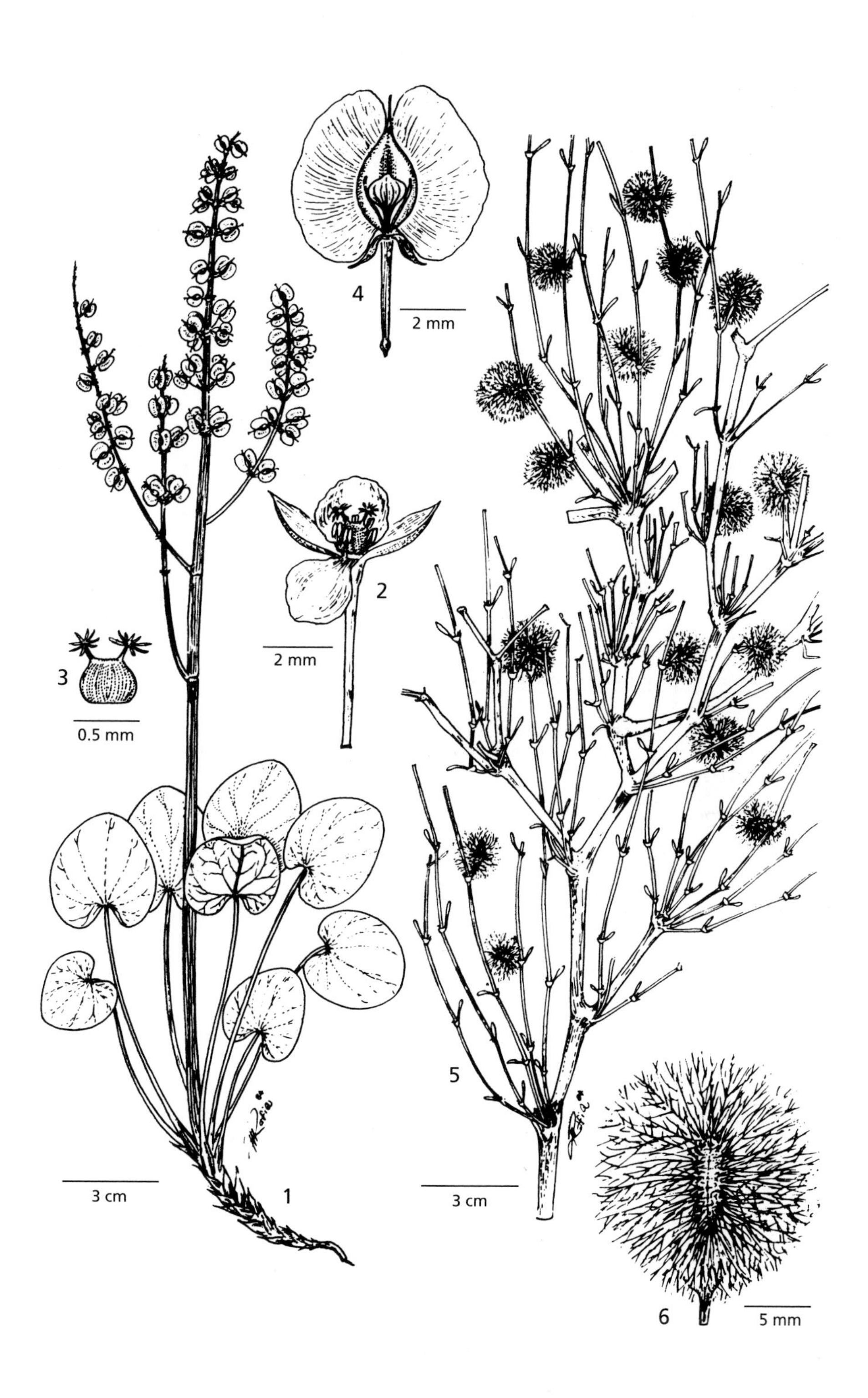

Fig. 53. **Oxyria digyna**. 1, habit; 2, flower; 3, pistil; 4, fruit. **Calligonum polygonoides**. 5, habit; 6, fruit. Reproduced with permission from Flora of Pakistan 205: f. 20, 2001. Drawn by M. Rafiq.

smooth (rarely rugose) ovate-oblong tubercles, margins with 4–9 acute teeth up to 2.5 mm. Nut 2.5-3 × 1.5 mm, trigonous, brown.

DISTRIB. **MRO**: Savan village on Jabel Karoukh, 1000–1500 m, *Nuri & Kass* 27277! Determined as *R. chalepensis* by K.H. Rechinger in 1998; this is an immature specimen and based on this single collection we cannot be sure if the species occurs in Iraq.

Syria, S Anatolia, Caucasus, Iran, C Asia to Manchuria.

SPECIES DOUBTFULLY RECORDED

A specimen from the alpine region of Mt. Halgurd on the frontier of Iran, collected and named as *Rumex patientia* L. on 26 June 1893 by J. Bornmüller, is barely in flower. K.H. Rechinger has annotated it "in statu florendi non certe determinabilis" (not definitely identifiable in flowering condition). Its lower leaves are rounded at the apex (acute in *R. patientia*). Another poor specimen from "Kurdish hills near Rayat, 6000 ft, gravelly soil", *A. Low* 206 (BM!) bears an ambiguous pencil note by K.H. Rechinger "subs. pat." As these are the only records for this species from Iraq the occurrence of *R. patientia* must be treated as doubtful; or they may represent material of 4. *R. ponticus,* collected from Mt. Halgurd, similar enough to have been treated as a variety of *R. patientia* (var. *kurdicus* Boiss.).

3. **OXYRIA** Hill

Veg. Syst. 10: 24 (1765)

Perennial herbs with thick rootstock. Leaves glabrous, mostly basal, long-petiolate, reniform to orbicular-cordate, somewhat fleshy. Inflorescence terminal, paniculate. Flowers bisexual. Perianth segments 4, greenish, outer pair narrower, spreading in fruit, inner pair clasping the fruit, becoming enlarged and tinged with red. Stamens 6. Stigmas 2. Fruit lenticular, glabrous, reddish, broadly winged, longer than inner perianth segments.

Oxyria (from Gr. οξύς, *oxus*, sharp-tasting.)

Monotypic, distributed in Eurasia and the Arctic regions.

Oxyria digyna (*L.*) *Hill*, Hort. Kew. 158 (1769); Cullen in Fl. Turk. 2: 269 (1967); Rech.f. & Schiman-Czeika, Fl. Iranica: 24 (1968); Qaiser in Fl. Pak. 126: 169 (2001); Breckle et al., Vasc. Pl. Afghan. Checkl.: 406 (2013).

Rumex digynus L., Sp. Pl. 337 (1753).
Oxyria elatior R.Br. in Wallich, Numer. List n. 1726 (1829).

Glabrous, tufted, perennial herb, stems 9–16 cm; rootstock clothed with persistent petiole bases. Leaves 2–3.5 × 1–3 cm, reniform to orbicular, mostly basal, becoming reddish with age; petiole 6–12 cm; ochrea membranous. Inflorescence narrowly paniculate, bearing clusters of 2–5 flowers. Flowers bisexual. Perianth segments 4, greenish to pinkish, without tubercles, outer pair spreading or reflexed, not accrescent; inner pair appressed to nut, enlarging slightly in fruit. Fruit ovoid, 3–5 mm, exceeding the inner perianth segments, broadly winged, cordate at base, apex notched, margin bright pink and becoming scarious, denticulate. Fig. 53: 1–4.

HAB. Stony mountain slopes, crannies of metamorphic rocks, by melting snow patches; alt. 2300–3600 m; fl. Jul.–Aug.
DISTRIB. Local in thorn-cushion zone of the high mountains of NE Iraq. **MRO**: Halgurd Dagh, *Guest* 2838!, 3068! *Gillett* 9601!, 12354! *Rawi & Serhang* 24356!, 24756!, 24818! *Rechinger* 11867; Qandil range, *Rawi & Serhang* 24394!, 24443a!, 24603!, 26777! *Rechinger* 11824.

Arctic Eurasia and North America, mountains of Europe and SW Asia eastwards to Himalaya, China and Korea; Turkey, Armenia, Caucasus, N Iran, Afghanistan, Pakistan.

4. **RHEUM** L.

Sp. Pl. 372 (1753)

A stout herb with thick rhizome. Stem hollow, sulcate, glabrous. Leaves basal and cauline, simple, broadly elliptic to reniform, margins sinuate; ochreae not ciliate. Inflorescence branched, paniculate, flowers in clusters. Flowers hermaphrodite. Perianth of 6 equal segments. Stamens (7–)9. Styles 3; stigmas inflated. Achenes trigonous, membranous-winged.

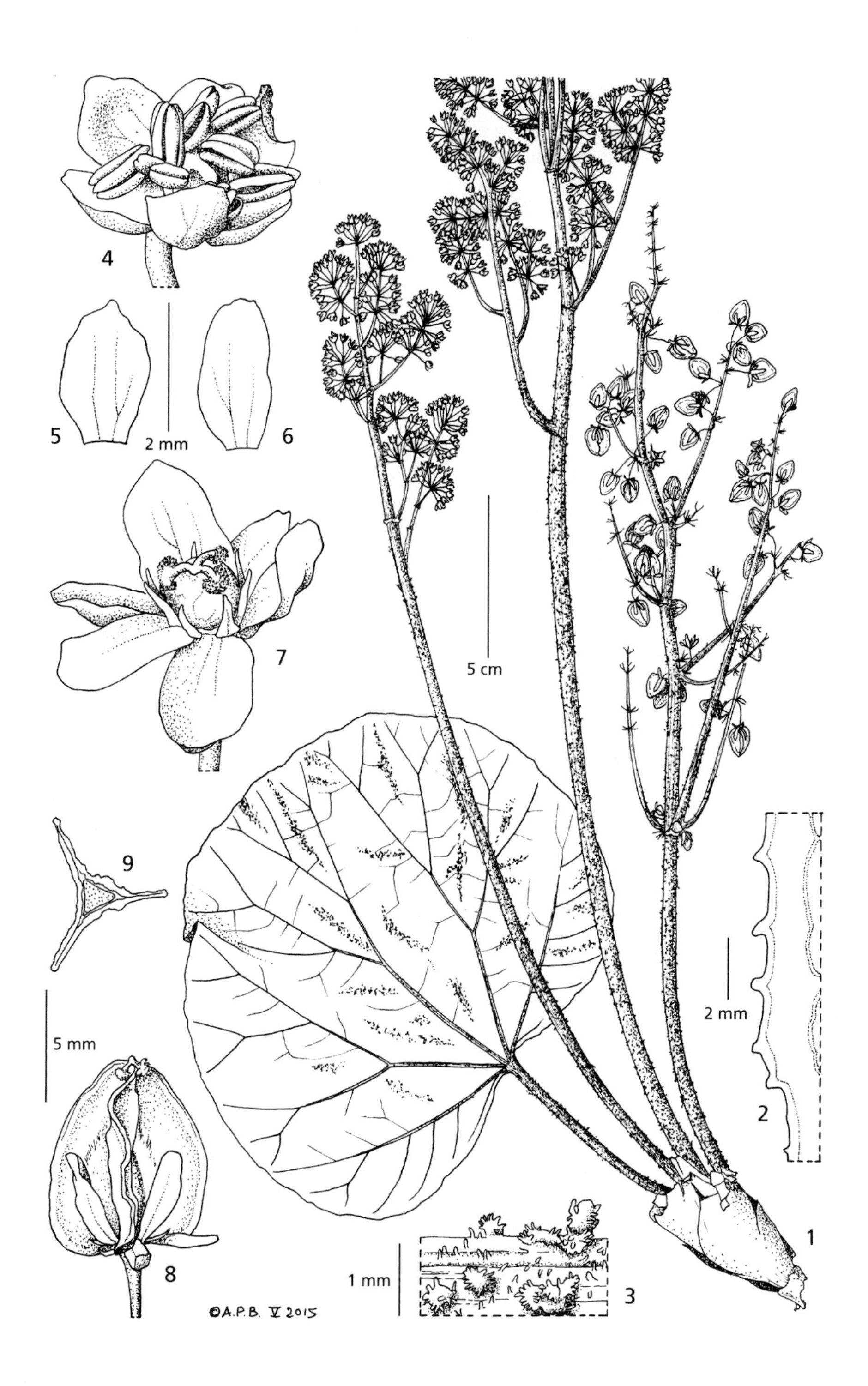

Fig. 54. **Rheum ribes**. 1, habit; 2, leaf margin; 3, indumentum on petiole; 4, young flower; 5, outer perianth segment; 6, inner perianth segment; 7, mature flower; 8, fruit lateral view; 9, transverse section of fruit. 1, 3–9 from *Davis* 44479; 2 from *Davis* 45626. © Drawn by A.P. Brown.

Rheum (from Gr. ρεύμα, *reuma*, stream, possibly because the plant was first introduced from the region of the river Volga, whose ancient name was Rha.)

About 50 species distributed in temperate and subtropical Asia; two species in Iraq.

Fruit 10–16 mm (including wings) ....1. *R. ribes*
Fruit 6–10 mm (including wings) ....2. *R. palaestinum*

1. **Rheum ribes** *L.*, Sp. Pl. 372 (1753); Jaub. & Spach, Ill. Pl. Or. 5: 470 (1856); Losinsk. In Fl. USSR 5: 497 (1936); Cullen in Fl. Turk. 2: 268 (1967); Rech.f. & Schiman-Czeika, Fl. Iranica: 27 (1968); Leonard, Contrib. Fl. et Veg. Deserts d'Iran 5: 84 (1985); Nyberg & Miller in Fl. Arab. Penins. & Socotra 1: 141 (1996); Qaiser in Fl. Pak. 130: 169 (2001); Breckle et al., Vasc. Pl. Afghan. Checkl.: 409 (2013).

A stout perennial herb with leafy stems (leafless in upper part), to 100 cm. Leaves 2–5, reniform to suborbicular, petiolate, lamina bullate, to 50 × 70 cm. Inflorescence a much-branched, widely spreading panicle. Pedicels jointed below the middle. Perianth segments white, brown and persistent in fruit. Fruit triquetrous, reddish-brown, with broad wings, 10–16 mm (including wings). Fig. 54: 1–9.

HAB. Stony places in mountains, dry loam in N.-facing *Quercus* forest; alt. 1220-2000 m; fl. May–Jun.
DISTRIB. Found occasionally in the upper forest zone of NE Iraq. **MRO**: Hawara Blinda, NE of Haji Umran, *Rawi, Nuri & Kass* 27783 p.p.!; Kuh Sefin nr Shaqlawa, *Bornmüller* 1773; Shaqlawa, *Haines* s.n.!; Rubar-i Sideka, *Haley* 205!; Hissar-i Rost, *Thesiger* 912! **MSU**: mt. Pira Magrun, *Haussknecht* s.n., *Rawi* 11531 & 12100, *Faris* 6057 & 6058;

RAIWAS (Kurd.-Pira Magrun, Rawi 11531 & 12100). Said to taste of currents (*Ribes*). The flowering stems are gathered for food in the Van province of SE Turkey, S Iran (Kuh-e Dena, fide Kotschy) and in Lebanon, so it is likely that this is also a seasonal delicacy in NE Iraq, though local information is lacking. It is regarded as a medicinal plant across much of its native range, especially the dried root. Probably under-recorded, as due to its awkward size the plant is difficult to press.

Palestine, Lebanon, Syria, Turkey (E Anatolia), Armenia, NW, S and E Iran, Afghanistan, Pakistan.

2. **Rheum palaestinum** Feinbr., Palestine J. Bot., Jerusalem Ser., 3: 117 (1944); Zohary, Fl. Palaest. 2: 58 (1972).

Stout perennial herb, to 50 cm. Ochrea somewhat triangular, acute at apex. Leaves 2–3, lying close to ground surface; petiole ± 5 cm; lamina bullate, reniform to suborbicular, up to 52 × 83 cm. Inflorescence a much-branched, spreading panicle, stems sulcate, glabrous. Perianth segments pink. Fruit pink 6–10 mm (including wings).

HAB. Gravelly, sandy, and loamy and clay soils, in the Western Desert; alt. 200-800 m; fl. Mar.–Apr.
DISTRIB. Rare and coming up only after good rain, in the Western Desert. **DWD**: nr. Baiji, *Rawi & Hamada* 33572! 60 km W of Rutba, *Rawi* 30490 (BAG); 5 km from Rutba-Nihaidain junct. *Al-Khayat, Omar & Adel* 55146 (BAG); Rutba, *Rawi* 23752 (BAG); 55 km S of Rutba, Ras Al-Emsad, *Musa* 2101 (AUH).

GHUTAIRFAN, AUTHNA (in Western Desert District). Recorded as not being grazed by livestock. In olden times water accumulated in the ochrea was drunk as a cure for abdominal diseases; extract of flowers has been reported to be used to treat intestinal worms.

Palestine, Jordan, ?Saudi Arabia.

5. **ATRAPHAXIS** L.

Sp. Pl. 333 (1753); Gen. Pl. ed. 5.1754

Erect often much-branched shrublets 60–80 cm with fibrous bark; older branches persisting as spines. Leaves elliptic, subsessile; ochreae brownish below, scarious above, bifid. Flowers hermaphrodite, borne in racemes. Perianth segments 4 or 5, whitish, inner accrescent in fruit forming broad scarious wings. Stamens 6–8, fused in lower part. Stigmas 2–3. Nuts triquetrous.

Atraphaxis (from Gr. α, *a*, lacking and τροφή, *trofi* nourishment.)

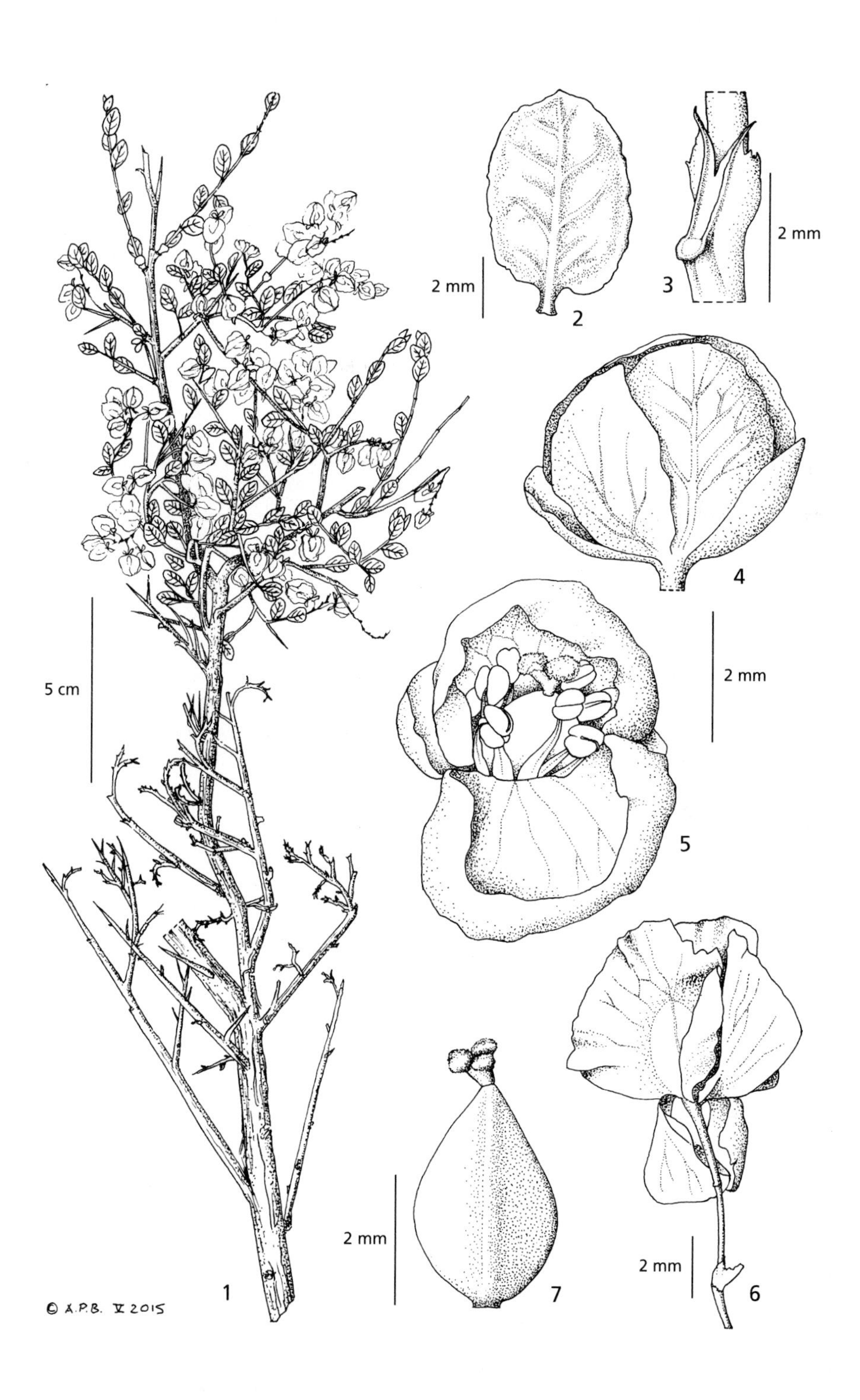

Fig. 55. **Atraphaxis billardieri**. 1, habit; 2, leaf abaxial surface; 3, ochrea; 4, flower lateral view; 5, flower view from above; 6, mature flower; 7, nut with style attached. 1–3 from *Breidy et al.* MSSKJB1819; 4–7 from *Khairallah et al.* MSSKAJ1952. © Drawn by A.P. Brown.

A genus of about 25 species distributed in N Africa, SE Europe to the Himalayas and Eastern Siberia; a single species in Iraq.

Atraphaxis billardieri *Jaub. & Spach*, Ill. Pl. Or. 2: 14, t. 111 (1844–46); Cullen in Fl. Turk. 2: 266 (1967);

Atraphaxis tournefortii Jaub. & Spach, Ill. Pl. Or. 2: 14, t. 111 (1844–46).

Intricately branched dwarf shrub 20–30 cm, with somewhat spiny branches; young bark greyish, older brownish. Leaves elliptic-oblong to ovate. Ochreae bicuspidate. Pedicels jointed in lower half. Perianth 5-merous, white. Stamens 8; Styles 3. Fruiting perianth pinkish white, becoming brown and scarious, outer two lobes shorter than the inner and reflexed. Fruit triquetrous, with sharp angles, to 0.8 cm. Fig. 55: 1–7.

HAB. Stony mountain slopes, sometimes on slate or serpentinite, *Quercus* forest; alt. 1360-1800 m; fr. Jun.–Aug.
DISTRIB. Occasional in the upper forest zone of NE Iraq. **MRO**: Halgurd Dagh, Gundar Sher Darband, *Gillett* 12389!; Jabal Karoukh between Saran and Kilkil, *Kass, Nuri & Serhang* 27333!; Dargala, 35 km NW by N of Rania, *Rawi, Nuri & Kass* 28889! **MSU**: Penjwin, *Rawi* 12239! *Rechinger* 10437!; mt. Kajan nr Penjwin, *Rawi* 22687! **FNI**: Mt. N of Mosul, *Dimmock* 0396! **FPF**: Jabal Hamrin, *Sutherland* 424!

Two varieties are sometimes distinguished, var. *billardieri* and var. *tournefortii* (Jaub. & Spach) Cullen, but as noted by Cullen (Fl. Turkey 2: 267, 1967) they are difficult to separate.

KARWAN KUGA (Kurd.-Penjwin, *Rawi* 12239).

C Greece, Crete, Turkey, Syria, Lebanon, Sinai, W Iran. According to Browicz (op. cit.) this species often occurs in badly degraded, overgrazed habitats.

6. **PTEROPYRUM** Jaub. & Spach

Ill. Pl. Or. 2: 7 (1844)

Virgate shrubs with greyish bark. Leaves fasciculate, entire, somewhat fleshy when young; ochreae very short. Flowers bisexual, in axillary fascicles. Perianth petaloid, 5-fid, the two outer lobes becoming deflexed, inner clasping fruit. Stamens 8, inserted at throat of perianth, filaments barbulate at base, anthers versatile. Styles 3, short. Fruit with 3 broad membranous wings, each wing bilobed with a rounded sinus above middle.

Pteropyrum, from Gr. ϖτέϱυξ, *pteryx*, wing and ϖυϱος, *pyros*, kernel).

A small genus of five species distributed in SW and S Asia; two species in Iraq.

1. Leaves ± linear, margins revolute, 6× or more as long as broad; pedicels scarcely articulated near the base . . . . 1. *P. aucheri*
   Leaves oblong-spathulate, ± flat or with margins revolute towards tip, less than 6× as long as broad; pedicels distinctly articulated near the base. 2
2. Rigidly branched shrub to 1.5 m; leaves oblong to obovate-spatulate, (2–)3–5 mm broad; flowers spicate in upper 1/3 of stems . . . . 2. *P. olivieri*
   Small shrub to 30 cm, branchlets whitish, papillose; leaves obovate, up to 20 × 10 mm; flowers in clusters on capillary, minutely papillate pedicels. . . . . 3. *P. naufelum*

1. **Pteropyrum aucheri** *Jaub. & Spach*, Ill. Pl. Or. 2: 8 (1844); Boiss., Fl.Orient. 4: 1002 (1879); Rech.f. & Schiman-Czeika in Fl. Iranica 56: 36 (1968); Leonard, Contrib. Fl. et Veg. Deserts d'Iran 5: 80 (1985); Qaiser in Fl. Pak. 205: 166 (2001); Breckle et al., Vasc. Pl. Afghan. Checkl.: 408 (2013).

Pteropyrum ericoides Boiss., Fl. Orient. 4: 1002 (1879).

Rigidly branched shrub to 1 m, young branches with greyish-white bark. Leaves glabrous, narrowly linear-lanceolate, to 2 mm broad, margins revolute. Inflorescence occupying the upper 1/2 of stems, spicate; flowers shortly pedicillate; Perianth white. Anthers red. Fruit green, wings scarlet. Fig. 56: 1–4.

HAB. River island, rocky slopes, dry conglomerate hills, sandy rocky soil, denuded *Quercus* woodland; alt. 300–1000 m; fl. May; fr. Jun.–Oct.

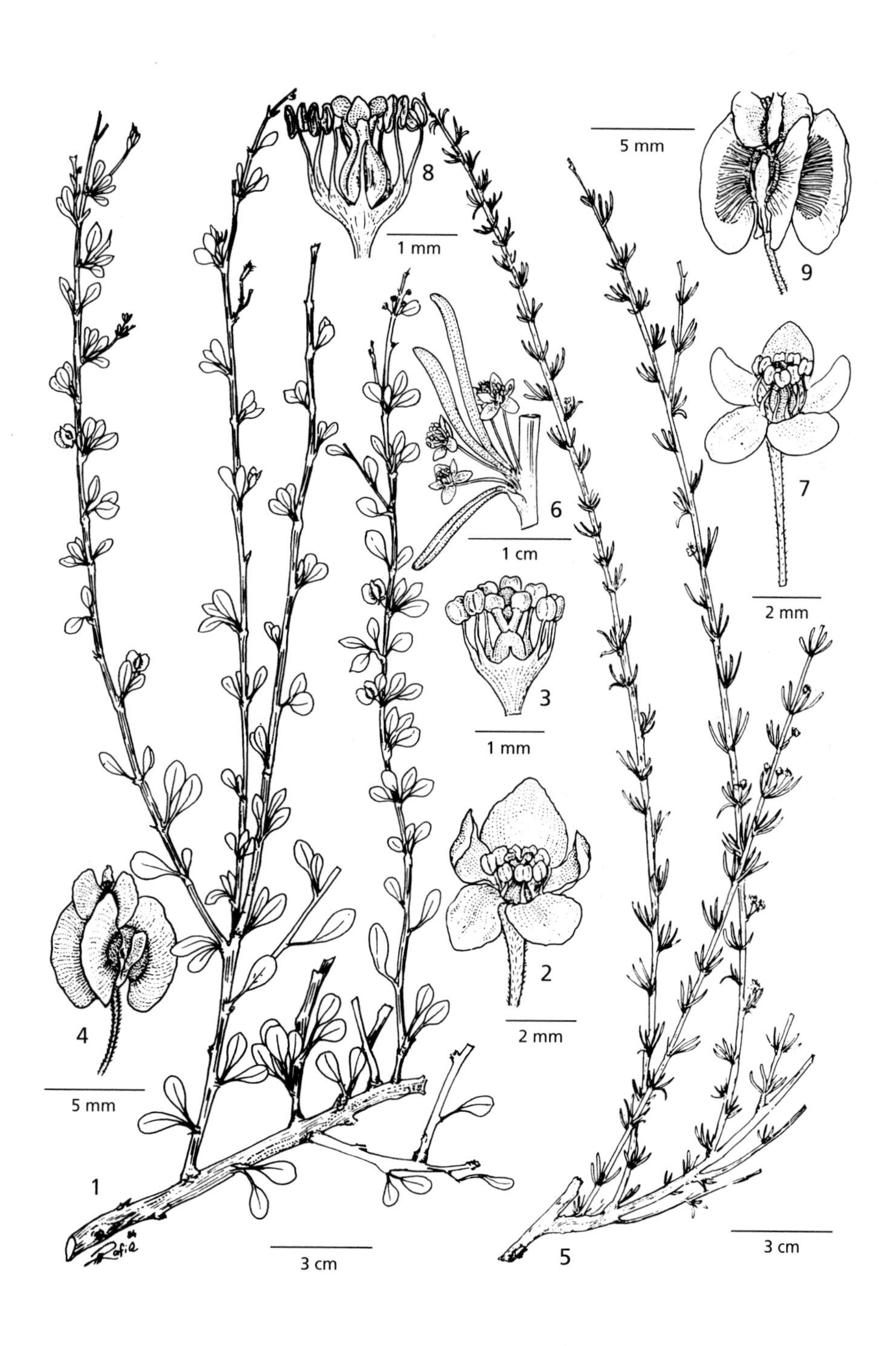

Fig. 56. **Pteropyrum aucheri**. 1, habit; 2, flower; 3, flower with perianth removed; 4, fruit. **Pteropyrum olivieri**. 5, habit; 6, flowering branch; 7, flower; 8, flower with perinath removed; 9, fruit. Reproduced with permission from Flora of Pakistan 205: f. 28, 2001. Drawn by M. Rafiq.

DISTRIB. Scattered in dry steppe areas of lowland Iraq. **MRO**: Khalakan, *Omar, Kaisi & Wedad* 49461! **MSU**: Serchinar nr Sulaimaniya, *Raddi* 5282!; Qara Dagh village, *Gillett* 7964!; Darband-i Khan, *Haines* W1535! **MSU/FPF**: Jalaula to Darbandikhan, *Barkley* 9268. **FNI**: islands in r. Zab nr Eski Kellek, *Gillett* 8203! **FAR**: S of Bastura Chai on Arbil-Shaqlawa road, *Gillett* 11290!; Arbil to Kirkuk, *Nabelek* 462; Bulaq, 20 km N of Arbil, *Jackson* 3530 **FPF**: Sangar nr Khanaqin, *Chakravarty & Rawi* 30760!; **FPF**: 3 km W of Sa'diya, *Kaisi* 42896!; Jabal Muwaila, nr Kuwait, ± 70 km N of Amara, *Guest, Rawi & Rechinger* 17582!; **FPF**: Jabal Ruhaila, 70 km N of Amara, *Rechinger* 14150. **LCA**: Baghdad, *Rogers* 0247!

Gillett (11290) notes that "I believe that this shrublet may once have been characteristic of moist steppe in N Iraq but has been mainly destroyed by overgrazing, cultivation and the hunt for fuel."

Iran, Afghanistan, Pakistan.

2. **Pteropyrum olivieri** *Jaub. & Spach*, Ill. Pl. Or. 2: 5 (1846); Cullen in Fl. Turk. 2: 268 (1967); Leonard, Contrib. Fl. et Veg. Deserts d'Iran 5: 82 (1985); Qaiser in Fl. Pak. 205: 169 (2001); Breckle et al., Vasc. Pl. Afghan. Checkl.: 409 (2013).

Pteropyrum gracile Boiss., Diagn. ser. 1, 12: 102 (1853); *P. olivieri* Jaub. & Spach var. *gracile* Boiss., Fl. Or. 4 : 1002 (1879); *P. noeanum* Boiss. ex Meisner in DC., Prodr. 14: 31 (1856).–Type: "Ahanneky ad ripam Dialae" (unknown locality on the bank of the r. Diyala), *Noë* 1002, Sep. 1851; "am Fluss Diala", *Noë* 1002, Aug. 1851 (syn. K!). Rech.f. & Schiman-Czeika in Fl. Iranica 56: 38 (1968).

Rigidly branched shrub to 1.5 m, with whitish bark. Leaves glaucous, oblong to obovate-spatulate, (2–)3–5 mm broad, margins scarcely revolute. Inflorescence occupying the upper $^1/_3$ of stems, spicate; flowers shortly pedicellate; Perianth white. Anthers reddish pink. Ovary green. Fruits oblong, pink, wings reddish. Nut brownish. Fig. 56: 5–9.

HAB. Sandy, gravelly and rocky desert, dry limestone & conglomerate hills; alt. 160–620 m; fr. Oct.
DISTRIB. Scattered in dry steppe areas of Iraq. **MSU**: Darbandikhan, *Haines* W535!; Jarmo, *Haines* W229!; Timar, mt. Qaradagh, *Karim, Hamid & Jasim* 40559; a few km N.E. of Mandali towards Khir Charmaga, *Kaisi & Hamad* 43566. **LEA**: Jabal Hamrin, nr Muqdadiya (Shahraban), *Guest* 15798! ?**LEA**: "Ahanneky ad ripam Dialae" (unknown locality on the bank of the r. Diyala), *Noë* 1002, Sep. 1851 (syntype of *P. noeanum*)!; ?**LEA**: sine loc., *Sutherland* 425, 426.

We agree with Rechinger (Fl. Iranica: 38, 1968) as he states that it is doubtful whether the specific distinction between *P. noeanum* and *P. olivieri* is justified. While our treatment recognises the two species in Iraq, further work is needed to establish whether or not the differences of leaves and bark represent merely edaphic variants.

Turkey (SE Anatolia), Iran, Afghanistan, Pakistan.

3. **Pteropyrum naufelum** *Al-Khayat*, Nordic J. Bot. 13 (1): 33, f.1 (1993). Type: Iraq, (District FPF), Hashima, on Iraq-Persian border, nr. Badra, Muthanna Province, 2 Dec. 1962, *Agnew, Hashimi & Hahad* (BUH holo; BAG iso).

Small shrub to 30 cm, much branched, with a brownish white, fissured bark; branchlets whitish, papillose. Ochrea short, truncate, hyaline, with 2 triangular auricles. Leaves think and leathery, obovate, up to 20 × 10 mm, margins scarcely revolute, papillose, greyish white. Flowers in clusters; pedicels capillary, minutely papillate, jointed above the base. Perianth of 5 unequal segments, ± 3 mm, obovate. Fruit cordate-ovoid, 5–8 × 5–7 mm, beaked, red becoming brown when mature.

HAB. Dry stony hillside; alt. 160–620 m; fl. & fr. Dec.–Feb.
DISTRIB. Localised. **FPF**: Anaiza, between Mandali & Badra, *Agnew, El-Hashimi & Hahad* 5143 (BUH); nr. Hashima, Tursuk-Badra rd., 2 Feb. 1962, *El-Hashimi* 28088 (BUH); Anaiza police station, 1 Dec. 1962, *El-Hashimi* 32494 (BUH).

Endemic to Iraq, but likely to occur in Iran.

7. **CALLIGONUM** L.

Sp. Pl. 530 (1753); Gen Pl. ed. 5, 235 (1754)

Much-branched shrubs, lacking spines; leaves borne on current year's flexible young twigs. Leaves simple, opposite, linear, caducous; ochrea short, membranous, scarious. Inflorescence axillary; pedicels slender, jointed. Flowers bisexual, short-pedicellate, in axillary clusters. Perianth 5-fid, not accrescent in fruit. Stamens 12–18; filaments fused at base. Ovary 4-angled, with 4 styles; stigmas capitate. Nuts trigonous, ellipsoid to ovoid, becoming woody; ribs with wings or bristles.

Ref. Yu. D. Soskov (1975). Novye serii, podvidy i gibridy v rode *Calligonum* L. (Polygonaceae). *Bot. Zhurn.* 60 (8): 1162–1163 (1975).

Calligonum (from Gr. καλλιοτος, *kalliotos*, most beautiful, and γονυ, *gonu*, knee.). The Ar. name of the genus is 'IRTAH.

About 80 species in Asia, Africa and SE Europe; three species in Iraq.

1. Fruits with 4 sinuate to denticulate membranous wings. . . . . 1. *C. tetrapterum*
   Fruits densely clothed with soft branched bristles. 2
2. Fruit oblong, with no more than 10 setae-branches on each wing . . . .2. *C. polygonoides*
   Fruit subglobose; more than 10 setae-branches on each wing . . . . 3. *C. comosum*

1. **Calligonum tetrapterum** *Jaub. & Spach*, Ill. Pl. Or. 2: t. 471 (1844–46). — Type: (?Syria/Iraq): de Bagdad à Alep, *Olivier* s.n. (P). Rech.f., Fl. Lowland Iraq: 177 (1964); Rech.f. & Schiman-Czeika in Fl. Iranica 56: 42 (1968); Mouterde, Nouv. Fl. Lib. et Syrie 2: 675 (1970); Nyberg & Miller in Fl. Arab. Penins. & Socotra 1: 142 (1996); Ghazanfar, Fl. Oman: 1: 85 (2003).

Shrub 30–60 cm, with stiff greyish-white main branches. Leaves deciduous, usually absent, arising from the younger shoots. Flowers arising in clusters at nodes of younger branches; pedicels usually slightly shorter than length of perianth. Perianth segments whitish with pink median stripe, oblong, obtuse, ± 3 mm. Fruits reddish or greenish, to 1 cm, with 4 slightly twisted denticulate wings, lacking bristles. Fig. 57: 3.

HAB. Gravelly and sandy desert, rocky hill, clay gravelly soil, slopes of sandy gypsum soil; fl. Feb.–May, fr. Apr.–Oct.
DISTRIB. Scattered, mainly in the desert zone of Iraq. **DLJ**: Thirthar depression, *Haines* W1459!; shore of lake Thirthar, *Musawi* 54002. **DWD**: Habbaniya, *Adqa & Ani* 1409!; 40 km W of Ramadi towards Rutba, *Kaisi & Hamad* 48930; 70 km N. of Rutba, *Chakr., Rawi, Khatib & Alizzi* 32924. **DSD**: Jabal Sanam, *Rawi & Gillett* 6149 p.p!; Falluja desert, *Agnew & Haines* s.n.!; 20 km from Busaiya, *Karim, Nuri, Hamid & Kadhim* 40343; **LCA**: 35 km W of Baghdad, *Agnew & Juma Brahim* 60/3! 10 km NW of Falluja, *Rawi* 20264.

Saudi Arabia, Jordan, Syria, Turkey, Iran, Afghanistan, Pakistan, India, C Asia.

2. **Calligonum polygonoides** *L.*, Sp. Pl. 530 (1753); Boiss., Fl. Orient. 4: 998 (1879); Pavlov in Fl. USSR 5: 568 (1936); Cullen in Fl. Turk. 2: 268 (1967); Rech.f. & Schiman-Czeika in Fl. Iranica 56: 43 (1968); Nyberg & Miller in Fl. Arab. Penins. & Socotra 1: 142 (1996); Boulos, Fl. Egypt. 1: 24 (1999); Qaiser in Fl. Pak. 205: 177 (2001).

Subshrub, to 1.2 m tall. Older branches with brownish-white bark; younger shoots green, 1–2 mm in diameter. Leaves up to 1 cm, soon deciduous and usually absent. Perianth segments oblong, obtuse, 3–5 mm, white. Anthers pink. Fruits oblong, red or greenish, 0.7–0.8 × 0.5–0.6 cm, clothed in branched ± rigid bristles 3–5 mm arising from 4 pairs of short wings. Fig. 53: 5–6.

HAB. Rocky gullies; alt. 30–50 m; fl. Feb.
DISTRIB. Rare in the desert zone of central Iraq. **DGA/LCA**: Adhaim district, N of Khalis, *Robertson* H.35. **DSD**: Jabal Sanam, *Rawi & Gillett* 6149! *Graham* 43! Ichrishi, 35 km E by N of Busaiya, *Guest & Rawi* 14180 p.p. **FPF**: Jabal Hamrin, *Sutherland* 422!

Egypt (Sinai), Arabian Peninsula, Palestine, Syria, Turkey (E Anatolia), Armenia, NW and C Iran.

3. **Calligonum comosum** *L'Herit.*, Trans. Linn. Soc. Lond. 1: 180 (1791); Boiss., Fl. Orient. 4: 1000 (1879); Rech.f., Fl. Lowland Iraq: 177 (1964); Rech.f. & Schiman-Czeika in Fl. Iranica 56: 44 (1968); Zohary, Fl. Palaest. 2: 68 (1972); Nyberg & Miller in Fl. Arab. Penins. & Socotra 1: 141 (1996); Qaiser in Fl. Pak. 205: 175 (2001); Ghazanfar, Fl. Oman 1: 84 (2003); Breckle et al., Vasc. Pl. Afghan. Checkl.: 405 (2013).

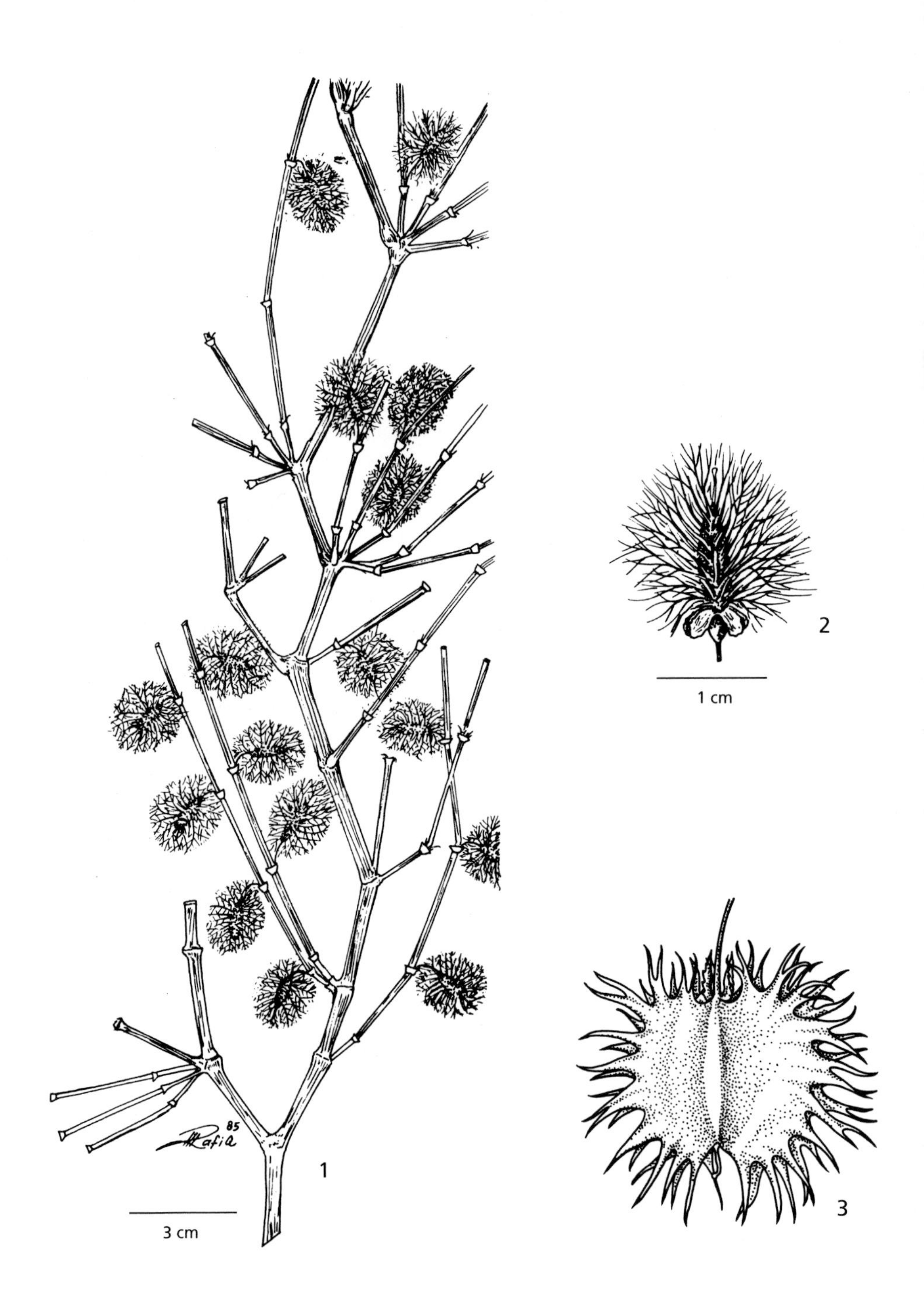

Fig. 57. **Calligonum comosum**. 1, habit × $^1/_2$; 2, fruit. **Calligonum tetrapterum**. 3, fruit × 4. 1, 2 Reproduced with permission from Flora of Pakistan 205: f. 28, 2001; 3 reproduced with permission from Flora of the Arabian Peninsula & Socotra 1: f. 25. 1996. Drawn by M. Rafiq.

Calligonum polygonoides L., Sp. Pl. 530 (1753) subsp. *comosum* (L'Herit.) Soskov, Nov. Sist. Vyssh. Rast. 12: 153 (1975); Boulos, Fl. Egypt 1: 24 (1999); *C. mejidum* Al-Khayat, Candollea 43(1): 273 (1988).

Subshrub, 0.5–1.5 m tall. Older branches with whitish bark, knotted at nodes; younger shoots green, 1–2 mm in diameter. Leaves 2–4 mm, soon deciduous and usually absent. Flowers fragrant. Perianth segments obtuse, ± 3 mm, greenish-white with a dark pink median stripe. Anthers pale purple. Stigmas whitish. Fruits ovoid, red or greenish, 10–15 × 05–8 mm, clothed in branched soft bristles 3–5 mm arising from 4 pairs of short wings. Fig. 57: 1–2.

HAB. Sandy desert; alt. 30–50 m; fr. Apr.–Oct.
DISTRIB. Rare in the desert zone of central Iraq. **DGA**: N of Khalis, *Robertson* 13486! **DLJ**: SE of Thirthar depression, *Haines* W1489! *Khayat & Omar* 54885 (type of *C. mejidum*); 70 km NNW of Falluja, *Rawi* 20264! **DSD**: Jabal Sanam, 1919, *Sharples* s.n.!; Falluja desert, *Haines* W193! Ichrishi, 35 km E by N of Busaiya, *Guest & Rawi* 14180 p.p.

C. tetrapterum and *C. comosum* cannot easily be distinguished unless fruits are present. The following specimens are therefore unassigned to a species: **DLJ**: Jazira nr Thirthar, *Emberger, Guest, Long, Schwarz & Serkahia* 15333!; **DSD**: Umm Qasr, *Khatib & Alizzi* 32669!

C. tetrapterum and *C. comosum* occupy similar territories across the Syrian Desert, though the latter also occurs in E Anatolia and Transcaucasia. Both occur in the Arabian Peninsula, but according to the map in Mandaville (p. 109, 1990) *C. tetrapterum* has a more northerly distribution. Both species are quite rare, perhaps because their suitability as fuel wood in desert regions has greatly reduced their populations, and their roots are also much sought after for turning into charcoal.

'IRTA (Busaiya, *anon.* 16445).

Egypt, Palestine, Jordan, Arabian Peninsula, Iran, Afghanistan, Pakistan.

8. **PERSICARIA** (*L.*) *Mill.*

Gard. Dict. Abrig. ed. 4, (1754); Ronse Decraene & Akeroyd in Bot. J. Linn. Soc. 98: 330 (1988)

Annual or perennial herbs. Stems usually erect. Leaves longer than wide, pinnately veined. Ochreae usually truncate, ciliate or fimbriate. Flowers usually bisexual, in cymes arranged in terminal or axillary panicles, racemes or spike-like or capitate inflorescences. Perianth-segments (4–)5, ± equal, petaloid, not winged or keeled. Stamens 5–8. Styles 2(–3). Nut trigonous, triquetrous or lenticular, enclosed in the persistent perianth or protruding from it for less than half its length.

About 150 species (inc. *Bistorta*), widespread and almost cosmopolitan; seven species in Iraq.

Persicaria, from the resemblance of the leaves to those of Peach (*Prunus persica*).

1. Annual, but sometimes rooting below 2
   Perennial, with well developed rootstock, sometimes semi-aquatic or aquatic 5
2. Plant densely pubescent; ochreae up to 20 mm, partly herbaceous . . . . . 2. *P. orientalis*
   Plant glabrous, subglabrous or glandular; ochrea usually less than 10 mm 3
3. Flowers greenish, in lax nodding racemes. . . . . 7. *P. hydropiper*
   Flowers pink or greenish-white, in dense spike-like racemes 4
4. Peduncles smooth, eglandular or nearly so, shiny . . . . 4. *P. maculosa*
   Peduncles rough with sessile yellow glands . . . . 5. *P. lapathifolia*
5. Leaves mostly basal, with winged petioles; stems unbranched . . . . 1. *P. bistorta*
   Leaves mostly cauline, with unwinged petioles; stems usually branched. 6
6. Petiole at least 2 cm; flowers in stout racemes; fruits lenticular. . . . . 3. *P. amphibia*
   Petiole up to 1 cm; flowers in slender racemes; fruit trigonous. . . . . 6. *P. salicifolia*

1. **Persicaria bistorta** (*L.*) *Samp.*, Herb. Port. 41 (1913).

Polygonum bistorta L., Sp. Pl. 360 (1753); Rech.f. & Schiman-Czeika in Fl. Iranica 56: 63 (1968); Li Anjen et al. in Fl. China 5: 296 (2003).
Bistorta officinalis Delarbre, Fl. Auvergne, ed. 2: 516 (1800).
Polygonum bistortoides (non Pursh) Boiss., Diagn. ser. 1, 5: 46 (1844).

subsp. **bistorta**

An erect, unbranched perennial 20–120 cm. Rhizome creeping, contorted, fleshy. Leaves mostly basal; basal leaves with lamina 10–25 × 1.5–3(–10) cm, ovate, obtuse to shortly

Fig. 58. **Persicaria bistorta**. 1, habit × 1/2; 2, flower dissected. Reproduced with permission from Flora of China 5: f. 245. 2003. Drawn by Zhang Chunfang.

acuminate, truncate at the base, paler and densely scabrid-puberulent beneath, petiole to 30 cm, winged in the upper part; cauline leaves sessile, triangular, acuminate, cordate at the base, the uppermost narrow. Ochreae up to 8 cm, obliquely truncate, ± laciniate, brown. Inflorescence cylindrical, 2–6 × 1–1.5 cm, solitary on an erect unbranched stem; flowers scented; peduncle and the short pedicels glabrous; bracts caudate to tricuspidate, ± 0.2 mm apart. Perianth-segments 5, 3–5 mm, reddish-pink. Longer stamens strongly exserted. Nut a little longer than the perianth, 3.3–4.2 mm, dark brown, glossy. Fig. 58: 1–2.

Hab. Damp areas of alpine grassland, rocky slopes and screes on metamorphic rock, by lake, 2700–3500 m; fl. Jun/July.
Distrib. Very local in the alpine zone of two mountains in NE Iraq. **mro**: Halgurd Dagh, *Gillett* 9557! 12361! *Guest & Ludlow-Hewitt* 2863! *Haley* 131!; N of Halgurd, near Bermasand lake, *Rawi & Serhang* 20214, 24790!; Ser Kurawa, *Gillett* 9734!

subsp. *carneum* (C. Koch) Greuter & Burdet, Willdenowia 19: 41 (1989), with oblong-globose inflorescences and distinctly pedicillate, deeper pink flowers, occurs in Armenia, Georgia and Turkey (NE Anatolia).

Turkey, NW Iran. Circumpolar, extending south into the higher mountains of Eurasia and in the Rocky Mountains.

2. **Persicaria orientalis** (*L.*) *Spach,* Hist. Nat. Vég. 10: 537 (1841).

Polygonum orientale L., Sp. Pl. 362 (1753); Rech.f. & Schiman-Czeika in Fl. Iranica 56: 57 (1968).

Robust, densely pubescent annual, with erect branched stems to 150 cm, branched. Basal leaves 8–20 × 2–3 cm, the base cordate, acuminate; petiole 3–5 cm. Ochreae 10–20 mm, thick, brown but with green leaf-like lobe at apex when young. Inflorescence of several dense drooping spike-like racemes 2–8 cm. Perianth segments 5, 3–4 mm, purplish-pink to reddish. Stamens 7–8. Fruits 2.5–3 mm, orbicular, lenticular, biconvex, black.

Hab. By streams; alt. to 1100 m; fl. Jun.–Jul.
Distrib. Occasional. **mam**: Baidaho nr Kani Masi, *Botany Staff* 43902! **mro**: Pishtashan, *Rechinger* 11030. **lca**: Baghdad, *Noë* s.n.!; ?**lea**: "Khan Sewil" (?Khan Jadida), *Sutherland* 418!

E and SE Asia; naturalized in S Europe and elsewhere.

3. **Persicaria amphibia** (*L.*) *Delarbre,* Fl. Auvergne ed. 2, 519 (1800); Grierson & Long in Fl. Bhutan 1(1): 161 (1983); Decraene & Akeroyd in Bot. J. Linn. Soc. 98: 366 (1988); Qaiser in Fl. Pak. 205: 28 (2001); Nyberg & Miller in Fl. Arab. Penins. & Socotra 1: 130 (1996); Breckle et al., Vasc. Pl. Afghan. Checkl.: 406 (2013).

Polygonum amphibium L., Sp. Pl. 361 (1753); Coode & Cullen in Fl Turk. 2: 272 (1967); Rech.f. & Schiman-Czeika in Fl. Iranica 56: 56 (1968).

Aquatic, usually glabrous perennial with creeping rhizome; stems 100–300 cm, producing adventitious roots at nodes, emergent (sometimes terrestrial) stems to 50 cm, erect, unbranched; floating leaves 7–15 × 2–4 cm, ovate-oblong, obtuse, truncate to subcordate at the base, acute; petiole 2–4 cm. Emergent leaves narrower, oblong-lanceolate, subacute, pubescent. Ochreae acute, hairy when young, ± ciliate. Inflorescence 2–5 × 0.8–1.5 cm, obtuse, dense. Peduncles stout. Perianth segments 5, 3.5–4 mm, bright pink, eglandular. Stamens 5, ± exserted. Styles 2. Nut 2–3 mm, lenticular, dark brown, glossy.

Hab. Ditches; alt. ± 1830 m; fl. Jul.
Distrib. Lower Forest zone of NE Iraq, recorded only twice. **mro**: Ser-i Hasan Beg, *Guest* 3025!

Yemen, Saudi Arabia, Turkey, Caucasia. Widespread in N Hemisphere: Temperate Eurasia south to the Mediterranean region and China, and North America south to Mexico.

4. **Persicaria maculosa** Gray, Nat. Arr. Br. Pl. 2: 269 (1821); Nyberg & Miller in Fl. Arab. Penins. & Socotra 1: 130 (1996); Boulos, Fl. Egypt 1: 26 (1999); Qaiser in Fl. Pak. 205: 36 (2001); Breckle et al., Vasc. Pl. Afghan. Checkl.: 406 (2013).

Polygonum persicaria L., Sp. Pl. 361 (1753); Rech.f., Fl. Lowland Iraq: 178 (1964); Coode & Cullen in Fl Turk. 2: 273 (1967); Rech.f. & Schiman-Czeika in Fl. Iranica 56: 56 (1968); Li Anjen et al. in Fl. China 5: 289 (2003).
Persicaria maculata (Rafin.)Á.Löve & D.Löve in Acta Hort. Gothob. 20: 164 (1956) non S.F.Gray; Decraene & Akeroyd in Bot. J. Linn. Soc. 98: 366 (1988).

A branched, erect or decumbent, mostly glabrous annual, 20–80 cm; stems reddish, swollen above the nodes. Leaves 3–10 × 0.5–3 cm, lanceolate or narrowly ovate-lanceolate, acute or acuminate, tapering to the base, usually with a large median black blotch; petiole up to 1 cm. Ochreae truncate, hairy, long-ciliate. Flowers in dense, spike-like racemes 1–4 cm. Peduncles smooth (rarely with a few yellow glands), shining. Perianth segments ± 3 mm long, bright pink, sometimes white, eglandular. Stamens usually 5, not exserted. Styles 2 or 3. Nut 2–2.5 mm, lenticular and biconvex or bluntly trigonous with concave faces, black, glossy. Fig. 59: 1–4.

HAB. Moist and wet places in mountains; alt. ± 1100 m; fl.
DISTRIB. **FAR**: Pushtashan, *Rechinger* 11030 (W).

Europe, NW Africa, SW Asia.

5. **Persicaria lapathifolia** (*L.*) *Delabr.*, Fl. Auvergne ed. 2, 519 (1800); Grierson & Long in Fl. Bhutan 1(1): 161 (1983); Qaiser in Fl. Pak. 205: 29 (2001); Nyberg & Miller in Fl. Arab. Penins. & Socotra 1: 132 (1996); Boulos, Fl. Egypt 1: 27 (1999); Breckle et al., Vasc. Pl. Afghan. Checkl.: 406 (2013).

Polygonum lapathifolium L., Sp. Pl. 360 (1753); Boiss., Fl. Orient. 4: 1030 (1879); Rech.f., Fl. Lowland Iraq: 178 (1964); Zohary, Fl. Palaest. 1: 56 (1966); Coode & Cullen in Fl. Turk. 2: 273 (1967); Rech.f. & Schiman-Czeika in Fl. Iranica 56: 57 (1968); Li Anjen et al. in Fl. China 5: 289 (2003).
Polygonum nodosum Pers., Syn. 1 (1805).
Polygonum obtusatum Steud. ex Meisn. in DC., Prodr. 14: 118 (1856).–Type: Islands in r. Tigris near Mosul, *Kotschy* 437 (K!, syn).

A branched, erect or procumbent, mostly glabrous annual 30–100 cm; stems greenish, often reddish and swollen above the nodes. Leaves 5–15 × 1–5 cm, linear- or ovate-lanceolate, acute or acuminate, tapering to the base, often with a large median black blotch; petiole short, 0.5–1.5 cm. Ochreae truncate, shortly ciliate. Flowers in dense spike-like racemes 1–4 cm. Peduncles rough with sessile yellow glands. Perianth segments 3–5, 2–3 mm, dull pink or greenish-white, ± yellow-glandular. Stamens 3–5, not included. Styles 2. Nut 2–3 mm, usually suborbicular and biconvex or trigonous, black or dark brown, glossy.

HAB. By water channels and in marshes; alt. s.l.–1550 m; fl. May.–Sep.
DISTRIB. Locally abundant. **MAM**: Sarsang, *Barkley* 9034. **MRO**: Walash, *Thesiger* 314; Haji Omran, *Dabbagh & Hamad* 46281! **FUJ**: islands in r. Tigris nr Mosul, *Kotschy* 437 (type of *P. obtusatum*)! **FNI**: Mindan bridge on r. Khazir between Mosul and Aqra, *Alizzi & Omar* 35292! **LEA**: 20 km N of Baquba, *Botany Staff* 47145A!; Diyala, *Sutherland* 416! **LBA**: Siba, opposite Abadan, *Guest* 25457! **LSM**: Mu'ail marsh nr Amara, *Hadač, Haines & Waleed* W. 1887!

SHAWASAR (Siba, *Guest* 25457).

Europe, Turkey, Armenia, Lebanon, Palestine, Jordan.

6. **Persicaria salicifolia** (*Brouss. ex Willd.*) *Assenov*, Fl. Reipubl. Popul. Bulg. 3: 243 (1966).

Polygonum salicifolium Brouss. ex Willd., Enum. Pl. 1: 428 (1809); Rech.f., Fl. Lowland Iraq: 178 (1964); Zohary, Fl. Palaest. 1: 56 (1966); Coode & Cullen in Fl. Turk. 2: 273 (1967); Rech.f. & Schiman-Czeika in Fl. Iranica 56: 59 (1968);

Glabrous perennial, 20–80 cm, stems erect, ascending, decumbent or partly submerged, branched, rooting at lower nodes. Ochreae to 2.5 cm, truncate, brownish, conspicuously ciliate. Leaves 5–15 × 0.8–2.5 cm, narrowly lanceolate, acute to acuminate, without a black blotch; petiole short, up to 1 cm. Flowers in terminal narrow racemes 3–8 cm. Peduncles long, glabrous. Perianth 2.5–3 mm, pink or whitish. Stamens 5–8. Fruits 2–3 mm, trigonous, dark brown or blackish, glossy.

HAB. Cultivated land, irrigation channels, river banks and marshes, mud flats, s.l.–20 m; fl.
DISTRIB. Mainly lowland alluvial areas of Iraq. **MSU**: Shaikh Sadiq, 10 km from Sulaimaniya, *Agnew, Hadač, Haines & Waleed* W. 1842! **FPF**: Sa'diya, *Kaisi* 47213! **LEA**: Ba'quba, *Omar, Mokhtar & Sahira* 37000!; Siba, nr Abadan, *Guest* 1627! **LCA**: Hurriya nr Baghdad, *Agnew, Hadač & Haines* W1683! **LSM**: Chabaish, *Dabbagh & Kaisi* 41894! *Thamer* 50105! *Khayat & Redha* 52318!; Hor al-Hammar, *Hamid* 38542!; Hor al-Kassara N of Al-Azair, *Rawi* 16537!; Al Fahad & Kabaish, *Karim & Hamid* 37795!; Kaba'ish, *Khayat & Radha* 52318! *Omar & Thamer* 47894! *Hamid* 38548! *Kaisi* 44033!; Al Mahar, 9 km from Garma Bani Sa'd, *Noori* 41352!; Kaba'ish, Fuhud, *Salman* 38578!; Garmat Ali, *Omar* 34507!; Amara marshes, *Guest* 1621!; Abu Safa 1 km from Chabaish, *Noori* 41333!; Qurna, *Alizzi & Omar* 34944!; 8 km E of Qurna, *Thamer, Wedad & Hana* 46768!; Hor as-Shaiba S. of Majaar, *Thamer* 46653!; Mdaina, nr Qurrah, *Hadač,*

Fig. 59. **Persicaria maculosa**. 1, habit; 2, perianth opened; 3, ovary and styles; 4, nut. **Persicaria hydropiper**. 5, habit; 6, leaf detail; 7, perianth opened; 8, ovary and styles; 9, nut. Reproduced with permission from Flora of Pakistan 205: f. 6, 2001. Drawn by M. Rafiq.

Haines & Waleed W1888!; Mdaina to Kaba'ish, *Rawi & Hamad* 34103!; Qala't Salih al-Ashimah, *Omar & Karim* 36795! **LEA**: Muhil marsh nr Amara, *Hadač, Haines & Waleed* W1889!; Abul-Khasib, *Alizzi & Omar* 34689! **LBA**: Siba (nr. Abadan), *Guest* 25457! *Guest* 1627!; Basra, *Omar & Thamer* 47886! *Gillett 10042*! *Guest* 303! *Alizzi & Omar* 34747!; Basra, Ma'qil, *Hadač, Haines & Waleed* W1881!; 10 km SE of Basra, *Rawi & Gillett* 5946!; Confluence of Tigris, Euphrates and Shatt al-Arab, *Gamal Abdin* 13!

SHAWASAR (Siba, Ir.- *Guest* 25457.), AS'AYAT-AR-RA'I (Siba, Ir.-*Guest* 1627). Eaten by all stock (*Gillett* 10042). Its leaves float on the water surface, but the plant has a very different emergent facies when growing in exposed muddy areas. Apparently the commonest *Persicaria* species in Iraq.

MRO: Berrog mt. on road to Qandil, *Rawi & Serhang* 23947! Probably an error, as the "Dry mountain slope" habitat and 1200 m altitude are unlikely for a species of lowland marshes.

Old World Tropics, Atlantic Islands, Mediterranean Europe, N Africa, Arabia.

7. **Persicaria hydropiper** (*L.*) *Delarbre*, Fl. Auvergne ed. 2: 518 1800; Grierson & Long in Fl. Bhutan 1(1): 162 (1983); Qaiser in Fl. Pak. 205: 38 (2001); Breckle et al., Vasc. Pl. Afghan. Checkl.: 406 (2013).

Polygonum hydropiper L., Sp. Pl. 361 (1753); Webb & Chater in Fl. Europ. 1: 79 (1964); Coode & Cullen in Fl. Turk. 2: 274 (1967); Rech.f. & Schiman-Czeika in Fl. Iranica 56: 60 (1968); Li Anjen et al. in Fl. China 5: 291 (2003).

An acrid, erect or suberect, branched subglabrous annual 20–40 cm, with stems often swollen above the nodes and rooting below. Leaves 3–7 × 0.8–1.5 cm, lanceolate, undulate, acute, narrowed to the base, asperous on the margins, almost glabrous, sessile or with short petiole. Ochreae inflated, glabrous or sparsely pubescent, shortly ciliate. Flowers in lax, nodding, spike-like racemes, leafy in the lower part with a single cleistogamous flower in each leaf-axil; peduncles glandular. Perianth segments 3–4, 2.5–3 mm, greenish or reddish, with numerous flat brownish or yellowish glandular dots. Stamens usually 4 or 6, not exserted. Styles 2(–3). Nut 2.5–3.5 mm, lenticular or subtrigonous, punctulate, dull, dark brown or black. Fig. 59: 5–9.

HAB. Damp and wet ground; alt. to 1000 m; fl. May.
DISTRIB. **MSU**: Damp ground at head-off from dam, 10 km towards Penjwin, *Agnew, Hadač, Haines & Waleed*, W1841!

Agnew et al. 1841 notes that the plant has "hot taste", a diagnostic feature of the species.

NW Africa, Temperate Asia, Pakistan and India eastwards to Japan; N America; widespread.

9. **POLYGONUM** L.

Sp. Pl. 359 (1753); Gen. Pl. ed. 2, 116 (1754)

Annual, biennial or perennial herbs. Stems branched, mostly more or less prostrate or only weakly erect. Leaves small, narrowed at the base, much longer than wide, pinnately veined. Ochreae bipartite or lacerate, more or less silvery or hyaline. Flowers small, solitary or few, subsessile or on short pedicels, in axils of leaves. Perianth segments 5, equal, often greenish but ± petaloid. Stamens 5–8. Styles (2–)3. Nut trigonous or lenticular, enclosed in the persistent perianth or protruding from it for less than half its length.

About 65 species, distributed in the N Temperate regions; eight species in Iraq.

Polygonum (from Gr. ϖολοί, *poloi*, many and γονυ, *gonu*, knee).

1. Annual, but sometimes rooting below 2
   Perennial, with woody stock 6
2. Perianth segments united for half or more their length 3
   Perianth segments united for much less than half their length 4
3. Leaves few, at least some more than 15 mm; perianth segments united for about 1/2 their length .... 2. *P. argyrocoleum*
   Leaves few, less than 15 mm; perianth segments united for about 2/3 their length .... 5. *P. polycnemoides*
4. Leaves linear-spatulate; nuts less than 1.5 mm, black .... 3. *P. corrigioloides*
   Leaves variable but not linear-spatulate; nuts more than 1.5 mm, brown 5

5. At least the upper bracts scarious, shorter than flowers; fruit often protruding slightly from perianth. . . . . 1. *P. patulum*
Bracts leaf-like, longer than flowers; fruit enclosed by perianth . . . . 4. *P. aviculare*
6. Branches erect, stiff; perianth segments united near base . . . . 1. *P. luzuloides*
Branches prostrate; perianth segments united for about 1/2 their length 7
7. Flowers 2–8 per node, longer than ochreae; perianth at least 3 mm. . . . . 6 *P. cognatum*
Flowers solitary per node, partly enclosed by ochreae; perianth not more than 2 mm . . . . 8. *P. paronychioides*

1. **Polygonum patulum** *Bieb.*, Fl. Taur.-Caucas. 1: 304 (1808); Kom., Fl. USSR 5: 630 (1936); Webb & Chater in Fl. Europ. 1: 78 (1964); Zohary, Fl. Palaest. 1: 55 (1966); Rech.f. & Schiman-Czeika in Fl. Iranica 56: 80 (1968); Qaiser in Fl. Pak. 205: 94 (2001); Li Anjen et al. in Fl. China 5: 285 (2003); Breckle et al., Vasc. Pl. Afghan. Checkl.: 408 (2013).

Polygonum bellardii non All. auctt., Fl. Pedem. 2: 207, t. 90, fig.2 (1785); Boulos, Fl. Egypt 1: 28 (1999).
P. equisetiforme (non Sm.) auctt.

Slender glabrous annual with erect stem 20–70 cm tall, branched from the base. Lower leaves 15–45 × 2–8 mm, lanceolate or oblong-lanceolate, acute, deciduous; upper leaves much smaller. Ochreae hyaline, laciniate, silvery, brown at the base. Flowers 1–3 at each node, mostly crowded towards the end of the branches, subsessile or on pedicels to 3 mm; upper bracts shorter than flowers, scarious. Perianth segments united near base, 2–3 mm, pink or greenish, with reddish margins. Nut 2–3 mm, as long as or slightly exserted from perianth, brown, smooth, shiny. Fig. 60. 1–5.

HAB. Gardens, fields, flooded ground, hillsides, clearings in *Quercus* forest, alt. 35–2600 m; fl. Jun.
DISTRIB. Widespread around Baghdad, occasional elsewhere. **MAM**: Jabal Khantur, *Rechinger* 10818! **MRO**: Halgurd Dagh, nr Nowanda, *Rechinger* 11337; **MRO**: Rost, *Haley* 175. **MSU**: Halabja, *Nuri & K. Hamad* 41203!; Penjwin, *Rechinger* 10459! **LCA**: Baghdad, *Rogers* 028! 0336! *Guest* 1102! *Graham s.n.* (1920)! Mahmudiya, *Guest* 923!; Abu Ghraib, *Alizzi* 34185! *Jenan* 36957!; Aziziya, *Robertson* 223!; Rustam, 18 km from Baghdad, *Rogers* 209! *Lazar* 1174a! *Guest* 1174! *Cowan & Darlington* 68!; Hillah, *Civilian Vet. Surgeon of Hillah* 2526!; Baghdad to Zafraniya, *Rechinger & Naib* 22! (fide Coode & Cullen).

MUWASSALAH (Ir.-nr Baghdad, *Guest* 1174, 2503). Said to be eaten by most animals (horses, sheep, cattle etc.).

S and SE Europe, Cyprus, Greece, Turkey, Syria, Lebanon, Palestine, Kuwait, N Iran and C Asia.

2. **Polygonum argyrocoleum** *Steud. ex Kunze*, Linnaea 20: 17 (1847); Rech.f., Fl. Lowland Iraq: 180 (1964); Rech.f. & Schiman-Czeika in Fl. Iranica 56: 82 (1968); Nyberg & Miller in Fl. Arab. Penins. & Socotra 1: 128 (1996); Qaiser in Fl. Pak. 205: 92 (2001); Li Anjen et al. in Fl. China 5: 407 (2003); Breckle et al., Vasc. Pl. Afghan. Checkl.: 407 (2013).

Polygonum chlorocoleum Steud. ex Boiss., Fl. Orient. 4: 1034 (1879).
P. noeanum Boiss., Bot. Zeitung (Berlin) 11: 734 (1853).–Type: "in arena insularum Tigridis prope Mossul" (on sand of islands in R. Tigris nr Mosul), *Kotschy* 440 (A, B, GOET, K!, HAL, MO, MPU, NY, P, US, iso).

Glabrous annual, stems erect or ascending 20–80 cm, branched from the base or above. Leaves few, deciduous, lower leaves 15–40 × 2–5 mm, sessile to petiolate (petiole up to 5 mm); leaves on branches much smaller. Ochreae hyaline, laciniate, silvery, brown at the base. Flowering branches almost leafless, with long internodes, bracts inconspicuous. Flowers crowded towards ends of the branches, 1 at each node, on pedicels 1.5–2 mm. Perianth segments united for half their length, 1.5–2.2 × ± 1 mm, greenish-white with white or pink margins. Nuts 1.5–2.3 × 1–1.5 mm, included in perianth, smooth, brown, shining. Fig. 60: 6–10.

HAB. Stony ground, roadside, irrigated alluvium, *Helianthus* field, *Phoenix* plantations, waste ground; alt. s.l.–2200 m; fl. Jun., fr. Nov.
DISTRIB. Throughout the alluvial plains and foothills of Iraq. **MJS**: Jabal Sinjar, *Kaisi* 49742! N. slope, *Kaisi, Khayat & Karim* 50935! **MAM**: Aradin (Amadiya valley), *Guest* 4998! **MRO**: Haruna, *Gillett* 9648! ; Sideka to Diana (about 12 km NE of Rowanduz), *Guest & Husham* (*Alizzi*) 15867!; Magar range, Haji Omran, *Rawi & Serhang* 24315!; Gali Ali Beg, *Omar, Kaisi & Wedad* 49545! **MSU**: 10 km from Sulaimaniya

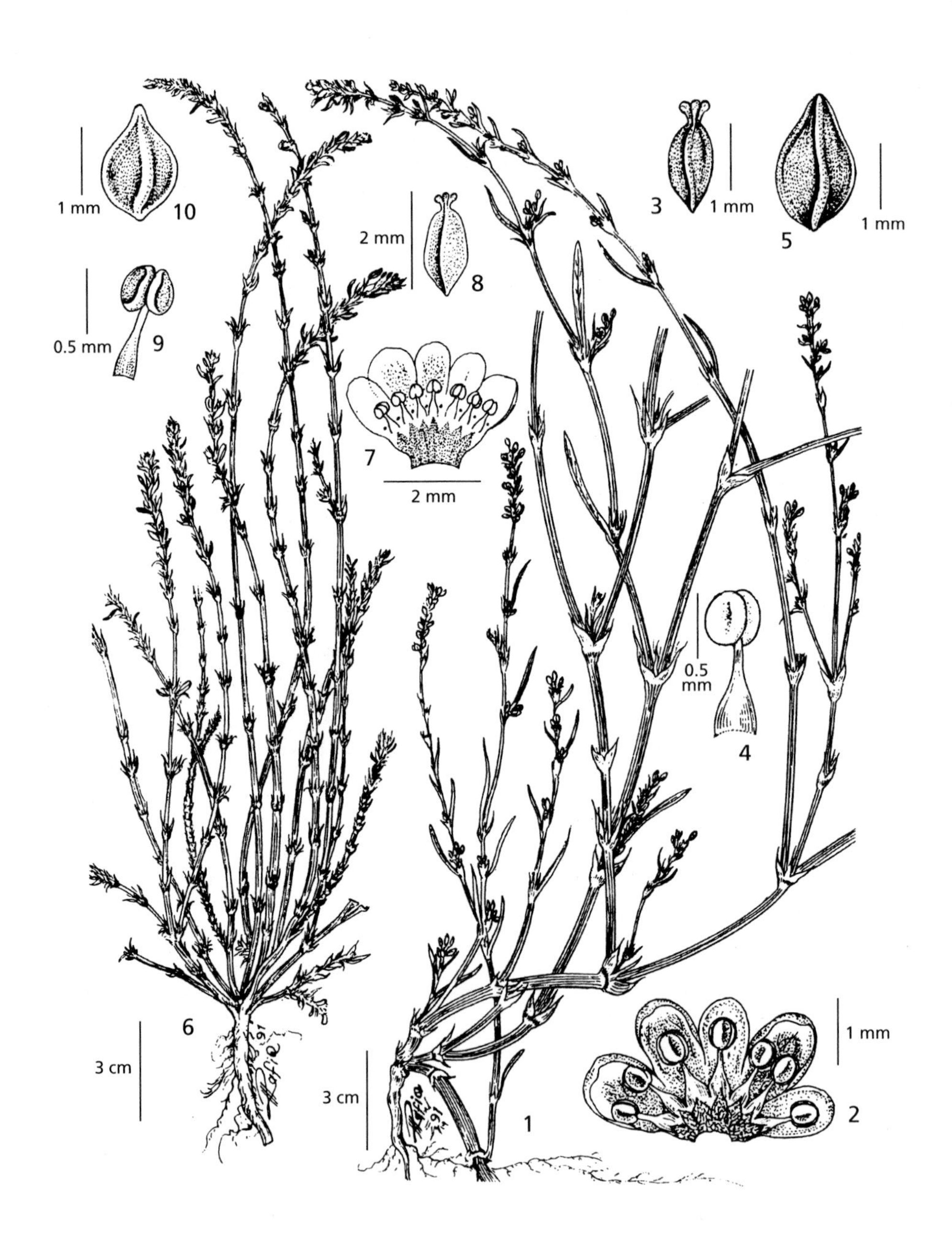

Fig. 60. **Polygonum patulum**. 1, habit; 2, perianth opened; 3, ovary and styles; 4, stamen; 5, nut. **Polygonum argyrocoleum**. 6, habit; 7, perianth opened; 8, ovary and styles; 9, stamen; 10, nut. Reproduced with permission from Flora of Pakistan 205: f. 14. 2001. Drawn by M. Rafiq.

to Chuwarta, *Kaisi & Khayat* 51149!; Penjwin, *Abd Majid Salim Agha* 5510!; Azmir, *Karim* 39352! ; 5 km from Saiyid Sadiq to Ahmad Awa, *Kaisi & Wedad* 49413! **FNI**: Nineveh, *Bornm.* 1780! **FPF**: Badra, *Ali Burhan Addin* 5527! *Naji Sufer* 5526!; 30 km N.E. of Baquba, *Botany Staff* 47150! **FUJ**: banks of the Tigris nr Mosul, *Hausskn.* s.n. (1867)! **DLJ**: Rawa, *Alizzi & Omar* 35360! *Gillett* 7069! **DWD**: Massad depression, Rutba, *Alizzi & Omar* 36184! Habbaniya, *Botany staff* 47195. **DSD**: 10 km S. of Zubair, *Rawi & Gillett* 6096! **LEA**: Ali al-Gharbi to Tib, *Kaisi & Khayat* 50537!; 3 km S. of Kut, *Barkley & Jumaa Brahim* 4091! Abul-khasib, *Omar* 34498! *Rawi* 25935!; Ba'quba, *Rechinger* 9742! **LCA**: Abu Ghraib, *Rawi* 10745! *Gillett* 8424! *Alizzi & Omar* 34271!; Za'faraniya, *Gillett* 10177! *Ahmad* 9848! *Aftan al-Rawi* 19922! *Haines* W40!; Shaikh Sa'd to Kut, *Gillett* 6740!; Shaikh Sa'd, *Rawi* 12537!; Baghdad, *Colvill* 71! *Noë* 344!; Baghdad to Za'faraniya, *Rechinger & Naib* 22!; Kiria (?Karrada) Mariam, *Haines* W221A!; Baghdad to Baquba, *Haines* W3! W298!; Baghdad, Chelebi's garden, *Haines* W62!; 64 km NW of Nasiriya towards Samawah, *Khatib & Alizzi* 32778!; nr Baghdad, Guest 174! *Lazar* 555!; 25 km S of Diwaniya, *Barkley, Safat & Agnew* 3745!; 50 km W of Baghdad, *Barkley (illegible!)*; 46 km SE of Musaiyib, *Thamer & Wedad* 47272! **LSM**: Musaida nr Amara, *Lazar* 72!; Musaida 15 km SW of Kut, *Dabbagh & Kaisi* 41937! **LBA**: Garma Bani Sa'd, *Rawi* 16567!; Basra, Ma'qil, *Guest, Rawi & Rechinger* 16714!; Basra, Guest 302! *Chapman* 10038!

KHARREIS (Ir.-Basra, *Chapman* 10038); 'AQEID (Ir.-Zafraniya, *Ahmad* 9848).

Arabia, Palestine, Iran, Caucasus, Pakistan, Afghanistan, Turkmenistan; introduced in North America.

3. **Polygonum corrigioloides** *Jaub. & Spach*, Ill. Pl. Or. 2: 34 (1845); Rech.f., Fl. Lowland Iraq: 179 (1964); Rech.f. & Schiman-Czeika in Fl. Iranica 56: 73 (1968); Nyberg & Miller in Fl. Arab. Penins. & Socotra 1: 129 (1996); Breckle et al., Vasc. Pl. Afghan. Checkl.: 407 (2013).

Glabrous, somewhat glaucous annual, stems 20–40 cm, slender, procumbent, branched. Ochreae dentate-lacerate, whitish. Leaves 4–15 × 1–2 mm, linear-spatulate, obtuse, the upper smaller. Flowers in clusters of 2–6 at each node, on pedicels 3–5 mm; perianth segments 1–2 mm, united only near the base, green or reddish. Nut ± 1mm, black, shiny.

HAB. Cultivated field weed, clay soil in ditch, margin of flooded area, mud banks; alt. ± 30 m. fl. Apr.–Jun.
DISTRIB. Mostly lowland. **DWD**: Ramadi, *Keith* (*Graham* 133)! *Graham* 197! **LEA**: nr Aziziya, *Khatib & Alizzi* 33485! *Guest* 13591!; by mouth of r. Diyala, *Sutherland* 419! **LCA**: Mahmudiya, *Guest* 2394!; between Baghdad and Babylon, *Bornmüller* 629!; 'near Baghdad', Nábělek (P). **LSM**: Qurna to Basra, *Rawi* 15032!; Mesopotamia, *Watson* s.n.!

?Syria, Yemen, Iran.

4. **Polygonum aviculare** *L.*, Sp. Pl. 362 (1753); Rech.f., Fl. Lowland Iraq: 180 (1964); Coode & Cullen in Fl. Turk. 2: 277 (1967); Rech.f. & Schiman-Czeika in Fl. Iranica 56: 79 (1968); Mouterde, Nouv. Fl. Liban et Syria 2: 675 (1970); Nyberg & Miller in Fl. Arab. Penins. & Socotra 1: 129 (1996); Boulos, Fl. Egypt 1: 28 (1999); Qaiser in Fl. Pak. 205: 95 (2001); Li Anjen et al. in Fl. China 5: 284 (2003); Breckle et al., Vasc. Pl. Afghan. Checkl.: 407 (2013).

Polygonum arenastrum (Boreau) auctt.; Mouterde, Nouv. Fl. Lib. et Syrie 2: 675 (1970).

A much-branched, glabrous annual with stems ascending to suberect, rarely prostrate, 20–100 cm. Leaves on main stem 20–40 × 3–18 mm, lanceolate to ovate-lanceolate, sometimes obovate or narrowly elliptical, acute to subobtuse, somewhat deciduous; leaves on branches smaller, narrower and more persistent; petiole up to 5 mm but usually shorter or absent. Ochreae ± 5 mm, laciniate, dull silvery-white, brownish at the base. Inflorescence of 1–3(–5)-flowered axillary clusters. Perianth segments ± 2 mm, united only near the base, greenish with broad pink or white margins. Fruit 2.5–3.5 × 1.5–1.8 mm, included in the perianth, minutely punctulate-striate, reddish-brown or dark brown, dull.

HAB. *Rhus-Quercus* forest, hillsides. alt. 100–2900 m; fl. Jun.–Sep.
DISTRIB. In the moist steppe, forest, and montane zones of N Iraq. **MAM**: Kani Masi, *Botany Staff* 43896; Bazingrah, *Rechinger* 11522. **MRO**: Halgurd Dagh, *Gillett* 12339! Jabal Khantur, *Rechinger* 10817. **MSU**: Sarchinar to Sulaimaniya, *Raddi* 5281!; Sulaimaniya, *Abd Majid Salim Agha* 5364!; Shaqlawa, *Barkley & Said* 3458!; Penjwin, *Rawi* 22545!; Pira Magrun, *Haussknecht* s.n. (1867)!; Qara Anjir, *Rechinger* 12526! **FKI**: Kirkuk, *Hausskn.* s.n.! **DWD**: 20 km S of Rutba, Dabbagh, *Taher & Tawfiq* 41787! **LCA**: Khan Bani Sa'd, *Barkley & Askari* 1747!; Baghdad, Tarmiya, *Haines* W67!; Baghdad, *Guest* 3226! *Rogers* 028!; Rustamiya, *Rogers* 266!; Tarmiya to Baghdad, *Luckman & Salah* s.n.!; Banks of r. Tigris, Adhimiya, *Shehbaz & Mousawi* s.n.!

Egypt, Yemen, Saudi Arabia, Palestine, Syria, Turkey, Armenia, Iran, Afghanistan, Pakistan; temperate Eurasia and North America. Widely introduced as a cosmopolitan weed.

5. **Polygonum polycnemoides** *Jaub. & Spach,* Ill. Pl. Or. 2: 30 (1845); Coode & Cullen in Fl. Turk. 2: 277 (1967); Rech.f. & Schiman-Czeika in Fl. Iranica 56: 74 (1968); Qaiser in Fl. Pak. 205: 106 (2001); Li Anjen et al. in Fl. China 5: 285 (2003); Breckle et al., Vasc. Pl. Afghan. Checkl.: 408 (2013).

Polygonum olivieri Jaub. & Spach, Ill. Pl. Or. 2 : 31 (1845).

Glabrous annual, stems 5–25 cm, prostrate, slender, rather fragile, much branched. Ochreae triangular, lacerate, membranous. Leaves dense, 5–12 × 0.5–2 mm linear, to narrowly lanceolate, sessile, acute, margins recurved. Flowers 1–2 at each node, on pedicels to 1 mm, in long slender leafy spikes. Perianth segments united for about 2/3 their length, 1.5–2 mm, greenish-white or pinkish. Stamens 5. Fruits ± 2 mm, enclosed by the perianth, punctulate-striate, dark brown, shiny. Fig. 61: 1–4.

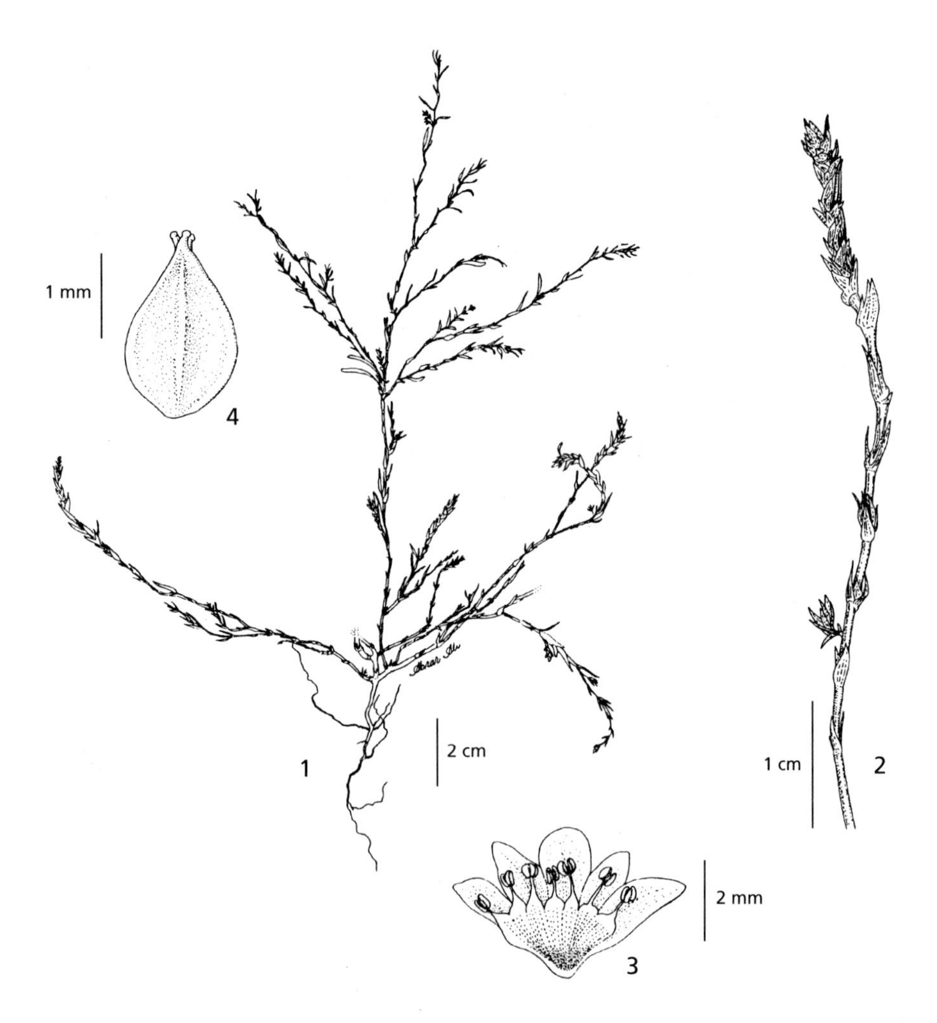

Fig. 61. **Polygonum polycnemoides**. 1, habit; 2, part of branch; 3, perianth opened; 4, ovary and styles. Reproduced with permission from Flora of Pakistan 205: f. 12. 2001. Drawn by M. Rafiq.

Hab. In coppiced *Quercus*, on limestone slate and serpentinite; alt. 1600–2740 m; fl. Jun.
Distrib. Local in the thorn-cushion zone of mountains in NE Iraq. **MRO**: Halgurd Dagh above Nowanda, *Rechinger* 11389. **MSU**: mountains nr Penjwin, *Rechinger* 10452!; Malakawa pass nr Penjwin, *Rechinger* 12324; spur of Jabal Hawraman N of Biyara, *Gillett* 11807!; Mt Hawraman, *Rawi* 22078!

The type specimen (P!) is labelled 'Mesop.[otamia] ad Baghdad', *no.* 2590, but is more likely to have been collected while Olivier & Bruguière were traversing the mountains en route to Kermanshah (Iran), as the species is otherwise unknown from lowland Iraq. Similarly the type of *P. olivieri* is labelled 'Mésopotamie', *Olivier* s.n.

Turkey (E Anatolia), Syria, Lebanon, Caucasus, N Iran, Turkestan, Afghanistan, Pakistan.

6. **Polygonum cognatum** *Meisn.*, Monogr. Polyg. 91 (1826); Coode & Cullen in Fl Turk. 2: 276 (1967); Qaiser in Fl. Pak. 205: 83 (2001); Li Anjen et al. in Fl. China 5: 284 (2003); Breckle et al., Vasc. Pl. Afghan. Checkl.: 407 (2013).

Polygonum alpestre C.A. Mey., Verz. Pfl. Cauc. Kasp. Meer 157 (1831); Rech.f. & Schiman-Czeika, Fl. Iranica : 65 (1968).
P. cognatum var. *alpestre* (C.A.Mey.) Meisn. in DC., Prodr. 14: 96. 1875.
P. ammanioides Jaub. & Spach, Ill. Pl. Or. 2: 28 (1844).

Glabrous, sprawling perennial with a woody stock, forming prostrate mats; stems densely leafy, to 40 cm. Leaves 10–35 × 2–10 mm, ovate to elliptic or oblong, usually subobtuse (or narrowly lanceolate and acute, var. *ammanoides* (Jaub. & Spach) Meissn.), rather fleshy, shortly but distinctly petiolate. Ochreae almost entire, hyaline. Flowers in dense leafy flattened spikes, in clusters of (2–)5–8 at each node, on pedicels usually shorter than perianth. Perianth segments united for half their length, 3-5 mm, green with broad white or pink margins, somewhat indurate in fruit. Fruits 2–3 mm, included in the perianth, smooth, black, glossy.

Hab. Rocky and stony ground; alt. 1200–3200 m.; fl. Jul.
Distrib. In the thorn cushion and upper montane zones of N & NE Iraq. **MAM**: Ser Amadiya, Gali Mazurka, *Dabbagh, Qaisi & Hamid* 46003!; Sheikh Adi nr Ain Sifni, *Field & Lazar* 755!; Sersang, *Haines* W484!; Alho, *Rawi* 8609B!; above Bazingrah, N of Zakho, *Rechinger* 11530!; Zawita, *Rechinger* 10937; Valley between Marsis and Zawita, *Rechinger* 12018. **MRO**: Halgurd Dagh, *Guest* 2864! above Nowanda, *Rechinger* 11406; Sermoka, *Rawi* 13734!; Ser Kurawa, *Gillett* 9773!; Ser-i Rost, *Haley* 99; Qandil mts., *Rawi & Serhang* 24483! mt. Malikh, *Rawi & Serhang* 24041!; Qandil mts., above lake Goam-i Kirmosaran, *Rechinger* 11781. **MSU**: Tainal, *Omar & Karim* 37920! **FNI**: Ain Sifni, *Field & Lazar* 755!

A polymorphic species with a wide distribution in Asia. Rechinger (l.c.) questions whether *P. cognatum* and *P. alpestre* are synonymous, pointing out that the former was described from Siberia. Type not seen.
KHANATITK (Alho, *Rawi* 8609B).

Syria, Lebanon, Turkey (E Anatolia), Caucasus, Iran, Pakistan, C Asia, Siberia.

7. **Polygonum luzuloides** *Jaub. & Spach,* Ill. Pl. Or. 2: 37 (1845); Coode & Cullen in Fl Turk. 2: 274 (1967); Rech.f. & Schiman-Czeika in Fl. Iranica 56: 65 (1968).

Polygonum setosum Jacq. subsp. *luzuloides* (Jaub. & Spach) Leblebici.–Type: *Aucher-Eloy* 2589 (P, lecto).

Perennial with woody stock and prostrate, rooting stems. Branches to 40 cm, erect, rigid, somewhat zigzag. Ochreae prominent, lacerate, silvery. Leaves linear to narrowly elliptic, 20–40 × 2–5 mm, acute, greyish-green, margins strongly revolute. Flowers 2-4 at each node, on pedicels the same length as perianth, in long, slender, branched inflorescences. Perianth segments c. 3 mm, united only near the base, white flushed with pink; fruits enclosed by the perianth, blackish, shiny.

Hab. On metamorphic and limestone rocks, mountain slopes; alt. 2000–3200 m; fl. Aug.
Distrib. Local in the thorn cushion zone of mountains in NE Iraq. **MAM**: Jabal Khantur, NE of Zakho, *Rawi, Tikriti & Nuri* 29014. **MRO**: Halgurd Dagh, *Gillett* 9544! above Nowanda, *Rechinger* 11369; SE of Serva (Zirva), *Gillett* 9691! **MSU**: Qandil mts, above lake Goam-i Kirmosoran, *Rechinger* 11139; Newrobar, Qandil range, *Rawi & Serhang* 24151; mt. Hawraman, *Rawi, Chakr., Nuri & Alizzi* 19788! 1867, *Haussknecht* s.n.!

Turkey (SE and E Anatolia), Armenia, NW and W Iran.

8. **Polygonum paronychioides** *C.A.Mey. ex Hohen.*, Bull. Soc. Nat. Mosc. 4: 356 (1838); Coode & Cullen in Fl Turk. 2: 275 (1967); Rech.f. & Schiman-Czeika in Fl. Iranica 56: 71 (1968); Qaiser in Fl. Pak. 205: 88 (2001); Li Anjen et al. in Fl. China 5: 283 (2003); Breckle et al., Vasc. Pl. Afghan. Checkl.: 408 (2013).

Polygonum mucronatum Royle ex Bab., Trans. Linn. Soc. Lond. 18: 115 (1841).

Densely tufted perennial with woody stock, much branched at base, stems 2–10 cm, prostrate, forming loose mats. Leaves dense, 3–12 × 1–1.5 mm, sessile, linear, ± mucronate, greyish-green, margins revolute. Ochreae conspicuous, hyaline, lacerate, silvery-white, partly enclosing the flowers. Flowers solitary in axils of lanceolate bracts, subsessile, in crowded inflorescences. Perianth segments 1.5–2 mm, united for half their length, pink. Stamens 8. Fruit included in the perianth, ± 2 mm, blackish, smooth. Fig. 62: 1–4.

HAB. Serpentine screes, hillside in *Rhus-Quercus* forest; alt. 1320–2700 m; fl. Jun.–Jul.

DISTRIB. Local in the upper forest and thorn-cushion zones of mountains in NE Iraq. **MRO**: Chiya-i Mandau, *Guest* 2710!; Halgurd Dagh, Goum Tawera, *Rawi & Serhang* 24732! **MSU**: Penjwin, Malakawa pass, *Rechinger* 10409! *Rawi* 22493!; nr Penjwin, *Rechinger* 12243; Penjwin, *Rawi* 12227!

Mountains of Caucasus, NE Turkey, Iran, Afghanistan, Pakistan, NW India, Turkmenistan, C Asia to W China.

SPECIES DOUBTFULLY RECORDED

A specimen labelled *Polygonum arenarium* Waldst. & Kit., Pl. Rar. Hung. 1: 69 (1801) (**LSM**: By r. Tigris below Amara, *W.E. Evans*, 11.11.1917, det. Cullen & Coode [E], may represent a casual occurrence of a species that occurs from SE & C Europe to the Caucasus.

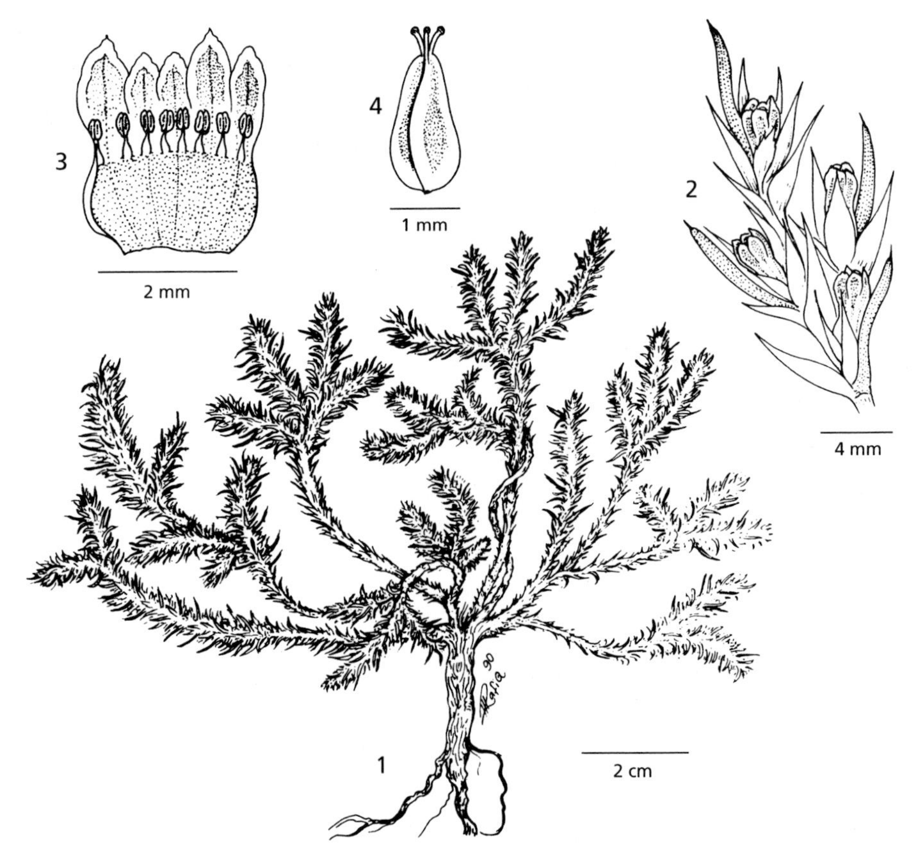

Fig. 62. **Polygonum paronychioides**. 1, habit; 2, fowering branch; 3, perianth opened; 4, ovary and styles. Reproduced with permission from Flora of Pakistan 205: f. 13. 2001. Drawn by M. Rafiq.

10. **FAGOPYRUM** Mill.

Gard. Dict. Abr. ed. 4 (1754)

Annual or perennial herbs. Stem single, erect, hollow, branched. Leaves sagittate or hastate, palmately veined, petiolate. Ochreae membranous, not ciliate. Flowers bisexual, heterostylous, in axillary and terminal umbels. Perianth segments 5, subequal, not enlarged in fruit. Stamens 8. Styles 3, long and slender. Nut sharply trigonous, much exceeding the perianth.

About 15 species distributed in Europe and temperate Asia; one species in Iraq.

Fagopyrum (from φαγο, *fago*, beech and ϖυϱος, *pyros*, kernel).

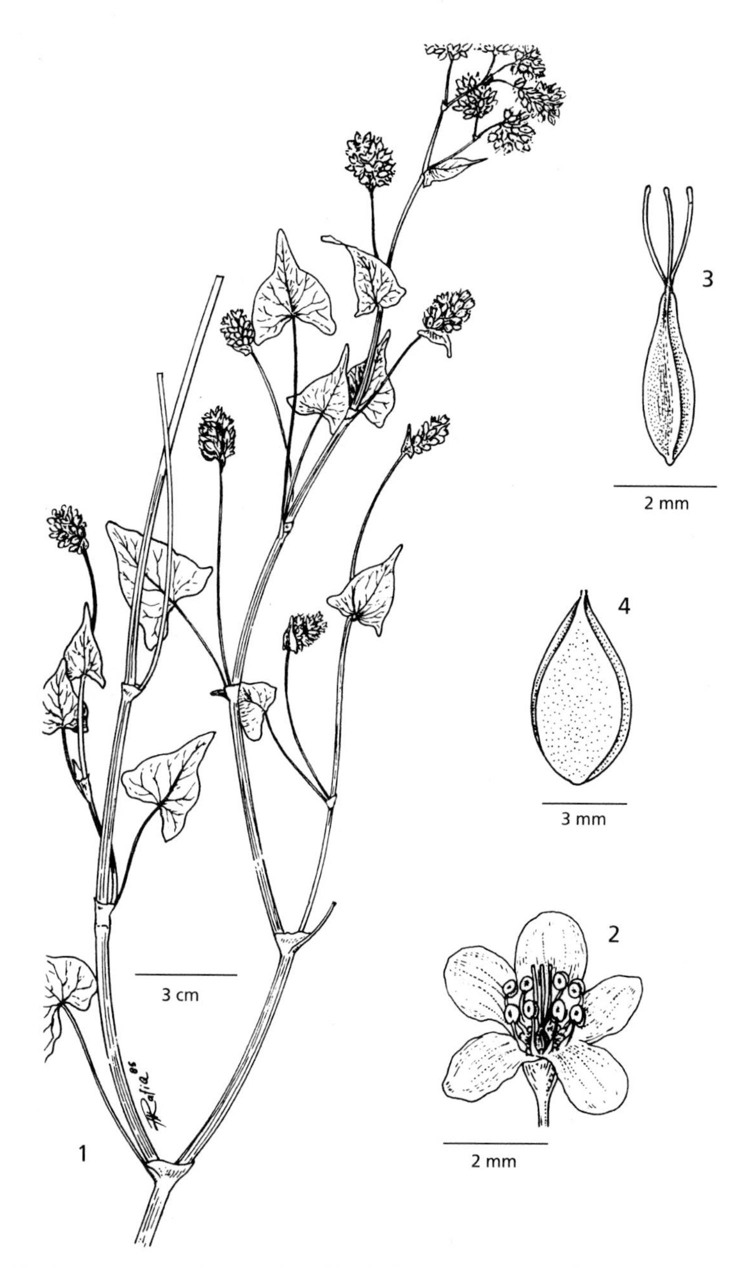

Fig. 63. **Fagopyrum esculentum**. 1, habit; 2, flower; 3, ovary and styles; 4, nut. Reproduced with permission from Flora of Pakistan 205: f. 17. 2001. Drawn by M. Rafiq.

Fagopyrum esculentum *Moench,* Methodus 290 (1794); Graham in Fl. Trop. East Africa Polygonaceae: 25 (1958); Rech.f. & Schiman-Czeika in Fl. Iranica 56: 83 (1968); Husain & Kasim, Cult. Pl. Iraq & their importance: 98 (1975); Qaiser in Fl. Pak. 205: 121 (2001); Li Anjen & Suk-pyo Hong in Fl. China 5: 323 (2003); Breckle et al., Vasc. Pl. Afghan. Checkl.: 406 (2013).

Polygonum fagopyrum L., Sp. Pl. 364 (1753).
Fagopyrum vulgare T. Nees, Gen. Fl. Mon. 53 (1835).

Annual 20–100 cm, stems becoming reddish with age. Leaves 3–8(–10) × 2–5(–7) cm, ovate-triangular, acuminate, cordate, somewhat hairy on the petioles and veins beneath; lower leaves petiolate, the upper amplexicaul. Ochreae 2–6 mm, not fringed. Flowers in short dense axillary panicles on long peduncles, mostly crowded into clusters at the stem apex. Perianth segments 3–4 mm, elliptical to obovate, whitish, pink or reddish. Nut considerably exceeding the perianth, 5–6 mm, trigonous with slightly concave faces, dull dark brown. Fig. 63: 1–4.

HAB. Irrigated alluvium; alt. to 40 m; fl. May.
DISTRIB. Found only once, probably as an escape from cultivation, in the Mesopotamian alluvial plain. **LCA**: Abu Ghraib, *Rawi* 10755!

Originally from C Asia, this crop is grown widely across temperate Eurasia for its floury seeds known as "buckwheat" (Eng.), "grechikha" (Russ., гречиха), الحنطة السوداء (Ar.).

Native to C and E Asia.

11. **FALLOPIA** Adans.

Fam. 2: 274 (1763) emend. Holub. in Folia Geobot. Phytotax. Praha 6: 171 (1971).
Bilderdykia Dumort., Fl. Belg.: 18 (1827)

Annual or perennial, prostrate, training or twining herbs or smal shrubs. Leaves petiolate, often with a gland at the base; ochrea truncate, deciduous. Flowers in loose axillary fascicles, terminal racemes or panicles. Perianth 5(–6), in 2 whorls, the outer segments keeled or membranous or winged on the back, accrescent in fruit. Stamens 8. Ovary trigonous; styles short. Nut trigonous, smooth or granulate, included in the perianth.

About nine species distributed in the N Temperate regions of Europe and Asia; a single species, possibly introduced, in Iraq.

Fallopia convolvulus (*L.*) *Á. Löve,* Taxon 19: 300 (1970); Ronse Decraene & Akeroyd in Bot. J. Linn.Soc. 98: 369 (1988); Qaiser in Fl. Pak. 205: 116 (2001); Li Anjen & Chong-wwok Park in Fl. China 5: 316 (2003); Breckle et al., Vasc. Pl. Afghan. Checkl.: 406 (2013).

Polygonum convolvulus L., Sp. Pl. 364 (1753); Kom, Fl. USSR 5: 694 (1936); Cullen & Coode in Fl. Turk. 2: 281 (1966); Rech.f. & Schiman-Czeika in Fl. Iranica 56: 51 (1968).
Bilderdykia convolvulus (L.) Dumort., Fl. Belg.: 18 (1827); Webb & Tutin in Fl. Europ. 1: 18 (1964).
Fagopyrum convolvulus (L.) H.Gross in Bull. Geogr. Bot. 23: 21 (1913).

Prostrate, scrambling or climbing annual with flexuous, twining stems 20–120 cm long. Leaves 2–8 × 1–4 cm, ovate, acute to acuminate, cordate-sagittate at the base, papillose on the petiole and veins beneath; petiole up to 3.5 cm. Ochreae obliquely truncate, more or less laciniate. Inflorescence pedunculate or subsessile, lax and interrupted; pedicels 1–3 mm, slender. Perianth segments 5, greenish-white or pinkish, the 3 outer obtusely keeled or narrowly winged in fruit. Nut 3–5 mm, black, dull, minutely punctulate. Fig. 64: 1–4.

HAB. Field weed; alt. ± 1800 m; fl. Jul.–Aug.

DISTRIB. A casual weed, possibly introduced with crop seed. **MRO:** Gunda Zhor valley E of Halgurd Dagh, *Gillett* 12317! **LCA:** Baghdad, Ma'amoon, *Sahira* C841!

Europe, NW Africa, Egypt and temperate Asia; naturalized in North America and other temperate regions.

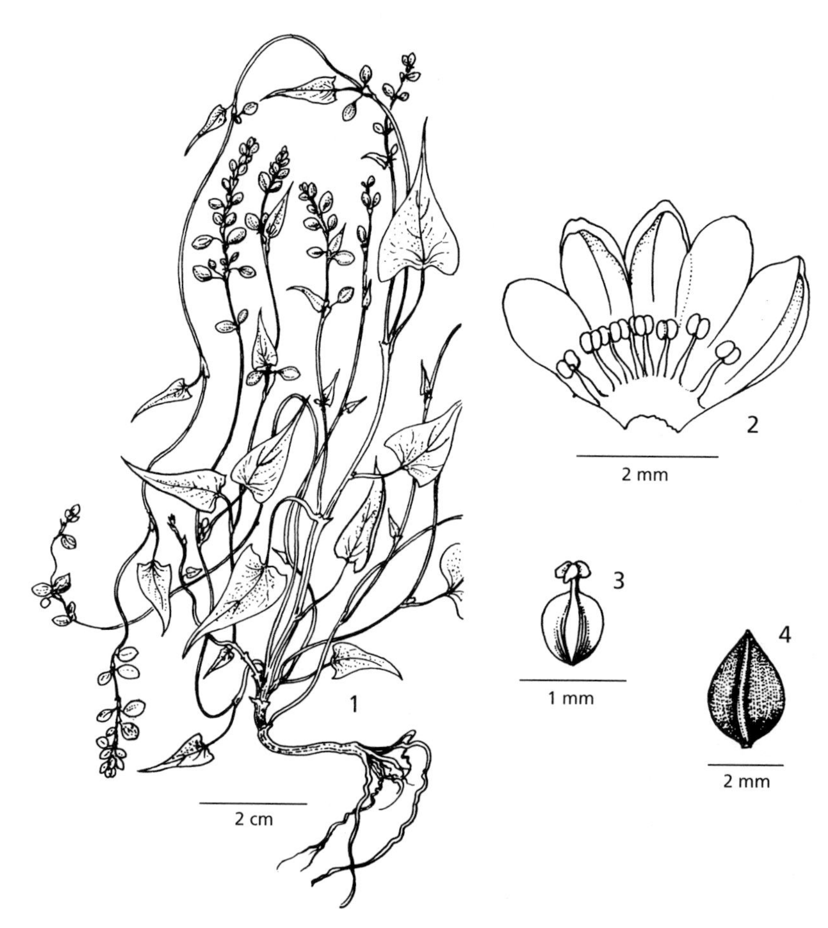

Fig. 64. **Fallopia convolvulus**. 1, habit; 2, flower opened; 3, ovary and styles; 4, nut. Reproduced with permission from Flora of Pakistan 205: f. 16. 2001. Drawn by M. Rafiq.

94. **ILLECEBRACEAE** R.Br.

Prodr. 413. 1810

Illecebraceae has been recognized as a separate family by Hutchinson in his classification (1959), however traditionally genera included in the Illecebraceae have been included in the Caryophyllaceae (Bentham & Hooker, 1862–1883 & Engler & Prantl 1903).

Most classifications divide Caryophyllaceae into 3 subfamilies: Alsinoideae, Silenoideae & Paronychioideae (a key to these subfamilies is given under Caryophyllaceae in this volume). Amongst the Paronychioideae, genera with hypogynous and perigynous flowers, flowers with or without corolla and fruit that is a capsule or nut (or utricle) are found. The group of plants with perigynous flowers, lacking corolla and a fruit that is a nut or utricle have been recognised by some authorities as a separate tribe, the Paronychieae. This tribe has sometimes been recognised as the family Illecebraceae.

Bittrich (1993: Caryophyllaceae. In: Kubitzki (ed.), The families and genera of vascular plants vol. 2: 214) gives a review of the subdivisions and relationships within the family Caryophyllaceae to show that amongst the subfamilies and tribes based on their morphological, and molecular characters, the exclusion of one of them cannot be justified.

Genera treated under the Illecebraceae by some authorities are included in the Caryophyllaceae in this Flora. [S.A. Ghazanfar & J.R. Edmondson – eds.]

95. CHENOPODIACEAE

A.P. Sukhorukov[13], P. Aellen[14], J.R. Edmondson[15] & C.C. Townsend[16]

Annual or perennial herbs, shrubs or trees. Leaves alternate or opposite, without stipules, petiolate or not; lamina simple, flattened, entire to pinnatisect, terete or reduced to scales. Flowers hermaphrodite or unisexual (plants monoecious or dieocious), solitary or arranged in spikes or cymes, often with bracts and bracteoles, with actinomorphic perianth consisting of 2–5 free or diversely connate segments; rarely without perianth or with 1 segment. At fruiting stage, perianth can change its consistence or shape that enhance anemochorous or zoochorous dispersal. Stamens 1–5, opposite to perianth segments, free or united at base into a ring. Ovary superior or semi-inferior, free, rarely adnate to perianth, with a single basal ovule. Stigmas 2 or 3–5. Fruit one-seeded, as a rule indehiscent, rarely dehiscent or raptured. Seeds horizontal or vertical (sometimes both positions at one individual); embryo straight, bent, annular or spirally coiled. Endosperm absent; perisperm abundant, traces or lost.

The taxonomy of many Chenopodiaceae has been changed in the recent decade, and the circumscription of each genus is given in accordance to the results of detailed molecular investigations.

About 100 genera and 1600 species distributed worldwide, but mostly in arid or semi-arid regions.

Key to genera

1. Plant covered with dendroid hairs (comprising of central row bearing in its surface shorter lateral outgrowths); leaves indurated . . . . 11. *Agriophyllum*
   Plants glabrous or with other indumentum types; leaves never indurated 2
2. Stem and leaves covered with stellate hairs; bracts enveloping the fruit with large simple and short stellate hairs; pericarp with scattered stellate indumentum. Subshrub with separated male and female flowers (plant monoecious) . . . . 9. *Krascheninnikovia*
   Stems and leaves glabrous or with other indumentum types; pericarp glabrous, papillate or with glandular trichomes 3
3. Leaves flat, (4–)5 and more mm wide, entire to pinnatisect 4
   Leaves terete or reduced to the scales, up to 5 mm wide 14
4. Plants aromatic, covered with simple, glandular hairs and subsessile yellow glands . . . . 2. *Dysphania*
   Plants non-aromatic or with unpleasant smell; yellow glands absent or occasional. 5
5. Plant densely pubescent with long simple hairs, sometimes bearing woolly indumentum. . . . . 10. *Bassia* s.l. (part)
   Dense simple hairs absent. 6
6. Leaves sessile, semi-terete, entire, fleshy, glaucous; seed perisperm absent or scanty; embryo spirally coiled. 7
   At least lower leaves petiolate, not fleshy; seed perisperm present (abundant); embryo annular or hoseshoe-shaped 8
7. Annuals; leaves 20–40 × 5–10 mm; perianth at fruiting stage forms prominent horizontal wings . . . . 16. *Bienertia*
   Shrubs or sometimes annuals; leaves up to 20(–30) × 5–6 mm; perianth without wings . . . . 15. *Suaeda* (part)

---

[13] Moscow State University
[14] Formerly Basel, Switzerland; died 1973
[15] Honorary Research Associate, Royal Botanic Gardens, Kew
[16] Royal Botanic Gardens, Kew

A.P. Sukhorukov would like to thank the Bentham-Moxon Trust, Russian Fund for Fundamental Research (project 14-04-00136: revision of the material in European herbaria) and Russian Scientific Foundation (project 14-50-00029: investigations of reproductive characters) for their financial support during the preparation of the manuscript.

8. Ovary semi-inferior; fruits often dehiscent by a lid; perianth of some fruits adnate to each other, but infrutescence do not bear spiny outgrowths. Glabrous or slightly pubescent annual or perennial herbs (in latter case with thickened root) . . . . 1. *Beta*
   Ovary superior; perianth mostly free, sometimes adnate to each other and form stellate-shaped glomerules with spiny outgrowths 9
9. Plant glabrous or with scattered long-pedicelled (to 0.5 mm) bladder hairs*, mostly dioecious; infrutescence as a rule spiny (except *Spinacia oleracea* known only in cultivation); stigmas 4–5. . . . . .7. *Spinacia*
   Long-pedicelled bladder hairs absent; stigmas 2–3. 10
10. The lowest leaves densely arranged and look like a rosette; perianth of 2–4 equal and small segments, at fruiting stage mostly juicy and red-coloured; fruit ovoid; pericarp adnate to the seed coat . . . . 8. *Blitum*
   No leaf rosette present; perianth not juicy 11
11. Plant monocious or dioecious, mostly farinose or green; fruit enclosed in two flattened or almost spherical, free or diversely connate valves, their dorsal part smooth or with diverse outgrowths; only a part of flowers may consist of equal (3–5) segments . . . . 6. *Atriplex*
   Flowers bisexual; perianth not accrescent 12
12. Leaves glabrous or glabrescent (bladder hairs mostly seen on youngest leaves); perianth free or united into a cup enveloping the fruit; seed embryo horizontal, vertical or in both positions on the same plant . . . . . .5. *Oxybasis*
   Plants with blader hairs, at least covering perianth and lower leaf surface; perianth segments almost free; seed embryo horizontal 13
13. Plant dark green; seeds with a prominent keel.. . . . . 4. *Chenopodiastrum*
   Plant greyish or green; seeds not keeled or only acutish . . . . .3. *Chenopodium*
14. Leaves (or leaves reduced to scales) and/or branches opposite 15
   Leaves and branches alternate 22
15. Leaves well-developed, from 5 mm long 16
   Leaves scale-like (except seedlings that might have longer leaves) 18
16. Leaves stout, not fleshy; only 3 of 5 perianth segments developing wings at fruiting stage; seed embryo vertical . . . . 26. *Girgensohnia*
   Leaves fleshy; all 5 segments developing equal or unequal wings 17
17. Annual branches (if perennial) green or greyish, with unpleasant smell when fresh; fruit up to 2 mm in diameter, their surface papillate; seed embryo vertical . . . . .27. *Anabasis* (part)
   Annual stems (if perennial) whitish, with no unpleasant smell; fruit more than 2 mm in diameter, smooth; seed embryo horizontal . . . . 24. *Seidlitzia*
18. Annuals; perianth segments (3) fused to the top; seeds with hooked hairs . . . . 14. *Salicornia*
   Small trees, shrubs or subshrubs; perianth segments free or connate only at base; seeds or fruits with no hooked hairs 19
19. Subshrubs forming dense mats; buds spherical; leaves with no tufts in axils; perianth segments 3, with no outgrowths. . . . . 13. *Halocnemum*
   Plants not forming dense mats; buds not spherical; leaves with tufts of hairs in axils; perianth segments 5 forming wings at fruiting stage 20
20. Small trees or tall shrubs; scale-like leaves mucronate; wings on perianth segments whitish or yellow. . . . . 28. *Haloxylon*
   Shrubs or subshrubs; leaves with no prominent mucro 21
21. Pericarp clearly fleshy in its upper part and reddish coloured; seed embryo vertical; annual and biennial shoots thick (2.5–5 mm). . . . . 27. *Anabasis*
   Pericarp almost not fleshy; seed embryo horizontal; annual and biennial shoots usually 1.5–3 mm . . . . 29. *Hammada*

---

* bladder hairs: modified epidermal hairs consisting of a stalk cell and a bladder cell. The bladder cell accumulates salt through osmosis from the mesophyll and eventually bursts; also called salt bladder.

22. Glabrous annuals to 25 cm with caducous lower leaves and succulent spherical bracts; inflorescence cone-like, very dense; perianth segments 3, with no projections at fruiting stage; seed coat near embryo with papillae-like outgrowths . . . . 12. *Halopeplis*
    Leaves always present; bracts and inflorescence of other shape, not cone-like; seed coat smooth (but pericarp can bear papillae) 23
23. Flowers with small hyaline bracteoles; seed coat at least in some seeds black and crustaceous; plants glabrous or very short pubescent, often turning black when dry . . . . .15. *Suaeda*
    Bracteoles present or absent, but not hyaline; seed coat not crustaceous. 24
24. Flowers with bracts (no transversal bracteoles are present); seeds with abundant perisperm; embryo horseshoe-shaped; leaves tapered at base . . . . 10. *Bassia*
    Bract and two bracteoles (similar in shape) present at each flower; seed perisperm absent; seed embryo coiled; leaves clearly broadened at base 25
25. Leaves very narrow (to 1 mm wide), 2–5 cm long, aggregated in the lower stem part; branches often thorny . . . . 25. *Noaea*
    Leaves wider, if narrow (1 mm) much smaller and not aggregated in lower part of stem 26
26. Plants glabrous or with short or minute straight papillae or short glandular hairs; simple hairs if present as tufts in leaf axils; at least leaves (and often bracts) with mucro or clearly acuminate 27
    Plants with simple (sometimes mixed with bladder) hairs that can cover different plant parts; straight papillae absent; leaf mucro present or absent 31
27. Leaves fleshy, setose with easily caducous mucro; 3(–4) of 5 perianth segments bearing wings; subshrubs with no tumble-weed habit; scattered glandular hairs can be present on stem and leaves . . . . .31. *Agathophora*
    Leaves stout, with persistent mucro or only slightly cuspidate; all 5 perianth segments bearing wings, or 1–2 of them with large thorn; plants of spherical or spreading habit; glandular hairs absent. 28
28. Tufts of hairs in the leaf axils well-visible with naked eye; perianth not winged 29
    Tufts of hairs as short hairs seen only at magnification or absent; perianth with wings 30
29. One or rarely two perianth segments bearing thorn; annuals or subshrubs . . . . .30. *Cornulaca* (part)
    Perianth with tubercles; subshrubs. . . . . 32. *Traganum*
30. Fresh plants often of unpleasant smell; perianth indurated at fruiting stage. . . . . .22. *Halothamnus*
    Plants with no unpleasant smell; perianth not indurated. . . . . .23. *Kali*
31. Leaves (at least some of them) with a persistent mucro or cuspidate 32
    Leaves not cuspidate or mucronate at apex. 33
32. One (rarely two) from 5 perianth segments (indurated basally) bear long straight spine; anthers with inconspicouos vesicle; annuals (sometimes with long thorns) or subshrubs . . . . .30. *Cornulaca* (part)
    Perianth segments at fruiting stage without spines, indurated at least in lower half; annuals; anthers with large vesicles. . . . . 21. *Halimocnemis*
33. Plants throughoutly covered with two types of indumentum consisting of simple hairs only): tangled hairs with no thickened base and longer straight hairs with conical podium; fruiting perianth with no projections or tubercles; annuals. . . . . 19. *Halocharis*
    Plant glabrous or hairy, sometimes with both short and large simple hairs in lower leaves but without conspicouos conical podium 34
34. Tufts of hairs present around the leaf base and especially in leaf axils; perianth with tubercles. . . . . .34. *Traganum* (some forms with obtuse leaves; see also 29)
    Plants hairy throughout and at leaf base; perianth with wings or smooth (without any appendages) 35

35. Leaves often gibbous at base, well-developed or often scale-like; staminal vesicles (appendages at the tips of stamens) very short or inconspicouos or rarely to 1 mm, but in latter case not clearly separated from the thecae. . . . . 17. *Caroxylon*
Leaves not gibbous or hardly gibbous, always well-developed; staminal vesicles prominent, large, clearly separated from the thecae 36
36. Annuals with decurrent leaves; wings of perianth segments 12–20 mm in diameter (in our area) . . . . 20. *Climacoptera*
Subshrubs; leaves not decurrent; wings absent or to 12 mm in diameter . . . 18. *Kaviria*

Key to the subfamilies

1. Leaves well-developed, flat or terete, of diverse shapes; seeds mostly black and/or brownish, with annular, horseshoe-shaped or almost straight embryo; perisperm present 2
Leaves terete or semi-terete, rarely flattened or reduced to the scales; seeds smooth or with hair-like or mammilate outgrowths; embryo annular, spirally coiled or bent 5
2. Plants covered with branched or dendroid hairs (with alternate outgrowths that can be long or short); pericarp flattened, dehiscent in a middle fruit portion or indehistent; seed coat thin . . . . Subfam. Corispermoideae
Plants glabrous or covered with diverse indumentum types (branched hairs absent), sometimes with stellate hairs (with several rows radiating from cell body); fruit indehiscent or dehistent by a lid or irregularely. 3
3. Plants glabrous or with scattered simple hairs on the leaves and stem; perianths of the neighbouring flowers often connate; ovary semi-inferior; fruit dehiscent by a lid (in our flora) . . . . . . Subfam. Betoideae
Plants with simple, bladder-like, glandular, stellate hairs or with yellow glands, very rarely glabrous; perianths of the neighbouring flowers not connate to each other; ovary superior; fruit indehiscent or irregularely dehiscent 4
4. Indumentum of simple hairs; leaves entire (flattened or terete), sessile or short-petiolate, perianth often with outgrowths (spines, wings, tubercles); embryo horseshoe-shaped. . . . . .Subfam. Camphorosmoideae
Indumentum of bladder, glandular or stellate hairs or with subsessile glands; simple hairs if present scattered and intermixed with bladder hairs and not visible under low magnification; leaves entire, dentate, lobate or pinnatisect; perianth smooth, with longitudinal keel or tuberculate; embryo annular or almost straight . . . . .Subfam. Chenopodioideae
5. Leaves terete or semi-terete, rarely flattened, but always entire; seeds smooth, black, red, brown or translucent; embryo spirally coiled; perisperm absent 6
Leaves reduced to opposite or alternate fleshy scales; seeds often with hair-like or mamillate outgrowths; perisperm present or absent; embryo annular or bent . . . . .Subfam. Salicornioideae
6. Plants glabrous or with scattered indumentum; pericarp easily raptured or scraped off the seed; seeds with dark crustaceous coat always present; seeds often heterospermous (in annuals) . . . . Subfam. Suaedoideae
Plants usually hairy (at least in leaf axils); pericarp not scraped off the seed (fruit is of diaspore type); seed coat translucent. . . . . Subfam. Salsoloideae

SUBFAMILY BETOIDEAE Ulbr.

1. BETA L.

Sp. Pl.: 222 (1753); Gen. Pl. ed. 5: 103 (1754)

Refs: P. Aellen, "Die orientalischen *Beta*-Arten", Ber. Schweiz. Bot. Ges. 48: 470–484 (1938).
G. Kadereit, S. Hohmann, J.W. Kadereit, "A synopsis of *Chenopodiaceae* subfam. *Betoideae* and notes on the taxonomy of *Beta*", Willdenowia 36: 9–19 (2006).

Annual, biennial or perennial herbs, rootstock slender to thick. Leaves alternate, simple, glabrous or with scattered simple hairs, basal forming a rosette (except annuals), long-petiolate; cauline leaves ± reducing, more shortly petiolate. Inflorescence spiciform with 1-several flowers in axils of upper leaves, which are frequently reduced and bracteiform. Perianth segments 5, deltoid to linear or spathulate, often concave, ± incurved, with an obscure or distinct longitudinal keel dorsally. Stamens 5. Ovary semi-inferior, fused with receptacle in fruit, fruiting receptacles of adjacent flowers frequently fused so that each cluster falls as a unit; stigmas 2–3, linear-oblong. Fruit indurate, seeds horizontal; embryo annular.

About 12 species extending from Macaronesia through the Mediterranean to temperate Eurasia; one species in Iraq.

Beta (the Lat. name for *B. vulgaris*, cited by Pliny the Elder).

Beta maritima L., Sp. Pl. ed. 2: 322 (1762); Boiss., Fl. Orient. 4: 898 (1879).

Beta vulgaris L., Sp. Pl. ed. 1: 222 (1753); Boiss., Fl. Orient. 4: 898 (1879); Hand.-Mazz. in Ann. Naturh. Mus. Wien 26: 23 (1912); Nábělek in Publ. Fac. Sci. Univ. Masaryk 105: 8 (1929); Guest in Dep. Agr. Iraq Bull. 27: 15 (1933); Iljin in Fl. SSSR 6: 35 (1936); Zohary in Dep. Agr. Iraq Bull. 31: 45 (1950); Blakelock in Kew Bull. 12: 477 (1958); Rawi in Dep. Agr. Iraq Tech. Bull. 14: 163 (1964); Rech.f., Fl. Lowland Iraq: 183 (1964); Zohary, Fl. Palaest. 1: 139 (1966); Aellen in Fl. Turk. 2: 299 (1967); Assadi in Fl. Iran 38: 19 (2001).

Annual, biennial or perennial with an erect to decumbent or procumbent strongly grooved and ridged stem 20–120 cm (more in luxuriant or cultivated forms), simple to much-branched, glabrous or rarely with scattered multicellular hairs. Root slender to thick and fleshy. Basal leaves with the lamina cordate-ovate to cuneate-rhomboid, vary variable in size and colour, ± long-petiolate with lamina narrowly decurrent along petiole; inflorescence leaves very small and bract-like or remaining larger than the subtended flower or flower-clusters. Inflorescence dense to lax (increasingly so below in fruit). Flowers solitary or in clusters of up to 8 which fall as a unit in fruit. Tepals broadly deltoid to linear-oblong, ± concave and cucullate, incurved, 1.5–3(–5) mm in flower, narrowly to broadly pale-margined, strongly to feebly carinate dorsally, ± accrescent in fruit and ± hardening and saccate at base or not. Receptacle indurate below. Stigma 0.5 mm. Seeds lenticular, horizontal, 2–2.5 mm, black, reticulate, ± immersed in the indurate receptacle, with a thinner apical pericarp.

HAB. Usually a weed of cultivation, especially where irrigated and the soil sometimes noted as ± saline, also in sandy loam bottom land, along riverbanks, on rocky ground, along roadsides and on saline alluvial soil by a lake; alt. 2–260 m; fl. Jan.–May (one specimen fl. in Nov.), fr. Apr.–May.
DISTRIB. Widespread in the Lower Mesopotamian Plain, rare in the foothills and desert: **MRO**: Shaqlawa, *Eig & Zohary* s.n. (HUJ!). **FUJ**: Baidha, *Chakr., Rawi, Khatib & Alizzi* 32103!; **FNI**: Eski Kellek, *Anderson* 41401!; **FPF**: Mandali, *Guest* 12649! **DWD**: 4 km W. of Hit, *Rawi & Gillett* 6864!; Fathat Imam: *Rawi* 31074! **DSD**: S of Zubair, *Rawi* 25872! **LEA**: Shahraban (Muqdadiya), *Paranjpye* in Graham s.n.!; Nahrwan, *Robertson* 77 (S/1031)! **LCA**: Nasiriya, *Karim & Hamid* 37790 !; Shamiya, *Qaimaqam of Shamiya* 2357!; Abu Ghraib, *Alizzi* 34182!; Baghdad, Karrada Mariam, *Haines* 789!; Baghdad, Rustam, *Lazar* 3908!, *Guest* 265!; 20 km E of Baghdad, *Alizzi & Omer* 34888! **LSM**: 37 km S of Ali Gharbi, *Rawi & Haddad* 25739!; 20 km S of Amara, *Rawi* 15019! 15040!; 5 km N of Al-Qurna, *Barucha, Rawi & Tikriti* 2932! **LBA**: Basra to Khadr al-Mai, *Bharucha & Abbas* s.n.!; 10 km S of Basra, *Rawi* 25912!; Abu al-Khasib, 15 km S of Basra, *Chakravarty & Rawi* 2987!

Distribution (of species): native on the coasts of S and W Europe, Macaronesia, the Mediterranean area, and in coastal regions and inland saline areas of SW Asia eastwards to NW India. Cultivated and occasionally naturalised or spontaneous elsewhere.

We have not been able to make much of the hierarchy of forms of this variable species as applied to the Iraqi material. Apart from the cultivated material cited below, all of the specimens seen appear (where basal parts have been collected) to be annual or perhaps occasionally biennial. Aellen, l.c., records subsp. *maritima* "L." var. *foliosa* (Aschers. & Schweinf.) Aellen and var. *glabra* (Del.) Aellen (which should be called var. *maritima*) for Iraq and has determined material at Kew as this subspecies. To separate subsp. *maritima* from subsp. *vulgaris* he utilises chiefly the perianth segments, stating that in subsp. *maritima* these are shortly triangular and only about equalling the radius of the receptacle – i.e. about reaching the style. This is true of the Iraqi material which he has named. However *B. vulgaris* "maritima" (what rank Linnaeus intended for this in ed. 1 of Sp. Pl. is not clear) was described from England and Belgium, and wild beet from the coasts of both those countries has linear-oblong perianth segments clearly longer than the rachis of the receptacle. Other authors have used the perennial habit of the wild sea beet to separate it at subspecific rank, but here too Thellung (in Syn. Mitteleur. Fl. 6 (1): 14 (1913)) allows a var. *annua*, and how this may be separated from an "annual or biennial" var. *foliosa* and var. *vulgaris* is quite obscure. Preparation of a regional flora does not allow time for revision of such a complex group; I am content to record Iraqi *B. vulgaris* with no attempt at further subdivision of the wild material. The var. *foliosa* seems to me to be only the extreme form of a range, and not worthy of recognition.

Mention must be made of the subsp. *lomatogonoides* Aellen, op. cit. : 478 (1938), based on two specimens, one form Iran, the other from Iraq. This is *Field & Lazar* 13 from LSM: Gatt al-Dwat (F!); although the fact is not stated in his revision, Aellen's determinavit label states "Beta vulgaris L. ssp. lomatogonoides Aellen, not typical", and we cannot separate the plant from much other Iraqi material, although the pale margins are perhaps slightly wider than most.

The cultivation of the various forms belongs to *Beta vulgaris* L. It is clear that the most commonly grown form in Iraq is var. **vulgaris**, or **Beetroot** that is originated from *B. maritima* due to selection, not known escaped from cultivation, and var. **cicla** L., the Leaf-Beet or Spinach-Beet, SILIQ or ZILIQ. Specimens apparently referable here are: LCA: Karrada, *Paranjpye* in Graham s.n.!; Rustam, *Guest* 249!, 1776!; Baghdad West, *Gillett* 10723!; College of Agriculture, Abu Ghraib, *Barkley, Hikmat Abbas & Juma'a Brahim* 1308!; LSM: 3 km E. of Amara on Fakka road, *Barkley* 6414!

The Beetroot, var. *rubra* (L.) Moq. is also grown, and known by the names SHUWANDAR, CHUKUNDAR, SUKARI or PANJAR. This is clearly grown to a much lesser extent, but one specimen may belong here: LCA: College of Agriculture, Abu Ghraib, *Ani* 687 !

SUBFAMILY CHENOPODIOIDEAE: TRIBE DYSPHANIEAE

2. DYSPHANIA R. Br.

Prodr. Fl. Nov. Holland.: 411 (1810)

Annual, subshrubs or shrubs covered with yellow subsessile glands (strongly aromatic) that are intermixed with simple and sometimes glandular hairs. Leaves alternate, petiolate, entire, dentate or pinnate. Inflorescence lax, much branched, or dense and spike-like. Cymes often reduced to 1 flower. Flowers hermaphrodite or unisexual. Perianth of 2–5 free or diversely connate segments, often keeled or cristate at the midrib. Stamens 2–5. Stylodia 2, free or connate at base. Fruits almost spheroidal, with length/thickness as 1.2–1.3:1, rarely ovoid and compressed. Pericarp very thin (as a rule 3–15 μm), 1–2-layered, hyaline, scraped off the seed, often with papillae or glandular hairs at its surface, sometimes the surface smooth, with reticulate or undulate sculpture. Seed coat 5–25 μm, smooth or undulate, outer seedcoat layer (testa) without stalactites. Embryo horizontal or vertical. In some groups spatial heterospermy is evolved.

Dysphania, from Gr. δυσ, *dys*, bad and φραίνομαι, *frainomai*, appear (due to the smell of plants).

About 50 species distributed worldwide but predominantly in the tropics and subtropics; one species, *D. botrys* is widespread in W and C Asia.

Leaves entire, dentate or sinuate; inflorescence dense, spike-like; perianth segments at least of some flowers connate to the middle, enclosing the fruit completely; fruit 0.8–1 mm in diameter; pericarp with glandular hairs in the upper part . . . . 2. *D. ambrosioides*
Leaves lobate or pinnatifid; infrorescence lax; perianth segments fused at base, open at fruiting stage but not falling off; fruit 0.6–0.8 mm in diameter; pericarp with small papillae throughout. . . . . 1. *D. botrys*

1. **Dysphania botrys** *(L.) Mosyakin & Clemants,* Ukr. Bot. Zhurn. 59(4): 383 (2002); Sukhorukov & Akopyan, Comp. Chenop. Cauc.: 17 (2013).

Chenopodium botrys L., Sp. Pl. ed. 1: 219 (1753); Boiss., Fl. Orient. 4: 903 (1879); Eig in Pal. Journ. Bot. Ser. 3: 120 (1945); Zohary in Dep. Agr. Iraq Bull. 31: 46 (1950); Blakelock in Kew Bull. 3: 480 (1957); Mouterde, Nouv. Fl. Lib. et Syr. 1: 410, f. 3 (1966); Aellen in Fl. Turk. 2: 300 (1967); Uotila in Fl. Iranica 172: 54 (1997) & in Fl. Pak. 204: 47 (2001); Assadi in Fl. Iran 38: 61 (2001); Zhu, Mosyakin & Clemants in Fl. China 5: 377 (2003).

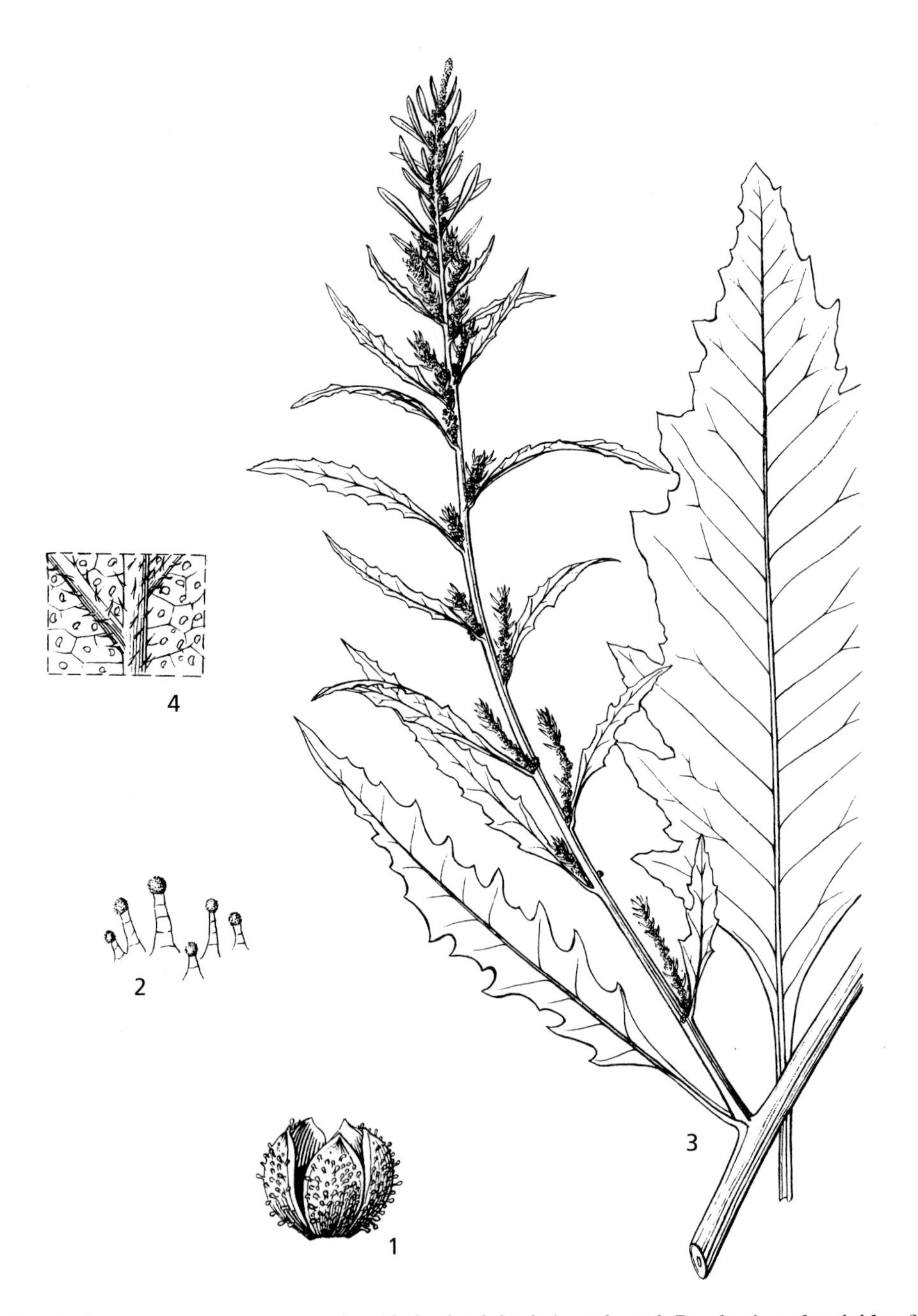

Fig. 65. **Dysphania botrys**. 1, perianth × 12; 2, glandular hairs enlarged. **Dysphania ambrosioides**. 3, habit × 1/4; 4, leaf portion abaxial surface detail. Reproduced with permission from Flora of China 5: fig. 307. Drawn by Cai Shuqin.

Annual, 10–50 cm, aromatic. Leaves 2–8 × 1–3 cm, ovate or oblong, pinnatisect or pinnatifid with 4–6 lobes on each side, with yellow glands and simple hairs. Inflorescence long, terminal, mostly pubescent with glandular hairs, composed of leafless cymes. Perianth segments acuminate, with a broad membranous margin, rounded on back, nearly keelless, with glandular and scattered simple hairs; no subsessile glands present. Fruit 0.6–0.8 mm; pericarp with short papillae. Seeds horizontal, smooth or with shallow pits. Fig. 65: 1–2.

HAB. Waste places, sandy and gravelly soil, roadsides, river-beds, cultivated land; alt. s.l.–1550 m; fl. & fr. Jun.–Sep.
DISTRIB. Locally common in the forest zone of N Iraq: **MAM**: nr Aradin, *Guest* 4999!; Merga Qus village NE of Zakho, *Rawi* 23525!; from Zawita to Atrush, *Al-Kaisi, Hamad & Hamid* 46081!; Jabal Khantur N of Zakho, *Rawi* 23326!; Sarsang, *Haines* W.515!; *Zohary* s.n. (HUJ!). **MRO**: Haji Omran, *Dabbagh* 46293!; Rowanduz, *Omar, Sahira, Karim & Hamid* 38508! *Eig & Feinbrun* s.n. (HUJ!); Rost, *Haley* 108 (BM!). **MSU**: Pir-i-Mukrun, *Eig & Duvdevani* s.n. (HUJ!); Sulaimaniya, *Zohary & Amdursky* s.n. (HUJ!); Sulaimaniya, *Hadač et al.* 4817 (G!). **MJS**: Penjwin, *Guest* 12962! **FNI**: Mindan bridge on Agra road, *Chapman* 26245!

The strong-smelling aromatic plant is avoided by insects, and so is often used as protection against moths.

Europe, temperate Asia, introduced in N America and Australia.

2. Dysphania ambrosioides (*L.*) *Mosyakin & Clemants*, Ukrainsk. Bot. Zhurn. 59 (4): 382 (2002); Zhu, Mosyakin & Clemants in Fl. China 5: 377 (2003); Sukhorukov & Akopyan, Comp. Chenop. Cauc.: 17 (2013).

Chenopodium ambrosioides L., Sp. Pl.: 219 (1753); Uotila in Fl. Pak. 204: 26 (2001).

Annual or short-lived perennial up to 120 cm, aromatic. Stems branching. Leaves short-petioled, oblong to lanceolate, mostly sinuate-dentate, upper leaves entire or nearly so, all with yellowish glands on lower surface. Flowers in dense clusters, forming elongated spike-like bracteose inflorescence; branches with short simple hairs. Perianth about 1.5 mm, completely enclosing fruit; tepals 4–5, united at base to the middle, not or slightly keeled. Stamens 4–5. Fruit 0.8–1 mm, subspherical. Pericarp tiny, in upper part with glandular hairs. Seed mostly horizontal, glossy. Fig. 65: 3–4.

HAB. Waste places, roadsides, river-beds, cultivated land; alt. to 500 m; fl. & fr. Jun.–Sep.
DISTRIB. Probably a widespread alien plant (as a ruderal), known in adjacent areas from 19th century. In Iraq recorded only once: **LSM**: Amara, *Chakravarty* 37764!

Central and South America, common weed in many tropical regions of Eurasia, especially in humid regions.

TRIBE CHENOPODIEAE

3. CHENOPODIUM L.

Sp. Pl.: 218 (1753); Gen. Pl. ed. 5: 103 (1754)

Annual herbs and shrubs, rarely small trees, covered with bladder hairs. Leaves petiolate, entire or more frequently lobed or dentate, very rarely terete and without petiole. Inflorescence composed of small cymose clusters arranged in paniculate racemes. All flowers hermaphrodite or part female, green or sometimes reddish. Perianth segments 5 (rarely 4), free or united at base. Stamens (2–4)–5, free or basally connate. Stigmas 2(–3), free. Fruit depressed-globular, falling off separately or together with perianth. Pericarp mostly thin, membranous, of 1–2(–3) parenchymatous layers, as a rule with small cylindrical or conical papillae (in dry fruits, the pericarp structure mostly looks as reticulate, and after soaking, the papillae retrieve their structure). In some members presently transferred to *Chenopodium* (*Rhagodia, Einadia*), the pericarp (at least in majority of flowers) appears fleshy (berry), coloured, and many-layered, but part of the fruits remain dry (heterocarpy). Seeds black or rarely brownish, with horizontal embryo; outer seed-coat layer (testa) of the black seeds with vertical stalactites.

About 100 species in recent circumscription distributed in all parts of the world; some are widespread or cosmopolitan. This genus is split into several genera of different systematic position within Chenopodioideae (see Kadereit & al., 2010; Fuentes-Bazan & al., 2012a, b) and distinct carpology (Sukhorukov & Zhang, 2013).

Chenopodium (from Gr. χῆνα, *chena*, goose, πόυς, *pous*, foot)

Our species are anthropogenic. They accompany human cultivation as weeds and occur in wadis and desert flats that have been landscaped by humans.

None of the species of *Chenopodium* in Iraq form distinct associations. Individual plants occur as isolated ephemerals, rarely in ± dense stands. They mostly colonise soils farmed by humans and survive repeated cultivation without much detriment. They occur on undisturbed soils predominantly in association, and despite the presence of, other native, widely distributed, perennial species. The chenopods of Iraq are primarily of Mediterranean species that extend their range into SW Asia. There are no endemic species.

Many of the species are eaten as spinach or the seeds used as meal, chiefly when the testa is thin and the seeds white. The distinction of the species presents no difficulties but ripe seeds are required to confirm identification.

1. Leaves as broad as long, the blades usually not exceeding 2.5–3 cm 2
   Leaves distinctly longer than broad, length of the blades usually more than 3 cm 3
2. Leaves 3-lobed, with broad toothed lobes, not or slightly foul-smelling . . . . 1. *C. opulifolium*
   Leaves broadly rhombic-ovate to rhombic, entire or with a simple, acuminate teeth; plant smelling of rotten fish . . . . 2. *C. vulvaria*
3. Leaves 3-lobed, terminal lobe clearly longer than the lateral ones . . . . . . 4. *C. ficifolium*
   Leaves not 3-lobed, entire or toothed or slightly lobate 4
4. Plant green . . . . 3. *C. album*
   Plant greyish or farinose 5
5. All leaves narrowly oblong or lanceolate, entire or slightly dentate . . . . 5. *C. novopokrovskyanum*
   Lower leaves toothed or slightly 3-lobate . . . . 6. *C. iranicum*

1. **Chenopodium opulifolium** *Schrad.* in Koch & Ziz, Cat. Pl. Palat. 6 1814); Boiss., Fl. Orient. 4: 901 (1879); Eig in Palest. J. Bot., Jerusalem ser., 3: 120 (1945); Tackholm, Stud. Fl. Egypt: 422 (1956); Rech.f., Fl. Lowland Iraq: 185 (1964); Mouterde, Nouv. Fl. Lib. et Syr. 1: 409 (1966); Zohary, Fl. Palaest. 1: 142 (1966); Aellen in Fl. Turk. 2: 302 (1967); Uotila in Fl. Iranica 172: 52 (1997); Assadi in Fl. Iran 38: 47 (2001); Sukhorukov & Akopyan, Comp. Chenop. Cauc.: 18 (2013).

Annual, 40–80 cm, loosely and divaricately branched, dark green, mostly grey-mealy, with red spot in axils of branches. Leaves 1–3 × 1–3 cm, lower and median rhombic-ovate, often broader than long, 3-lobed, the lateral lobes short and broad, middle lobe broadly triangular to shortly ovate, obtuse or subacute, subentire or toothed, mucronate. Upper leaves narrower, toothed. Inflorescence irregularly branched, often contracted, glomerules becoming mealy. Perianth segments broadly keeled on back. Seeds black, rounded at margin.

HAB. Waste places, roadsides, cultivated land; alt. 500–1500 m; fl.& fr. Jun.–Aug.
DISTRIB. **MAM**: Sarsang, *Haines* W554!; Sharanish, *Rawi, Nuri & Tikriti* 29053!; Shaikh Adi, *Guest* 3671! **MRO**: Pushtashan, *Rawi & Serhang* 24236!; Baradost, *Thesiger* 328 (BM!). **MSU**: Qara Dagh, *Buthaina Makki* 534! **FNI**: Nineveh to Mosul, *Chapman* 25380!

This species prefers warm and dry countries. The dense mealy indumentum and small, broad, 3-lobed leaves with a broad rounded middle lobe are characteristic for this species.

S and E Europe, N Africa, SW and C Asia.

2. **Chenopodium vulvaria** *L.*, Sp. Pl. ed. 1: 220 (1753); Boiss., Fl. Orient. 4: 901 (1879); Nábělek in Publ. Fac. Sci. Univ. Masaryk 105: 8 (1929); Zohary in Dep. Agr. Iraq Bull. 31: 45 (1950); Tackholm, Stud. Fl. Egypt: 422 (1956); Rech.f., Fl. Lowland Iraq: 184 (1964); Mouterde, Nouv. Fl. Lib. et Syr. 1: 409 (1966); Zohary, Fl. Palaest. 1: 141 (1966); Aellen in Fl. Turk. 2: 303 (1967); Uotila in Fl. Iranica 172: 42 (1997) & in Fl. Pak. 204: 33 (2001). Assadi in Fl. Iran 38: 19 (2001); Sukhorukov & Akopyan, Comp. Chenop. Cauc.: 20 (2013).

Annual, 20–100 cm. Stems branched from base, often decumbent, in young parts mealy, having a strong stale-fish smell. Leaves 1–2.5 × 1–2.5 cm, rhombic or ovate, entire or with a single tooth-like angle on each margin at broadest part, mealy beneath. Inflorescence small, terminal and axillary. Perianth segments rounded on back. Seeds black, rounded at margin. Fig. 66: 1–4.

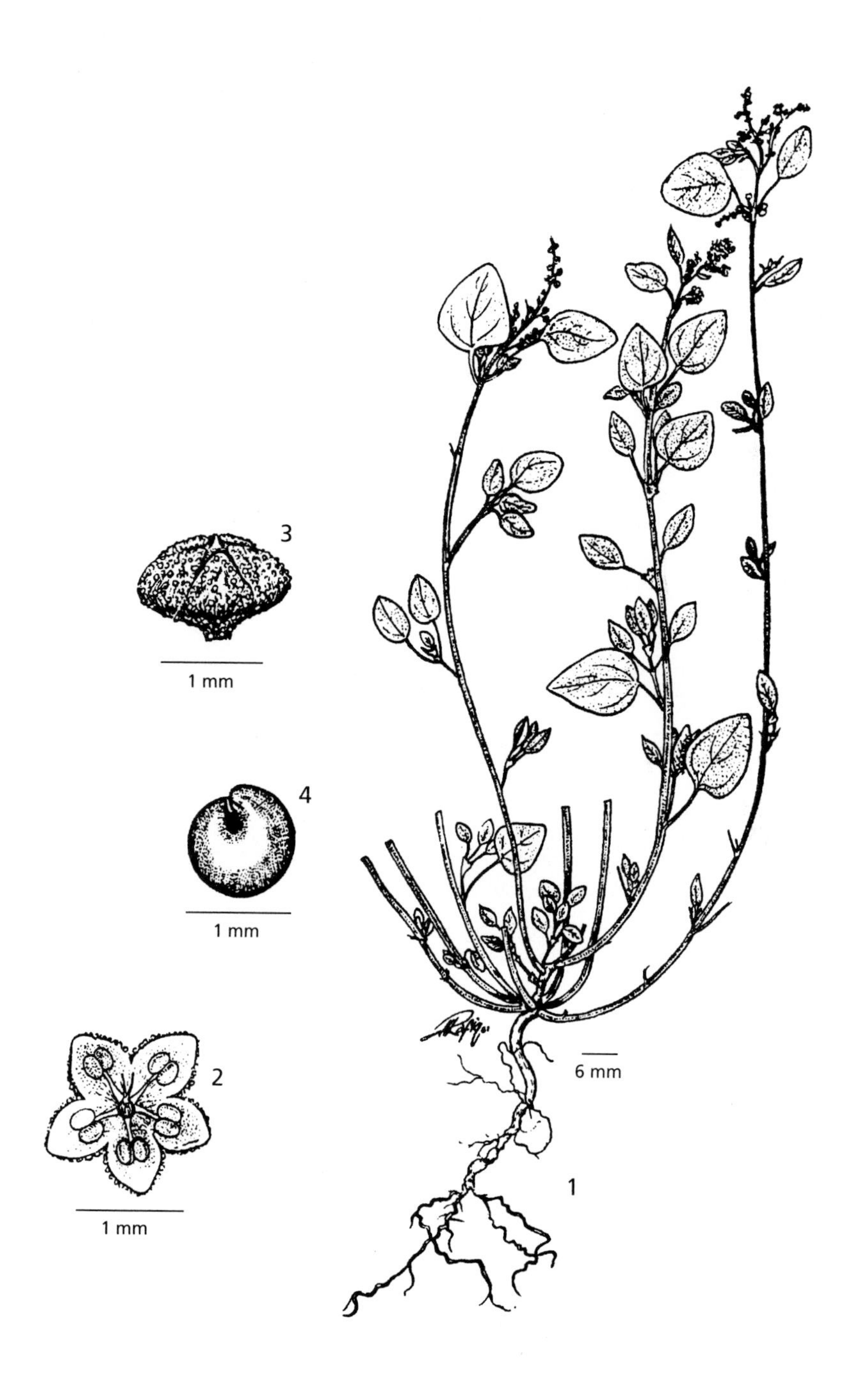

Fig. 66. **Chenopodium vulvaria**. 1, habit; 2, flower; 3, perianth with fruit; 4, seed. Reproduced with permission from Flora of Pakistan 204: f. 6. 2003. Drawn by M. Rafiq.

HAB. Waste places, nitrate-rich soils, roadsides, rarely as a weed of gardens and fields; alt. 250–1000 m; fl.& fr. May–Aug.
DISTRIB. Occasional in the Mesopotamian plain and foothills of Iraq: **MAM**: Zawita, *Guest* 4888! **FNI**: Nineveh, *Chapman* 25385! **LCA**: Baghdad, *Lazar* 3418!, *Haines* W.1361!; nr Baghdad, *Rechinger* 11705!

The species has a characteristically unpleasant smell of rotten fish, which it does not lose even when growing on sandy ground low in nitrate. However, some populations in Eastern Mediterranean can be almost without smell when fresh (Sukhorukov, obs.).

Eurasia, N Africa; introduced in N America, Australia and New Zealand.

3. **Chenopodium album** *L.*, Sp. Pl. ed. 1: 219 (1753); Boiss., Fl. Orient. 4: 901 (1879); Eig in Palest. J. Bot. Jerusalem ser., 3: 119 (1945); Zohary in Dep. Agr. Iraq Bull. 31: 45 (1950); Tackholm, Stud. Fl. Egypt: 422 (1956); Blakelock in Kew Bull. 3: 479 (1957); Rech.f., Fl. Lowland Iraq: 185 (1964); Mouterde, Nouv. Fl. Lib. et Syr. 1: 409 (1966); Zohary, Fl. Palaest. 1: 142 (1966); Aellen in Fl. Turk. 2: 304 (1967); Uotila in Fl. Iranica 172: 48 (1997) & in Fl. Pak. 204: 40 (2001); Assadi in Fl. Iran 38: 42 (2001); Zhu, Mosyakin & Clemants in Fl. China 5: 383 (2003); Sukhorukov & Akopyan, Comp. Chenop. Cauc.: 18 (2013).

Annual, 20–200 cm, erect, many-branched, ± mealy. Leaves 2–8 × 1–5 cm, rhombic, deltoid or ovate-lanceolate, entire or denticulate, lower leaves often 3-lobed (middle lobe tapering to apex), mealy beneath. Flowers in glomerules arranged in terminal and axillary panicle-like inflorescence. Perianth segments with rounded keel (± sharp when dry) on back. Seeds obtuse on margins, nearly smooth with shallow radial furrows. Fig. 67: 1–4.

HAB. Waste places, cultivated land; alt. 0–1500 m; fl.& fr. May–Sep.
DISTRIB.: **MRO**: Rayat, *Guest* 13031! **FNI**: Wadi Aloka, *Barkley & Al-Ani 9009*! **DLJ**: Balad, *Rawi, Nuri & Tikriti* 28947! **LEA**: between Muqdadiya and Ba'quba, *Rechinger* 8048! **LCA**: Baghdad airport, *Rechinger* 8013!; 12 km NW of Falluja, *Rawi* 25466!; Baghdad, Chelebi's garden, *Haines* W.43!

The species is of Eurasian origin but is now cosmopolitan. Extremely polymorphic; it develops many geographical races and shows great adaptability to different ecological conditions.

4. **Chenopodium ficifolium** *Sm.*, Fl. Brit. 1: 276 (1800); Boiss., Fl. Orient. 4: 901 (1879); Nábělek in Publ. Fac. Sci. Univ. Masaryk 105: 8 (1929); Zohary in Dep. Agr. Iraq Bull. 31: 45 (1950); Tackholm, Stud. Fl. Egypt: 422 (1956); Blakelock in Kew Bull. 3: 480 (1957); Rech.f., Fl. Lowland Iraq: 185 (1964); Aellen in Fl. Turk. 2: 305 (1967); Uotila in Fl. Iranica 172: 40 (1997); Assadi in Fl. Iran 38: 39 (2001); Zhu, Mosyakin & Clemants in Fl. China 5: 383 (2003); Sukhorukov & Akopyan, Comp. Chenop. Cauc.: 19 (2013).

Annual, to 80 cm, loosely branched, ± mealy. Leaves 2–6 × 1–3 cm, 3-lobed, lateral lobes in lower part of leaf short, entire or sinuate; middle lobe long, narrow, with ± parallel margins, irregularly sinuate-dentate. Flowers in small glomerules in ± terminal panicles. Perianth green, becoming later yellow-brown or black-brown, enclosing fruit; segments on back slightly carinate. Pericarp scraped off the seed. Seeds small, rounded or slightly keeled, testa with narrow elongate pits, without radial furrows between pits. Fig. 67: 5–6.

India, Nepal, SE Asia (Myanmar, Thailand), China, Korea, Japan, Taiwan, alien in temperate Eurasia.

All plants from Iraq, insofar as they have ripe fruits and are therefore determinable, belong to the following subspecies:

subsp. **blomianum** (*Aellen*) *Aellen* in Hegi, Ill. Fl. Mittel-Eur. ed. 2, 3 (2): 624 (1960).
C. blomianum Aellen in Bot. Notiser 203 (1928).

Testa of seeds with radial furrows between pits; pits are less deep than in the type subspecies and margins are often curved.

HAB. Waste places, as a weed in gardens and other cultivated land; alt. s.l.–500 m; fl.& fr. May–Aug.
DISTRIB. Mainly collected in the Baghdad area, occasional elsewhere: **FUJ**: nr Mosul, *Kotschy* 450 (BM!). **DWD**: Ramadi, *Regel* s.n.! **LCA**: Abu Ghraib, *Alizzi, Thaib & Tenan* 32574!; Rustam, *Lazar* 3435!; Zafariniya nr Baghdad, *Al-Kadi & Rechinger* 15846!; Baghdad, Al-Khir bridge, *Rechinger* 8004!; Karrada nr Baghdad, *Regel* s.n.; Baghdad, *Regel* s.n.! *Lazar* 331!; **LSM**: Musaida, *Field & Lazar* 81!; Amara, *anon.* 42496!; Amara, *Evans* s.n. (E!).

Distribution of subspecies: probably the same as type subspecies.

Fig. 67. **Chenopodium album**. 1, habit × 1; 2, leaf × 1 ; 3, perianth × 15; 4, seed × 20. **Chenopodium ficifolium**. 5, habit flowering branch × 1; 6, leaf × 1. Reproduced with permission from Flora of China 5: fig. 312. Drawn by Cai Shuqin.

5. **Chenopodium novopokrovskyanum** *(Aellen) Uotila*, Ann. Bot. Fennici 30: 192 (1993); Uotila in Fl. Iranica 172: 40 (1997) & in Fl. Pak. 204: 37 (2001); Sukhorukov & Akopyan, Comp. Chenop. Cauc.: 19 (2013).

C. *album* L. subsp. *novopokrovskyanum* Aellen, Trudy Rostov. Oblastn. Biol. Obsch. 2: 3 (1938).

Very similar to *C. album* in most characters, but with lanceolate leaves that are mealy below or on both sides.

DISTRIB. LCA: Baghdad, Rustam, *Rogers* 0256!, *Haines* W.43!

Caucasus, Iran, Eastern Mediterranean, Afghanistan.

6. **Chenopodium iranicum** *(Aellen) Hamdi & Malekloo*, Iran. J. Bot. 16(1): 69 (2010).

C. album L. subsp. *iranicum* Aellen, Not. Roy. Bot. Gard. Edinb. 28(1): 30 1967; ?*C. zerovii* Iljin, Fl. USSR 6: 650 (1936).

Similar to *C. album*, but leaves farinose or greyish on both sides, ovate or oblong, dentate or slightly 3-lobate. Perianth greyish, with broad white margins and a keeled midrib.

HAB. Found on irrigated places. 0–1500 m.
DISTRIB. Overlooked species that is known mostly from Iran, but collected also in the neighbourung countries including Iraq: MRO: Surade, *Rawi* 26518!; nr Rayat, *Eig & Zohary* s.n. (HUJ!). FUJ: nr Mosul, *Hohenacker* 450b (LE!); LCA: Abu Ghraib, *Rechinger* 8121!; Baghdad, *Guest* 3234!; Baghdad, Karaftiya estate, *Guest* 3233!; Baghdad, Adhimiya, *Shehbaz & Mousawi* (E!, LE!).

?South Ukraine, Turkey, Iran. The type of *C. zerovii* as well as authentic specimens are not found in LE. Further work is necessary to confirm the identity of *C. iranicum* and *C. zerovii.*

4. **CHENOPODIASTRUM** S.Fuentes, Uotila & Borsch

Willdenowia 42 (1): 14 (2012)

Annuals, glabrous or covered with bladder hairs, sometimes with scattered yellowish glands. Leaves triangular or rhombic, petiolate, entire, dentate or lobate, rarely pinnatisect. Inflorescences spreading, mostly leafless. Flowers bisexual, with 5 free or basally connate segments not changing at fruit. Stamens 5. Stylodia 2. Fruits 1.3–2.5 mm in diameter; pericarp 1–2-layered, with conical or cylindrical papillae (forming alveolate or reticulate surface when dry). Seed black, with keel or not, its surface alveolate or punctate, sometimes with deep combs. Seedcoat testa with vertical or oblique oriented stalactites; latent structural heterospermy expressing in the diverse testa thickness is observed in some species. Seed embryo horizontal.

At least 7 species native in Eurasia, N America and Australia, but Eurasian group *C. hybridum* is poorly investigated so far. One species in Iraq. In morphological characters, *Chenopodiastrum* is very close to *Chenopodium* s.str.

Chenopodiastrum, from *Chenopodium* (q.v.) and Lat. *-astrum*, partly resembling.

1. **Chenopodiastrum murale** *(L.) S. Fuentes, Uotila & Borsch*, Willdenowia 42 (1): 14 (2012); Sukhorukov & Akopyan, Comp. Chenop. Cauc.: 22 (2013).

Chenopodium murale L., Sp. Pl. ed. 1: 219 (1753); Boiss., Fl. Orient. 4: 902 (1879); Eig in Pal. J. Bot., ser. 3: 120 (1945); Zohary in Dep. Agr. Iraq Bull. 3: 479 (1950); Tackholm, Stud. Fl. Egypt: 422 (1956); Blakelock in Kew Bull. 3: 479 (1957); Rech.f., Fl. Lowland Iraq: 186 (1964); Mouterde, Nouv. Fl. Lib. et Syr. 1: 409 (1966); Zohary, Fl. Palaest. 1: 142 (1966); Aellen in Fl. Turk. 2: 302 (1967); Uotila in Fl. Iranica 172: 38 (1997); Assadi in Fl. Iran 38: 36 (2001); Uotila in Fl. Pak. 204: 26 (2001).

Annual, 10–70 cm, erect, much-branched, dark green, slightly mealy. Leaves 2–6 × 1–5 cm, broadly deltoid-ovate, upper narrow and long-acuminate, mucronulate, cuneate to rounded at base, coarsely and irregularly toothed, rarely subentire; teeth acute and ± incurved. Inflorescences terminal and axillary, of loosely branched cymes; glomerules small. Perianth segments slightly keeled, with a distinct swelling below apex. Pericarp tightly adjoining to the seed coat. Seeds with prominent keel, testa densely covered with small pits. Fig. 68: 1–7.

HAB. Waste places, roadsides, seashores, riversides, in fields and gardens, on fine gravel soil mixed with sheep manure; alt. s.l.–300 m; fl.& fr. Apr.–Aug.

Distrib. Widespread but scattered in the desert and plains of Iraq: **msu**: Qaradagh, *Graham* s.n.! **dwd**: Ramadi, *Regel* s.n.!; **dsd**: 85 km NW of As-Salman, *Rechinger* 9423!, *Guest, Rawi & Rechinger* 19209!, 19221!; Kuwaibda, *Eig & Zohary* s.n. (HUJ!); Safwan, *Botany Staff* 24447! **lea**: Kumait, *Al-Dabagh & Al-Kaisi* 41915! Ba'quba, *Regel* s.n.!; Ma'qil nr Basra, on island in Shatt al-Arab, *Rechinger* 8438!; **lca**: Baghdad airport, *Haines* s.n.!; 37 km S of Ali Gharbi, *Rawi & Haddad* 25740! garden behind college, Baghdad, *Regel* s.n.!; nr Baghdad, *Lazar* 553!; 65 km S of Basra, *Chakravarty, Rawi, Khatib & Tikriti* 29984! Adhamiya nr Baghdad, *Rechinger* 11702!, *Regel* s.n.!; Abu Ghraib, *Alizzi, Thaib & Tanan* 32576!; Khalis, *Barkley* 2487! 12 km NW of Falluja, *Rawi* 25467! ; **lsm**: Qal'a Salih, 40 km S. of Amara, *Rechinger* 8853!; Qal'a Salih, *Guest, Rawi & Rechinger* 17407! Amara, *Evans* s.n. (E!). **lba**: Basra, *Rawi & Gillett* 5984! Kuwaibda nr Basra, Eig & Zohary s.n. (HUJ!); Abu Hassib nr Basra, *Regel* s.n.!, *Rawi* 25930!

Cosmopolitan with unknown origin (E Africa?), introduced in N America and Australia.

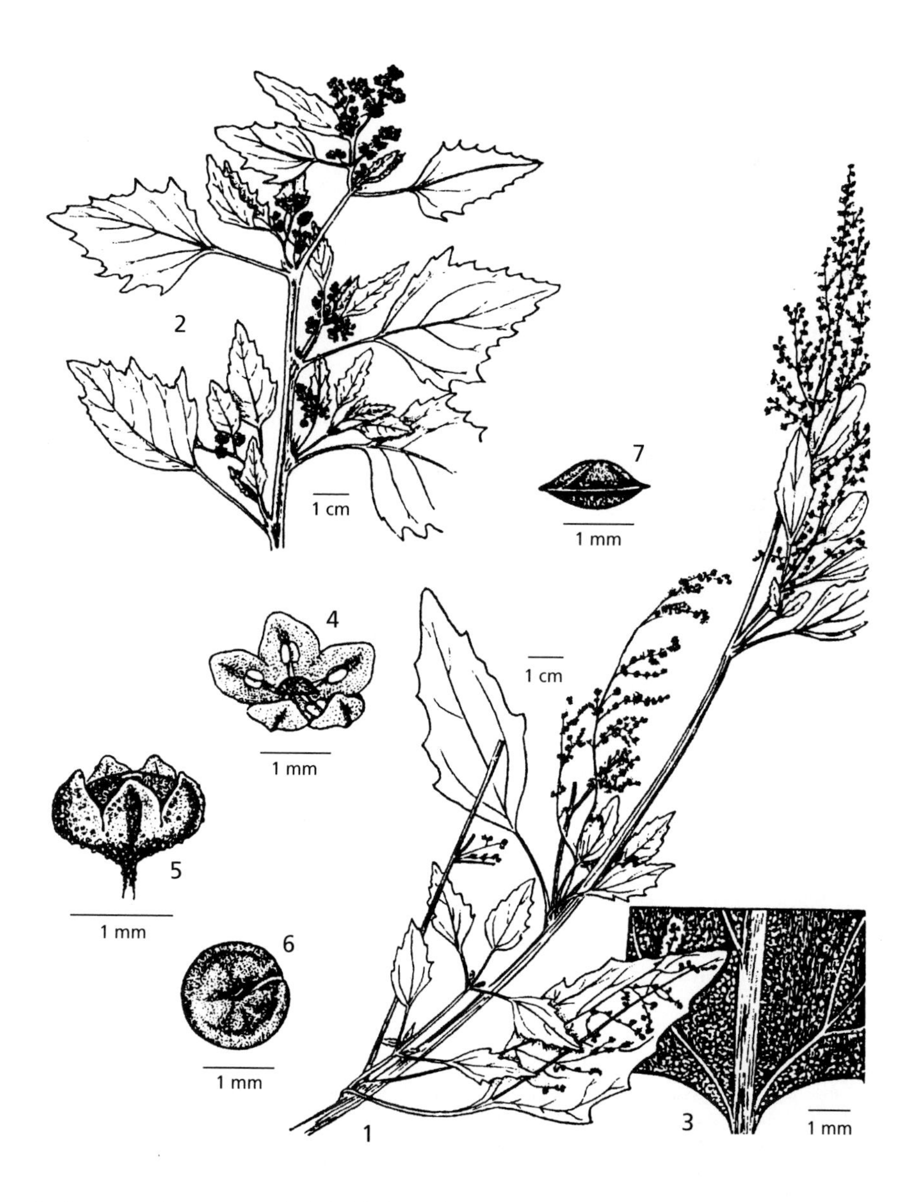

Fig. 68. **Chenopodiastrum murale**. 1, habit; 2, habit fruiting; 3, leaf detail abaxial surface; 4, flower; 5, perianth with fruit; 6, seed; 7, seed side view. Reproduced with permission from Flora of Pakistan 204: f. 4. 2003. Drawn by M. Rafiq.

5. **OXYBASIS** Kar. & Kir.

Bull. Soc. Imp. Nat. Mosc. 4: 738 (1841)

Annual herbs, branched from the base or with a single stem, glabrous or covered with bladder hairs. Leaves entire, undulate or lobate, of rhombic, triangular or oblong. Inflorescence cylindrical with lateral branches mostly appressed to the stem, flowers in dense glomerules. Perianth of 2–5 free or diversely connate hyaline or greenish segments (in some species both perianth forms are present). Flowers bisexual or (lateral ones) sometimes female. Stamens 1–5. Stylodia 2. Pericarp thin or rarely if several so consists of equal layers, smooth, mamillate or rarely papillate. Seeds mostly small (to 1.2 mm) in diameter, red or black. Embryo horizontal or vertical, not rare both embryo positions present at one individual (spatial heterospermy). Structural (latent) heterospermy expressing in the seedcoat testa is common by almost all representatives. Outer seedcoat layer (testa) with stalactites.

Oxybasis, from Gr. ὀξύς, *oxus*, sharp and – βάσησ, *basis*, basis or foundation.

At least 12 species distributed in the temperate regions of Eurasia and America. Two species in Iraq.

Plant to 100 cm, stem straight, without reddish colour; perianth segments free; pericarp papillate (reticulate when the fruits dry); seed black . . . . 2. *O. urbica*
Plant to 50 cm, often reddish; perianth segments in a part of flowers connate to the top; pericarp smooth; seed red . . . . 1. *O. chenopodioides*

1. **Oxybasis chenopodioides** (*L.*) *S. Fuentes, Uotila* & *Borsch* in Willdenowia 42 (1): 15 (2012); Sukhorukov & Akopyan, Comp. Chenop. Cauc.: 20 (2013).

Blitum chenopodioides L., Mantissa 2: 170 (1771). *Chenopodium botryoides* Sm. in Sowerby & Sm., Engl. Bot. 32: t. 2247 (1811); Brenan in Fl. Europaea 1: 94 (1964); *Oxybasis minutiflora* Kar. & Kir. in Bull. Soc. Imp. Nat. Mosc. [4]: 739 (1841). *Chenopodium crassifolium* Hornem., Hort. Hafn.: 254 (1813); Aellen in Mag. Bot. Lapok 25: 55 (1926). *Blitum rubrum* L. var. *subintegrum* Boiss., Fl. Orient. 4: 905 (1879). *Chenopodium chenopodioides* (*L.*) *Aellen* in Ostenia: 98 (1933); Aellen & Iljin in Fl. SSSR 6: 50 (1936); Aellen in Hegi, Ill. Fl. Mittel-Eur. ed. 2, 3 (2): 608 (1960) & in Fl. Turk. 2: 301 (1967); Uotila in Fl. Iranica 172: 33 (1997); Assadi in Fl. Iran 38: 58 (2001); Zhu, Mosyakin & Clemants in Fl. China 5: 380 (2003).

Annual herb, 10–50 cm, glabrous or sparsely covered with bladder hairs, erect or prostrate, with branches spreading from base. Leaves to 9 × 7 cm, fleshy or not, broadly triangular-rhomboid, apex acute or rounded, base broad and finely long-cuneate and attenuate, entire or sharply sinuate-dentate. Glomerules in distinct, rather distant groups in a much-branched terminal nearly leafless inflorescences consisting of few-flowered cymes. Perianth of terminal flower (male or female) in each glomerule with 4–5 almost free segments; segments of the lateral (female) flowers connate almost to apex, closely enclosing the fruit like a sac. Seeds horizontal (terminal flowers) or vertical (lateral flowers), red, with minute pits. Fig. 69: 1–2.

HAB. In salt marshes on the alluvial plain; alt. 0–200 m; fl.& fr. Jul.–Sep.
DISTRIB. Rare in the Mesopotamian plain: **LCA**: Babylon nr Baghdad, *Noë* 1105 & *Noë* 335 (LE!); Baghdad, Jadriya, *Haines* W.1830!; 30 km S of Falluja, *Haines* W.1817!

Coasts of the Atlantic and Mediterranean; inland salt marshes of Europe, Asia, N, E and S Africa and N America.

2. **Oxybasis urbica** (*L.*) *Fuentes, Uotila* & *Borsch*, Willdenowia 42 (1): 15 (2012); Sukhorukov & Akopyan, Comp. Chenop. Cauc.: 21 (2013).

Chenopodium urbicum L., Sp. Pl. ed. 1: 218 (1753); Boiss., Fl. Orient. 4: 902 (1879); Eig in Pal. Journ. Bot. J. ser. 3: 120 (1945); Zohary in Dep. Agr. Iraq Bull. 31: 46 (1950); Blakelock in Kew Bull. 3: 480 (1957); Mouterde, Nouv. Fl. Lib. et Syr. 1: 409 (1966); Aellen in Fl. Turk. 2: 303 (1967); Uotila in Fl. Iranica 172: 36 (1997); Assadi in Fl. Iran 38: 33 (2001); Zhu, Mosyakin & Clemants in Fl. China 5: 381 (2003).

Fig 69. **Oxybasis chenopodioides**. 1, habit × 1/2; 2, perianth detail. Reproduced with permission from Flora of China 5: fig. 310. Drawn by Cai Shuqin.

Annual herb, 30–100 cm, stiffly erect, lower branches ascending, farinose only when young. Leaves 4–15 × 3–12 cm, lower triangular-ovate or oblong-rhomboid, truncate or cuneate at base, usually dentate, teeth often long and hooked; upper narrower, mostly entire; all leaves fleshy, dark green, somewhat shiny. Inflorescence with erect and stiff, dense and mostly leafless branches appressed to the stem; glomerules small and distant. Perianth segments not or scarcely keeled on back, with broad membranous margins, subglabrous, in ripe fruit horizontally spread. Seeds not completely enclosed by perianth, with shallow grooves or punctate pits.

Hab. Waste places, damp places in gardens, fields; alt. 1000–1500 m; fl.& fr. Jun.–Aug.
Distrib. Rare in foothills of N Iraq: **mro**: nr Rayat, *Guest & Husham* (*Alizzi*) 15885! *Eig & Zohary* s.n. (HUJ!); Walash, *Thesiger* 304 (BM!).

Europe, Turkey, Syria, Iran, Palestine, Central Asia, S Siberia, China and the Far East. Introduced in N America and Australia.

Oxybasis species likely to be found in Iraq, especially in northern part:

Oxybasis rubra (L.) Fuentes & Borsch (=*Chenopodium rubrum* L.). The species is easily distinguished from *O. chenopodioides* by the free perianth segments in all flowers. It is not found yet in Iraq in wet habitats, but scattered records from riversides are present from Turkey, Syria, and Iran. *Oxybasis glauca* (L.) Fuentes & Borsch (=*Chenopodium glaucum* L.) has prostrate or ascending stems and leaves farinose beneath. Scattered records are known in Turkey, Iran, and Arabia.

6. **ATRIPLEX** L.

Sp. Pl. 2: 1052 (1753); Gen. Pl. ed. 5: 472 (1754)

Annual herbs, subshrubs or shrubs, covered with bladder and (almost unvisible) simple hairs of white or brown colour. Leaves alternate or opposite, flat (but in some cases folded on ventral side), with C_3 or C_4 photosynthetic pathway. In the latter case, the bundles have a chlorenchyma sheath that looks like dark-green ornament near the veins. Flowers bisexual (rare) and unisexual (plants monocious or dioecious), arranged in clusters forming loose or dense inflorescences. Male flowers with (3–4)5 hyaline segments connate at base.

Stamens 4–5. Female flowers usually without actinomorphic perianth, enclosed in a bract-like cover consisting of two valves (termed in literature pre–2011 as "bracteoles") developed from accrescent and flattened perianth segments, their dorsal part often carry outgrowths or appendages. In addition to the female flowers enclosed in valves, three species also have female flowers with actinomorphic perianth of (3–4)5 segments (and therefore this flower type lacks the flattened valves). Stylodia 2, rarely 3. Seeds rounded (never keeled), elliptic or ovoid, with perisperm. Embryo vertical (in the seeds originated from flowers covered in valves), and horizontal when female or bisexual flowers with actinomorphic perianth are present. Many annuals feature conspicuous or (rarely) cryptic heterospermy.

Ref.: Aellen P., "Die *Atriplex*-Arten des Orients", Bot. Jahrb. 70 (1): 1–66 (1939).
Sukhorukov A.P., "Zur Systematik und Chorologie der in Russland und benachbarten Staaten (in den Grenzen der ehemaligen UdSSR) vorkommenden *Atriplex*-Arten (*Chenopodiaceae*)", Ann. Naturhist. Mus. Wien 108 B: 307–420 (2006).

About 260 species in temperate and warmer regions (the most species-rich genus within Chenopodiaceae after splitting of *Chenopodium* and *Salsola*). The monophyly of *Atriplex* including *Obione, Senniella, Blackiella, Pachypharinx* etc. (except *Halimione*) is proven after results of Kadereit et al. (2010). The Americas and Australia are extremely rich in species. They can be categorized as thermophile, and partly halophile and nitrophile. In Iraq they do not form their own associations; their occurrence is mostly sporadic (native or escaped after cultivation).

While the male flowers resemble those of most *Chenopodium* species in having a 5-lobed perianth, the female flowers (with the exception of a few species) lack a perianth but their fruit is provided with 2 enlarged valves of a diverse shape.

The species growing in steppe and desert are eagerly eaten by grazing cattle. Australian shrubby species are occasionally sown as green crops.

Atriplex (from Lat. *a*-, 'not', *triplex*, 'in threes', cited by Pliny, may refer to the fact that unlike *Rumex* which has three perianth segments (valves) *Atriplex* has only two segments. This derivation is speculative. Nehmé (2000) offers an alternative derivation, namely that Lat. *Atriplex* is a corruption of the Gr. Ατράφαξις (*atraphaxis*), its Gr. name.

1. Shrubs or subshrubs 2
   Annuals or short-lived perennials 5
2. Stems prostrate or ascending; leaves lanceolate or oblong, up to 1 cm wide; valves mostly mauve or reddish .... 13. *A. semibaccata*
   Stems erect; leaves broader; valves never mauve 3
3. Plant 1–3 m; leaves large, up to 6 × 5 cm, with a cuneate base; valves connate only basally 4
   Plant up to 1m; leaves smaller, at most 2 × 2 cm (except young plants with longer leaves), shortly cuneate to truncate at base; valves connate to ¾ .... 2. *A. leucoclada*
4. Plant dioecious; female inflorescence leafy; male inflorescences leafless, very branched and spreading .... 1. *A. nummularia*
   Plant monoecious; inflorescences bracteose or aphyllous .... 3. *A. halimus*
5. Flowers trimorphic: male flowers enclosed in perianth of 4–5 segments, bisexual flowers (also included in similar perianth) forming horizontal seed and, and female flowers (like other *Atriplex*) in accrescent valves bear vertical seeds; leaves green or purple-brown, ± glabrous, sometimes the uppermost slightly farinose; cultivated plants .... 4. *A. hortensis*
   Female flowers monomorphic, all enclosed in valves; fruits bear vertical seed 6
6. Compass plant (leaves turning with their margins to the sun); leaves broadly triangular-hastate, green or silvery green; valves green, not sclerified, orbicular-cordate or rhombic, sessile or with minute (to 2 mm) stalk, usually entire 7
   Leaves mostly folded on the ventral side, at least below farinose; valves of other color, consistency or shape 8
7. Valves free, orbicular, smooth .... 6. *A. micrantha*
   Valves connate to their middle part or more, rhombic or rounded, mostly with two small outgrowths at dorsal part .... 5. *A. davisii*

8. Small (to 30 cm), much branched plant with unicoloured whitish leaves; valves in the female flowers spongy and soft, connate to the top, almost orbicular . . . . 12. *A. holocarpa*
Leaves mostly bicoloured (green or greyish at ventral side and white or grey below); valves never spongy 9
9. All valves with stalk to 20(–30) mm, fruit-containing part orbicular, 10–20 mm in diameter; fruiting time April to July 10
All or almost all valves sessile, of other shape; fruiting time August to October 11
10. Small plant to 30 cm with short life circle (fruiting time April to May); stem very branched from the base; leaves alternate; valves connate marginally only at base . . . . 9. *A. dimorphostegia*
Plant to 100 cm, branched in upper part, (fruiting time June to July); all or almost all leaves opposite; valves connate superficially, with indurated central (fruit-containing) part. . . . . 11. *A. flabellum*
11. Leaves to 3(–4) cm, triangular-cordate; inflorescence mostly leafy; valves deltoid-campanulate, to 4 × 3 mm . . . . 10. *A. belangeri*
Leaves not cordate; inflorescence leafy in its lower part or aphyllous; valves rhombic or triangular 12
12. Stem with peeling epidermis; branches slender, often filiform at their ends; at least majority of valves to 5 mm . . . . 7. *A. lasiantha*
Stem with epidermis not peeling; branches not slender; valves larger, to 8 mm. . . . . 8. *A. tatarica*

1. **Atriplex nummularia** *Lindl.* in Mitchell, Journ. Trop. Austral.: 64 (1848); Bentham, Fl. Austral. 5: 170 (1870); Aellen in Bot. Jahrb. 70 (1): 377, t. 2 f. G 1–2 (1939); Blakelock in Kew Bull. 3: 475 (1957); Assadi in Fl. Iran 38: 71 (2001); Zhu, Mosyakin & Clemants in Fl. China 5: 361 (2003).

Atriplex asphaltitis Kasapligil in J. Arnold Arb.: 47: 160 (1966); Sukhorukov in Willdenowia 42 (2): 177 (2012).

Woody, much-branched dioecious shrub up to 3 m. Leaves (including petiole) 3–5 × 2–5.5 cm, ovate-deltoid or orbicular, often some broader than long, entire or sinuate-dentate, sessile or petiolate (petiole 5–10 mm), long cuneate at base, obtuse or emarginate at apex. Male inflorescences leafless or with short bracts, very branched, panicle-like; female flowers in leafy or bracteose clusters. Valves 4–7 × 3–8 mm, semiorbicular, rounded-cordate, erose-dentate, in central and lower parts slightly spongy and hardened, smooth or with small outgrowths at back.

HAB. Cultivated and escaping from cultivation in semi-deserts and similar places; alt. to 300 m; fl. Apr.–Jun.
DISTRIB. Upper Mesopotamian plain: LCA: Abu Ghraib, *Regel* s.n.!; *Rechinger* 15847 (W!); Abu Ghraib agricultural school, *Gillett* 10160!; Daltawa, *Guest* 3313 (HUJ!).

Native to W and S Australia, New South Wales and Victoria. This shrubby and strongly woody species differs from *A. halimus* for instance in its dioecious character, leafy female inflorescences and (not rare) valves having appendages on their surface.

2. **Atriplex leucoclada** *Boiss.*, Diagn. ser. 2, 12: 95 (1853); Boiss., Fl. Orient. 4: 915 (1879); Handel-Mazzetti in Ann. Naturh. Mus. Wien 26: 141 (1912); Nábělek in Publ. Fac. Sci. Univ. Masaryk 105: 9 (1929); Aellen in Bot. Jahrb. 70 (1): 21 (1939); Eig in Pal. J. Bot., ser. 3: 123 (1945); Blakelock in Kew Bull. 3: 474 (1957); Rech.f., Fl. Lowland Iraq: 189 (1964); Zohary, Fl. Palaest. 1: 146 (1964); Boulos in Fl. Arab. Penins. & Socotra 1: 246 (1996); Hedge in Fl. Iranica 172: 80 (1997); Assadi in Fl. Iran 38: 95 (2001); Freitag et al. in Fl. Pak. 204: 56 (2001); Ghazanfar, Fl. Oman 1: 39 (2003).

Subshrub, up to 50 cm, often forming tumble-weed. Leaves short (except lower oblong leaves that can be to 4 cm), 1–2 × 1–2 cm, triangular-deltoid, shortly cuneate or truncate or hastate-auriculate at base, irregularly sinuate-dentate, middle and upper sessile. Flowers in many-branched bracteose inflorescences or in solitary or few-flowered clusters in axils of leaves. Valves up to 7 × 4 mm, those of solitary flowers campanulate, slightly lobate-dentate dorsally or denticulate (3–7 teeth), mostly with appendages; valves of clustered flowers deltoid, in upper part richly lobate, dentate, with irregular, partly linear lobes, mostly with muricate or horn-like appendages. Seeds 2.5 × 2 mm, dark brown. Fig. 70: 1– 8.

HAB. Deserts, waste places, roadsides, steppe, abandoned fields, saline depressions; alt. s.l.–1000 m; fl. Aug.–Sep., fr. Sep.–Dec.

DISTRIB. Mainly in the foothills and desert regions of Iraq: **FUJ**: Mosul, *Anders* 2425 (W!); ruins of Hadhr, *Barkley & Haddad* 9168!; **FUJ/DLJ**: Wadi Thirthar, *Guest* 3520!; **FKI**: Abu Ghraib, *Sahira* 39491!; **FPF**: 5 km W of Mandali, *Rechinger* 9611!; by Iranian frontier 54 km SE from Mandali towards Badra, *Rechinger* 12745!; **DLJ**: Balad, *Rawi, Nuri & Tikriti* 28946! **DWD**: Bahr al-Milh, 30 km W by S of Karbala, *Rawi & Rechinger* 16162!, *Rechinger* 8341!; Rutba to Ramadi, *Tikriti & Hazim* 29696! *Rechinger* 2157 (W!);

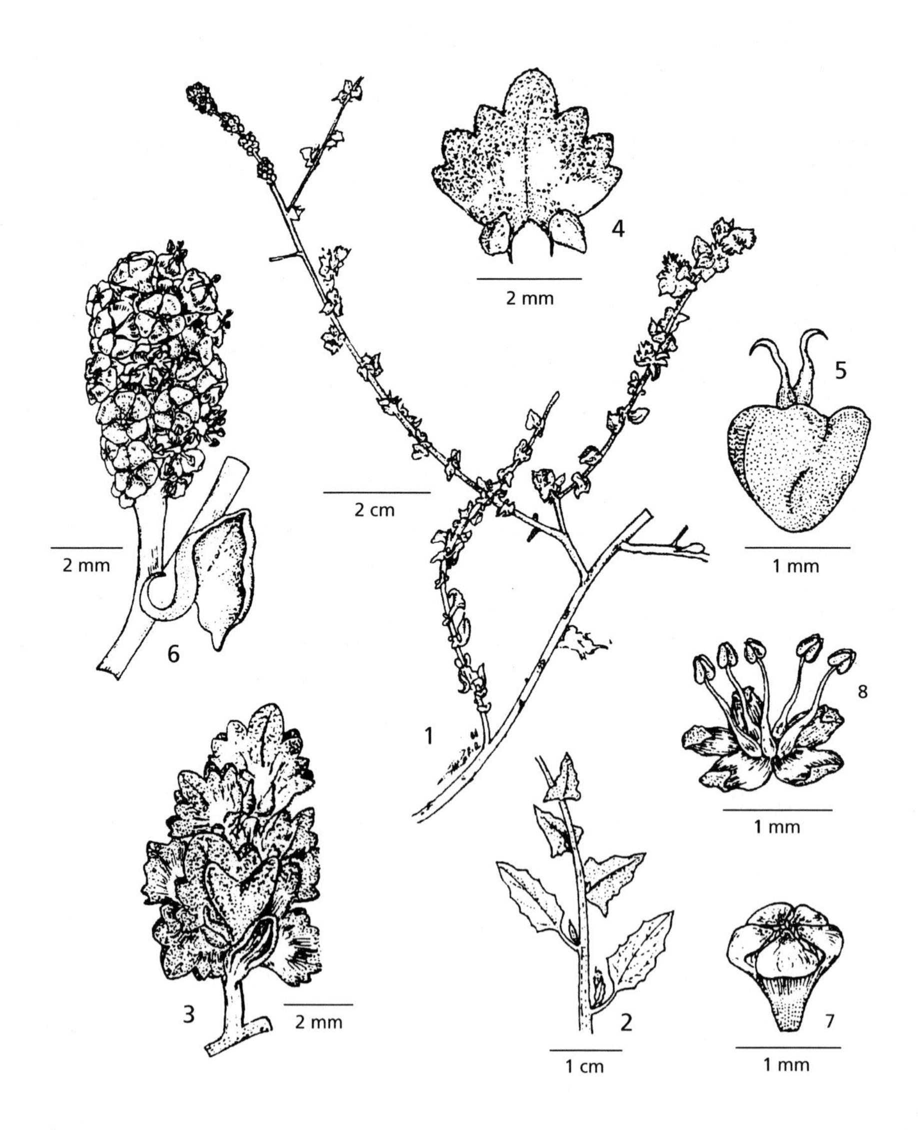

Fig. 70. **Atriplex leucoclada**. 1, habit; 2, lower portion of stem with leaves; 3, part of female inflorescence; 4, bracteoles enclosing female flower; 5, female flower with bracteoles removed; 6, male inflorescence; 7, male flower; 8, male flower opened. Reproduced with permission from Flora of Pakistan 204: f. 8. 2003. Drawn by M. Rafiq.

20 km S of Rutba, *Al Dabbagh & al.* 41782!; Umm al-Junaina, NW of Rutba, *Rawi* 26879!; Ukhaidhir, *Regel* s.n.! **DSD**: 95 km E of As-Salman nr Jalib Bakur, *Rechinger* 9337!; Saudi frontier at As-Salman, 135 km SW of As-Salman, *Rechinger* 9387!; 55 km NW of Shabicha, *Guest, Rawi* & *Long* 14052!; 145 km W of Ramadi, *Rawi* 26838!; nr Al-Aidaha, *Guest, Rawi & Rechinger* 19138!; **LEA**: Aziziya, *Guest* 3570!; Dujaida nr Kut, *Guest* 13105a!; 20 km NE of Baquba, *anon.* 47145!; 25 km SW of Fakka, *Barkley & Abbas-al-Ani* 8928 (E!); Dujaida nr Kut, *Guest* 13105!; Kut al-Amara, *Rechinger* 8960 (W!); **LCA**: Nasiriya, Kharja & Brainat, *Rawi* 26875!; Karrada nr Baghdad, *Regel* s.n.!; nr Abu Ghraib, *Haines* W. 1410!, *Guest* 13585! *Haines* W.10!; Baghdad, *Noë* 559!; by Baghdad airport, *Rechinger, Haines & Khudairi* 26!; road to Zafariniya, *Rechinger* 8341!; **LSM**: 15 km N of Amira, *Rawi & Alizzi* 32414!; nr Amara, *Evans* s.n. (E!). **LBA**: nr Basra, *Rechinger* 15453 (W!); 50 km from Basra to Nasiriya, *Alizzi* 26620!

The differences of *A. leucoclada* from from *A. turcomanica* (Moq.) Boiss. (*A. leucoclada* var. *turcomanica* (Moq.) Zoh.) are still not clear. However, in Caucasus, North Iran and Turkmenistan, this plant is suffruticose and does not form the tumble-weed habit.

One of the most indicative plants in deserts of Eastern Mediterranean and Iran.

Iran, Palestine, Arabian Peninsula, Egypt, Syria, Pakistan.

3. **Atriplex halimus** *L.*, Sp. Pl. 2: 1052 (1753); Boiss., Fl. Orient. 4: 916 (1879); Aellen in Bot. Jahrb. 70 (1); 11 (1939); Zohary, Fl. Palaest. 1: 145 (1964); Aellen in Fl. Turk. 2: 306 (1967); Hedge in Fl. Iranica 172: 84 (1997); Boulos in Fl. Arab. Penins. & Socotra 1: 244 (1996); Freitag et al. in Fl. Pak. 204: 69 (2001).

Monoecious shrub to 3 m; stems erect, branched. Leaves 1–6 × 1–4 cm, alternate, rhombic or ovate, mostly entire or with two small lateral lobes, both sides farinose. Inflorescence leafless or bracteose, dense, with intermixed male and female flowers. Valves at female flowers to 5 mm, semi-orbicular, sessile, slightly sclerified in lower part, entire ot dentate, without appendages on the dorsal part (in our area). Seeds 1–2 mm in diameter, dark brown or red.

Hab. West deserts, wadis, sometimes cultivated in agricultural stations; alt. 0–200 m; fl. Jun.–Jul., fr. Aug.–Oct.
Distrib. **DWD**: between Wadi Sarahan & Rutba, *Eig, Feinbrun & Zohary* s.n. (HUJ!).

Widely distributed in the Mediterranean area with some radiations to the neighbouring regions. Cultivated as a hedge plant in many arid regions and for fodder. The occurrence in Tropical E Africa or Mauritius belongs to other species (*A. brenanii* Sukhor. and *A. aellenii* Sukhor., respectively).

4. **Atriplex hortensis** *L.*, Sp. Pl. ed. 1: 1053 (1753); Boiss., Fl. Orient. 4: 907 (1879); Aellen in Bot. Jahrb. 70 (1); 27, t. 2, f. A (1939); Aellen in Fl. Turk. 2: 308 (1967); Hedge in Fl. Iranica 172: 68 (1997); Assadi in Fl. Iran 38: 76 (2001); Freitag et al. in Fl. Pak. 204: 64 (2001); Sukhorukov in Ann. Naturhist. Mus. Wien 108 B: 337 (2006); Sukhorukov & Akopyan, Comp. Chenop. Cauc.: 25 (2013).

Annual, up to 2.5 m, robust, glabrous or only the young shoots can be mealy. Leaves large, more than 10 cm, ovate or triangular, slightly cordate, weakly toothed or entire, upper leaves ovate to oblong-lanceolate. Inflorescence panicle-like, with intermixed flowers. Male and bisexual flowers with a 5-lobed perianth and 5 stamens; the latter ones bear fruits with horizontal, mostly black seeds of minor diameter (1.5–1.7 mm). Valves of female flowers entire, rounded, to 20 mm in diameter, reddish turning to brownish colour, membranous at fruiting stage, marginally free but united in basal-median part with the consequence that the fruit base is located near their central part. Seeds of the fruits in valves black, ± 2 mm in diameter, or yellow brown, 2–3.5 mm in diameter; testa thick or thin. Fig. 71: 1–3.

Hab. Cultivated in irrigated gardens and fields; alt. 0–1000 m; fl. Jul.–Aug., fr. Aug.–Oct. Not known as a weed.
Distrib. Widely cultivated and known in Kurdistan (**FAR**: Arbil) as ornamental plant near hotels.

The origin of the species is still unknown, but it grows in some natural habitats in Turkey. The leaves have been used for many centuries as spinach, chiefly in farms and local gardens. There are several colour forms: green, pink and purple.

5. **Atriplex davisii** *Aellen* in Not. Roy. Bot. Gard. Edinb. 28 (1): 30 (1967); Sukhorukov in Fedd. Repert.: 118 (3–4): 73 (2007); Sukhorukov & Danin in Fl. Medit. 19: 19 (2009).

Annual to 200 cm. Stem branched from the base, lower branches horizontally spreading or ascending. Leaves green or greyish below, often shiny, triangular-hastate, 3–6 × 2–4 cm, entire or weakly toothed. Both male and female flowers intermixed, in glomerules forming loose, usually aphyllous inflorescence. Valves rhombic or rounded, small (to 5 mm in diameter), only a little larger than the fruit, sessile or with short (up to 2 mm) peduncle, smooth or mostly with two dot-like appendages in upper part. Seeds heteromorphic: black seeds 1–1.5 mm, brownish seeds 1.6–2.2 mm in diameter.

HAB. Wet places (streams, moist ruderal habitats).
DISTRIB. **FUJ**: Mosul, *Safar* s.n. (LE!).

In habit very similar to *A. prostrata* Boucher ex DC., but the latter species has triangular valves united in their basal part. The occurrence in Iraq is possible in the part of the country.

Turkey, Eastern Mediterranean to Sinai, Cyprus, Greece, Iran, Afghanistan.

6. **Atriplex micrantha** *C.A. Mey.* in Ledeb., Ic. Pl. Fl. Ross. 1:11 (1829); Hedge in Fl. Iranica 172: 66 (1997); Zhu, Mosyakin & Clemants in Fl. China 5: 362 (2003); Sukhorukov in Ann. Naturhist. Mus. Wien 108 B: 360 (2006); Sukhorukov & Akopyan, Comp. Chenop. Cauc.: 28 (2013).

A. heterosperma Bunge, Reliq. Lehm.: 272 (1851); Iljin in Weeds SSSR 2: 119 (1934) & in Fl. SSSR 6: 93 (1936); Aellen in Bot. Jahrb. 70 (1): 30 (1939); Aellen in Hegi, Ill. Fl. Mitteleur. 3 (2): 69 (1960); Mouterde, Nouv. Fl. Lib. et Syr. 1: 414 (1966); Assadi in Fl. Iran 38: 76 (2001).

Annual herb, erect, up to 1.5 m, much-branched. Leaves 3–10 × 2–7 cm, triangular-hastate, upper leaves hastate or oblong-ovate, entire or sinuate-dentate, often fleshy, glabrous or silvery and shiny. Inflorescence loose, panicle-like. Flowers in lax clusters. Valves at fruiting stage 1.5–6(–8–12) × 1.5–6 mm, rounded to ovate, ± cordate at base, entire, green, later scarious, main nerves separated to near the base, without appendages on the back, sessile or with very short stalk. Seeds of two sorts: some ± 2.5 mm in diameter, flat, brownish, with a soft testa (from the flowers with larger valves); others ± 1 mm in diameter, black, with a hard testa (mostly originated from the fruits with small valves). Fig. 71: 4–5.

HAB. Wasteland, roadsides, riversides, wadis; alt. s.l.–1000 m; fl. Jul.–Aug., fr. Aug.–Oct.
DISTRIB. Mesopotamian plain: **FNI**: Nineveh plantation to Mosul, *Chapman* 26076!, 25381!; **DWD**: Ramadi, *Regel* s.n.! **DSD**: Julaida, *Regel* s.n.!; Khadar-al-Mai, *Rawi, Khatib & Tikrity* 29118! **LCA**: Mahmudiya nr Baghdad, *Haines* W2133!; *Haines* 159!; Abu Ghraib, *Gillett* 8419!, *Haines* 9!; *Fawzi & Ali* 39506! *Rawi* 16330! *Alizzi* 34142!; *Barkley & Abbas* 6160!; nr Baghdad, *Omer* & *Sahira* 34841! **LSM**: Qal'a Salih, *Guest, Rawi & Rechinger* 17406! 88_{54} (W!); nr Amara, *Evans* s.n. (E!).

All specimens seen at fruiting stage have extremely small (to 5 mm) valves. Such plants are found from South Caucasus through E Turkey to Iraq and Syria and can represent a new taxon.

E and SE Europe (spreading into CW Europe, as alien), S Siberia, W China, SW Asia (Caucasus, Syria, Palestine, Iran), C Asia (Kazakhstan, Uzbekistan, Kyrghyzstan, Turkmenistan, Afghanistan). Common in the Asian steppes and deserts.

7. **Atriplex lasiantha** *Boiss.*, Diagn. ser. 2, 12: 95 (1853); Aellen in Bot. Jahrb. 70 (1): 44 (1939); Blakelock in Kew Bull. 3: 475 (1957); Zohary, Fl. Palaest. 1: 149 (1966); Mouterde, Nouv. Fl. Lib. et Syr. 1: 416 (1966); Aellen in Fl. Turk. 2: 308 (1967); Hedge in Fl. Iranica 172: 76 (1997); Freitag et al. in Fl. Pak. 204: 65 (2001); Sukhorukov & Akopyan, Consp. Chenop. Cauc.: 30 (2013).

A. tataricum L. var. *virgatum* Boiss., Boiss., Fl. Orient. 4: 911 (1879) p.p.; Fl. Syr. ed. 2, 2: 434 (1933) p.p.
A. autrani Post, Fl. Syr.: 682 (1896); Dinsmore in Fl. Syr. ed. 2, 2: 434 (1933).
A. tatarica var. *tenera* Eig ("*tenerum*") syn. nov.
A. mesopotamica Eig (nomen in herb. HUJ!).

Annual herb, 0.1–1 m, many- and long-branched, richly lepidote; stem and branches with peeling epidermis, slender in their upper parts (to 0.5 mm in diameter). Leaves up to 7 × 2.5 cm, elongate-deltoid, ovate or rhombic, partly toothed; petiole up to 1 cm, upper leaves

Fig. 71. **Atriplex hortensis**. 1, habit fruiting branch × 1/2; 2, female flower × 2; 3, utricle with fruiting bracts × 1. **Atriplex micrantha**. 4, habit fruiting branch × 1/2; 5, utricle with fruiting bracts × 1. Reproduced with permission from Flora of China 5: fig. 299. Drawn by Cai Shuqin.

subsessile. Inflorescence long and laxly branched, with some female flowers in leaf-axils forming clusters; valves in clusters of 3–7 mm diameter, somewhat straggly, lower partly with bracts; valves in the main inflorescence rhomboid-deltoid to elongate-deltoid, to 5(–7) mm, often 3-lobed, entire or toothed, with or without nodular appendages on the back. Fig. 72: 1–4.

HAB. Dry steppe, waste places, cultivated land; alt. s.l.–1800 m; fl. May–Jul., fr. Aug.–Oct.
DISTRIB. This species seems to be locally common in the mountains or deserts: **MAM**: Gara, *Kotschy* s.n. (BM!); Kara Dagh, *Kotschy* 353!; *Rawi* 9269!; *Haines* 15841 (W!); Sarsang, *Haines* W460!; Sulaf, *Eig & Feinbrun* s.n. (HUJ!); Sharanish, *Rechinger* 10859 (W!); Jabal Khantur N of Zakho, *Rechinger* 10859! **MRO**: mt. Serin on road to Qandil, *Rawi & Serhang* 23990!; Rayat, *Eig & Zohary s.n.* (HUJ!); **MSU**: Qopi Qaradagh, *Haines* W1085!; Pir-i- Mukrun, *Eig & Duvdevani* s.n. (HUJ!); Sulaimaniya, *Zohary & Amdursky* s.n. (HUJ!); Avroman, *Haussknecht* s.n. (BM!). **FUJ**: Mosul, *Haussknecht* s.n. (BM!). **LCA**: Baghdad, *Haines* W.1825!; 13 km E Ali Gharbi, *Eig & Zohary* s.n. (HUJ!); **LSM**: nr Azair, *Guest, Rawi & Rechinger* 17663!

SW Asia (Caucasus, Turkey, Syria, Palestine, Jordan, Iran), C Asia (Afghanistan, Turkmenistan, Uzbekistan).

8. **Atriplex tatarica** *L.*, Sp. Pl. ed. 1: 1053 (1753); Boiss., Fl. Orient. 4: 910 (1879) excl. var. *virgata* Boiss.; Aellen in Bot. Jahrb. 70 (1): 42 (1939); Eig in Palest. J. Bot., Jerusalem ser., 3: 121 (1945); Zohary in Dep. Agr. Iraq Bull. 31: 46 (1950); Tackholm, Stud. Fl. Egypt: 425 (1956); Blakelock in Kew Bull. 3: 474 (1957); Rech.f., Fl. Lowland Iraq: 188 (1964); Mouterde, Nouv. Fl. Lib. et Syr. 1: 413 (1966); Zohary, Fl. Palaest. 1: 148 (1966); Aellen in Fl. Turk. 2: 309 (1967); Hedge in Fl. Iranica 172: 75 (1997); Zhu, Mosyakin & Clemants in Fl. China 5: 365 (2003); Sukhorukov in Ann. Naturhist. Mus. Wien 108 B: 376 (2006); Sukhorukov & Akopyan, Comp. Chenop. Cauc.: 29 (2013).

A. tatarica var. *desertorum* Eig, Palest. J. Bot., Jerusalem ser., 3: 122 (1945).
? *Atriplex olivieri* Moq., Chenop. Monogr.: 52 (1840) & in DC., Prodr. 13 (2): 91 (1849); Aellen in Bot. Jahrb. 70 (1): 37 (1939); Blakelock in Kew Bull. 3: 476 (1957); Aellen in Fl. Turk. 2: 311 (1967).

This name is still unresolved, and there are no strict morphological characters that can separate it from *A. tatarica* or *A. lasiantha*. The first author (A.S.) rather thinks that *A. olivieri* is more closely related to *A. lasiantha* and probably a variety growing under favorable conditions and forming bigger individuals.

Annual herb, erect or prostrate, usually strongly crusted. Leaves 3–8 × 2–7 cm, variously shaped, triangular-deltoid, hastate to lanceolate, usually irregularly sinuate-lobed or toothed. Flowers in axillary clusters or terminal, often on an elongate spike-like or paniculate inflorescence. Bracteoles to 8 × 7 cm, orbicular to deltoid-rhombic, often 3-lobed, reticulately veined, with or without dorsal appendages. Fig. 73: 1–8.

HAB. Waste places, roadside, sandy soils, salt flats, weed in gardens and fields; alt. s.l.–1000 m; fl. May–Aug., fr. Jul.–Oct.
DISTRIB. Alluvial plains, foothills and desert regions of Iraq: **DWD**: Nukhaib, *Rawi* s.n.!; **DSD**: 65 km W by N of Baniya, *Guest* 14148!; Ur, *Regel* s.n.!; **LEA**: between Amara and [illegible], *Rechinger* 8960!; 20 km E of Amara, *Rawi* 15041!; 20 km N of Amara, *Rawi* 15010!; **LCA**: Hilla, *Civilian Veterinary Surgeon of Hilla* s.n.!;

SW and C Asia, Mediterranean area, alien in N America. Very polymorphic; some varieties require further investigations.

9. **Atriplex dimorphostegia** *Kar. & Kir.* in Bull. Soc. Nat. Mosc. 15: 438 (1842); Boiss., Fl. Orient. 4: 909 (1879); Aellen in Engl. Bot. Jahrb. 70 (1): 34 (1939); Eig in Palest. J. Bot., Jerusalem ser. 3: 121 (1945); Zohary in Dep. Agr. Iraq Bull. 31: 46 (1950); Tackholm, Stud. Fl. Egypt: 425 (1956); Blakelock in Kew Bull. 3: 475 (1957); Rech.f., Fl. Lowland Iraq: 188 (1964); Zohary, Fl. Palaest. 1: 147 (1966); Boulos in Fl. Arab. Penins. Socotra 1: 247 (1996); Hedge in Fl. Iranica 172: 70 (1997); Freitag et al. in Fl. Pak. 204: 69 (2001); Assadi in Fl. Iran 38: 84 (2001); Zhu, Mosyakin & Clemants in Fl. China 5: 364 (2003); Sukhorukov in Ann. Naturhist. Mus. Wien 108 B: 388 (2006).

A. bracteosa Trautv., Act. Hort. Petrop. 1 (1): 17 (1871); *A. transcaspica* Bornm. & Sint. in sched.; Ulbrich in Engl. & Prantl., Nat. Pflanzenfam. ed. 2, 16c: 514 (1934), nomen.

Annual herb, up to 50 cm; stem erect, branches slender, ascending, silvery. Leaves short petiolate, up to 4 × 3.5 cm, lamina broadly ovate to deltoid, truncate or cordate at base, entire or crisp, white-crusted beneath, nearly glabrous above, upper leaves only a little reduced, usually appressed to the stem. Flowers 5–12 together in clusters in short axillary inflorescences,

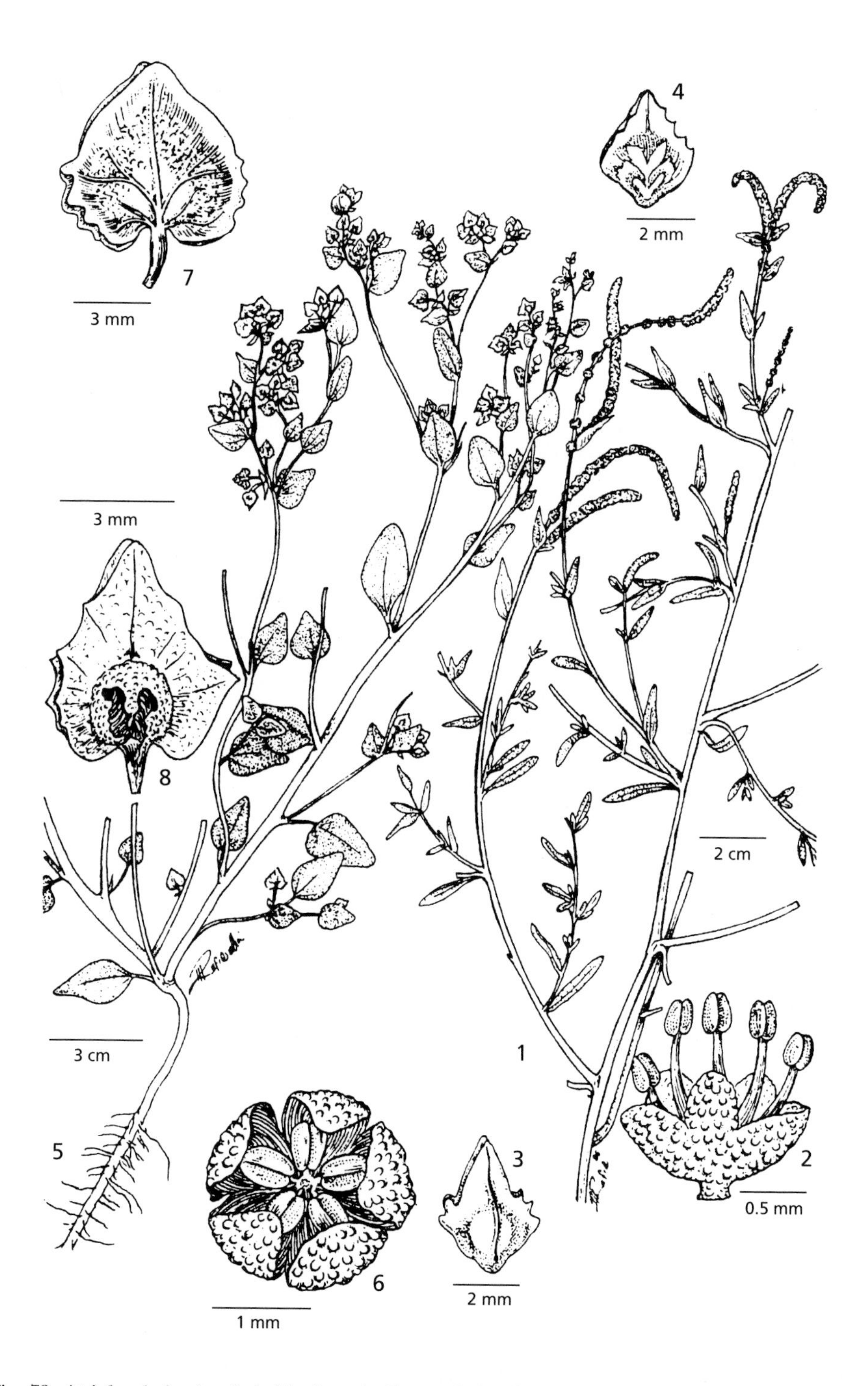

Fig. 72. **Atriplex lasiantha**. 1, habit; 2, male flower; 3–4, valves variation. **Atriplex dimorphostegia**. 5, habit; 6, male flower; 7–8, valves. Reproduced with permission from Flora of Pakistan 204: f. 11. 2003. Drawn by M. Rafiq.

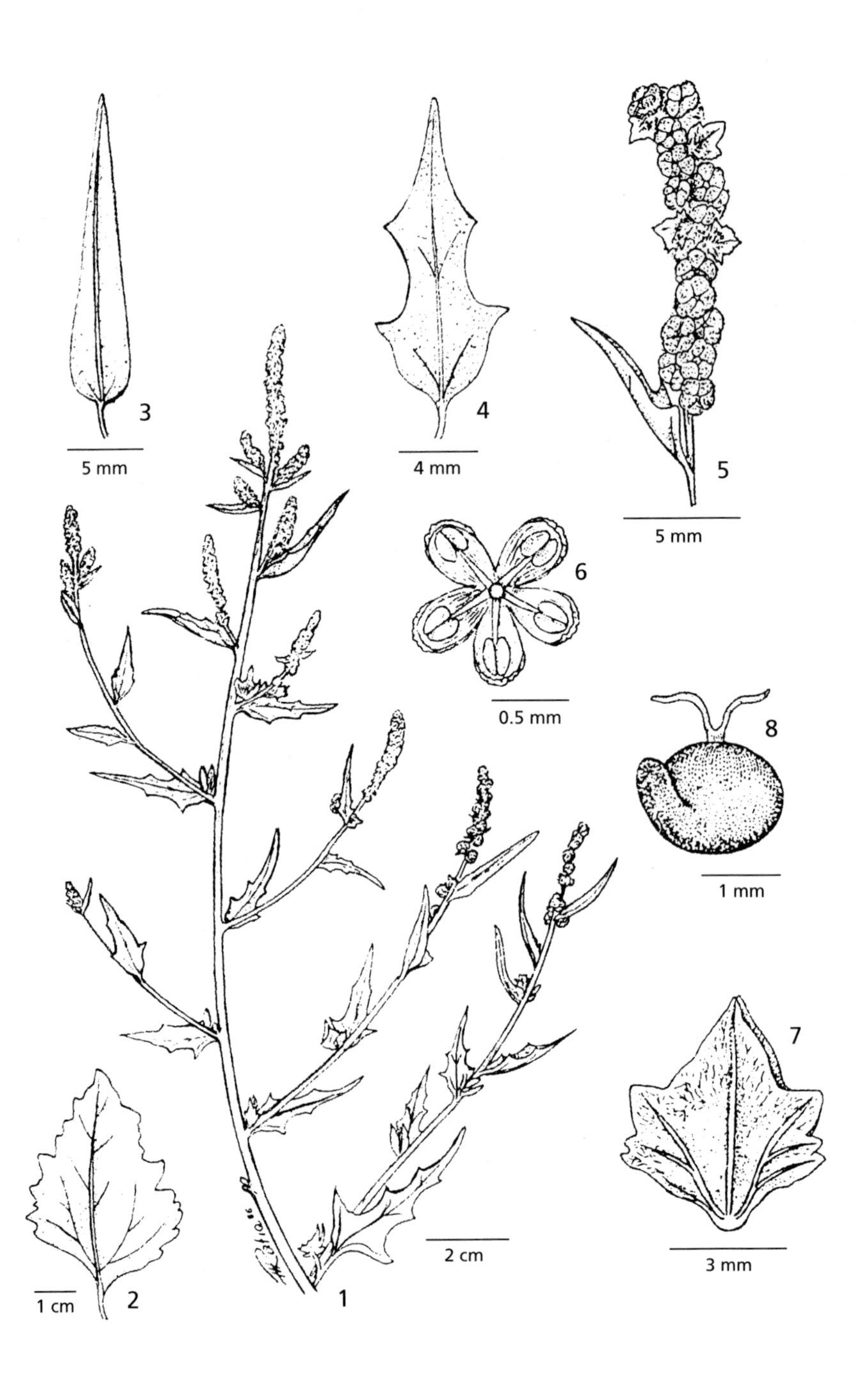

Fig. 73. **Atriplex tatarica**. 1, habit; 2–4, leaf variation; 5, male inflorescence with a few female flowers; 6, male flower; 7, valves; 8, female flower. Reproduced with permission from Flora of Pakistan 204: f. 10. 2003. Drawn by M. Rafiq.

lower clusters with female flowers. Valves stalked (stalk 3–30 mm), 10–13 mm broad, rounded-cordate, sinuate-denticulate, mostly with crispate appendages on back, distinctly reticulate-nerved. Seeds black, reddish-black, ± 1 mm, and brown, 1.5–2.5 mm. Fig. 72: 5–8.

HAB. Loamy or sandy deserts, saline soils; alt. s.l.–350 m; fl. & fr. Mar.–May.
DISTRIB. In desert and lowland alluvial regions: **DWD**: E of Shithatha, *Rawi* 199971!; **DSD**: 55 km NW of Shabicha, *Guest* 14052! **LCA**: Imam Ibrahim between Khan Mahawil & Suwaira, *Haines* W.1770!; Nahr Hamdaniya nr Kuwairish, *Handel-Mazzetti* 879!; *Nábělek* 1107!; Babylon, *Handel-Mazetti* 879 (WU!).

E Mediterranean region (Palestine, Syria), N Africa (Morocco to Egypt), Arabia, Iran, Turkmenistan, Afghanistan, Pakistan, Uzbekistan, Tajikistan, Kazakhstan, China (Xinjiang).

10. **Atriplex belangeri** (*Moq.*) *Boiss.* 4: 913 (1879); Assadi in Fl. Iran 38: 87 (2001); Sukhorukov in Ann. Naturhist. Mus. Wien 108 B: 394 (2006).

Obione thunbergiifolia Boiss. & Noë in Boiss., Diagn. ser. 2, 4: 74 (1859); Lectotype: Iraq: Kutt am Tigris, May 1851, *Noë* 993 (G!, isolecto K! LE! selected by Sukhorukov in Chenop. Carp. & Syst. 2014, p.349):
Atriplex thunbergiifolia (Boiss. & Noë) Boiss., Fl. Orient. 4: 911 (1879); Aellen in Bot. Jahrb. 70 (1): 36 (1939); Zohary in Dep. Agr. Iraq Bull. 3: 476 (1957); Blakelock in Kew Bull. 3: 476 (1957); Rech.f., Fl. Lowland Iraq: 188 (1964); Mouterde, Nouv. Fl. Lib. et Syr. 1: 414 (1966).

Annual herb 20–150 cm, erect, branched or not. Leaves 2–4 × 1–3 cm, shortly petiolate, rounded- to triangular-hastate, truncate or slightly cordate, entire or dentate. Clusters many-flowered, in axils of leaves and bracts over the whole plant, the upper in long mostly simply spicate-like or shortly branched inflorescences. Valves 4–6 × 2–3 mm, deltoid-campanulate, broadest towards the apex, slightly 3-lobed, with a basal tube without appendages, with 3–5 main nerves.

HAB. Riversides, swampy places, rarely as a weed in irrigated fields; alt. to 300 m; fl. Jun–Aug., fr. Aug.–Oct.
DISTRIB. Probably rare: **FPF**:10 km SW from Mandali, *Rechinger* 13399a (W!). **LEA**: Kut on the Tigris, *Noë* 943! (type of *Obione thunbergiifolia*); Tigris, *Noë* 506 (LE!); **LCA**: Khalis, *Barkley* 2489! **LSM**: Amara, *Evans* s.n. (E!).

Iran, S Kazakhstan, Turkmenistan, Uzbekistan, Afghanistan, Tajikistan.

11. **Atriplex flabellum** *Bunge* in Boiss., Fl. Orient. 4: 912 (1879); Hedge in Fl. Iranica 172: 73 (1997); Sukhorukov in Ann. Naturhist. Mus. Wien 108 B: 408 (2006).

Obione flabellum (Bunge) Ulbrich in Engler & Prantl, Nat. Pflanzenfam., ed. 2, 16c: 506 (1934); Blakelock in Kew Bull. 3: 475 (1957); Aellen in Verh. Naturf. Gesell. Basel 49: 136, f. 5–6 (1938).

Annual herb, to 100 cm tall, with many whitish branches. Leaves 4–7 × 4–7 cm, triangular-hastate to rounded-cordate, mostly sharply sinuate-dentate, crusted beneath, nearly all opposite or upper leaves alternate. Male flowers in glomerules, forming terminal panicles; female flowers in axils of leaves or bracts, 1–3 together or single in the male glomerules. Valves large, orbicular, 1 × 1 cm, with stalk to 15 mm, rounded, sinuate-cordate at base, with 6–12 radiate ridges, connate superficially near the margin; appendages absent or nodule-like. Seeds red, 1.5–2 mm, and brown, 2–4 mm.

HAB. Steppe of plains and mountains, as a weed in cultivated land, on waste and ruderal places; alt. s.l.–1400 m; fl. & fr. May–Jul.
DISTRIB. Reported from **LCA**: nr Baghdad, *Lazar* 156 (F, n.v.) after Aellen (1938) and Blakelock (1957).

Iran, Kazakhstan, Uzbekistan, Turkmenistan, Tajikistan, Afghanistan, China (Xinjiang).

12. **Atriplex holocarpa** *F. Muell.*, Repert. Pl. Babbage's Exped. 19 (1859).

?*Atriplex spongiosa* F. Muell., Trans. Proc. Phil. Inst. Victoria 2: 74 (1858); Hedge in Fl. Iranica 172: 86 (1997); *Senniella spongiosa* (F. Muell.) Aellen, Bot. Jahrb. 68: 417 (1938) var. *holocarpa* (F. Muell.) Aellen, l.c.. 418 (1938).

Annual or short-lived perennial, up to 30 cm; stem branched from the base forming dense habit. Leaves long-petiolate, blades to 5 × 3 cm, triangular to rhomboid, entire to serrate. Inflorescence leafy. Clusters of both male and female flowers. Valves suborbicular or ovoid, 0.8–1.2 cm long, spongy and soft, completely fused, shortly apiculate. Seeds brown, 1.5–2 mm.

HAB. In disturbed places; fl. Mar.–Nov.
DISTRIB. **LCA**: Baghdad, *Plitman* s.n. (HUJ!).

Introduced from Australia as an ornamental plant, escaped, and became naturalized in many parts of Eastern Mediterranean, North Africa, northwards to Iran and Uzbekistan. Other records are highly predictable.

13. **Atriplex semibaccata** R.Br., Prodr. Fl. Nov. Holl. 1: 406 (1810).

Subshrub, very branched at base, with prostrate or ascending shoots up to 1 m. Leaves 2–5 × 0.6–1 cm, oblong to lanceolate, entire or sinuate-dentate to lobate. Inflorescence leafy. Valves rhombic, to 5 mm, connate in lower half, reddish or mauve coloured, sclerified, entire or toothed, mostly without tubercles. Seeds 1–2 mm in diameter, brown.

HAB. Saline soil, dry rocky slopes, screes, waste places; alt.: to 250 m; fl. & fr. Jul.–Sep.
DISTRIB. Widespread alien plant in the deserts that has become common in many adjacent areas. **LCA**: Abu Ghraib, *Tikriti* 33181!

Native in Australia, cultivated and escaped in Mediterranean area, S Africa, subtropical parts of America, Polynesia.

EXCLUDED SPECIES

Atriplex nitens Schkuhr (now *A. sagittata* Borkh.) cited in Rechinger, Flora of Lowland Iraq, belongs to other species (mostly to *A. micrantha*). *A. nitens* is a temperate species widely distributed in the steppes and forest-steppes of Eurasia. Its southern limit is in Caucasus. In the vegetative stage, *A. sagittata* is easily recognized due to its folded (not "compass") leaves (Sukhorukov, 2006).

A. hastata L. (now *A. prostrata* Boucher ex DC.). The occurrence of *A. prostrata* is possible in Iraq, but the specimens seen belong to other, morphologically closely related species *A. micrantha* and *A. davisii.* The most southern findings of *A. prostrata* are collected Eastern Mediterranean (Syria and Lebanon), or is indicated near Urmia Lake (W Iran), but this species is very rare in these areas. Some specimens from adjacent territories also belongs to *A. patula,* species widely distributed in the temperate Eurasia and the Mediterranean that can be found scattered in moist places, and *A. laevis* distributed mainly in Central Asia with very dense inflorescences (naturalized in Turkey).

A. aucheri Moq., specimens seen belong to *A. micrantha.*

TRIBE ANSERINEAE Dumort.

7. SPINACIA L.

Sp. Pl. 1027 (1753); Gen. Pl. ed. 5: 452 (1754)

Ref.: P. Aellen, "Beitrag zur Kenntnis von *Spinacia* L.", Ber. Schweiz. Bot. Ges. 48: 485–490 (1938).

Dioecious, rarely monoecious annual or biennial plants; stem and leaves glabrous or with scattered, relatively long-stalked bladder hairs; leaves alternate, hastate to pinnatifid or runcinately pinnatisect. Flowers in sessile clusters, forming a spiciform or (the males) paniculate inflorescence. Female flower clusters axillary, the flowers without a "normal" (actinomorphic) perianth but with 2 sclerified valves fused completely around ovary and considerably accrescent in fruit, each frequently with a dorsal spine; style very short or obsolete, stigmas 4–5, long and filiform. Male flower clusters disposed along axils and slender branches of inflorescence, with 4(–5) perianth segments and 4(–5) free, slender filaments; anthers roundish, rather tumid, thecae finally widely gaping. Fruits indehiscent; seeds vertical, with brownish seed coat; endosperm copious; embryo annular.

Three species, two wild in SW and C Asia, the third (*S. oleracea*) cultivated, probably also temperate Asiatic in origin.

Spinacia (derived from the Medieval Lat. name for spinach)

Glabrous; leaves hastate, entire to pinnatifid; female flowers mostly in clusters of 6 or more; all fruits of each cluster free from each other, spinous or not . . . . 1. *S. oleracea*
Scattered long-stalked bladder hairs usually present; leaves pinnatisect; female flowers mostly in clusters of fewer than 6; all fruits of each cluster fused into a single hard spiny cover . . . . 2. *S. tetrandra*

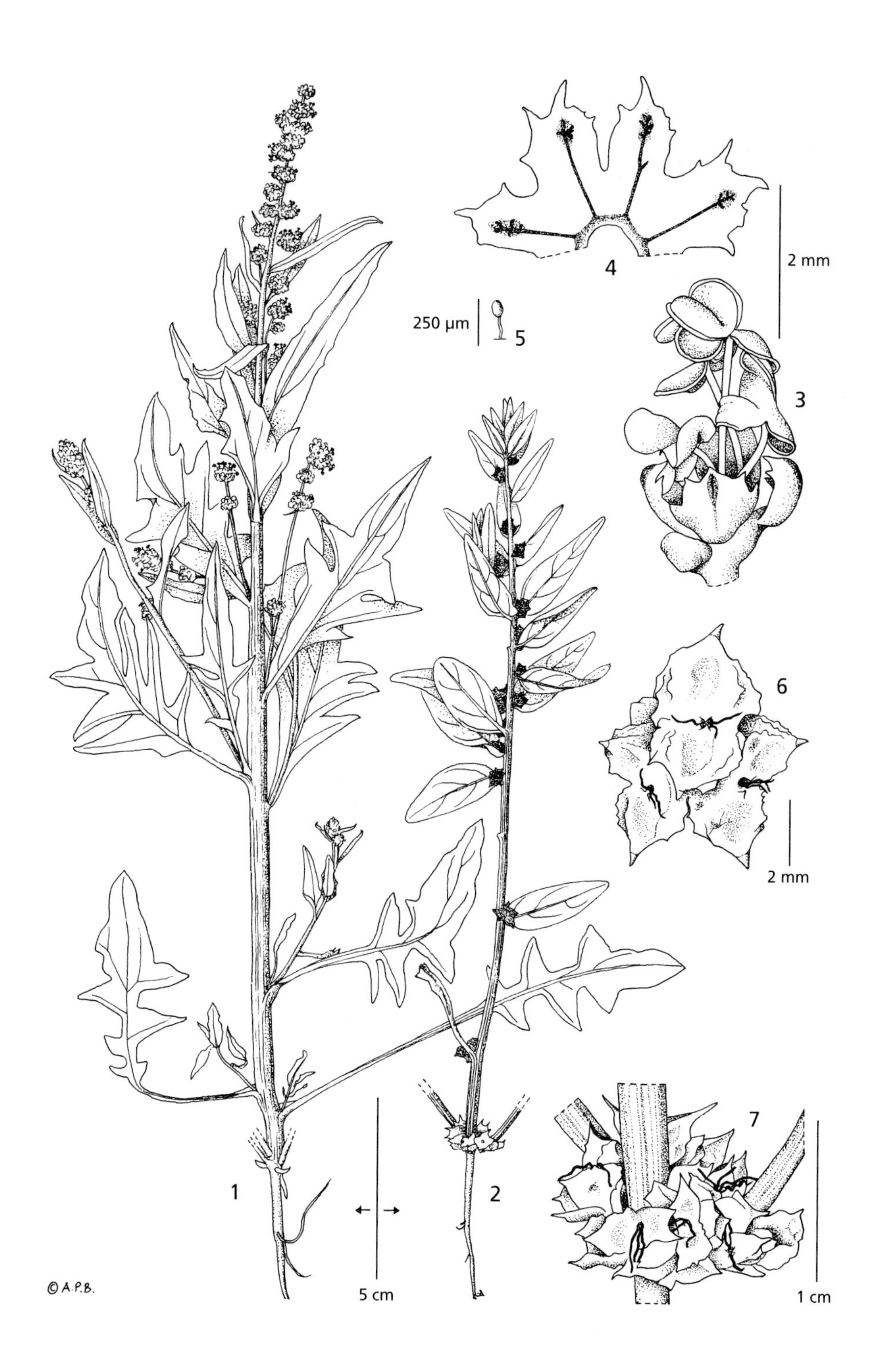

Fig. 74. **Spinacia tetrandra**. 1, habit male plant; 2, habit female plant; 3, male flower lateral view with all anthers dehisced; 4, perianth of male flower opened; 5, bladder hair enlarged; 6, group of axillary fruits 7, fruit cluster at base of stem. 1 from *Ikodze* 552; 2, 6–7 from *Rawi* 30506; 3–5 from *Jacobs* 6641. © Drawn by A.P. Brown.

1. **Spinacia oleracea** *L.*, Sp. Pl. ed. 1: 1027 (1753); Uotila in Fl. Iranica 172: 62 (1997); Assadi in Fl. Iran 38: 68 (2001).

Annual or biennial herb, 25–50 cm, glabrous, light green, simple or sparingly branched at base or above. Basal and lower stem leaves long-petiolate, deltoid-hastate to runcinate-pinnatifid with terminal lobe up to 10 × 8 cm, deltoid-hastate and very much larger than the 1(–2) pairs of laterals; upper leaves considerably reducing, those of upper part of inflorescence in female plants petiolate, acute, lanceolate or narrowly hastate, those of male inflorescences bract-like or absent. Female flowers in clusters of mostly 6–10, valves with or without a dorsal process, each ± 3 mm in diameter (excluding spines if present), stigmas 1–3(–4.5) mm. Male flowers in panicle-like inflorescences, lower clusters becoming more remote; perianth segments 4(–5), oblong-ovate to oblong-obovate, 1.5–1.7 mm, obtuse, entire or almost so; filaments 4(–5), 2.5–3 mm; anthers 1 mm.

Two varieties are separated based on the presence or absence of the spines of the valves: spine up to 3 mm (var. **oleracea**), or valves smooth and rounded (var. **inermis** (Moench) Metzg.).
Known only in cultivation.

HAB. Cultivated in fields.
DISTRIB. Rarely collected. LCA: Baghdad, Za'faraniya, 15.1.1960, cultivated, *Haines* s.n.!

Origin unknown. Spinach is widely cultivated in the world as an important leaf vegetable.

2. **Spinacia tetrandra** *Stev.* in Mem. Soc. Imp. Nat. Mosc. 2: 182 (1809); Rech.f., Fl. Lowland Iraq: 186 (1964); Aellen in Fl. Turk. 2: 305 (1967); Uotila in Fl. Iranica 172: 61 (1997); Assadi in Fl. Iran 38: 69 (2001); Sukhorukov & Akopyan, Comp. Chenop. Cauc.: 32 (2013).

Annual herb, 10–40 cm, with scattered long-stalked bladder hairs or glabrescent, simple or branched from base with several simple stems, remains of fruit "burr" frequently conspicuous at top of rootstock in female plants. Basal and lower cauline leaves sinuose to pinnatifid with broad lobes, or pinnatisect with narrow segments, terminal lobe not so considerably larger than the 2–4 pairs of laterals as in *S. oleracea*; upper leaves considerably reducing, those of upper part of female inflorescence sessile-amplexicaul, oblong, entire or rarely slightly lobed, those of male inflorescence sessile-amplexicaul, oblong, entire or rarely slightly lobed, those of male inflorescence present only in basal clusters, bract-like. Female flowers in clusters of 3–5(–6), valves with a dorsal process; stigmas (2–)3–6 mm. Male flowers in panicles, lower clusters becoming more remote; perianth segments oblong, 1.5–1.7 mm, commonly with a few usually large acute teeth; filaments 4, 2.5–3 mm; anthers ± 1 mm. All fruits of each cluster fused into a single spiny "burr' 9–12 mm in diameter including the broad-based 2.5–3 mm spines; surface of fruit smooth or slightly rugose. Fig. 74: 1–7.

HAB. Semideserts, wadis, riversides, disturbed places; alt. s.l.–1300 m; fl. & fr. Apr.–Jun.
DISTRIB. MAM: Mantufa, *Field & Lazar* 109!; FUJ: Mosul, *Maresch* 21 (WU!); DWD: 15km S of Rutba, *Rawi & Khatib* 32161!; Wadi Tibul, *Rawi* 30499!; Anah, *Barkley & Abbas* 753!; Wadi Al-Ajrumiya, *Rawi* 30506!

Caucasus, E Turkey, Iran, Syria.

8. **BLITUM** L.

Sp. Pl.: 4 (1753)

Annual or perennial herbs, glabrous or covered with scattered bladder or glandular hairs. Leaves long-petiolate, mostly broad (triangular, hastate or trullate), basal ones often with short internodes forming leaf rosettes. Inflorescences as a rule leafy. Flowers in dense glomerules, bisexual or sometimes unisexual. Perianth of 1–5 free or insignificantly concrescent segments or reduced, green, but in the fruiting stage sometimes fleshy and red-coloured, or indurated. Stamens 1–5, usually equal to the perianth segments. Stigmas 2–3. Fruits with 1–2(–3)-layered, smooth, mamillate or rarely papillate pericarp, free or tightly adjoining to the seed coat. Seed round or ovoid, red or reddish black, without keel or with 2 dull keels. Seed smooth or alveolate, the cells without stalactites, rarely cells containing hair-like outgrowths. Seed embryo horizontal or vertical, sometimes both embryo positions present on an individual (spatial heterospermy).

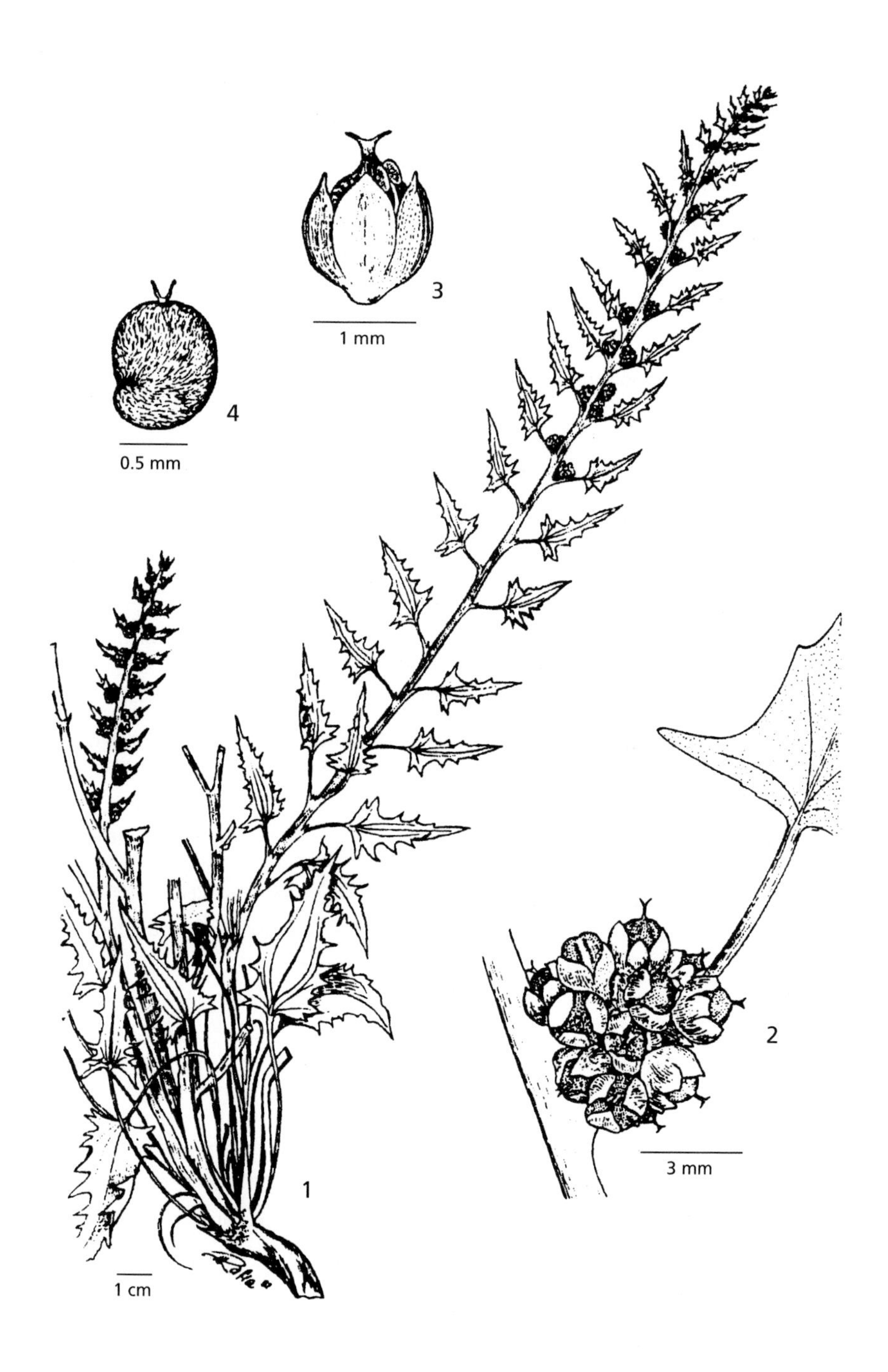

Fig. 75. **Blitum virgatum** subsp. **virgatum**. 1, habit; 2, glomerule; 3, perianth with fruit; 4, seed. Reproduced with permission from Flora of Pakistan 204: f. 11. 2003. Drawn by M. Rafiq.

Blitum, from Gr. βλητόυ, *bliton*, worthless. About 12 species in recent circumscription distributed in the temperate, montanious or subarctic regions of the World. One species in Iraq.

1. **Blitum virgatum** *L.*, Sp. Pl. ed. 1: 4 (1753); Boiss., Fl. Orient. 4: 905 (1879); Eig in Palest. J. Bot., Jerusalem ser., 3: 120 (1945); Zohary in Dep. Agr. Iraq Bull. 31: 46 (1950); Sukhorukov & Akopyan, Comp. Chenop. Cauc.: 32 (2013).

Chenopodium foliosum (Moench) Aschers., Fl. Prov. Brandenb. 1: 572 (1863); Blakelock in Kew Bull. 3: 479 (1957); Mouterde, Nouv. Fl. Lib. et Syr. 1: 411 (1966); Aellen in Fl. Turk. 2: 301 (1967); Uotila in Fl. Iranica 172: 29 (1997) & in Fl. Pak. 204: 18 (2001); Assadi in Fl. Iran 38: 29 (2001); Zhu, Mosyakin & Clemants in Fl. China 5: 379 (2003). *Morocarpus foliosus* Moench, Meth.: 342 (1794).

Branches straight or ascending; inflorescence leafy near the top; seed 1–1.2 mm, ovoid .... a. *B. virgatum* subsp. *virgatum*.
Branches prostrate; inflorescence bracteose; seed 0.6–0.8 mm, rounded .... b. *B. virgatum* subsp. *montanum*.

a. **B. virgatum** subsp. **virgatum**.

Annual or short-lived perennial, 10–100 cm, erect or with ascending stems, glabrous or with scattered bladder hairs and a few glandular hairs. Lower leaves as long as broad, to 10 cm, triangular to hastate or rhomboid, mostly strongly dentate, long-petiolate; upper leaves narrow or hastate, cuneate at base. Inflorescence leafy to the top. Flowers in sessile, globose glomerules. Perianth segments 2–5, at first green, later red and fleshy in fruit or remaining membranous in plants from dry localities. Fruit 1–1.2 mm, ovoid. Pericarp adherent to the seed coat. Seeds vertical, dark red, grooved, densely punctate-pitted. Fig. 75: 1–4.

HAB. Mountain slopes, among rocks and in stony steppes; alt. 2000–3500 m; fl. Jun.–Aug., fr. Aug.–Oct.
DISTRIB. Alpine zone of N Iraq: **MRO**: Mergan nr Bardanas, *Rawi* 24350!; Algurd Dagh, *Rawi* 24901!; *Guest & Ludlow-Hewitt* 2942 (HUJ!). **FAR**: Arbil, *Rechinger* 1172!

Europe, Asia, N Africa.

The fleshy perianth is recorded to be eaten by the Assyrians (*Guest & Ludlow-Hewitt* 2942); in other parts the fleshy perianths are used as rouge cosmetic. The leaves have been used in Europe in former times as pot-herbs.

b. **B. virgatum** subsp. **montanum** *(Uotila) S. Fuentes, Uotila & Borsch* in Willdenowia 42 (1): 17 (2012).

Chenopodium foliosum (Moench) Asch. subsp. *montanum* Uotila in Ann. Bot. Fenn. 30: 190 (1993).

Stems prostrate; lower leaves smaller, to 5–7 cm, entire or near so; glomerules bracteose, also smaller; fruits 0.6-0.8 mm in diameter, as long as wide.

HAB. Mountain slopes, among rocks; alt. 2300–3500 m; fl. Jun.–Aug., fr. Aug.–Oct.
DISTRIB. **MRO**: Qandil range, *Rawi & Serhang* 26751!; Halgord (Algurd) Dagh, *Guest* 2862! **FAR**: Arbil, *Rechinger* 1172! (together with the type variety)

Turkey, Iran.

TRIBE AXYRIDEAE G. Kadereit & Sukhor.

9. **KRASCHENINNIKOVIA** Gueldenst. nom. conserv.

Nov. Comm. Ac. Sci. Petrop. 16: 551 (1772)

Ceratoides Gagnebin, Acta Helv., Phys.-Math. 2: 59 (1755) nom. rejic.;
Eurotia Adans., Fam. Pl. 2: 260 (1763) nom. illegit.;
Ceratospermum Pers., Syn. Pl.: 2 (2): 551 (1807)

Shrubs or subshrubs. Stem and leaves covered with stellate hairs turning fulvous when dry. Leaves alternate, short-petiolate, entire, linear, oblong or ovate, green. Flowers unisexual (plants monoecious); male flowers agglomerated in dense spike-like inflorescences that terminate the branches, perianth hyaline, of 3–5 almost free segments covered with stellate, but easily caducous indumentum; female flowers below in the bract's axils, enclosed in the

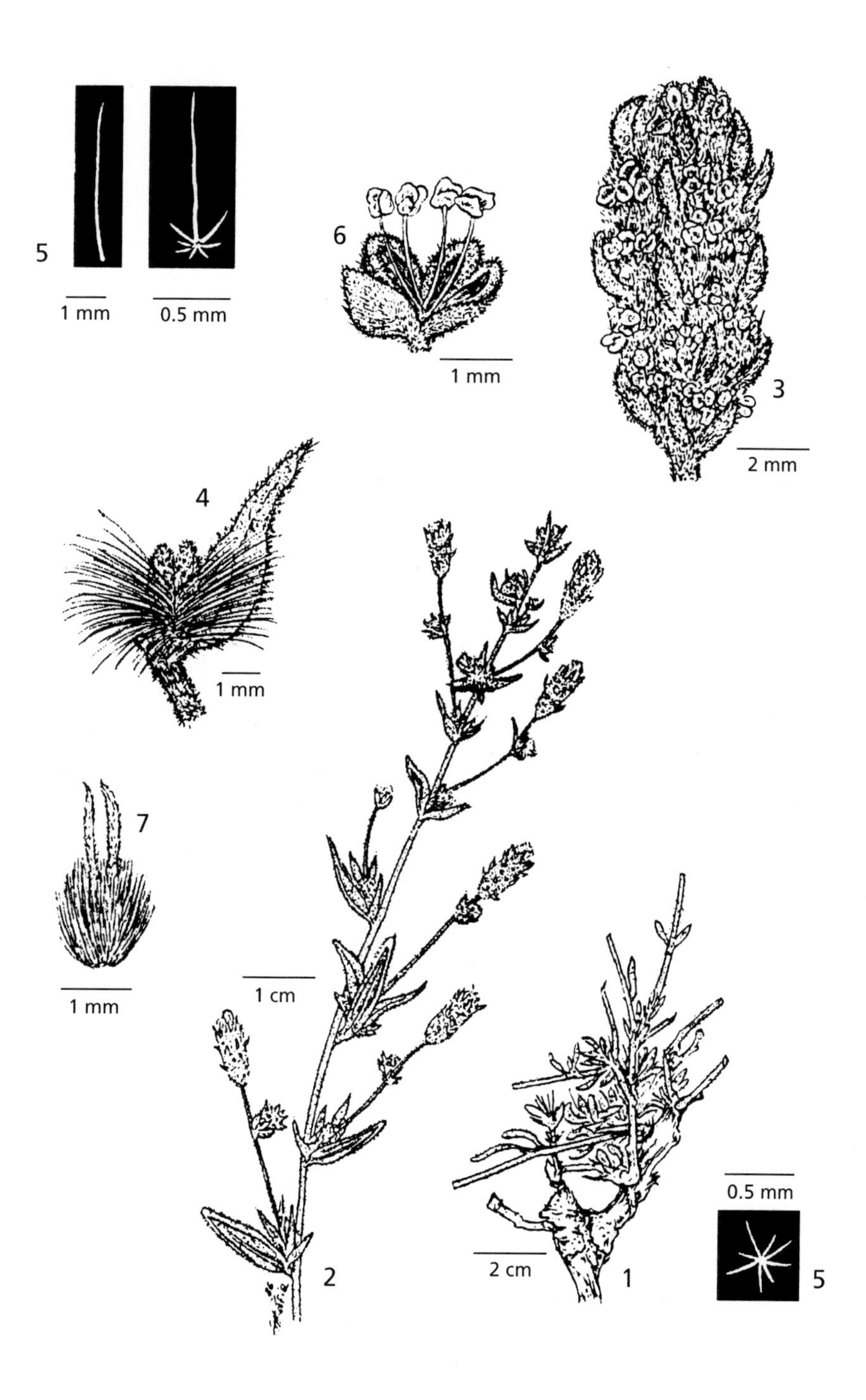

Fig. 76. **Krascheninnikovia ceratoides**. 1, habit basal portion; 2, habit flowering and fruiting stems; 3, inflorescence; 4, bracts; 5, hairs enlarged; 6, male flower; 7, female flower. Reproduced with permission from Flora of Pakistan 204: f. 13. 2003. Drawn by M. Rafiq.

cover consisting of 2 connate accrescent bracts covered with long simple hairs and much smaller stellate hairs Stylodia 2. Fruit with hyaline, very thin pericarp having scattered stellate hairs. Seed with thin seed coat, with perisperm and vertical horseshoe-shaped embryo.

According to the recent investigations (Heklau & Röser, 2008), the genus comprises a single species that is divided into several subspecies.

Krascheninnikovia, in honour of S.P. Krascheninnikov, Russian botanist (1713–1755).

1. **Krascheninnikovia ceratoides** *(L.) Gueldenst.*, Nov. Comm. Ac. Sci. Petrop. 16: 555 (1772); Aellen in Fl. Turk. 2: 313 (1966); Hedge in Fl. Iranica 172: 88 (1997); Freitag et al. in Fl. Pak. 204: 71 (2001); Sukhorukov & Akopyan, Comp. Chenop. Cauc.: 33 (2013).

Axyris ceratoides L., Sp. Pl.: 979 (1753); *Ceratospermum papposum* Pers., Syn. Pl. 2(2): 552 (1807); *Ceratoides papposa* (Pers.) Botsch. et Ikonn., Nov. Syst. Vyssh. Rast. 6: 267 (1970).

Description of species as for genus. Fig. 76: 1–7.

HAB. Mountain zone of Iraq; alt. 1000–3500 m. fl. Jul.–Sep., fr. Sep.–Oct.
DISTRIB. **MRO**: Rowanduz, *Safar* s.n. (LE!).

Widely distributed in the steppes, deserts or mountain regions of temperate Eurasia; scattered in North Africa and N America (*K. ceratoides* subsp. *lanata*).

SUBFAMILY CAMPHOROSMOIDEAE Ulbr.

10. BASSIA All.

Mel. Phil. Math. Soc. Turin 3: 177 (1766)

Kochia Schrad., Neues Journ. Bot. 3, 3–4: 85 (1809);
Panderia Fisch. & C.A. Mey., Ind. Sem. Horti Petrop. 2: 246 (1835)

Annual herbs or subshrubs with simple, alternate, flattened and entire or terete leaves having diverse types of kranz-anatomy. Inflorescence spiciform, leafy, formed of few-flowered axillary clusters. Flowers solitary or 2–5 in clusters, hermaphrodite or unisexual (mostly female), of 5 free segments or connate to the middle lobes that develop in the fruiting stage wing-like, spiny or tuberculate outgrowths; rarely perianth unchanged. Ovary round; style very short, with 2–3 filiform stigmas. Fruit round or ovoid, compressed, with smooth pericarp tightly adjoining to the seed coat; seeds with horizontal or vertical horseshoe-shaped embryo; perisperm abundant.

As previously suggested by Townsend (the first draft of the present manuscript), the genus *Panderia* is extremely close to *Bassia* and some ex-*Kochia* species, from which it differs virtually only in the vertical seeds; and since in such a typical chenopodiaceous genus as *Suaeda* vertical and horizontal disposition of seeds may occur even in the same plant, this scarcely seems a strong character. In addition, *Panderia* and some *Kochia* have the same leaf atriplicoid anatomy as *Bassia* in its recent circumscription. On other hand, the close relaltioship between distinct groups of *Bassia* and *Kochia* is proposed by Sukhorukov (2003). Formal reduction of *Panderia* and majority of *Kochia* to *Bassia* is undertaken by Freitag & Kadereit (2011).

A genus of about 20 species found in Europe, Asia and N Africa. The Australian representative formerly considered as *Bassia* are transferred to other endemic genera.

Bassia, in honour of Ferdinando Bassi (1710–1774), an Italian botanist and prefect of the Bologna Botanical Garden.

1. Subshrub growing in the mountanious region; perianth segments with the wings. . . . . 6. *B. prostrata*
   Annuals. 2
2. Flowers and fruit concealed in a dense, white, woolly indumentum of simple flexuose hairs, spines at the perianth segments very short and sunken in the fleece; leaves dark-green, hairy but not woolly. . . . . 1. *B. eriophora*
   Flowers and fruit ± densely pilose or even glabrous but not woolly, spines or tubercles of fruiting perianth clearly visible. 3
3. Lower leaves to 8 cm; perianth glabrous or its lobes ciliate only, winged or tuberculate at fruiting stage . . . . 5. *B. scoparia*

Lower leaves usually shorter; perianth pubescent throughout or in upper part, winged or spiny at fruits, rarely tuberculate 4

4. Plants to 50 cm, divaricately branched; spines of fruiting perianth straight, to 4 mm, covered with long hairs at least in lower half; fruiting time in late spring-early summer........................ 2. *B. muricata*

Perianth at fruiting stage tuberculate, winged or with hooked spines; fruiting time in autumn.. 5

5. Plant to 2–2.5 m, forming tumble-weed habit; leaves succulent; perianth with small wings or tubercles; seed embryo horizontal .... 4. *B. indica*

Plant to 1 m; leaves not succulent; perianth uncinate or winged (in the latter case seed embryo vertical) 6

6. Perianth spherical, without tubus, the spines uncinate; seed embryo horizontal.. 3. *B. hyssopifolia*

Perianth tubiform, winged or tuberculate but not uncinate; seed embryo vertical.. 7

7. Small montainous plants to 15 cm; leaves linear, to 2 mm broad; perianth tuberculate at fruiting stage...........................8. *B. monticola*

Plants from 15 to 100 cm; leaves much broader; perianth short-winged at fruiting stage ....7. *B. pilosa*

1. **Bassia eriophora** (*Schrad.*) *Aschers.* in Schweinf., Beitr. Fl. Aethiop.: 187 (1867); Rech.f., Fl. Lowland Iraq: 190 (1964); Zohary, Fl. Palaest. 1: 153 (1966); Boulos in Fl. Arab. Penins. Socotra 1: 250 (1996); Hedge in Fl. Iranica 172: 101 (1997); Assadi in Fl. Iran 38: 128 (2001); Freitag et al. in Fl. Pak. 204: 84 (2001); Zhu, Mosyakin & Clemants in Fl. China 5: 388 (2003).

Kochia eriophora Schrad., Neues J. Bot. 3 (3–4): 86, t. 3 (1809); *K. latifolia* Fresen., Mus. Senckenb. 1: 179 (1834).

Erect or decumbent or ascending annual, 6–40 cm, simple or much-branched and bushy with ascending branches when well-grown. Stem and branches whitish, ± densely white-lanuginose throughout. Leaves flat, lanceolate to lanceolate-oblong, obtuse to subacute, 7–20 × 1.5–5 mm, thinly to moderately villous, much diminishing along branches and inflorescence or, where stem leaves are short, scarcely so. Inflorescences spiciform, the few-flowered clusters concealed in a dense woolly indumentum, dense to somewhat distant. Flowers hermaphrodite or female or male. Perianth of female and hermaphrodite flowers small, ± 1 mm diameter, villous, fused in basal $^{1}/_{3}$ (which is hyaline with 5 green nerves), lobes acute with green veins, incurved, each bearing a dorsal tubercle which is variably accrescent to a spine of up to 1 mm in fruit, the villous, straight and stout spines also concealed in fleece; ovary rounded, depressed; style very short, with brownish filiform stigmas 2–3 mm. Filaments white, delicate, 3–3.5 mm; anthers oblong, 1.0–1.5 mm. Fruit ± 2 mm diameter; seed compressed-ovoid, greenish-brown with a pale centre. Fig. 77: 1–4.

Hab. Gravelly, sandy, clayey and silty soils, tolerating some salinity; alt. 0–400(–680) m; fl. Mar.-May, fr. Apr.–Jun.

Distrib. Widespread in the desert and semi-desert alluvial plains of Iraq: **FPF**: Fakka, 70 km E of Amara, *Rawi* 25806!; S of Mandali to Kalal Tahlas Nukhta, *Guest* 788!; 16 km S of Badra, *Alizzi* 18225!; 25 km E of Badra, *Rawi* 20780! **DLJ**: 18 km W of Khamran, *Alizzi & Husain* 33784!; Baka police station, *Alizzi & Husain* 33834!; **DLJ/LCA**: Is-haqi, *Robertson* RB 45! **DWD**: nr Shithatha, *Rawi* 19953!; Rutba, Faidhat al-Masad depression, *Alizzi & Omar* 36188!; K.3 pumping station, *Barkley & Palmatier* 1075!, *Hamid* 39065!; Ukhaidhir palace, *Haddad* 9898!; 3 km E of K.3, *Chakravarty, Rawi, Khatib & Alizzi* 32958!; Habbaniya, *Rawi & Alizzi* 34463!; 18 km E of Ana, *Barkley, Juma'a Brahim & Ani* 551!, 601!; Qutaina (?Qutniya), 90 km N of Shabicha, *Rawi* 14826!; between Ramadi and Hit, *Hamid* 38990!; Habbaniya, *Qaisi* 44251!; 18 km S of Hit, *Barkley & Juma'a Brahim* 4300!; between Abukemal & Ramadi, *Handel-Mazzetti* 667 (WU!); **DWD/DLJ**: Haditha, *Chakravarty, Rawi, Khatib & Alizzi* 32936! **DSD**: 30 km W by N of Busaiya, *Guest, Rawi & Long* 14161!; Zubair, *Rechinger* 8627!, *Rawi & Gillett* 6045!, *Thamer* 47378!; nr Ashuriya well, 55 km NW of Shabicha, *Guest, Rawi & Long* 14053!; 30 km from Salman to Samawa, *Qaisi, Hamad & Hamid* 48296!; 5 km from Salman to Samawa, *Qaisi, Hamad & Hamid* 48128!; Shabicha, *Karim, Nuri, Hamid & Kadhim* 39987!; nr Samawa, *Hossain & Safaji* 92 (B!); 5 km S by E of Zubair, *Guest, Rawi & Rechinger* 16790!; ± 15 km SE of Ashuriya, *Guest, Rawi & Rechinger* 19392!; Imam Ali, W of Zubair, *Rawi & Walter* 16694!; Salman to Al Ain, Al Hilwa, *Karim, Nuri, Hamid & Kadhim* 40120!; 80 km W of Shabicha, *Chakravarty, Rawi & Tikriti* 30074!; E of Wadi Ubaiyidh, *Rawi & Gillett* 6410!; nr Saudi border, about 70 km SW of Nukhaib, *Rawi* 31066!; Ur, *van der Maesen* 1688 (W!), *Regel* 110 (B!). **LEA**: Mashiriya canal, 10 km W

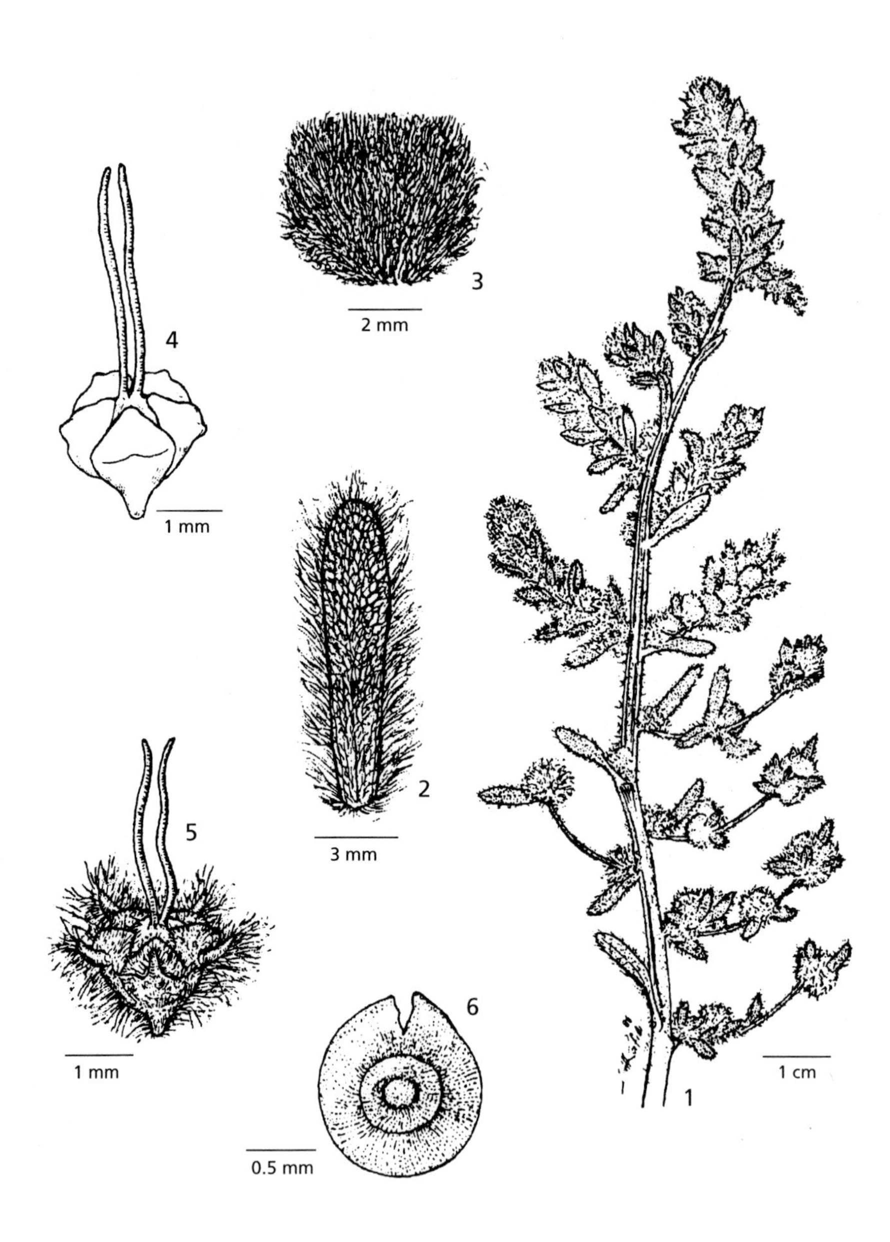

Fig. 77. **Bassia eriophora**. 1, habit; 2, leaf; 3, flower concealed in woolly indumentum; 4, female flower; 5, female flower with hairs removed; 6, seed. Reproduced with permission from Flora of Pakistan 204: f. 15. 2003. Drawn by M. Rafiq.

of Ba'quba, *Anderson* 2017!; 5 km NE of Baquba, *Stutz* 1343 (W!). **LCA**: 5 km W of Tib, *Thamer* 47691!; Abu Ghraib, *Khalid* 37586!; Baghdad to Falluja, *Omar & Wedad* 47503!; Karrada Mariam, *Haines* W110!; 30 km S of Mahmudiya, *Rawi* 20143!; Sudur, *Barkley, Abdul Wahab & Oraha* 6819! **LCA/DWD**: Hindiya barrage, *Rawi & Walter* 18372!; **LCA/DSD**: between Basra & Zubair, *Botany Staff* 135!, 8627!, 33631! **LBA**: Tanuma, *Thamer & Abdel Jabar* 50083!; 'Mesopotamia', *Watson* s.n.!; Tigris, ? Kittam, *Noë* s.n.!

Functionally male flowers do not seem to have been noticed in this species before, but such certainly appear to be present. Some few plants, indeed, appear to be entirely male. Such individuals tend to be rather slender, and narrow, in both leaf and inflorescence.

Morphologically *B. eriophora* is similar to *Bassia eriantha* (*Londesia eriantha*), but with no close relationship to each other (Akhani & Khoshravesh in Phytotaxa 93, 1: 1–24, 2013). The latter taxon has e.g. ovate leaves and smooth perianth lobes, and it is distributed in Irano-Turan (Turkmenistan).

ALAICH-AL-GHAZAL (Babylon, *Guest*). QUTAINAH (Baghdad, *Lazar*), nr Hit (*Rawi & Gillett*). KHATHRAF (Wadi Thirthar, *Rawi & Gillett*). HALŪLĀN (Ashuriya, *Guest et al.*).

Iran, Pakistan, Arabia, Palestine, N Africa.

2. **Bassia muricata** (*L.*) *Aschers. & Schweinf.*, Beitr. Fl. Aethiop. 1: 187 (1867); Rech.f., Fl. Lowland Iraq: 190 (1964); Zohary, Fl. Palaest. 1: 147 (1966); Boulos in Fl. Arab. Penins. Socotra 1: 247 (1996); Hedge in Fl. Iranica 172: 101 (1997); Assadi in Fl. Iran 38: 126 (2001); Ghazanfar, Fl. Oman 1: 41 (2003).

Salsola muricata L., Mant. 1: 54 (1767); *Salsola monobractea* Forssk., Fl. Aeg.-Arab.: 55 (1775); *Kochia muricata* (L.) Schrad., Neues Journ. Bot. 3 (3–4): 86 (1809).

Erect or decumbent annual, 10–40(–50) cm, with long ascending branches from base upwards or in smaller forms simple or with few simple stems from base. Stem and branches pale-stramineous, wiry, ± densely lanuginose throughout or older parts sometimes glabrescent. Leaves flattened, linear to linear-lanceolate, ± acute, densely hairy with long, upwardly-directed hairs, those of main stem and lower part of branches 5–16(–20) × 1–2 mm, those of upper part of branches and in inflorescence slightly to considerably diminishing. Inflorescences spiciform, flowers (hermaphrodite and female) in groups of 2–3, lower distant in fruit or remaining ± approximate. Perianth ± 1.5 mm diameter, green, deltoid-ovate, rather blunt, incurved, each with a dorsal spine which is accrescent to 4 mm in fruit, straight, and long-pilose at least in lower half. Ovary rounded, depressed; style very short with 2 brownish filiform stigmas 1.5–2 mm. Filaments white, delicate, 1 mm; anthers roundish, 0.5 mm. Fruit 1.5–1.75 mm in diameter.

HAB. Sand dunes and flats; alt. 0–700 m; fl. Apr.–Jun., fr. Jun.–Jul.
DISTRIB. Frequent but local in the Southern Desert region of Iraq: **DWD**: Wadi Sarahan & Rutba, *Eig, Feinbrun & Zohary* s.n. (HUJ!); Habbaniya, *Regel* s.n. (G!); **DSD**: Al-Ichrishi, 35 km E by N of Busaiya, *Guest & Rawi* 14192!; Al-Waksa well, 85 km W by N of Shabicha, *Guest, Rawi & Rechinger* 19213!; 55 km SE by E of Busaiya, *Guest & Rawi* 14270!; 389 km W of Baghdad, *Eig & Zohary* s.n. (HUJ!); Zubair, 1953, *Regel* s.n. (G!); **DSD/LCA**: Baghdad to Karbala road, *Haines* W.663! **LEA**: Muqdadiya, *Rechinger* 9738 (B!, W!, WU!).**LCA**: Imam Ibrahim, *Hadač* 1685 (G!).

?KHUDHRADH; eaten by camels but not relished by sheep (*Guest & Rawi* 14192).

N Africa, Palestine, Cyprus, Syria, Arabia, S Iran.

3. **Bassia hyssopifolia** (*Pall.*) *O. Kuntze*, Rev. Gen. 2: 547 (1891); Volkens in Pflanzenfam. 3, 1a: 70 (1893); Rech.f., Fl. Lowland Iraq: 190 (1964); Boulos in Fl. Arab. Penins. Socotra 1: 250 (1996); Hedge in Fl. Iranica 172: 100 (1997); Assadi in Fl. Iran 38: 126 (2001); Freitag et al. in Fl. Pak. 204: 82 (2001); Zhu, Mosyakin & Clemants in Fl. China 5: 387 (2003); Sukhorukov & Akopyan, Comp. Chenop. Cauc.: 38 (2013).

Kochia hyssopifolia Pall., Reise Russ. Reichs 1: 491 (1771); *Willemetia lanata* Maerkl., J. Bot. 3, 1: 330 (1801) nom. illegit.; *Kochia hyssopifolia* (Pall.) Roth, Neue Beitr.: 176 (1802); *Echinopsilon hyssopifolium* (Pall.) Moq. in Ann. Sc. Nat. ser. 2, 2: 127 (1834); *E. lanatum* (Maerckl.) Moq. in Soyer-Willem., Ann. Sci. Nat. Ser. 2, 2: 127 (1834) nom. illegit.; *E. hyssopifolium* (Pall.) Moq., Chenop. Monogr. Enum.: 87 (1840); *E. caspicum* Bunge in Mém. Sav. Étr. Petersb. 7: 455 (1851).

Erect annual, (10–)30–100 cm, with long ascending branches from base upwards or in dwarfed forms simple. Stem and branches whitish, wiry, glabrescent in older parts, increasingly clothed upwards with patent or crisped whitish hairs. Leaves flat, oblong or ovate, ± acute, hirsute or villous, those of main stem and lower part of branches 5–20 ×

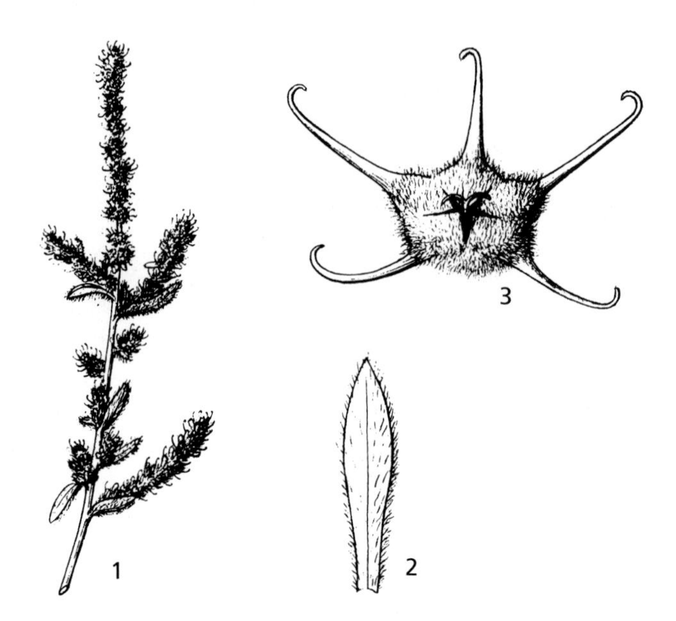

Fig. 78. **Bassia hyssopifolia**. 1, fruiting branch × 1/2; 2, leaf × 3; 3, perianth with wings × 20. Reproduced with permission from Flora of China 5: fig. 314. Drawn by Cai Shuqin.

2–10 mm, diminishing upwards along branches and much reduced in inflorescence. Inflorescences spiciform; flowers solitary or in groups of 2–3 which become distant below in fruit. Flowers hermaphrodite and female. Perianth long-pilose, fused at its half, lobes rounded, incurved, membranous and green-vittate, each bearing a dorsal spine which is seen at the flowers and rapidly accrescent, ± glabrous and strongly inwardly uncinate at tip. Ovary rounded, depressed; style very short, with 2 brownish filiform stigmas ± 1 mm. Filaments white, delicate, 1 mm; anthers roundish. Fruit ± 1.5 mm diameter, compressed-ovoid, greenish to black with a paler centre. Fig. 78: 1–3.

HAB. Saline flats, roadsides, bank of Washash drain; alt. 0–1000 m; fl. & fr. Sep.–Nov.
DISTRIB. Occasional and rather local in the desert region of Iraq: **DSD**: nr Karbala, *Rechinger* 8310 (E!, W!); **LCA**: Diwaniya, *Guest* 3565!; Baghdad, Al Khur bridge, *Rechinger* 8002!; Hilla, *Fawzi* 39528!; nr Musaiyib on road to Karbala, *Guest, Rawi & Rechinger* 16146!; Baghdad, *Haines* W1815! **LSM**: nr Amara, *Evans* 303 (E!); **LBA**: Basra, *Handel-Mazzetti* 3136 (WU!).

?SUĀD or ARID (*Guest* 3565); ?KHANAIQ or HUMMAIDH (*Guest et al.* 16146).

Spain, SE and E Europe, C Asia, Turkey, Iran, Kuwait, Arabia, Afghanistan, Pakistan, China (Xinjiang).

The species appears to be not uniform within the range: the forms from Spain with almost glabrous perianth are described as *B. reuteriana* Guerke; besides, such forms with glabrescent perianth and lanceolate leaves are known in Xinjiang.

4. **Bassia indica** (Wight) A.J.Scott in Fedd. Repert. 89: 108 (1978); Boulos in Fl. Arab. Penins. Socotra 1: 251 (1996).

Kochia indica Wight, Ic. Pl. Ind. Or. 5, 2: 5, t. 1791 (1852); Zohary, Fl. Palaest. 1: 153 (1966); Hedge in Fl. Iranica 172: 109 (1997); Freitag et al. in Fl. Pak. 204: 87 (2001).
Kochia griffithii Bunge ex Boiss., Fl. Orient. 4: 924 (1879).
Bassia joppensis Bornm. & Dinsmore in Bornm., Repert. Sp. Nov. (Fedd. Repert.) 17: 274 (1921).

Very branched annual to 2.5 m forming tumble-weed habit; stems reddish, hairy but in the lower part glabrescent; lower leaves to 5 cm, tapering at base, upper leaves to 3 cm, all very succulent; inflorescence large, with remote glomerules; flowers in clusters with tufts of hairs at base; perianth pubescent, forming wing-like or tuberculate appendages.

HAB. Clayey soils; ruderal places; wadis.

Distrib. Widespread alien common in SE Eastern Mediterranean and NE Africa. **LCA**: Abu Ghraib, *Janan al-Mokhtar* 38162!; *Omar* 38518!; Hinaidi, *Jackson* 328 (BM!). **LBA**: Basra, *Haines* s.n. (E!).

India, Pakistan, Iran, Arabia, Kuwait, Eastern Mediterranean, N Africa, Cyprus.

5. **Bassia scoparia** (L.) A.J.Scott in Fedd. Repert. 89 (2–3): 108 (1978); Boulos in Fl. Arab. Penins. Socotra 1: 251 (1996); Sukhorukov & Akopyan, Comp. Chenop. Cauc.: 38 (2013).

Chenopodium scoparium L., Sp. Pl. ed. 1: 221 (1753).
Kochia scoparia (L.) Schrad. in Neues J. Bot. 3, 3–4: 85 (1809); Aellen in Fl. Turk. 2: 316 (1967); Assadi in Fl. Iran 38: 141 (2001); Freitag et al. in Fl. Pak. 204: 90 (2001); Zhu, Mosyakin & Clemants in Fl. China 5: 364 (2003). *K. sieversiana* (Pall.) C.A. Mey. in Ledeb., Fl. Alt. 1: 415 (1829). *K. scoparia* β [var.] *densiflora* Turcz. ex Moq. in DC., Prodr. 13(2): 131 (1849). *K. parodii* Aellen in Verh. Naturf. Ges. Basel 50: 151 (1939). *K. densiflora* (Turcz. ex Moq.) Aellen in Mitt. Basler Bot. Ges. 2, 1:13 (1954). *K. scoparia* subsp. *densiflora* (Turcz. ex Moq.) Aellen in Hegi, Ill. Fl. Mitteleur., ed. 2, (2): 710 (1961). *Bassia scoparia* subsp. *densiflora* (Turcz. ex Moq.) Cirujano & Velayos, An. Jard. Bot. Madrid 44(2): 577 (1987).

Erect annual, 20–100(–150) cm, with long ascending branches from base upwards. Stem and branches whitish, striate, sometimes reddish-tinged, moderately to densely clothed throughout with short to rather long, soft, whitish hairs. Leaves flat, narrowly oblong-lanceolate to lanceolate or linear, 3-nerved, those of main stem and lower parts of branches 20–50(–80) × 1.5–7(–10) mm, apex acute, gradually attenuate and subpetiolate at base, ± densely pilose when young, when mature glabrous to thinly pilose on upper surface, more densely finely appressed-hairy below; inflorescence leaves rather abruptly reducing. Inflorescences spiciform, foliose, lax to rather dense, axis with fine appressed to rather long, spreading, brownish hairs; flowers solitary or 2–5, unisexual or hermaphrodite, surrounded by basal tufts of hair or not. Perianth of female and hermaphrodite flowers ± 2 mm in diameter, fused in basal ²/₃, free apical lobes incurved, blunt, greenish (sometimes reddish-tinged), ciliate-margined or glabrous, each with a basal dorsal umbo which develops in fruit into a hollow, bluntly conical sac or rarely a wing; ovary rounded, depressed; style very short, stigmas ± 1 mm. Filaments delicate, ± 1.5 mm; anthers shortly oblong, 1 mm, often reddish-tinged. Fruit ± 2 mm diameter, compressed-ovoid, dark brown, ± 1.5 mm. Fig. 79: 1–6.

Hab. In the mountain zone; mountain slopes mostly on gravelly substrates; alt. 1000–2500 m; fl. .
Distrib.: **MRO**: Haji Omran, *Rawi & Serkahia* 24308! 24945!

Possible origin of species is Central Asia (E Kazakhstan, Kyrghyzstan, W China, Mongolia), alien in many temperate regions of Eurasia, N & S America.

f. **trichophylla** *Schinz & Thell.* (within *Kochia*) in Verz. Säm. Bot. Gart. Zürich: 10 (1909).

Plant forming a dense, ovoid "bush", densely foliose with linear, mostly 1-nerved leaves. Widely cultivated as decorative plant elsewhere.

Hab. & Distrib. Escaped from cultivation, as ephemerophyte; collected around Baghdad: **LCA**: nr Baghdad, *Lazar* 532 (Field Museum 786194)!; Abu Ghraib, *Alizzi* 34180!

Grown in gardens for the beautiful reddish colour which it develops on senescence – hence its English name of "Burning Bush". Cultivated elsewhere including neighboring territories (Arabia, Kuwait, Palestine).

6. **Bassia prostrata** (L.) A.J. Scott in Fedd. Repert. 89 (2–3): 108 (1978); Boulos in Fl. Arab. Penins. Socotra 1: 251 (1996); Sukhorukov & Akopyan, Comp. Chenop. Cauc.: 37 (2013).

Salsola prostrata L., Sp. Pl.: 222 (1753).
Kochia prostrata (L.) Schrad. in Neues J. Bot. 3, 3–4: 85 (1809); Aellen in Fl. Turk. 2: 316 (1967); Hedge in Fl. Iranica 172: 105 (1997); Assadi in Fl. Iran 38: 141 (2001); Freitag et al. in Fl. Pak. 204: 86 (2001); Zhu, Mosyakin & Clemants in Fl. China 5: 384 (2003); *K. suffruticulosa* Less. in Linnaea 9: 202 (1835).

Subshrub to 20–70 (–100) cm, basally branched. Stems reddish or whitish, covered with simple curved hairs, often glabrescent. Leaves to 25 × 2 (–2.5) mm, linear or linear-lanceolate, semi-terete or slightly flattened, tapering at base, softly villose or pubescent.

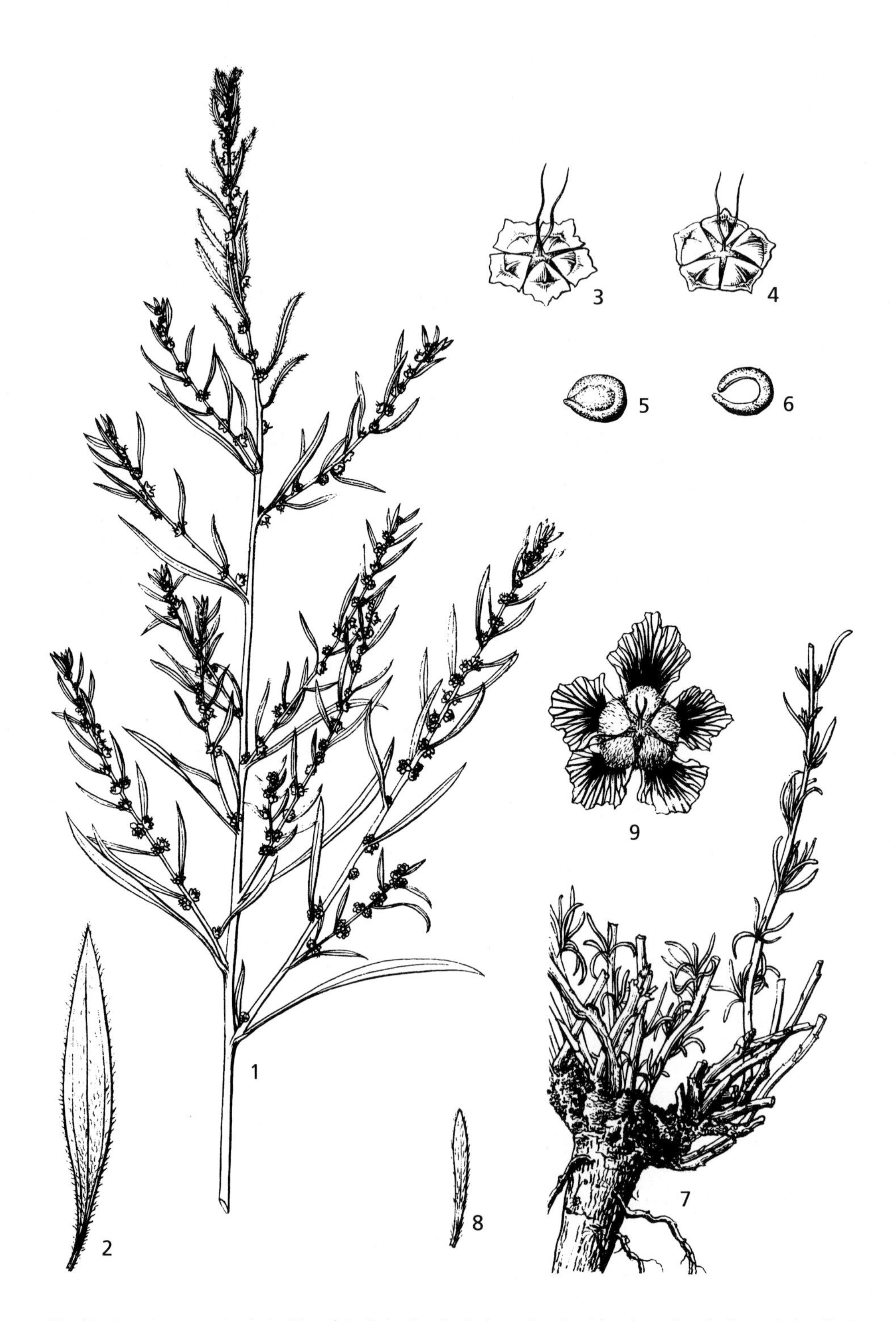

Fig. 79. **Bassia scoparia**. 1, habit × 1/2, 2, leaf × 2; 3–4, perianth with wings detail; 5, seed detail; 6, embryo detail. **Bassia prostrata**. 7, habit lower part × 1/2; 8, leaf × 1; 9, perianth detail. Reproduced with permission from Flora of China 5: fig. 313. Drawn by Cai Shuqin.

Bracts small, equal or slightly longer than flower clusters. Perianth of 5 basally connate segments, villose throughout, at fruiting stage with white or brownish to reddish wings 3.5–6 mm in diameter Anthers ± 1 mm. Fruits brownish, 2–2.5 mm in diameter. Fig. 79: 7–9.

Hab. Rocks, hill slopes at alt. 1000–3000 m; fl. & fr. Aug.–Oct.
Distrib. **mro**: nr Rowanduz, *Safar* s.n. (LE!).

Another subshrubby species, *B. arabica* (Boiss.) Maire & Weiller (=*Chenolea arabica* Boiss.; ≡*Chenoleoides arabica* (Boiss.) Botsch.) may be found in the southern deserts of Iraq. It is well distinguished by its terete fleshy leaves and perianth bearing small tubercles at fruiting stage. It reaches Jordan and Syria in the northern parts of its range.

Widely distributed in the steppes and deserts of Eurasia. Steppes and semi-deserts of Europe, Central Asia, Iran, Turkey, Afghanistan, N Himalaya.

7. **Bassia pilosa** (Fisch. et C.A. Mey.) Freitag & G.Kadereit in Taxon 60(1): 73 (2011); Sukhorukov & Akopyan, Comp. Chenop. Cauc.: 37 (2013).

Panderia pilosa Fisch. & C.A. Mey., Ind. Sem. Hort. Petrop. 2: 46 (1835); Rech.f., Fl. Lowland Iraq: 189 (1964); Hedge in Fl. Iranica 172: 96 (1997); Freitag et al. in Fl. Pak. 204: 69 (2001).
Panderia turkestanica Iljin in Bull. Jard. Bot. Acad. Sci. URSS 30: 364 (1932); *P. iraqensis* Sukhor. in sched.
Kochia noeana Boiss., Fl. Orient. 4: 919 (1879) nom. invalid. in syn. –Syntypes: Iraq, Baghdad, *Noë* (G! K!).

Annual herb, tall (to 1 m), much-branched from base upwards, branches divaricate or ascending, white to reddish, slightly angular, striate, with a mixed indumentum of short, patent or crisped hairs and long hairs which may be ± flexuous or commonly, at least in the lower internodes of the stem, horizontally spreading; often partly glabrescent with age. Leaves flat, narrowly oblanceolate to narrowly elliptic or linear, 8–15(–20) × (1.5)2–5 mm on main stem and lower part of branches but reducing and oblong to ovate towards ends of branches, narrowed at each end, apex acute to subacute, with long and/or shorter ± appressed to patent hairs. Inflorescence narrow and spiciform along stem and branches, foliose, indumentum of axis similar to that of stem; flowers in groups of mostly 1–5 in axils of reduced floral leaves, hermaphrodite or (usually less commonly) female, rarely functionally male. Perianth pilose, especially above, that of all types of flower similar, with a shortly campanulate tube 1–1.25 mm; lobes incurved, cucullate, 0.5-0.75 mm, with broad pale margin and a green ± deltoid vitta, with a small dorsal umbo when in flower. Filaments of hermaphrodite flowers narrowly ribbon-like, 2.5–3 mm; anthers shortly oblong, ± 1 mm. Styles of female and hermaphrodite flowers very short, ± 0.5 mm, slightly dilated below the 1.25–1.5 mm filiform shortly pilose stigmas. Functionally male flowers with ovule abortive or absent, stigmas abortive and short, "burnt"-looking. Fruiting perianth with dorsal umbo of each segment variably developed from a short, conical process to a distinct, lingulate or flabellate, longitudinally veined wing. Fruit distinctly compressed, 1.2–1.75 mm; seed compressed, shortly elliptic, brown to greenish-brown, 1.1–1.6 mm, with vertical embryo.

Hab. Waste places, irrigated often subsaline areas, abandoned fields; alt. 30–45+ m; fl. Jul.–Aug., fr. Sep.–Oct.
Distrib. Locally common in the Baghdad area. **fpf**: Jabal Hamrin, *Sutherland* 483 (BM!). **lea**: Ba'quba, by the bridge, *Rechinger* 8069! **lca**: Jadriya, *Haines* s.n.!; Abu Ghraib, *Gillett* 11549!, 12488!, 12495!, 12503!, *Barkley & Hikmat Abbas* 3870B!; Baghdad, *Barkley* 3514!, *Colvill* s.n.!, *Guest* 3228!, 3230!, *Haines* W.1460!, *Noë* s.n. (type of *Kochia noeana*)!; Baghdad airport, *Barkley & Safwat* 3404!, *Haines* s.n.!, *Rechinger* 33626!; Zafariniya, *Sahira & Sakin* 34773!; Mahmudiya, *Haines* W.7132!; nr Falluja, *Guest & Rawi* 15959!

The reasons for reducing *P. turkestanica* to synonymy have been discussed elsewhere [Townsend in Kew. Bull. 35 (2): 292 (1980), see also Kadereit & Freitag, 2011]. However, almost all specimens collected in Iraq are distinguished by the tall and much branched stem forming tumble-weed habit; leaves turn black when dry. Such exemplares were noted as *Kochia noeana* as synonym to *Panderia pilosa* and called as *Halostigmaria maris-mortui* (Eig in HUJ) or *Panderia iraqensis* (Sukhor. in sched. K & E); these need further investigation concerning their taxonomic status (A.S.). The distribution of this unusual race is localized in Iraq, Syria and Jordan.

?HĀMIDH (*Guest* 3228); ?HAITAM (*Guest & Rawi* 15959).

Turkey, Iran, Palestine, Syria, Lebanon, Afghanistan, Pakistan, Central Asia.

8. **Bassia monticola** (*Boiss.*) *O. Kuntze*, Revis. Gen. Pl. 2: 547 (1891).

Kochia monticola Boiss. ex Moq. in DC., Prodr. 13 (2): 133 (1849).
Pentodon barbatus Ehrenb. in Boiss., Fl.: 4 (2): 925 (1879) nom. inval. (in syn.)

Annual herb to 15 cm, basally branched; stems ascendent, rarely almost straight, whitish and glabrescent, leaves (except basally located leaves) remote, 10–15 × 1–2 mm, with brachyblasts in their axils; inflorescence dense, bracts to 6 mm, slightly exceeding the flower clusters; perianth 2–5 mm, in its base with tufts of simple hairs, perianth tube white, almost glabrous or slightly pilose, teeth at the top hairy, with green dots that bear the tubercles or bulges at fruiting stage. Fruit compressed, ± 1.5 mm, with easily ruptured white pericarp; seed with vertical embryo.

HAB. Alpine vegetation; 1900–2500 m; fl. Jul.–Sep., fr. Sep.–Oct.
DISTRIB. **FAR**: Makhmur, *Ehrenberg* 236 (LE!).

A neglected species that was poorly known. However, it is very remarkable due to dwarf habit, small linear leaves, and bulging perianth at fruiting stage. The vertical embryo position is stated here for the first time seen on the material in HUJ.

Lebanon, Syria, Iran, E Turkey.

SUBFAMILY CORISPERMOIDEAE Raf.

11. **AGRIOPHYLLUM** Bieb.

Fl. Taur.-Cauc. 3: 6 (1819)

Ref.: P. Aellen, *Agriophyllum* in "Ergebnisse einer botanisch-zoologischen Sammelreise durch den Iran". Verh. Naturf. Ges. Basel 63: 254–262 (1952).

Stiff annual herbs, at least the younger parts clothed in a ± dense indumentum of dendroid hairs, or glabrous. Leaves alternate (except the lowermost which may be opposite), sessile (in ours), with prominent whitish veins, spine-tipped. Flowers in heads, hermaphrodite, sessile in leaf axils. Perianth segments (1–5) free, small, hyaline. Stamens 1–3(–5). Ovary becoming hard in fruit with 2 persistent styles and 2 horns of similar size. Seed vertical; albumin present; embryo circular.

About 7 species found mainly in sandy desert areas; one species in Iraq.

Agriophyllum, from Gr. αγϱιος, *agrios*, wild, φύλλον, *phyllon*, leaves).

1. **Agriophyllum minus** *Fisch. & C.A. Mey.* ex Ledeb. Fl. Ross. 3(2): 755 (1851); Boulos in Fl. Arab. Penins. Socotra 1: 249 (1996); Hedge in Fl. Iranica 172: 70 (1997); Assadi in Fl. Iran 38: 84 (2001); Zhu, Mosyakin & Clemants in Fl. China 5: 364 (2003); Ghazanfar, Fl. Oman 1: 41 (2003).

Much-branched annual, becoming stiff and frangible, with a variably dense clothing of dendroid hairs. Stems 12–25 cm, leafy; leaves linear-lanceolate, to 30 × 6 mm, cuneate at base, spine-tipped, sessile, with prominent veins on lower surface. Inflorescence branched, flowers sessile in axils of upper spinose leaves. Perianth segments very small, oblong. Fruit ± 4 × 2 mm, flattened, winged below and laciniate-toothed; styles hard, persistent, with 2 horns arising basally and equalling the styles in length. Seed broadly elliptic, 1.5 × 1 mm. Fig. 80: 1–7.

HAB. Sand dunes; alt. ± 5 m; fl. Apr., fr. ?Jun.
DISTRIB. Rare, only recorded once from semi-desert areas of Iraq: **LCA**: Nasiriya, 25.4.1973, *D.C.P. Thalen & Samey* 40438!

Iran, Afghanistan, Arabia, C Asia, W China.

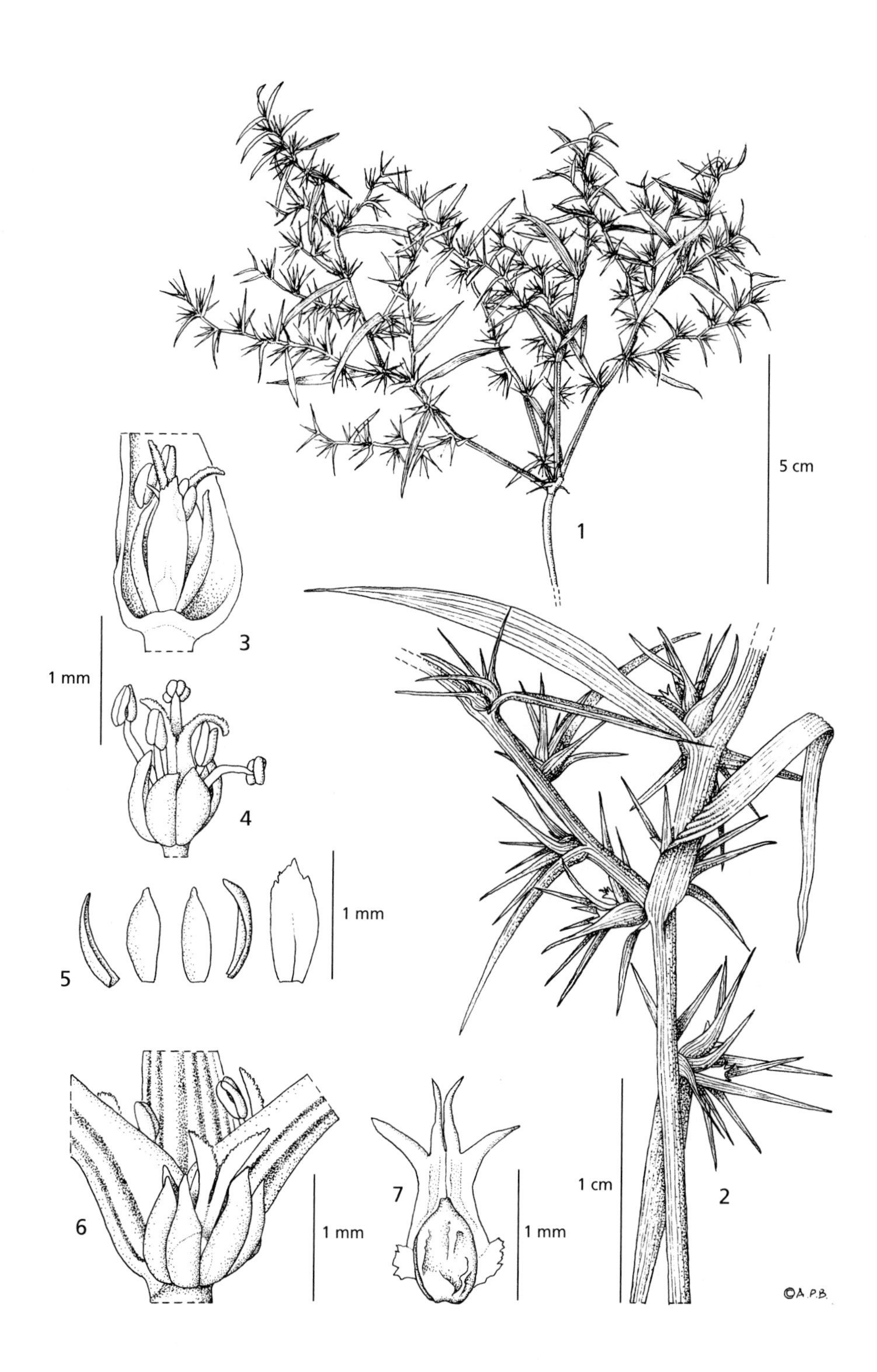

Fig. 80. **Agriophyllum minus**. 1, habit; 2, lower part of stem detail; 3, flower in situ in axil of leaf; 4, flower; 5, tepals; 6, older flower in situ showing young flowers behind in leaf axil; 7, fruit. All from *Mandaville* 634. © Drawn by A.P. Brown.

SUBFAMILY SALICORNIOIDEAE Kostel.

12. **HALOPEPLIS** Bunge ex. Ung.-Sternb.

Vers. Syst. Salic.: 105 (1866)

Plants glabrous, halophytic, annual herb or subshrubs, succulent with alternate branches. Leaves alternate, very thick and fleshy, subglobular to cordiform, ± clasping stem. Inflorescence spike-like, terminal on stem and branches; flowers usually in 3-flowered cymes in leaf axils, hermaphrodite, united with each other and with the bracts. Perianth flattened, obscurely and bluntly 3-toothed, formed of 3 almost completely united segments, almost closed. Stamens 1–2, anthers not appendiculate. Ovary ovoid-pyriform, compressed; style short; stigmas 2, filiform. Fruit with ruptured pericarp in its basal part; seeds yellowish brown, with mamillate seed coat; perisperm present, embryo arcuate.

Two to three species from Europe to C Asia; one species in Iraq.

Halopeplis (from Gr. ἅλας, *hals*, salt, and *Peplis*, (Gr. name of *Peplis*).

Halopeplis pygmaea (*Pall.*) *Bunge ex Ung.-Sternb.*, Vers. Syst. Salic.: 105 (1866); Rech.f., Fl. Lowland Iraq: 191 (1964); Assadi in Fl. Iran 38: 166 (2001); Zhu, Mosyakin & Clemants in Fl. China 5: 355 (2003); Sukhorukov & Akopyan, Comp. Chenop. Cauc.: 43 (2013).

Salicornia pygmaea Pall., Ill. Pl.: 9, t. 2 f. 2 (1803).

Annual herb, (3–)5–20 cm, sparingly to much-branched from base upwards, glaucous to deep green, frequently suffused with red or purple. Leaves very fleshy, subglobose, sessile, clasping stem and margins decurrent along it; lower internodes longer than leaves, upper shorter. Bracts with a broad, conspicuous scarious border to ± 0.4 mm wide. Inflorescences 0.6–3.4 × 3–5 mm, obtuse. Cymes mostly 3-flowered. Flowers strongly dorsiventrally compressed, apex semi-lunar, central flower larger than laterals. Stamens 1(–2), anthers oblong, ± 0.5 mm, exserted. Stigmas very short, ± 0.25 mm, usually included. Seeds 0.8–1 mm, rounded to shortly reniform, testa cells form narrowly cylindrical, truncate papillae at least near embryo. Fig. 81: 1.

HAB. Wet salt marsh, margin of salty lakes; alt. ± 35 m; fl. Oct.–Nov.
DISTRIB.: Rare in Desert region of Iraq: **DWD**: Hit, *Guest* 3547!; Bahr al-Milh, 10 km NW of Shithatha, *Guest, Rawi & Rechinger* 16182!, *Rechinger* 8345!

We are by no means convinced of the distinctness of this species from *H. nodulosa* (Del.) Bunge ex Ung.-Sternb (*H. amplexicaulis* (Vahl) Ung.-Sternb. ex Ces., Passer. & Gib.), which ranges in the Mediterranean area. Neither habit nor internode length (as used to separate the two taxa in *Flora Europaea*) are very helpful. Boissier (Boiss., Fl. Orient. 4: 934–935, 1879) gives *H. pygmaea* as "staminibus binis" and *H. amplexicaulis* as "stamine unico"; but Iljin in Fl. SSSR states of *H. pygmaea* "stamen mostly one". Unger-Sternberg emphasises the seed character – the seeds papillose all over in *H. pygmaea* but only over the embryo in *H. amplexicaulis.* Our material is characterized with the seeds papillose throughout (Sukhorukov, l.c., SEM photo). Certainly it may be said that in "eastern" *Halopeplis* of this affinity the lateral faces of the seed are papillose and in "western" material they are smooth. But there is variation even here. Some Egyptian material named as *H. amplexicaulis* has the papillae confined to the outer line of the embryo; other plants have them covering the embryo and encroaching onto the lateral faces. Likewise, some Italian material ("in locis salsis proper Tarantum, Jul. 1875, *H. Groves* s.n.) has the papillae reduced to blunt, conical tubercles. However, herbarium specimens are notoriously unsatisfactory to deal with in this group and formal reduction is

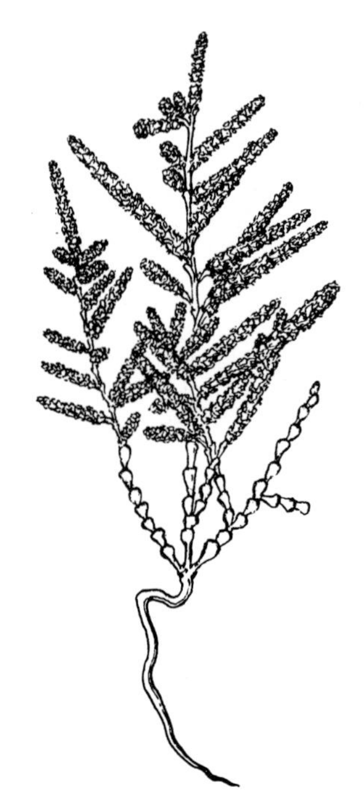

Fig. 81. **Halopeplis pygmaea**. Habit flowering plant. Reproduced with permission from Flora of China 5: fig. 293.

scarcely possible without a careful study of a range of fresh material. On seed character and rather denser habit, the Iraqi plant is best referred to *H. pygmaea*. *H. nodulosa*, however, is the earlier name if in the future the two should be combined.

THILAIDH (*Guest* 3547).

Iran, C Asia.

13. HALOCNEMUM Bieb.

Fl. Taur.-Cauc. 3: 3 (1819)

Small halophytic glabrous shrub with shoots of current year's growth jointed, succulent, with opposite leaves and branches. Leaves connate around stem, all reduced to fleshy cups, lamina virtually obsolete. Inflorescence spike-like, terminal on stem and branches; bracts free; flowers 3 together in leaf axils, hermaphrodite. Perianth segments 3, free or usually shortly fused below. Stamen solitary, anterior, connective not appendiculate. Ovary ovoid; style short but distinct; stigmas 2, filiform. Fruit ovoid, enclosed within perianth, with a delicate pericarp; seeds with mamillate testa, perisperm present, embryo arcuate, vertical.

Lectotype of the genus: *H. strobilaceum.*

A genus of two species, but some populations of *H. strobilaceum* in Arabia are distinct in some characters.

Halocnemum (from Gr. ἅλς, *hals*, salt, κνεμις, *knemis*, sheath).

Halocnemum strobilaceum (*Pall.*) *Bieb.*, Fl. Taur.-Cauc. 3: 3 (1819); Rech.f., Fl. Lowland Iraq: 191 (1964); Zohary, Fl. Palaest. 1: 155 (1966); Boulos in Fl. Arab. Penins. Socotra 1: 252 (1996); Hedge in Fl. Iranica 172: 126 (1997); Boulos, Fl. Egypt 1: 386 (1999); Freitag et al. in Fl. Pak. 204: 103 (2001); Assadi in Fl. Iran 38: 169 (2001); Zhu, Mosyakin & Clemants in Fl. China 5: 356 (2003); Sukhorukov & Akopyan, Comp. Chenop. Cauc.: 44 (2013).

Salicornia strobilacea Pall., Reise Russ. Reichs 1: 481 (1771).
[*Arthrocnemum glaucum* (non (Del.) Ung.-Sternb.): Guest in Dep. Agr. Iraq Bull. 27: 10 (1933)]

Low glabrous shrub, forming ± rounded hummocks 15–50 cm tall, much-branched, old growth frequently rather gnarled and contorted with a whitish to pale brown cortex; rootstock stout and tough. Young stems and branches ± straight and erect, jointed, those of current year green, succulent, joints globular to cylindrical with a cup-shaped pair of connate leaves at top of each. Leaves reduced to fleshy concave cups. Flowers on short lateral branchlets on current year's shoots and also at shoot tips, in axillary groups of 3. Perianth small, lobes hyaline, broadly oblong, 1–1.25 mm, truncate at apex, central lobe flattened dorsally, laterals gibbous and often green at tip. Stamens ± 2 mm, filaments delicate; anthers oblong, ± 0.75 mm. Ovary ovoid; style ± 0.5 mm, stigma ± equalling it or longer. Fruit ± 1.25 mm, very delicate, compressed-ovoid; seeds light brown, 0.75–1 mm, compressed-ovoid to bluntly triangular, minutely verruculose. Fig. 82: 1–5.

HAB. Saline flats; alt. s.l.–150 m; fl. & fr. Mar.–Nov.
DISTRIB. Common in lowland *sabkha* areas, particularly in the Southern Desert. Due to the very large number of specimens gathered, only one to three typical sheets per area have been cited: **FUJ/DLJ**: Wadi Thirthar nr Sabkha Umm Rahal, *Guest* 3544!; **FPF**: nr Tursaq, *Rawi* 12577!; Haj Yusuf, 10 km E of Mandali, *Rawi* 20687!; 8 km NE of Gassan, *Weinert & Mousawi* s.n. (LE!); **DLJ**: 80 km NW of Rawa, *Rawi, Khatib & Alizzi* 31921!, 31924!; **DWD**: Shithatha, *Rawi & Gillett* 6504!; 4 km W of Hit, *Rawi & Gillett* 6865!; between Basra and Umm Qasr, *Hossain* s.n.!; **DSD**: Shu'aiba, *Rogers* 03!; Zubair, *Rawi & Gillett* 6048!; 15 km ENE of Jaliba station, *Guest, Rawi & Rechinger* 16141!; 17 km W of Samawa, *Rechinger* 8160 (E!); Umm Qasr, *Rechinger* 14527 (W!). **LEA**: 7 km E of Abul-Khasib, *Barkley & Hikmat Abbas-al-Ani* 6458! **LSM**: nr Amara, *Evans* 215 (E!); 6 km E of Amara, *al-Ani & Barkley* 8922 (W!); shore of Hor al-Hammar, *Rechinger* 8254! **LBA**: 50 km S of Basra, *Khatib & Alizzi* 32649!;

This species is one of the most spectacular features of the scenery on and around salt flats, particularly in autumn when the plants become reddish in colour.

THELETH (*Rawi & Gillett* 6048); THIL(L)AIDH (*Guest* 3544, *Guest et al.*, 16141); THELLAITH (*Rawi* 16586); HAMTH (*Rawi & Gillett*); ?RIMTH (*Rawi & Gillett* 6865).

Spain, E Mediterranean, N Africa, Egypt, Sinai, Turkey, Caucasus, Iran, Arabia, Turkmenia, Afghanistan, Pakistan, China (Xinjiang and Kansu).

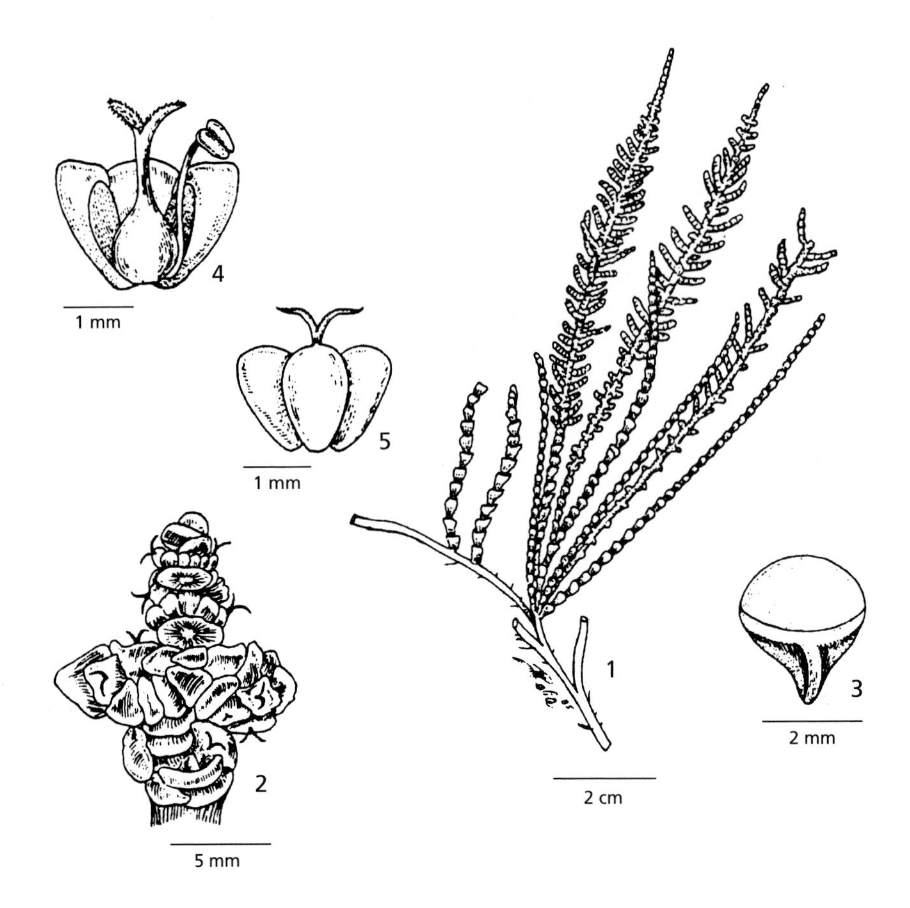

Fig. 82. **Halocnemum strobilaceum**. 1, habit; 2, part of inflorescence; 3, bract; 4, hermaphrodite flower; 5, female flower. Reproduced with permission from Flora of Pakistan 204: f. 18. 2003. Drawn by M. Rafiq

14. **SALICORNIA** L.

Sp. Pl.: 3 (1753); Gen. Pl. ed. 5: 4 (1754)

Halophytic glabrous annuals with succulent, jointed stems and branches; branches opposite. Leaves connate to form a cup around stem and branches, lamina obsolete. Inflorescence a spike, terminal on stem and branches, formed of axillary, opposite, (1–)3-flowered cymes sunk in a pair of connate, fleshy bracts, flowers fused to each other and the bracts. Perianth cup-like almost closed, indistinctly 3-lobed, formed of 3 almost completely united segments, apex flattened and usually deltoid or rhomboidal. Stamens 1–2, anthers exserted, connective not appendiculate. Ovary ovoid; style short, included in perianth, with 2 filiform stigmas exserted through the central orifice. Fruit included, membranous; seeds brown, with hooked hair-like outgrowths; perisperm absent, radicle incumbent.

Possibly about 15 species, occurring in saline areas in both hemispheres.

The plant is now much sought after by gourmet chefs as an accompaniment to cutting-edge cuisine.

Salicornia (possibly from Celtic *sal*, near to and Lat. *corni*, of horn. This derivation is speculative, as the first syllable could be a reference to salt. According to Fl. SSSR vol. 6, Linnaeus may have based the name on *salicor*, the Fr. name for this plant.

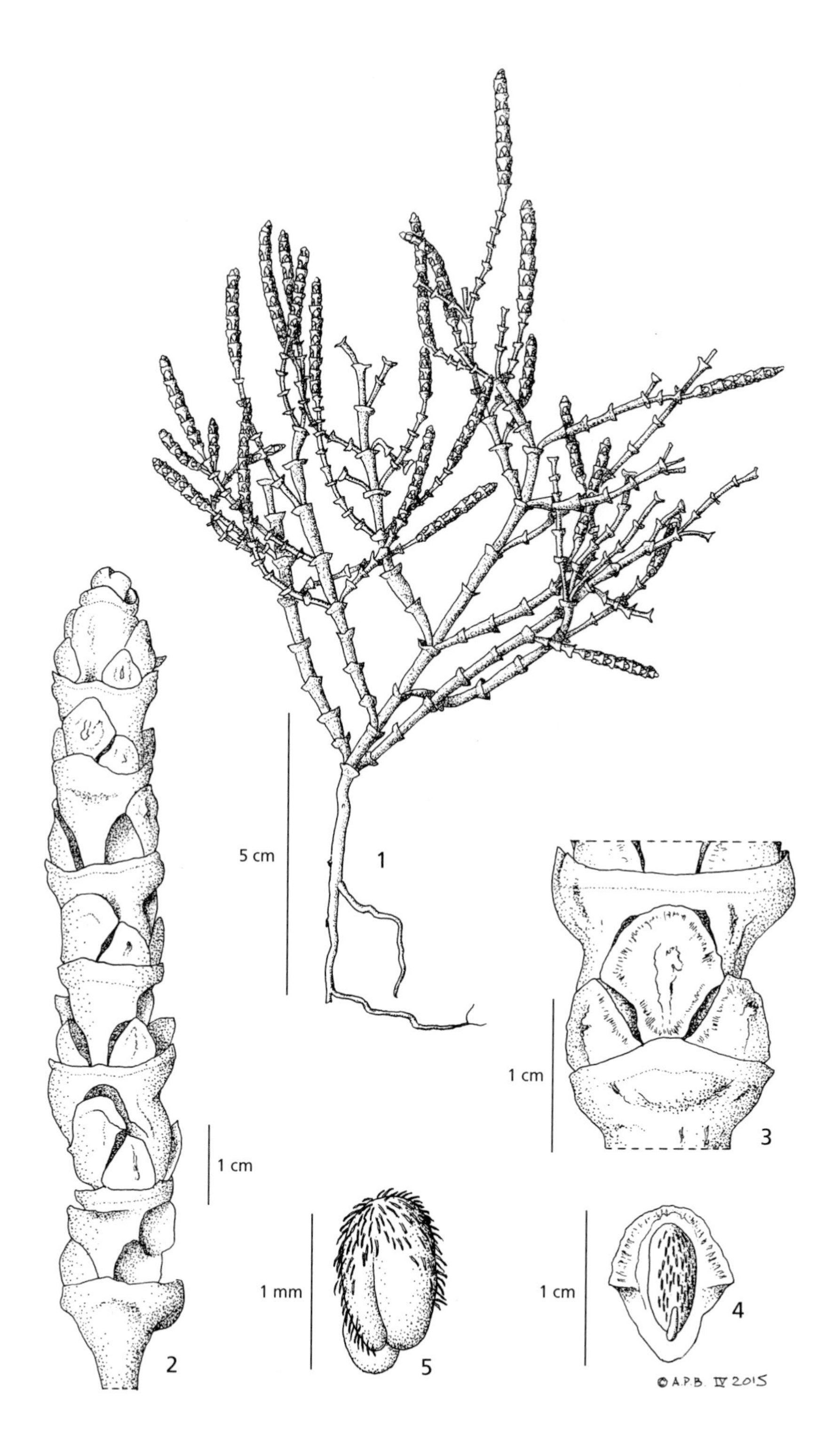

Fig. 83. **Salicornia perennans**. 1, habit; 2, inflorescence at fruiting stage; 3, inflorescence with flowers detail; 4, underside of flower showing seed; 5, seed. 1 from *Guest, Rawi & Rechinger* 16181; 2–5 from *Safwat & Barkley* 3959. © Drawn by A.P. Brown.

Salicornia perennans *Willd.*, Sp. Pl. 1: 24 (1797); Sukhorukov & Akopyan, Comp. Chenop. Cauc.: 44 (2013).

S. europaea auctt. non L., Sp. Pl. 3 (1753); Zohary, Fl. Palaest. 1: 155 (1966); Boulos in Fl. Arab. Penins. Socotra 1: 253 (1996); Hedge in Fl. Iranica 172: 130 (1997); Boulos, Fl. Egypt 1: 110 (1999); Assadi in Fl. Iran 38: 180 (2001).
S. herbacea auct. non L., Rech.f., Fl. Lowland Iraq: 192 (1964).

Erect annual herb, 10–30(–40) cm, usually considerably branched and bushy from base upwards, branches mostly spreading at an angle of 44–60°, the lowest as long as main stem in the best-developed forms. Plant reddish-green in fruit, tips of branches more distinctly red. Mature internodes of main stem and branches mostly 1–1.2 cm. Bracts with a narrow scarious border. Inflorescences obtuse, fertile segments 5–12(–16). Cymes 3-flowered visible part of lateral and central flowers subequal in size; base of central flower forming an angle of ≤ 90°; central flower ± 1.5 mm, larger than laterals. Stamen solitary, anthers ± 0.5 mm. Stigmas 0.8–1 mm. Seeds 1–1.2 mm, compressed-ovoid, brown, with thin hair-like outgrowths. Fig. 83: 1–5.

The taxonomy of the genus *Salicornia* is critical, and not at all well known even in W Europe, where it has been studied fairly effectively. In the Middle East it has scarcely been investigated at all. Thus the identity of the Iraqi plants partially remains questionable in respect to the presence of the morphologically cryptic species, and other taxa, especially *S. persica* Akhani can be encountered in our area.

HAB. Salt flats and salt marshes; alt. 0–10 m; fl. Sep.–Nov.
DISTRIB. Rather local in desert and subdesert areas of Iraq: **DLJ**: Thirthar, *Safwat & Barkley* 3959!; 40 km N of Baiji, *E. & F. Barkley* 3689!; **DWD**: nr Shithatha, *Rawi & Serkahia* 16304!, *Rawi* 26916!; Hit, *Guest* 3546!; Bahr al-Milh, 10 km N of Shithatha, *Guest, Rawi & Rechinger* 16181!, *Rawi* 26921!; W. shore of Bahr al-Milh, *Rechinger* 8346! **DSD**: nr Karbala, *Rechinger, Guest & Rawi* s.n. (E!). **LCA**: nr Musaiyib, on road to Karbala, *Guest, Rawi & Rechinger* 16148!

?HUMMAIR or ?HUMMAIDH (Guest *et al.* 16148): "eaten only by camels"; ?THILAIDH (anon. 3546, Hit): "eaten by camels".

Europe, W Asia.

SUBFAMILY SUAEDOIDEAE Ulbr.

15. **SUAEDA** Forssk. ex Scop.,

Intr. Hist. Nat.: 333 (1777) nomen conserv.

Dondia Adans., Fam. Pl. 2: 261 (1763); *Schanginia* C.A. Mey. in Ledeb., Fl. Altaica 1: 394 (1829); *Schoberia* C.A. Mey. in Ledeb., Ic. Pl. Fl. Ross. 1: 11 (1829); *Chenopodina* Moq. in DC., Prodr. 13(2): 159 (1849); *Brezia* Moq. in DC., Prodr. 13(2): 167 (1849); *Belovia* Moq. in DC., Prodr. 13(2): 168 (1849).

Halophytic plants, annuals to tall shrubs or trees, glabrous or with younger parts finely hairy. Leaves simple, alternate, fleshy, terete or semiterete, obtuse to acute or with a short terminal bristle. Inflorescences spiciform and ± leafy, rarely forming a terminal panicle, consisting of 1- to few-flowered axillary clusters (more rarely clusters set on petioles of inflorescence-leaves), subtended by scarious hyaline bracteoles. Flowers hermaphrodite, or plants monoecious or diocious. Perianth segments 4–5, slightly or considerably fused below, somewhat accrescent but not indurate in fruit, ± fleshy, often cucullate. Stamens 5; connective not appendiculate; staminodes present or not in female flowers of monoecious species. Ovary free or more rarely immersed in receptacular tissue; style very short or absent, with 2–4 stigmas. Fruit ± compressed with a delicate pericarp; seeds vertical or horizontal (sometimes in the same species: spatial heterospermy), in annual species showing seasonal variation (seeds with firm, shining black testa in summer, seeds with more delicate, dull, paler testa in autumn); perisperm scanty or absent; embryo spiral.

One of the most difficult genera in the family for the taxonomist, owing to the fact that many of the available characters are difficult to observe in the herbarium.

A cosmopolitan genus of about 100 species; seven species in Iraq.

Suaeda (from Ar. *suida* or *soued*, the Egyptian vernacular name of this plant, according to Nehmé 2000 and Provençal 2010).

It should be noted that all of the binomials in *Suaeda* employed by Forsskål (1775), although accompanied by full and careful descriptions, are illegitimate since no generic description of *Suaeda* was given, the genus itself dating from 1777, as cited above.

1. Monoecious tree of 2–4 m, rarely entirely female; female flowers with or without staminodes, male flowers with a columnar pistillode with a round, flattened apex and minute stylar remnants .... 3. *S. monoica*
   Annuals or shrubs, much shorter; flowers hermaphrodite, or hermaphrodite and female, if ovary with a central column then this bearing perfect styles and fertile ovary 2
2. Shrubby, ± blackening when dried; stems often with scattered hairs............... 3
   Annuals 4
3. Leaves flattened .... 2. *S. vermiculata*
   Leaves semiterete .... 1. *S. fruticosa*
4. Ovary immersed in and adnate to perianth base, seeds thus inferior to perianth segments; ovary produced into a style-like column from the ± concave top of which arise the 2–3 perfect stigmas .... 7. *S. aegyptiaca*
   Ovary free, seeds ripened amid perianth segments; ovary without a style-like column above, at most a very short neck produced.................. 5
5. Flower clusters slightly distant from leaf-axils due to concrescence of leaf basis and floral axis. Vegetative leaves thin, to 1(1.5) mm wide when dried .... 4. *S. altissima*
   Flower clusters axillary. At least older leaves wider, 2–4 mm 6
6. Leaves homogenous when dry, without whitish margin, hypodermis not present at cross-sections of fresh leaves. Anthers small (c. 0.25 mm), roundish. Dark seeds with reticulate ultrasculpture .... 5. *S. anatolica*
   Leaves with whitish margin, hypodermis present at cross-sections of fresh leaves; anthers larger, oblong; dark seeds glossy and smooth . .6. *S. carnosissima*

1. **Suaeda fruticosa** *Forssk. ex J.F.Gmel.*, Syst. Nat. ed. 13, 2: 503 (1791); Zohary, Fl. Palaest. 1: 159 (1966); Mandaville, Fl. Eeastern Saudi Arabia: 83 (1990); Akhani & Podlech in Fl. Iranica 172: 148 (1997); Assadi in Fl. Iran 38: 187 (2001); Freitag et al. in Fl. Pak. 204: 107 (2001).

Erect or more rarely decumbent shrub up to 2 m. Stem and branches glabrous except occasionally when very young, older cortex whitish to pale yellow and cracking, youngest parts purplish and glaucous-pruinose, soon smooth and green. Leaves fleshy, semiterete or terete, 5–25 × 1–2 mm, decreasing rapidly in the inflorescence. Inflorescence spiciform, of mostly (1–)3–5-flowered axillary clusters of hermaphrodite or female greenish flowers. Perianth segments shortly fused to each other below, (1–)1–1.5 mm, gibbous dorsally, concave within, narrowly pale-margined. Ovary pyriform, free except at extreme base, slightly contracted below apical rim; stigmas 3, filiform, densely papillose, ± 0.75 mm. Stamens ± 1 mm, finally exserted; anthers shortly oblong. Pericarp easily ruptured; seed black, 1.5 mm, glossy; embryo vertical.

HAB. Saline soils; deserts and desert wadis; alt. 0–200 m; fl. & fr. Jul.–Oct.
DISTRIB. Saline plains in the deserts. At least majority of specimens previously identified as *S. vermiculata* belong hereto: **DWD**: N of Shithatha, *Rechinger, Guest & Rawi* 24! *Rechinger* 8336 (W!); Bahr al-Milh, *Barkley & Safat* 3817 (W!); **DSD**: between Ur and Al Busaiya, *Rechinger* 8195!; Umm Qasr, *Guest, Rawi & Rechinger* 16881!; Chankula river, *Eig & Zohary* 2008 (HUJ!); **LEA**: N bank of river opposite Ashar, *Gillett* 10039!; Kut al-Imara, *Rechinger* 9106 (W!). **LCA**: Diwaniya, *Bharucha, Rawi & Tikr.* 29234! *al-Ani & Barkley* 8880 (W!); Darraji, *Guest* 2395! **LSM**: Amara, *Hazim & Nuri* 30611! *al-Ani & Barkley* 8924 (W!); nr Azair, *Guest, Rawi & Rechinger* 17383! **LBA**: Basra, *Haines* 825! *Rechinger* 8424 (W!).

The late Paul Aellen named *Hadač, Haines & Waleed* W.1887 as "*Suaeda* cf. *catenulata* Bornm. & Gauba" and designated *Guest, Rawi & Rechinger* 16710A as type of an unpublished new species *S. rawii.*

Egypt, Palestine, Saudi Arabia, Yemen, Afghanistan, Pakistan, N Africa.

2. **Suaeda vermiculata** *Forssk. ex J.F. Gmel.*, Syst. Nat. ed. 13, 2: 503 (1791); Rech.f., Fl. Lowland Iraq: 193 (1964); Zohary, Fl. Palaest.1: 160 (1966); Boulos in Fl. Arab. Penins. Socotra 1: 257 (1996); Boulos, Fl. Egypt 1: 113 (1999); Mozaffarian, Trees & Shrubs of Iran: 175 (2005); Assadi in Fl. Iran 38: 189 (2001).

S. mesopotamica Eig in Palest. J. Bot., Jerusalem ser., 3: 127 (1945). —Type: Iraq (Syrian desert): 130 km W of Rutbah plain, 11.10.1933, *A. Eig* 1045 (holo. HUJ-20778!); Rech.f., Fl. Lowland Iraq: 194 (1964).

Much-branched decumbent to erect shrub, 0.3–1.2 m, cortex of branches whitish or greyish to stramineous, striate; young twigs ± densely clothed with short, multicellular hairs, striate or sulcate, stem and branches glabrescent. Leaves very fleshy, rotund to narrowly linear-oblong, flat to concave on upper surface, strongly convex below, those of main stem and branches below inflorescence 5–18 × 3–5 mm, decreasing rapidly in the inflorescence. Inflorescence spiciform, of mostly (1–)2–4(–10)-flowered axillary clusters of hermaphrodite greenish yellow or purplish flowers. Perianth segments only shortly fused to each other, (0.75–)1.25–1.5 mm, thick and fleshy, convex dorsally, concave and darker within, tips incurved and ± cucullate, ± pale-margined. Ovary urceolate-pyriform, only slightly fused to receptacular tissue; stigmas (2–)3–4, 0.75–1 mm. Stamens 2 mm, finally exserted; anthers shortly oblong, 0.75–1 mm. Fruits membranous, ± 1.5 mm, included in the frequently somewhat accrescent perianth; seeds vertical, roundish, compressed, black, shining, very faintly reticulate. Fig. 84: 1–3.

HAB. Fields, waste ground, river banks, usually with saline soils, alt. 0–50 m; fl. Apr.–Jun., fr. Jun.–Dec.
DISTRIB. Common in the deserts and on the alluvial plain: **DWD**: 3 km S of Shithatha, *Rawi & Chakravarty* 29786!; 3 km S of Rahalliya, *Chakravarty & Rawi* 29806!; Bahr al-Milh, *Guest, Rawi & Rechinger* 16166! **DSD**: Salaibiyat al Hamr 35 km S by W of Ur, *Guest, Rawi & Rechinger* 16044A!; 10 km NW of Rumaitha, *Rawi & Serkahia* 16256!; nr Najaf, *Rawi* 19974!; nr Zubair, *Rawi & Gillett* 6012! **LEA**: Ma'qil, *Haines* W. 1887! **LCA**: Nasiriya, *Civil Vet. Surgeon* 2531A!, *Rawi* 19967!, *Rawi & Serkahia* 16306!; 30 km W of Darraji between Nasiriya and Samawa, *Rawi & Alkas* 16235!; **LSM**: Tigris near Esra's tomb (Al Azair), *Evans* s.n.!; 30 km E of Amara, *Bharucha, Rawi & Tikriti* 29294!; *Guest* 1635!; Garma Bani Sa'd, Rawi 16630! **LBA**: Basra, *Guest, Rawi & Rechinger* 16710A!; Basra, *Haines* W.825!

Suaeda vermiculata is closely related to *S. fruticosa*, and possibly conspecific; the identification of the material from Iraq is done using of only one character: flatteness degree of the leaves.

HAMUDH – eaten by sheep and camels (*Civil Vet. Surgeon* 2531); TAHĀMA (*Guest et al.* 16710A) ?TAKHME (*Gillett* 10039), who noted that the plant is eaten by sheep, but not by cattle on account of its salt content; when young, GŌGALLA (*Guest et al.* 16710A, Basra).

Egypt, Palestine, Sinai, Iran, Afghanistan, Arabian Gulf coast from Kuwait to Oman, Saudi Arabia, Yemen, N Africa (Libya to Morocco), Macaronesia (Canary Is.), NE tropical Africa (Eritrea to Somalia).

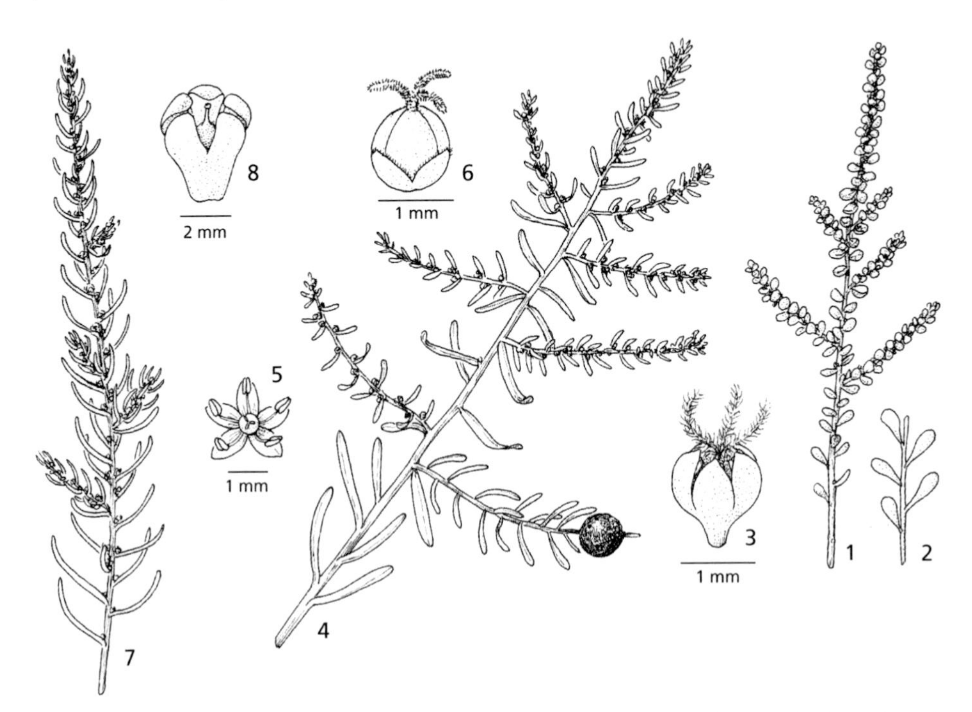

Fig. 84. **Suaeda vermiculata**. 1, habit portion flowering and fruiting branch; 2, leaves; 3, flower. **Suaeda monoica**. 4, habit portion flowering and fruiting branch; 5, male flower; 6, female flower. **Suaeda aegyptiaca**. 7, habit portion flowering and fruiting branch; 8, female flower. Reproduced with permission from Flora of Egypt 1: plate 22. Drawn by M. Tebbs.

3. **Suaeda monoica** *Forssk. ex J.F. Gmel.*, Syst. Nat. ed. 13, 2: 503 (1791); Boulos in Fl. Arab. Penins. Socotra 1: 256 (1996); Hedge in Fl. Iranica 172: 70 (1997); Akhani & Podlech in Fl. Iranica 172: 152 (1997); Boulos, Fl. Egypt 1: 112 (1999); Assadi in Fl. Iran 38: 187 (2001); Freitag et al. in Fl. Pak. 204: 112 (2001); Ghazanfar, Fl. Oman 1: 44 (2003).

Large shrub, usually 2–4 m in SW Asia, to ± 6 m in the tropics, much-branched, glabrous (or very faintly pubescent when young), cortex of branches pale green to white or stramineous, ridged. Leaves linear to linear-oblong, obtuse to subacute, those of main stem and lower parts of branches 1.2–3(–3.5) × 2–3 mm, decreasing above and in inflorescence. Inflorescence monoecious or rarely female only, of 1–7-flowered axillary clusters, flowers green. Male flowers with perianth segments fused in about the basal $^1/_3$, ± 2 mm long, incurved and cucullate above; stamens 5, included, ± 2 mm; anthers oblong, ± 1 mm; columnar rudimentary ovary present, 1 mm, with a round, flattened apex and minute stylar remnants. Female flowers 1.5–2 mm, perianth segments fused in basal $^1/_3$, minute staminodes sometimes present; ovary free except at extreme base, compressed-ovoid; stigmas 3–4, 0.8–1.4 mm. Fruit membranous, 1.5–2 mm, included; seeds vertical, compressed-ovoid to subrotund, 1.25–1.75 mm, black, shining and almost smooth. Fig. 84: 4–6.

HAB. saline waste land; alt. ± 6 m; not in flower or fruit.
DISTRIB. Very rare; only recorded once in the Southern Marshes district: **LSM**: half-way between Qurna and Basra, *Rawi* 15036!

Egypt, Palestine, Syria, Arabia, Pakistan, NE Africa; also reported from W Africa (Cape Verde), but it seems to be doubtful.

4. **Suaeda altissima** (*L.*) *Pall.*, Illustr. Pl.: 49 (1803); Akhani & Podlech, Fl. Iranica 172: 139 (1997); Assadi in Fl. Iran 38: 207 (2001); Sukhorukov & Akopyan, Comp. Chenop. Cauc.: 47 (2013).

Chenopodium altissimum L., Sp. Pl. ed. 1: 221 (1753); *Schanginia altissima* (L.) C.A. Mey., Verz. Pflanz. Cauc.: 159 (1831); *Salsola altissima* (L.) L., Sp. Pl. ed. 2: 324 (1762)

Annual herb, (0.2–)0.4–2 m, much-branched from base upwards, glabrous or youngest parts with short evanescent hairs, frequently blackening on drying; stem and branches terete, striate, frequently ± reddish or purplish. Leaves filiform, terete, obtuse or subacute, those of main stem mostly 10–30 × 0.5–1 mm, those of branches gradually reducing, inflorescence leaves small, ascending to incurved, slightly to rather considerably exceeding flower-clusters. Flowers hermaphrodite or female, in dense mostly 3–8-flowered clusters set on leaf-petiole, both clusters and individual flowers shortly stalked, forming an elongate spiciform inflorescence along stem and branches. Perianth segments free or fused below, 1–1.25 mm, somewhat fleshy, convex dorsally, incurved and somewhat cucullate above, green with pale margins. Ovary urceolate, only slightly fused to receptacular tissue; stigmas (2–)3(–4), short, 0.25 mm. Stamens 1.5 mm, anthers exserted, shortly oblong, 0.7 mm. Fruits membranous, 1–1.5 mm; seeds vertical or also horizontal (spatial heterospermy), round, compressed, black, shining, faintly reticulate.

HAB. & DISTRIB. Probably not common in Iraq. **LCA**: "Baghdad aestate in ruderatis", *Noë* 1099!; Daltawa, *Guest* 2451!; Hillah, *Graham* 271 (BM!); "Ad ripam Tigridis, in deserto", May 1851, *Noë* 123!; "am Tigris", May 1851, *Noë* 999!;

In vegetative stage *S. altissima* is similar to the thin-leaved forms of *S. aegyptiaca*, but the leaves are straight (*S. aegyptiaca* have often inwardly curving leaves).

S Europe (from Spain to S Russia), Turkey, Caucasus, Iran, Afghanistan, Jordan, S Siberia, C Asia.

5. **Suaeda anatolica** (*Aellen*) *Sukhor.*, Chenop. Carp. & Syst.: 348 (2014).

Suaeda prostrata subsp. *anatolica* Aellen, Not. Roy. Bot. Gard. Edinb. 28: 32 (1967); *Suaeda maritima* auct. non (*L.*) *Dum.*, Fl. Belg.: 22 (1827); Rech.f., Fl. Lowland Iraq: 183 (1964).

Glabrous annual herb, 15–75 cm, green or ± purplish-suffused with few to numerous simple or branched, stems. Leaves fleshy, linear, semiterete or upper leaves ± flattened on upper surface and rounded below, mucronate at tip, those of main stems (4–)7–25 × 0.5–2(–3) mm, rapidly reducing in the inflorescences, uppermost usually about twice as long as flower-clusters. Flowers solitary or more frequently in clusters of 2–4, hermaphrodite or rarely female. Perianth obtuse to subacute, free or almost so, green with a narrow pale margins, deeply concave; tube short; at fruiting stage star-shaped, one of the segments forms bigger horn-shaped outgrowth, the other segments with shorter equal outgrowths. Ovary free except at extreme base, broadly pyriform, slightly narrowed above to a short neck; stigmas 2, filiform, purplish, ± 0.6 mm, attached to outer upper edge of neck. Stamens ± 1 mm, anthers finally slightly exserted, very small, ± 0.25 mm, roundish. Fruit very thin-walled and delicate, included; seeds horizontal, compressed-rotund, 0.8–1 mm, black or brownish-black, faintly to conspicuously reticulate.

Hab. Saline depressions and ditches; alt. not recorded; fl. & fr. Oct.–Nov.
Distrib. Occasional in the alluvial plains. **dwd**: Shithatha, *Agnew & Haines* W.2113! **lca**: Mahmudiya nr Baghdad, *Haines* W.2134 (K!, E!).

HAMUDH, "grazed by sheep and more specifically by camels. Local herdsmen say that the camels get scab if the plant is absent from their diet" (Nasiriya, *Abd-ar-Rizaq Barbuti* 2531);

Turkey, Syria, Iraq.

6. **Suaeda carnosissima** *Post* in Fl. Syr.: 687 (1896); Aellen in Fl. Turk. 2: 327 (1967).

[*S. setigera* (non (DC.) Moq.) Boiss. Fl. Orient. 4: 942 (1879); et auctt. al., p.p.]; Zohary, Fl. Palaest. 1: 161 (1966); *S. birandii* Aellen, nomen in sched.; *S. guestii* Aellen, nomen in sched.

Annual herb, 17–35 cm, considerably branched from base upwards, glabrous except in younger parts of inflorescence axis and younger perianths, where whitish floccose hairs are frequently present; stem and branches whitish to stramineous, somewhat angled. Leaves on main stems and branches lanceolate or oblong-lanceolate, 4–12 × 2–3 mm, obtuse to acute and mucronate, very fleshy with outer cells pale and vesicular, green centrally, much-shrivelled on drying; floral leaves gradually to abruptly diminishing, uppermost shorter than to exceeding flower-clusters. Inflorescence spiciform, of (1–)2–5(–8)-flowered separated clusters. Bracteoles small, hyaline, deltoid to deltoid-ovate, 0.5–0.75 mm, dentate, ± concealed by the tightly set, sessile flower clusters. Perianth segments shortly connate below, gibbous dorsally, yellowish-green to green, very fleshy, with broad white margin and a dorsal keel which in all but the largest flowers is produced into a blunt or ± falcate appendage which overtops the segment. Ovary compressed-ovoid, free except at extreme base, with a very short funnel-shaped apex in which are set the 2–3, filiform, ± 0.75 mm, stigmas. Stamens ± 1.75 mm, included; anthers shortly oblong, ± 0.75 mm. Fruit very thin-walled, 1–1.25 mm, concealed by incurved perianth segments; seeds little compressed when ripe, ± 1 mm, black, very shining, absolutely smooth under a hand lens, vertical or occasionally horizontal.

Hab. Waste places, *hasswa* desert, fallow fields; alt. 0–300 m. fl. Aug.–Oct., fr. Oct.–Dec.
Distrib. Occasional in plains and deserts: **dwd**: Shithatha, *Guest, Rawi & Rechinger* 16173! **dsd**: 10 km NW of Rumaitha, *Rawi & Serkahia* 16257!; on Kerbala road, *Haines* 2119! **lca**: Baghdad, *Noë* 311 (as *S. setigera*)!; Nasiriya, Rawi 16241!; Hilla, *E. Berkley, Haddad & F. Barkley* 3705!; **lsm**: near Amara, *Evans* s.n.!; Amara, *Evans* s.n.!; Qal'at Sali, *Rawi & Tikrity* 24990!

The type of this species, in Beirut, has not been seen, it is reported that no type material is in AUB, Lebanon (Nada Sinnu, pers. comm.). The Iraqi material has been named by comparison with material cited by Aellen in *Flora of Turkey*, and one specimen (Amara, *Evans* s.n.) was determined as *S. carnosissima* by Aellen himself. *Suaeda cochlearifolia* Wol. in Denkschr. Kaiserl. Akad. Wiss., Math.-Naturwiss. Kl. 51 (2): 275 (1886) described from North Iran (Holotype: Persia borealis, in vinetio prope Chanabad, 27.07.1882, *Th. Pichler*, W–13699!) cannot be synonymized with *S. carnosissima* or *S. acuminata* undertaken in some treatments.

Turkey, Syria.

7. **Suaeda aegyptiaca** (*Hasselq.*) *Zohary* in J. Linn. Soc. Bot. 55: 635 (1957); Zohary, Fl. Palaest. 1: 161 (1966); Boulos in Fl. Arab. Penins. Socotra 1: 256 (1996); Akhani & Podlech, Fl. Iranica 172: 135 (1997); Boulos, Fl. Egypt 1: 112 (1999); Assadi in Fl. Iran 38: 200 (2001); Freitag et al. in Fl. Pak. 204: 118 (2001); Ghazanfar, Fl. Oman 1: 43 (2003).

Chenopodium aegyptiacum Hasselq., It. Palaest.: 460 (1757); *Suaeda baccata* Forssk. ex J.F. Gmel., Syst. Nat. ed. 13, 2: 503 (1791); *Schanginia baccata* (Forssk.) Moq., Chenop. Monogr. Enum.: 119 (1840); [*Schanginia hortensis* (non (Forssk.) Moq.) Boiss., Fl. Orient. 4: 945 (1878), saltem quoad pl. iraq.]; *Schanginia aegyptiaca* (Hasselq.) Aellen in Rech.f., Fl. Lowland Iraq: 195 (1964).

Annual herb, 15–60 cm tall, much-branched from base upwards, procumbent to erect, glabrous or young parts with fine, floccose, white hairs and later glabrescent; stem and branches white, terete, former up to 1 cm in diameter in large plants. Young plants rather weak, with fleshy filiform leaves mostly c. 1–2 × 0.1 cm. Leaves on mature stems c. 10–20(–30) × 1–3 mm, straight or ± inwardly curving, obtuse, fleshy; branch and floral leaves diminishing in size, the latter often rapidly so. Flowers hermaphrodite, yellowish-green, sessile or very shortly pedicellate, in a spiciform inflorescence of few- or many-flowered ± globular clusters which become increasingly distant below. Perianth tube adnate to ovary, turbinate, 0.5–0.75 mm; tepals 1.5–1.75 mm, green with scarious margins, incurved, very concave, cucullate at apex and ± gibbous dorsally, elliptic-oblong, with a slender acute tip at least when in flower. Ovary tapered into a shortly cylindrical, 0.7–1.25 mm beak, which is truncate-concave at tip, the 2–3 filiform, 1–1.5 mm stigmas arising from the concavity. Stamens 1.5 mm, ± included; anthers shortly oblong, 1 mm. Fruit very thin-walled, 1 mm, immersed in the somewhat accrescent to very spongy receptacle; fruiting perianth lobes somewhat to very strongly gibbous-inflated. Seeds horizontal across top of receptacle or more commonly vertical and deeply embedded within it, 1 mm, black and shining, with a distinct though faint reticulate pattern. Fig. 84: 7–8.

Hab. Salt ground, waste and uncultivated land; alt. 15–100 m; fl. May–Sep., fr. Sep.–Nov.
Distrib. Plains and deserts: **MJS**: El Chattunije, *Handel-Mazzetti* 1635 (W!); **FUJ**: Qaiyara, *Handel-Mazzetti* 3112 (W!); **FPF**: 5 km W of Mandali, *Rechinger* 13398 (W!); **DLJ**: nr Balad, *Baltaxe* 13478!; Sumaicha, *Rechinger* 8091 (W!); by spring in Wadi Thirthar, *Guest* 3550!; Thirthar Lake, *Barkley & Abbas* 3628 (W!); **DWD**: Bahr al-Milh, *Rawi & Rechinger* 16180! **DSD**: Karbala, *Rechinger* 8309 (W!); SW of Samawa, *Rechinger* 8162B (W!); Zubair, *Rechinger* 8634 (W!); 18 km W by S of Samawa, *Guest, Rawi & Rechinger* 16020!; nr Karbala, *Rechinger* 8309! **LEA**: Kut al-Imara, *Lazar* 3487!; Dabouni nr Kut, *Guest* 191!; Ba'quba bridge, *Rechinger & Khudairi* 32!; Kut al-Imara, *Gillett* 6590!; **LCA**: Nasiriya, *Abd-ar-Rizaq Barbuti* 2531!; Abu Ghraib, *Haines* W.66!; *al Jumaily* 389!; Hillah to Diwaniya, *Guest* 3169!; Diwaniya, *Gillett* 9925! *al-Ani & Barkley* 8879 (W!); **LSM**: Amara, 26.10.1917, *Evans* s.n.!; by Tigris above Amara, *Evans* s.n.!; Musaiyib nr Karbala, *Rechinger* 8309!; 30 km E of Amara, *Bharucha, Rawi & Tikriti* 29291!; 10 km N of Amara, *Eig & Zohary* s.n. (HUJ!); Garma Bani Sa'd, *Rawi* 16630 (W!); **LSM/LEA**: Shatt Al-Arab, *Abdin* 7! **LBA**: Basra, *Graham* 250 (BM!).

Vernacular names: TARTAI', *Lazar* 3487 (Kut), HAMUDH (*Guest* 3550 from Wadi Thirthar), SHAWATĀN (*Guest et al.* 16020 from nr Samawa), SHUATAN (*Gillett* 9925 from Diwaniya); known as a fodder plant.

Cyprus, Syria, Palestine, Iran, Yemen, Arabian Gulf from Kuwait to Oman, NE Africa (Egypt, Ethiopia, Somalia).

Excluded species:

Suaeda maritima auct. non L.: Rech.f., Fl. Lowland Iraq: 194 (1964). The distribution of this species lies in the Mediterranean and Atlantic Europe. All material under this name is referable to S. *anatolica.*

Suaeda salsa auct., non (L.) Pall. The occurrence of this species is not documented in Iraq, but belongs to other species (*S. aegyptiaca, S. prostrata* subsp. *anatolica*). The southern range border of *S. salsa* is the Black Sea area and Caucasus; *S. salsa* is known in the steppes and semi-deserts of Eurasia.

16. **BIENERTIA** Bunge ex Boiss.

Boiss., Fl. Orient. 4: 945 (1879)

Fleshy glabrous annual with erect or ascending stems becoming woody at base. Leaves fleshy, sessile, terete or flattened, glabrous or covered with scattered long-stalked bladder hairs. Inflorescence paniculate. Flowers borne on axillary shoots bearing hermaphrodite or female flowers; bracts present, minute. Perianth segments 5, fleshy, fused to half the length, becoming membranous, connivent and developing a circular wing. Stamens 5. Stigmas (2–)3, free. Fruit flattened; first seeds shiny, black, later seeds dull, brown. Embryo horizontal, coiled, perisperm lacking.

The genus contains three species, two of which have been described only recently from Iran and adjacent countries.

Bienertia (commemorating Dr Theophil Bienert (1833–1873), the Estonian botanist who was an early explorer with Alexander Bunge of the flora of Iran.)

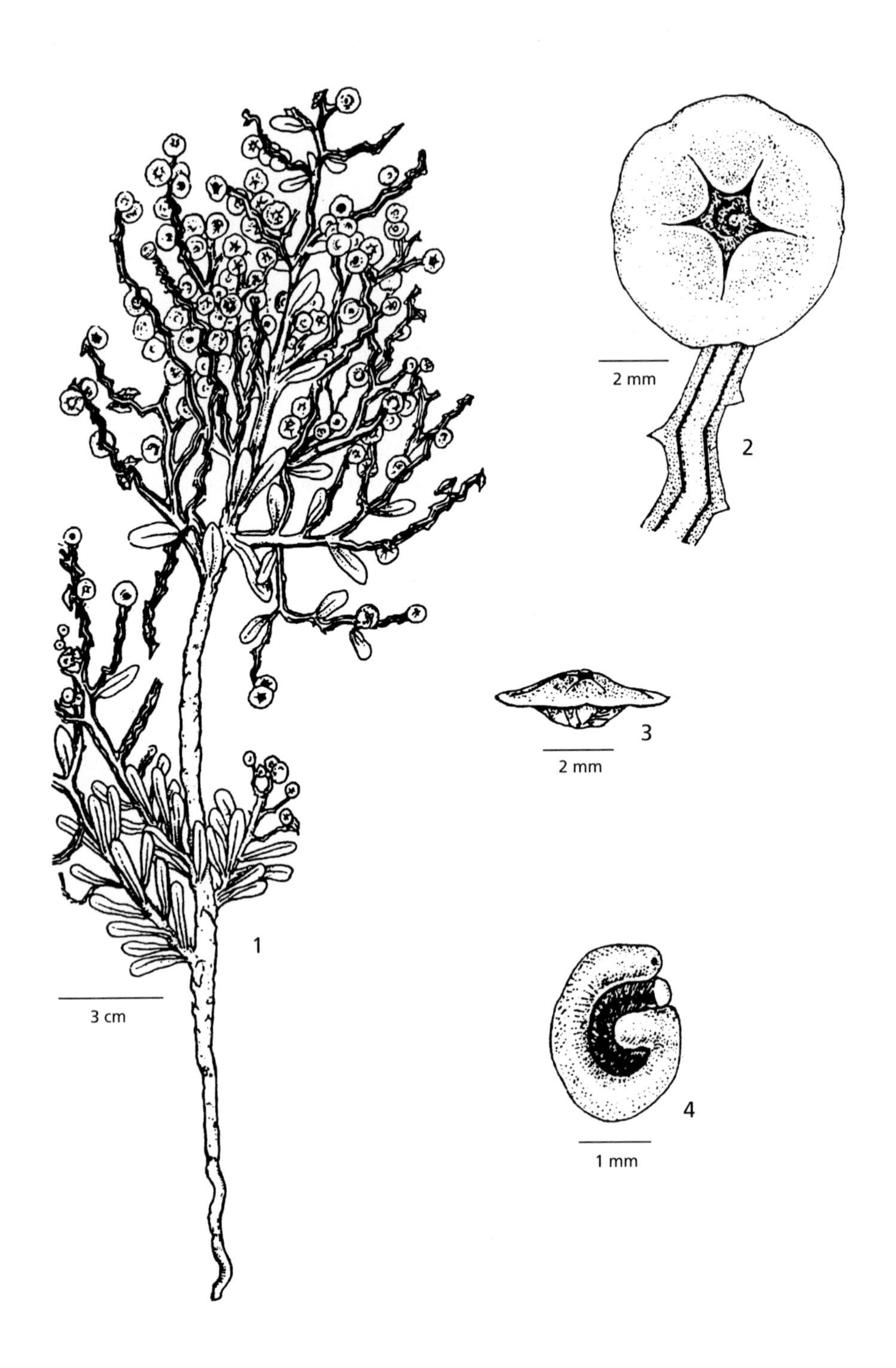

Fig. 85. **Bienertia sinuspersici**. 1, habit; 2, fruiting perianth; 3, fruiting perianth lateral view; 4, seed. Reproduced with permission from Flora of Pakistan 204: f. 24. 2003. Drawn by M. Rafiq.

1. **Bienertia sinuspersici** *Akhani* in Syst. Bot. 30 (2): 291 (2005).

B. cycloptera auct. non *Bunge ex Boiss.*, Rech.f., Fl. Lowland Iraq: 196 (1964); Boulos in Fl. Arab. Penins. Socotra 1: 259 (1996); Omar, Vegetation of Kuwait: 51 (2000); Freitag et al. in Fl. Pak. 204: 126 (2001).
Suaeda guestii Aellen in sched. (G!).

Fleshy annual, to 100(–150) cm, green or yellowish-green. Stems branched almost from base, erect or ascending with often drooping lateral shoots. Leaves glaucous, linear-oblong, 20–40 × 5–10 mm, fleshy. Perianth segments membranous, incurved, unwinged at flowering time; converging in fruit and expanding into a circular wing 7–7.5 mm across. Fruit adnate to perianth segments, flattened, dark bluish-green and somewhat translucent. Seeds black and brown, 1.8–2.3 mm. Fig. 85: 1–4.

Hab. Saline flats, gravelly soil, 2–120 m; fl. Oct., fr. Nov.–Dec.
Distrib. Widespread in desert regions of Iraq: **fpf**: 4 km from Mandali, *Al-Kaisi & K. Hamad* 43564!; **dwd**: ± 50 km NE of Ramadi, *Guest, Rawi & Nuri* 13550!; nr Shithatha, *Rawi & Serkahia* 16300!, *Rawi* 26925!; between Shithatha and Rahhaliya, *Rawi & Alkas* 16139!; Bahr al-Milh, E of Shithatha, *Guest, Rawi & Rechinger* 16185, *Rechinger* 8333!; Rahhaliya, *Weinert, Al-Mousawi* s.n. (E!). **dsd**: 27 km W by S of Samawa, *Guest, Rawi & Rechinger* 16022!; between Busaiya and Nasiriya, *Rawi & Alizzi* 32431!; Salaibiyat 25 km S of Ur, *Guest, Rawi & Rechinger* 16037!, *Rechinger* 8177!; W of Samawa, *F. Barkley, F. Safat & E. Barkley* 3727!; Ur, *Schwabe* s.n. (B!). **lea**: 5 km S of Abul-Khasib, *Khatib & Alizzi* 32648!; **lca**: 30 km W of Darraji between Nasiriya and Samawa, *Rawi & Alkas* 6234!; Sheikh Saad, *Rawi & Haddad* 25528! **lba**: 40 km SE of Basra towards Fao, *Rawi* 25894!; Abu Khasib, 22 km S of Basra, *Juma'a Brahim* 6203!; **lba/dsd**: between Basra and Zubair, *Alizzi & Omar* 35849!, 35850!;

Iran, United Arab Emirates, Saudi Arabia, Kuwait, Qatar.

Subfamily Salsoloideae Raf.: Tribe Caroxyleae Akhani & Roalson

17. **CAROXYLON** Thunb.,

Nov. Gen. Pl. 2: 37 (1782)

Annual herbs or small shrubs, often with unpleasant smell, usually covered with simple articulate, malpighian or bladder (vesicular) hairs. Stem and branches green or then white, villous when young and glabrescent. Leaves terete, often shortened or reduced to the scales, more or less gibbous at base, not mucronate. Bracts shorter than bracteoles or of equal size. Flowers axillary, solitary. Perianth segments 4–5, hyaline, ovate or triangular, with whitish, yellow or red coloured wing-like projections, rarely smooth or with small tubercles. Stamens 4–5, anthers divided almost to the top, with short appendices, rarely with prominent vesicles that are not clearly separated from the thecae. Stigmas 2. Fruit with membranous pericarp; seed with horizontal, rarely vertical embryo.

The precise number of species is still unknown, but it seems to be that this genus appears one of the most taxonomically diversified in Salsoloideae, and can comprise at least 70 representatives distributed in arid regions of Eurasia, N, E and S Africa.

Caroxylon, from Lat. *caro*, flesh and Gr. ξύλον, *xylon*, wood.

1. Indumentum of simple and bladder ("mealy") hairs; annuals 2
   Only simple hairs present 3
2. Lower leaves (always caducous at flowering stage) to 10 mm; bracts and bracteoles oblong or ovate, to 2 mm wide; wings at fruiting perianth 5–9 mm across.................... 5. *C. inerme*
   Lower leaves 10–30 mm; bracts and bracteoles suborbicular, 3–4 mm; wings at fruiting perianth 9–12 mm across .... 6. *C. jordanicola*
3. Leaves well-developed, to 15 mm; perianth wings at fruiting stage reddish or pink turning brown at dispersal stage; subshrubs 4
   Leaves (except lowest caducous leaves to 15 mm) to 5 mm long; wings at fruiting perianth white or yellowish, only sometimes pinkish 5
4. Leaves not or slightly gibbous at base, somewhat flattened; anther appendages relatively large, 0.8–1.4 mm; mountainous dwarf shrub to 40 cm .... 10. *C. canescens*

Leaves gibbous at base, needle-like; anther appendages much smaller than 0.8–1.4 mm; shrub to 60 cm growing on deserts plains. . . . . 4. *C. vermiculatum*

5. Shrubs with significant stem lignification; wings at fruiting perianth 3–5 mm across 6
   Annuals or subshrubs with basally lignified stem; wings at fruiting perianth 6–12 mm across. 7
6. Plant greyish; leaves and perianth silvery pubescent; often globular "cotton" galls are present on the branches . . . . 7. *C. cyclophyllum*
   Plant green; leaves glabrous or shortly pubescent; perianth glabrous or ciliate only; no globular galls on the branches present . . . . 8. *C. imbricatum*
7. Annuals 8
   Subshrub with caudex, stem stout, thick . . . . 9. *C. dendroides*
8. Perianth segments glabrous or slightly ciliate . . . . 2. *C. nitrarium*
   Perianth segments pubescent or villous 9
9. Anthers 0.5-0.7 mm; fruit 1.6–1.7 mm . . . . 3. *C. volkensii*
   Anthers 0.8–1.3 mm; fruit ± 2 mm . . . . 1. *C. incanescens*

1. **Caroxylon incanescens** *(C.A. Mey.) Akhani & Roalson* in Int. J. Pl. Sci. 168(6): 947 (2007).

Salsola incanescens C.A. Mey. in Eichw., Pl. Casp.-Cauc. 2: 35 (1833); Iljin in Fl. SSSR 6: 254 (1936); Rech.f., Fl. Lowland Iraq: 199 (1964); Freitag in Fl. Iranica 172: 200 (1997); Assadi in Fl. Iran 38: 278 (2001); Freitag et al. in Fl. Pak. 204: 163 (2001).

Annuals with unpleasant smell; stem very branched, covered with curved hairs, partially glabrescent; lower and middle leaves villose, caducous, to 20 mm, upper leaves and bracts scale-like, to 3 mm, slightly keeled at mibrib; perianth segments (5) pubescent with curved simple hairs that is persistent at the fruiting stage, 1.5–2 mm long, hyaline; stamens 5; anthers 0.8–1.3 mm, their appendage almost unnoticeable; wings at fruiting perianth white or yellowish, 8–11 mm in diameter; fruit ± 2 mm.

HAB. Irrigated fields, sandy loam, sandy saline soil, waste ground; alt. 0–80 m. Common plant in the plain; fl. Jul.–Oct., fr. Oct.–Dec.

DISTRIB. Widespread in the alluvial plains and deserts of Iraq: **FKI**: Hawija, *Barkley & Haddad* 3929L! **FPF**: S of Mandali, *Haddad & Agnew* 3H! **DLJ**: Sumaicha, *Handel-Mazzetti* 3120 (WU!). **DWD**: Bahr al-Milh, 30 km W of Karbala, *Guest, Rawi & Rechinger* 16164! *Hillcoat* 8343 (E!); 20 km NE of Ramadi, *Guest, Rawi & Nuri* 13541!; 30 km N of H3, *Serkahia* 15130!; Ukhaidhir, *Barkley & Safat* 3792 (E!); Habbaniya, *Hillcoat* 8305 (E!); 5 km N of Rahhaliya, *Weinert* & *Mousawi* s.n. (LE!). **DSD**: 40 km NW of Karbala, *Rawi & Serkahia* 16289! **LEA**: 30 km NE of Baquba, *anon.* 47151!; Diyala nr Kut, *Guest* 13104!; **LCA**: Baghdad, *Rechinger* 8019!; Abu Ghraib, *Robertson* 13477!; 30 km W of Darraji between Nasiriya and Samawa, *Rawi & Alkas* 16240!; Dujaila project, *Rawi* 12540!, 12541! **LSM**: nr Amara, Evans s.n. (E!).

Local name RIMITH (*Rawi* 12540), GHIDHĀM (*Guest, Rawi & Rechinger* 16164). Good for fodder as summer grazing (*Serkahiya* 15130, *Rawi* 12541).

Egypt east through the Arabian Peninsula (Yemen, Saudi Arabia) to Iran, Afghanistan, Pakistan; Anatolia, northern Caspian lowland.

2. **Caroxylon nitrarium** (*Pall.*) *Akhani & Roalson*, Int. J. Pl. Sci. 168(6): 947 (2007); Sukhorukov & Akopyan, Comp. Chenop. Cauc.: 55 (2013).

Salsola nitraria Pall., Ill. Pl.: 23 (1803); Freitag in Fl. Iranica 172: 196 (1997); Freitag et al. in Fl. Pak. 204: 161 (2001); Zhu, Mosyakin & Clemants in Fl. China 5: 404 (2003).

S. spissa M. Bieb. in Mem. Soc. Nat. Mosc. 1: 103 (1811).

S. pseudonitraria Aellen in Anz. Math.-Naturw. Kl. Oesterr. Akad. Wiss.: 27 (1967).—Type: Iraq: Basra, desertum meridionale (Southern desert), Naziriya, 10 km SE Ur in fosca secus viam, 6.11.1956, *K.H. Rechinger* 8178 (E!, W!, iso).

Nitrosalsola nitraria (Pall.) Tzvel. in Ukr. Bot. Zhurn. 50 (1): 80 (1993).

Annuals up to 50 cm with unpleasant smell; stem very branched, covered with curved hairs, glabrescent; lowermost leaves woolly, caducous, to 20 mm, upper leaves and bracts scale-like, to 3 mm, slightly keeled at mibrib; perianth segments (5) glabrous or slightly ciliate, 1.5–2 mm long, hyaline; stamens 5; anthers 0.8–1 mm, their appendage almost unnoticeable; wings at fruiting perianth white or yellowish, 5–8 mm in diameter; fruit ca. 2 mm.

Hab. Clayey salty substrates, disturbed areas; 0–1000 m; fl. Aug.–Sep.; fr. Oct.–Nov.
Distrib.: **lca**: Baghdad, *Hillcoat* 8142 (E!); Naziriya, *Rechinger* 8178 (E!, type of *Salsola pseudonitraria*).

One of the most widely distributed *Caroxylon* species in Irano-Turanian region: from S Russia through Central Asia eastwards to W China, Iran, Turkey, Afghanistan, Pakistan.

3. **Caroxylon volkensii** (*Asch. & Schweinf.*) *Akhani & Roalson* in Int. J. Pl. Sci. 168(6): 948 (2007).

Salsola volkensii Asch. & Schweinf., in Asch. et Schweinf., Ill. Fl. Egypt. 130 (1887); Rech.f., Fl. Lowland Iraq: 183 (1964); Boulos in Fl. Arab. Penins. Socotra 1: 163 (1996).

Annual to 40 cm, dark green, villose with denticulate hairs (having small protuberances). Stems erect, indurated, much branched. Leaves 3–8 × 1–2 mm, alternate, fleshy, semi-terete or linear, dilated at base, hairy. Bracts fleshy, more or less orbicular; bracteoles as long as or longer than bracts, more or less fleshy, orbicular. Flowers solitary, arranged in loose or dense spike-like inflorescences. Perianth segments connivent, oblong-lanceolate to ovate, acute, villose. Fruiting perianth with wings 5–8 mm in diameter; wings broadly obovate, imbricated, often turning blackish when dry.

Hab. Deserts, wadis or as ruderal; fl. Jul.–Sep.
Distrib. **dsd**: Chibritiya, *Hazim* 32490!; 150 km S of As-Salman, *Rawi & Tikrity* 24980!; "Mesopotamia", *Schwabe* s.n. (B!).

Africa, Arabia.

4. **Caroxylon vermiculatum** (*L.*) *Akhani & Roalson* in Int. J. Pl. Sci. 168(6): 948 (2007).

Salsola vermiculata L., Sp. Pl.: 223 (1753); Rech.f., Fl. Lowland Iraq: 202 (1964); Freitag in Fl. Iranica 172: 205 (1997); *S. villosa* Del. ex Roem. & Schult., Syst. Veg. ed. 15, 6: 232 (1820); Boulos in Fl. Arab. Penins. Socotra 1: 276 (1996).
S. vermiculata L. var. villosa (Del. ex Roem. & Schult.) Moq., Chenopod. Monogr. Enum.: 141 (1840).
S. palaestinica Botsch. in Bot. Zhurn. 60(4): 503 (1975).

Shrub up to 60 cm, without an unpleasant smell, covered with articulated and short spinulose hairs. Young shoots often reddish. Lower leaves to 15 mm, terete or semi-terete, needle-like, pubescent; upper leaves imbricate, scale-like with broader base, obtuse; their axiles with brachyblasts. Bracts ovate, 2–3 mm, bracteoles almost as long as bracts, suborbicular. Flowers solitary, perianth 2–2.5 mm, segments almost free, hairy, wings at fruiting perianth 8–12 mm across, near the base (or completely) pink, in upper parts sometimes white or yellow turning brown at dispersal stage. Fig. 86: 1–5.

Hab. Rocky, gravelly and sandy-clay hillsides, silty soil; alt. 0–650 m; fl. Jun.–Sep.; fr. Aug.–Nov.
Distrib. Widespread in the western Desert, scattered in other parts: **mam**: nr Zakho, *Regel* s.n.!; **mam/fuj**: N of Mosul, *Dinsmore* 396! **fuj**: 53 km W of Balad Sinjar, *Eig & Zohary* s.n. (HUJ!). **far**: Jabal Qara Choq, *Gorrie* 13494!, 13494A!, 13495!, 13497! **fki**: above Injana, *Gillett* 12486!; Qaraghan (Jalaula), L. Dimmock (comm. Rogers) K. 107!; Jabal Hamrin, *Sutherland* 472!, *Barkley* 3868! *Haines* 1366! *Evans* 26 (E!); Khanaqin, *Rogers* 406! *Haines* s.n. (E!); nr Mandali, *Hadač et al.* 4609! **dwd**: 260 km NW of Ramadi on road to Rutba, *Rawi* 20948!; 15 km S of Rutba, *Khatib & Hakim* 32443!; between Rutba and Ramadi, *Barkley, Agnew & Hikmat Abbas* 109!; *Rechinger* 9855 (W!); 70 km N of Rutba, *Rawi & Nuri* 27146!; between Rutba and Nukhaib, *Serkahia & Tikriti* 15125! **lea**: Shahraban (Muqdadiya), *Guest* 15793!; **lsm**: Amara, *Evans* s.n. (E!).

Local name RŪTHA (*Serkahia & Tikriti* 15125, between Rutba and Nukhaib).

From Macaronesia through the Mediterranean area to Turkey, Arabia, and Iran.

5. **Caroxylon inerme** (Forssk.) *Akhani & Roalson* in Int. J. Pl. Sci. 168(6): 947 (2007).

Salsola inermis Forssk., Fl. Aegypt.-Arab. 57 (1775); Boiss., Fl. Orient. 4: 955 (1879); Rech.f., Fl. Lowland Iraq: 200 (1964); Zohary, Fl. Palaest. 1: 171 (1966); Boulos in Fl. Arab. Penins. Socotra 1: 269 (1996); Freitag in Fl. Iranica 172: 205 (1997); Assadi in Fl. Iran 38: 287 (2001). *S. villosa* Roem. & Schult. nom. illegit. non Del.; *Bassia pulverulenta* H. Lindb. in Act. Soc. Sci. Fenn., ser. B: 2 (7): 12 (1946).

Annual up to 40 cm, densely covered with bladder hairs intermixed with scattered simple hairs. Stem branched from the base with ascending shoots. Lowermost leaves to 1 cm, caducous, very densely arranged, simple hairs to 3 mm long, with inconspicous or tiny conical base; other leaves drastically reduced to scales 1–2 mm, acute. Bracts ± 2

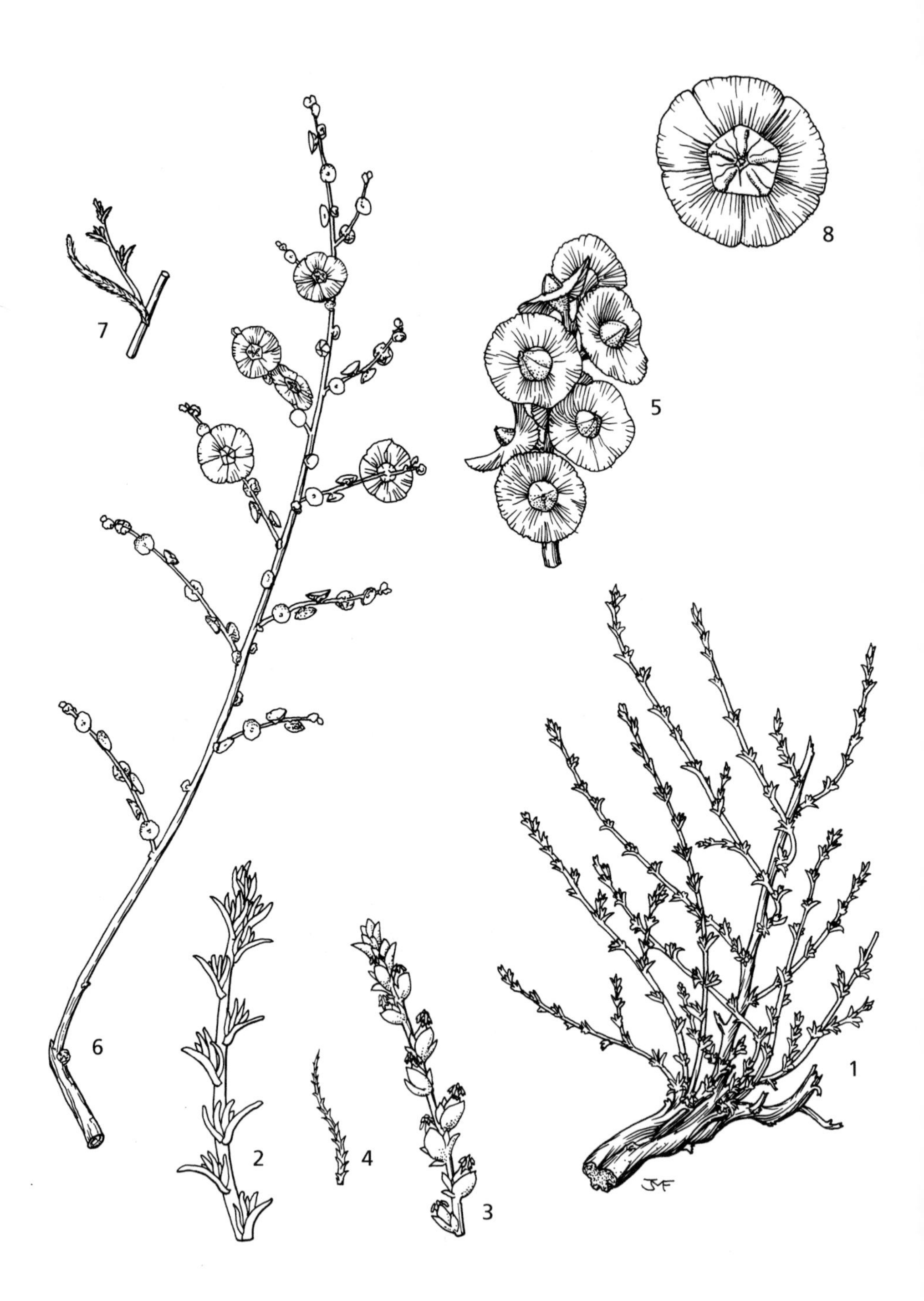

Fig. 86. **Caroxylon vermiculatum**. 1, habit × 1; 2, sterile branch × 1; 3, flowering branch × 2; 4, hair detail × 20; 5, fruits × 2. **Caroxylon jordanicola**. 6, habit × 1; 7, basal leaves × 1; 8, fruit × 3. Reproduced with permission from Flora of Arabia 1: fig. 50. 1996. Drawn by JMF.

mm, lanceolate to ovate-triangular; bracteoles about as long as bracts. Perianth 1.5–2 mm, segments with bladder hairs. Wings at fruiting perianth 4–8 mm across, pale purple turning brown at dispersal stage. Fruit ± 2 mm. Fig. 87: 1–4.

HAB. Clay plain, rocky hillsides, sandy and silty ground; alt. 0–700 m.; fl. Aug.–Oct., fr. Oct.–Nov.
DISTRIB. Widespread in Desert zones, occasional in the alluvial plains of Iraq: **MSU**: Sulaimaniya, *Fawzi* 39530! **MJS**: nr Jabal Sinjar, *Handel-Mazzettti* 1641 (WU!). **FUJ**: 68 km S. of Hadhr, *Barkley & Haddad* 3942!; **FKI**: 80 km W of Kirkuk, *Barkley* 2543!; Hawija in valley of R. Little Zab, *Barkley & Haddad* 3830A! **FKI/FUJ**: above Fatha ferry, *Barkley & Juma'a Brahim* 2544! 2545! **FPF**: 30 km W of Mandali, *Haddad* 10011!; 25 km SE of Mandali, *Rawi* 20722!; Ali al-Gharbi *Eig & Zohary* s.n. (HUJ!); N of R. Dujaila at W end of Jabal Hamrin, *Barkley* 3833! Jabal Hamrin, *Haines* 1455 (E!). **DLJ**: 39 km N of Baiji, *Barkley* 3688! **DWD**: Wadi al-Ajramiya, 73 km N of Rutba, *Rawi & Nuri* 27189!; 75 km W of Rutba, *Rawi* 21181!; 4 km S of Shithatha, *Barkley & Hikmat Abbas* 1918!; 43 km N of Rutba, *Rawi & Khatib* 32239! 30 km N of Rutba, *Chakravarty, Rawi, Alizzi & Khatib* 31494!; 7 km S of Shithatha, *Rawi* 26905!; 165 km W of Ramadi, *Rawi* 26848!; 75 km W of Rutba, *Rawi* 14708!; 50 km NE of Ramadi, *Rawi & Nuri* 13554! **DSD**: Salman to Aidha, *Fawzi, Hazim & H. Hamid* 38790!; nr Samawa, *Rawi & Rechinger* 16008B!; **LEA**: nr Shahraban (Muqdadiya), *Haines* W.1455!; 15 km E of Khan Bani Sa'd, *Barkley, Askari & Inman* 1772B!; 5 km NE of Ba'quba, *Stutz* 1344 (W!). **LCA**: Dujaila nr Kut, *Guest* 13108!; **LSM**: 40 km W of Amara, *Bharucha, Rawi & Tikrity* 29311!; 40 km E of Amara, *Rawi & Haddad* 25755! **LBA**: Basra, *Rechinger* 8151 (E!).

Turkey, Iran, Syria, Palestine, Saudi Arabia, Kuwait, Cyprus, North Africa.

6. **Caroxylon jordanicola** (*Eig*) *Akhani & Roalson* in Int. J. Pl. Sci. 168(6): 947 (2007).

Salsola jordanicola Eig in Palest. J. Bot., Jerusalem ser., 3 (3): 130 (1945); Rech.f., Fl. Lowland Iraq: 200 (1964); Zohary, Fl. Palaest. 1: 171 (1966); Boulos in Fl. Arab. Penins. Socotra 1: 269 (1996); Freitag in Fl. Iranica 172: 220 (1997); Assadi in Fl. Iran 38: 285 (2001).

Annual to 40 cm, semi-spherical, with simple articulated hairs and white bladder hairs, glabrescent; lower leaves 1–3 cm, caducous, other leaves to 2.5 mm, with broad, almost orbicular base. Bracts broadly ovate, to 3 mm. Perianth segments 2–3 mm long, ovate, with bladder hairs, at fruiting stage with mauve or white wings of 9–12 mm across. Fig. 86: 6–8.

HAB. Sandy, gravelly and stony deserts; alt. 0–450 m; fl. Apr.–Jul., fr. Jun.–Nov.
DISTRIB. Mainly in the Desert zone of SE Iraq: **MSU**: Sulaimaniya, *Fawzy* 39547! **FPF/LEA**: between Khanaqin and Ba'quba, *Evans* s.n.!, *Hadač et al.* 2992!, *Haines W.* 1362!; **FPF**: 30 km W of Mandali, *Haddad* s.n.!; 25 km SE of Mandali, *Rawi* s.n.!; Tursaq, *Barkley* 2419!; Ali al-Gharbi at Iranian frontier, *Petter* s.n.!; 20 m N of Badra, *Rechinger* 12736! **DLJ**: Tharthar Lake, *Barkley & Abbas* 3623 (W!); Sumaicha, *Rechinger* 8097 (W!); nr Balad, *Baltaxe* 1349! **DWD**: 165 km W of Ramadi, *Rawi* 26847!; 45 km NE of Ramadi, *Guest, Rawi & Nuri* 13526!, 13533!; 90 km S of Baghdad, *Rawi* 26884!; Massayih, Kharja and Braimat, NW of Rutba, *Rawi* 26875A!; 4 km S of Shithatha, *Barkley & Abbas* 1912 (W!). **DSD**: above L. Habbaniya, *Barkley & Safwat* 3963!; 30 km W of Darraji, *Rawi & Alkas* 16250!; between Busaiya and Zubair, *Sarkahia & Tikriti* 15138, 15166!; 28 km W by S of Samawa, *Guest, Rawi & Rechinger* 16024B!; al-Batin nr Jarishan, *Guest, Rawi & Rechinger* 16133!; 5 km from Ma'niya, *Hazim* 32518!; 180 km W of Basra, *Rawi & Tikriti* 24987!; nr Samawa, *Rawi & Rechinger* 16009B!; nr Busaiya, *Rechinger* 8194 (HUJ!); 8 km from Ramadi, *Qaisi, Hamad & H. Hamid* 43578!; 10 km SE of Ur, *Rechinger* 8173 (W!). **LEA**: nr Debouni (Kut), *Guest* 3469 (HUJ!); between Fakka & Msaieda, *Rawi* 14982!; Kut al-Imara, *Rechinger* 14068 (W!). **LCA**: Falluja, *Rechinger* 139 (E!); E of Falluja, *Rechinger* 33619!; 139!; Sudur, *Botany Staff* 47171!

GHIDHĀM (Guest et al. 16138B, Kuraiz al-Mali); KHIDRĀF (*Guest et al.* 13526, nr Ramadi), KHIDHRĀF (*Guest et al.* 13533, nr Ramadi).

Iran, Syria, Palestine, Egypt, Arabia (Kuwait).

7. **Caroxylon cyclophyllum** (*Baker*) *Akhani & Roalson* in Int. J. Pl. Sci. 168(6): 948 (2007).

Salsola cyclophylla Baker in Kew Bull. [without number]: 340 (1894); Rech.f., Fl. Lowland Iraq: 202 (1964); Greuter, Burdet & Long, Med-Checklist ed. 2, 1: 308 (1984); Boulos in Fl. Arab. Penins. Socotra 1: 272 (1996); Freitag in Fl. Iranica 172: 211 (1997); Freitag et al. in Fl. Pak. 204: 164 (2001).

Shrub up to100 cm forming spherical habit; branches ascending, silvery due to presence of adpressed simple hairs, almost without unpleasant smell. Stems often with specific globular "cotton" galls to 15 mm across; leaves minute, 1–2 mm, hairy. Bracts and bracteoles similar to leaves. Perianth segments ± 1.5 mm, with dense silvery hairs, at fruiting stage with white or yellowish wings 3–4 mm across. Anthers 0.8 mm, with almost unnoticeable appendage at the top. Fruiting perianth 4–4.5 mm in diameter.

HAB. Clayey and gravelly substrates; alt. 0–500 m; fl. Aug.–Oct.; fr. Nov.–Dec.

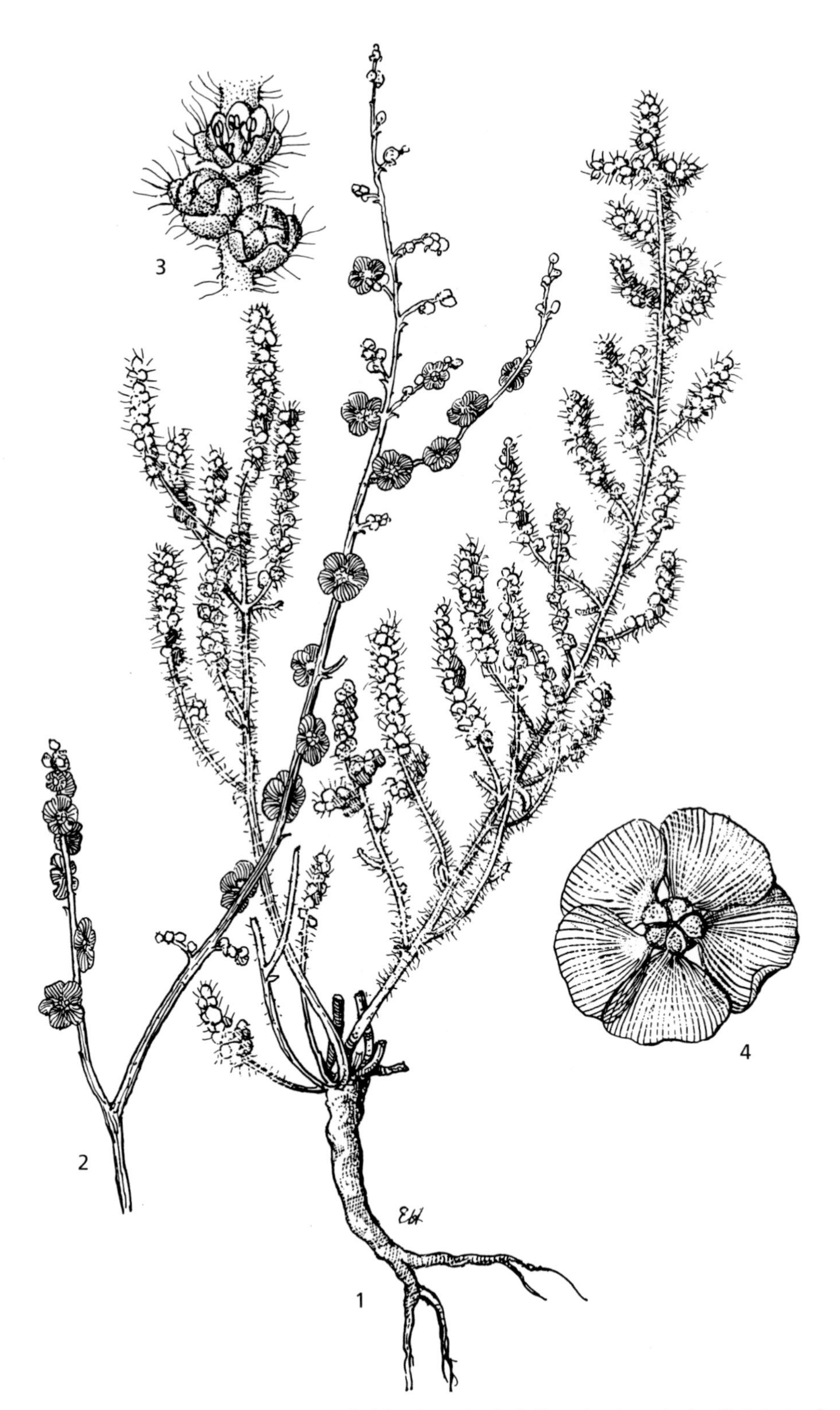

Fig. 87. **Caroxylon inerme**. 1, habit × 1; 2, fruiting branch × 1; 3, flowering branch, detail; 4, fruit × 2. Reproduced with permission from Zohary, Flora Palaestina 1: Plates, f. 249. 1966. Drawn by Ester Huber. © The Israel Academy of Sciences and Humanities.

DISTRIB. Desert zone of Iraq: **DWD**: between Shithatha & Rahhaliya, *Rawi & Alkas* 16325! **DSD**: Khadhar al-Ma'i, *Guest* 15246! *Rechinger* 138 (BM!); Samawa, *Hillcoat* 8158!; *Guest, Rawi & Rechinger* 16024!; Zubair, *Rawi & Alizzi* 32429!; Busaiya, *Haddad* 9548!

NE Africa, Arabia, Palestine, Syria, S Iran, S Pakistan.

8. **Caroxylon imbricatum** (*Forssk.*) *Moq.* in *DC.*, Prodr. 13(2): 177 (1849).

Salsola imbricata Forssk., Fl. Aeg.-Arab. 57 (1775); Boulos in Fl. Arab. Penins. Socotra 1: 275 (1996); Freitag in Fl. Iranica 172: 193 (1997); Freitag et al. in Fl. Pak. 204: 157 (2001).
Chenopodium baryosmum Roemer et Schultes, Syst. Veg. ed. 15, 6: 269 (1820).
Salsola baryosma (Roem. et Schultes) Dandy in Andrews, Flow. Pl. Anglo-Egypt. Sudan 1: 111 (1950).
Caroxylon imbricatum (Forssk.) Akhani & Roalson in Int. J. Pl. Sci. 168(6): 947 (2007) nom. superfl.;

Branched shrub to 200 cm. Leaves short, to 1.5 mm (except leaves of young plants that can be to 15 mm long), glabrous or covered with scattered barbellate hairs, at margins with white stripe. Bracts similar to stem leaves or bracteoles, imbricated. Perianth to 2 mm long, glabrous or ciliate, at fruiting stage with white wings of 3–5 mm across. Fruit ± 1.5 mm in diameter.

HAB. Clayey and gravelly substrates; alt. 0–500 m; fl. Sep.–Oct; fr. Nov.–Dec.
DISTRIB. Desert zone of Iraq: **DWD**: 10 km E of Ramadi, *Barkley & Barkley* 3960!; Ukhaidhir, *Barkley, Haddad & Ani* 6239!; Rahhaliya, *Guest, Rawi & Rechinger* 16190!; 2 km SE of Rahhaliya, *Weinert & Mousawi* s.n. (LE!); Shithatha, *Rawi & Alkas*, 16298!, *Rechinger* 8330 (BM!). **DSD**: between Ur & Al-Busaya, *Rechinger* 8193 (W!); Samawa, *Rechinger* 8164 (W!). **LCA**: 30 km W of Darraji, *Rawi & Alkas* 16231!; Nasiriya, *Guest, Rawi & Rechinger* 16031!; Hillah, *Graham* 416!; Falluja, *Barkley, Abbas & Maule* 140!; Babylon, *van der Maesen* 1609 (W!); Baghdad, *Barkley & Safwat* 3403C (W!). **LBA**: Basra, *Haines* 1886!; *van der Maesen* 1617 (W!).

Widely distributed species in Saharo-Arabian and Turanian Region, from N Africa through Eastern Mediterranean, Arabian Paninsula, Iran to W India.

9. **Caroxylon dendroides** (*Pall.*) *Tzvelev* in Ukr. Bot. Zhurn. 50 (1): 81 (1993).

Salsola dendroides Pall., Ill.: 22 (1803); Aellen in Fl. Turk. 2: 333 (1967); Gabrielian & Fragman-Sapir, Fl. Transcauc. & adjacent areas: 10 (2008).
Salsola dendroides Pall. subsp. *trichantha* Botsch. in Nov. Sist. Vyssh. Rast. 11: 33 (1974).
Salsola trichantha (Botsch.) Botsch. in Nov. Sist. Vyssh. Rast. 19: 81 (1982).

Unpleasant smelling subshrubs to 150 cm with thick and branched caudex; stem insignificantly lignificated (to 10 cm), thick at the base (up to 5–7 mm), as the leaves covered with dense indumentum of curved simple hairs, later glabrescent. Lower leaves short, to 7–10 mm, caducous, middle and upper leaves and bracts to 3 mm, gibbous, glabrous or with scattered hairs. Bracts equal to the bracteoles or slightly shorter, with median keel. Perianth segments (5) glabrous or slightly pubescent, at fruiting stage with white or pinkish wings of 7–10 mm across. Anthers 1–1.3 mm. Fruit ± 2 mm, pericarp translucent, slightly fleshy in upper fruit part.

HAB. Clayey deserts, alt. 0–1000 m; fl. Aug.–Oct., fr. Oct.–Dec.
DISTRIB. Desert zone of Iraq: **FKI**: Fatha, *Agnew & Haines* s.n. (E!). **LCA**: Baghdad, *Haines* 131 (E!).

Widely distributed in Irano-Turan. SE European Russia, Turkey, Iran, Central Asia, Afghanistan, Pakistan, W India.

10. **Caroxylon canescens** *(Moq.) Akhani & Roalson* in Int. J. Pl. Sci. 168(6): 947 (2007).

Salsola canescens (Moq.) Boiss. nom. illegit., non Pers. (1805); Boiss., Fl. Orient. 4: 963 (1879); Rech.f., Fl. Lowland Iraq: 202 (1964); Freitag in Fl. Iranica 172: 230 (1997); Assadi in Fl. Iran 38: 310 (2001); Freitag et al. in Fl. Pak. 204: 170 (2001).
Noaea canescens Moq. in DC., Prodr. 13 (2): 208 (1849). —Lectotype: [Iraq] In cacumine m. Gara Kurdist. orientem versus frequens, 1843, *Th. Kotschy* 346 (LE, iso – K-000899548; 000899548! W-0046462!).
Salsola boissieri Botsch., Bot. Zhurn. 53: 1442 (1968).
Caroxylon boissieri (Botsch.) Freitag, Vasc. Pl. Afghanistan: 264 (2013) nom. superfl.
Salsola chrysoleuca Aellen in sched.

Subshrub to 40 cm, with branched caudex and many stems, some of them do not produce flowers (sterile); leaves to 1 cm, inflorescence loose, bracts 5–8 mm, equal to bracteoles or slightly larger; not or slightly keeled, bracteoles of the upper flowers larger than subtending bract; perianth pubescent, at fruiting stage develops wings 8–11 mm across; anther appendages large, 0.8–1.4 mm, but not clearly separated from the thecae; fruit ± 3 mm; seed with vertical or oblique embryo.

HAB. Mountain slopes; alt. 1200–2500 m. fl. & fr. Jul.–Sep.
DISTRIB. Occasional in the mountain zone: **MAM**: Hawraz (Hauris), *Haley* 181!; Mt. Gara, *Kotschy* 346!; N of Zakho, nr Sharanish, *Rechinger* 10951! **MRO**: Mt. Qandil, *Rechinger* 11790!; Helgord Dagh, *Gillett* 9623!; Qandil range, New Robar valley, *Rawi* 24160!; nr Rayat, *Zohary* s.n. (HUJ!). **MSU**: Mt. Hauraman (Avroman), *Rawi, Nuri* & *Alizzi* 19753! *Haussknecht* s.n. (BM!).

This species is set apart from other *Caroxylon*, especially by large anther appendages and vertical seed embryo.

Turkey, Lebanon, Syria, Iran, Afghanistan, C Asia.

SPECIES EXPECTED TO OCCUR IN IRAQ

Caroxylon tetrandrum (*Salsola tetrandra*) is likey to be found in the southern deserts of Iraq. It is known in Saudi Arabia, Jordan, Palestine and Syria. It is easily recognized by the (sub)opposite branches and leaves and presence of tubercles on the perianth segments at fruiting stage.

18. **KAVIRIA** Akhani

in Int. J. Pl. Sci. 168 (6): 948 (2007)

Annuals or small shrubs, mostly covered with simple curved hairs sometimes intermixed in plant's basal part with longer bristle-like hairs, rarely glabrous; branches spiny or not; leaves relatively short, usually to 2 cm, semi-terete or flattened, alternate, spirally arranged or distichous, basally not gibbose, at the tip obtuse, without hypodermis at cross-sections, Flowers solitary, supported with bract and two bracteoles similar in shape; perianth of 5 free segments, membraneous, at fruiting stage winged or not. Anthers long (1.5 mm or more), divided (almost) to the tip, with small or prominent vesicle. Fruit dry; seed embryo horizontal or vertical, without perisperm.

Kaviria, from Persian *kavir*, salt desert (= Ar. *sabkha*).

Genus of about 10 species, but its richness not completely clarified.

1. Plant woolly or grey (lower leaves can be glabrescent); perianth without wings; seed embryo vertical .... 1. *K. lachnantha*
   Plant pubescent but not woolly; perianth with wings; seed embryo horizontal 2
2. Plant not thorny; bract base ovate .... 3. *K. tomentosa*
   Plant thorny at fruiting stage; bract base rounded .... 2. *K. azaurena*

1. **Kaviria lachnantha** (*Botsch.*) *Akhani* in Int. J. Pl. Sci. 168 (6): 948 (2007).

[*Salsola canescens* (non (Moq.) Boiss.) Aellen in Rech.f., Fl. Lowland Iraq: 202 (1964)].
S. tomentosa (Moq.) Spach subsp. *lachnantha* Botsch., Bot. Zhurn. 53: 1448 (1968).
S. lachnantha (Botsch.) Botsch., Nov. Syst. Pl. Vasc. 17: 124 (1980); Boulos in Fl. Arab. Penins. Socotra 1: 276 (1996); Assadi in Fl. Iran 38: 307 (2001).

Subshrub, to 50 cm tall, woolly. Cauline leaves to 22 × 2 mm, grey to white, caducous; bracts to 5 mm; perianth not accrescent, ± 3.5 mm, woolly pubescent in its upper part and (almost) glabrous in lower part, without any projections; anthers ± 1.7 mm, their appendages ± 1 mm, seed embryo vertical.

HAB. Gypsiferous hills, gravelly slopes, sandy deserts; alt. 0–580 m; fl. Jun.–July; fl. Aug.–Oct.
DISTRIB. Rare in the forest zone, widespread in the steppe and desert areas: **MSU**: Sulaimaniya, *Fawzi* s.n.! **FKI**: above Injana, *Gillett* 12479!; Tuz Khurmatu, *Rechinger* 10610 (WU!). **FPF**: Koma Sank police station on Iranian border, *Rawi* 20597!; Jabal al-Muwaila, nr Kuwait border, *Guest, Rawi & Rechinger* 17622!; Jabal Hamrin, *Rechinger* 8081 (E!, LE!); Mirjani, *Sutherland* 470 (BM!). **DLJ**: 50 km N by W of Falluja, *Guest, Rawi* & *Rechinger* 4525!, 14526!; 50 km NE of Ramadi, *Guest, Rawi & Nuri* 13534!; nr Balad, *Baltaxe* 13483!; 23 km N of Baiji, *Barkley* 3837!; Thirthar Lake, *Barkley & Abbas* 3622 (LE!). **DWD**: W of Falluja, shore of L. Habbaniya, *Guest* 15195!; Shithatha to Rahhaliya, *Rawi & Alkas* 16137!; nr L. Thirthar, *Barkley & Abbas* 3627!; Thirthar, *Barkley & Abbas* 3622A!; 30 km S of Ramadi, *Guest, Rawi & Rechinger* 16198! *Rechinger* 8355 (E!); nr Anaiza, *Agnew & Barkley* 2457!; 38 km E of Husaiba, *Weinert &*

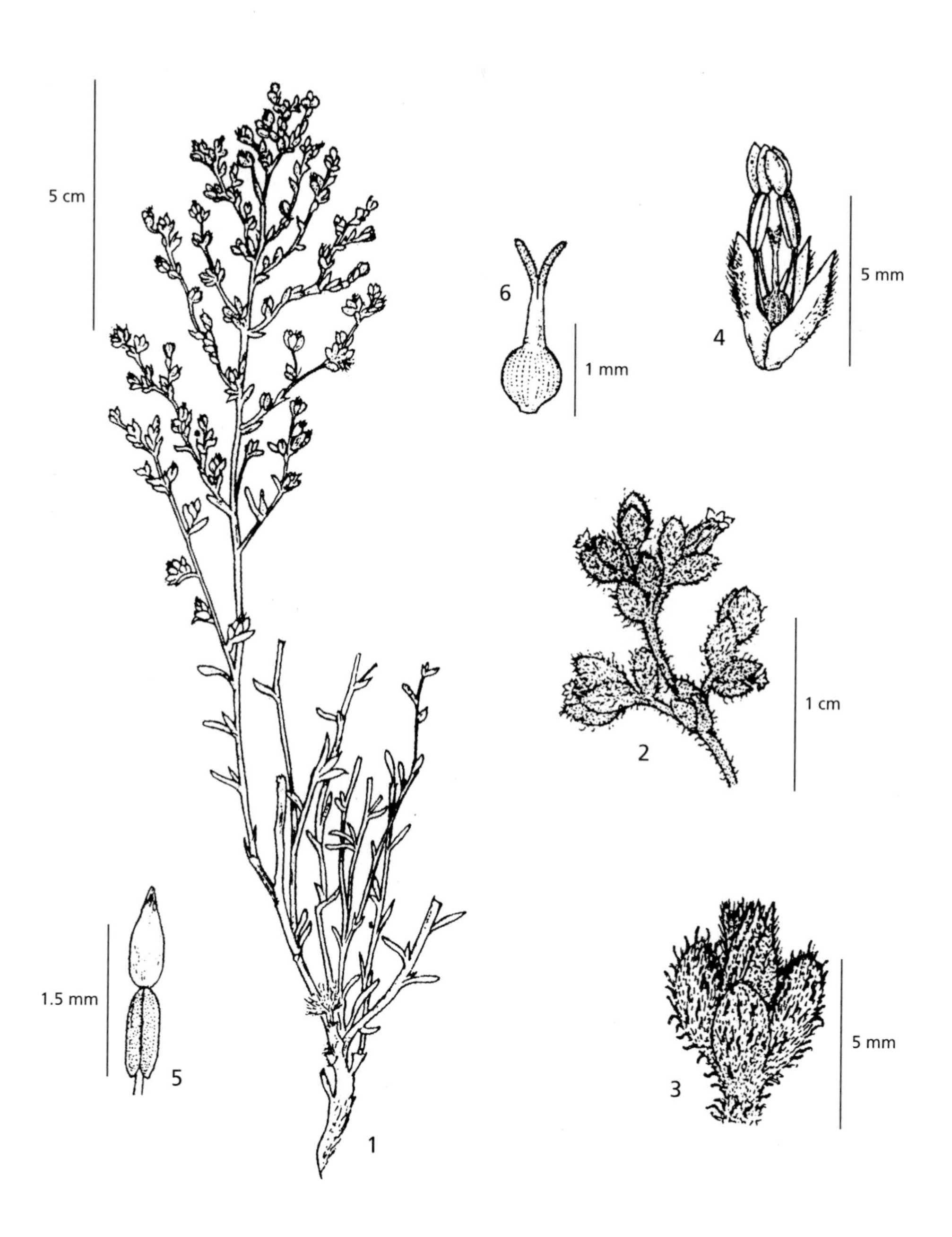

Fig. 88. **Kaviria tomentosa**. 1, habit; 2, inflorescence apical portion; 3, flower with bract and bracteoles; 4, flower; 5, anther with appendage; 6, ovary with style and stigmas. Reproduced with permission from Flora of Pakistan 204: f. 31. 2003.

Mousawi s.n. (LE!). **LEA**: nr Shahraban (Muqdadiya), *Haines* W.1367! **LCA**: Iskandariya, *Haines* W.751 (E!, LE!); *Rechinger & Haines* 8271 (LE!).

SW Iran, N Arabian Peninsula.

2. **Kaviria azaurena** (Mouterde) *Sukhor.*, Chenop. Carp. & Syst.: 348 (2014).

Salsola spinosa Pabot in Khatib, Contrib. Syst. Chenop. Syrie: 146 (1959) nom. illegit. non Lam. (1779).
S. azaurena Mouterde, Nouv. Fl. Liban & Syrie 1: 433 (1966).
Caroxylon azaurenum (Mouterde) T.A. Theodorova, Turczaninovia 14(3): 75 (2011).

Subshrub or shrub to 50 cm; caudex stout, prominent; stem very branched with almost horizontal spreading shoots often terminating with thorns after leaf fall; leaves to 2 cm long, semiterete or slightly flattened, to 2 mm broad, covered with barbellate simple hairs. Bract deflexed, at least 5 mm long, similar with the bracteoles; bracteoles base orbicular, with translucent hyaline margins. Perianth segments almost free, membraneous, slightly unequal, 3.5–4 mm long at anthesis, acutish to obtuse, covered with short hairs; anthers ± 1.5 mm long, with prominent (± 1.5 mm) triangular appendage (vesicle). Wings at fruiting perianth 10–12 mm across. Seed embryo horizontal.

HAB. gypsum soils; alt. 0–200 m.
DISTRIB. Scattered in deserts and plains of Iraq. **FKI**: Fatha-Ferry, *Agnew & Haines* 5814 (LE!); **FKI/FPF**: W of Jabal Hamrin, *Barkley* 3832 (LE!). **DLJ**: S Jazira, *Guest, Rawi & Nuri* 13531(LE!); Thirthar Lake, *Weinert* s.n. (LE!); *Safwat & Barkley* 3953 (LE!). **DWD**: 23 km W of Rahhaliya, *Rawi* 26934!; 6 km SW of Shithatha, *Barkley & Abbas* 1924 (LE!).

Syria.

3. **Kaviria tomentosa** (*Moq.*) *Akhani* in Int. J. Pl. Sci. 168 (6): 948 (2007).

Salsola tomentosa (Moq.) Spach in Kotschy, Exsicc. Pl. Alepp. Kurd. Moss. no 346: (1843); Freitag in Fl. Iranica 172: 234 (1997); Freitag et al. in Fl. Pak. 204: 172 (2001).
Salsola aurantiaca Bunge ex Boiss., Fl. Orient. 4: 963 (1879).
Caroxylon tomentosum (Moq.) Tzvelev, Ukr. Bot. Zhurn. 50 (1): 80 (1993).

Subshrub to 50 cm, with many stems arisen from the caudex; leaves to 10 mm, semiterete, covered with simple flexuose hairs; inflorescences lax in lower part, more condensed at the tips; bracts to 5 mm, with translucent hyaline margins; perianth segments ca. 3mm, hairy; anthers with prominent trtiangular vesicle of 1 mm long. Wings at fruiting perianth 8–10 mm across, pink turning brown at dispersal; fruit ca. 1.5 mm. Fig. 88: 1–6.

HAB. Steppes and semi-deserts; 500–1500 m; fl. Jul.–Sep.; fr. Sep.–Nov.
DISTRIB. Probably rare. **FPF**: 5 km S of Badra, *Zakirov & Shermatov* 471 (MW!); Jabal Hamrin, *Sutherland* 472 (BM!).

Iran, Caucasus, Afghanistan, Tajikistan, Pakistan.

19. HALOCHARIS Moq.,

in DC., Prodromus 13 (2): 201 (1849)

Annuals, branched, with prostrate or ascending stems, thickly clothed with barbellate hairs and straight hairs with conical base. Leaves linear, alternate but appearing whorled, somewhat succulent, sessile. Flowers hermaphrodite, borne in clusters in axils of leaves, occasionally solitary, arising from a tuft of crimped hairs. Tepals 5, free, membranous, not accrescent. Stamens 5; connective brightly coloured and inflated at apex; anther vesicles sessile or with short stalk. Style ± deeply divided, sometimes subentire, equal to stigmas. Fruit enclosed in persistent perianth. Seeds lacking perisperm; embryo spiral, vertical.

About 10 species distributed mainly in SW & C Asia. *Halocharis* share with *Halanthium* the terminal inflated staminal vesicles, the adaptive significance of which is unclear but can be connected with entomophyly. This organ provides useful characters to differentiate the species .

Halocharis (from Gr. ἅλς, *hals*, salt and χάρις *charis*, grace)

Plant hispid; anther appendages oblong, 1–2 mm .... 1. *H. sulphurea*
Plant not hispid; anther appendages ovate, 0.6-0.7 mm................ 2. *H. brachyura*

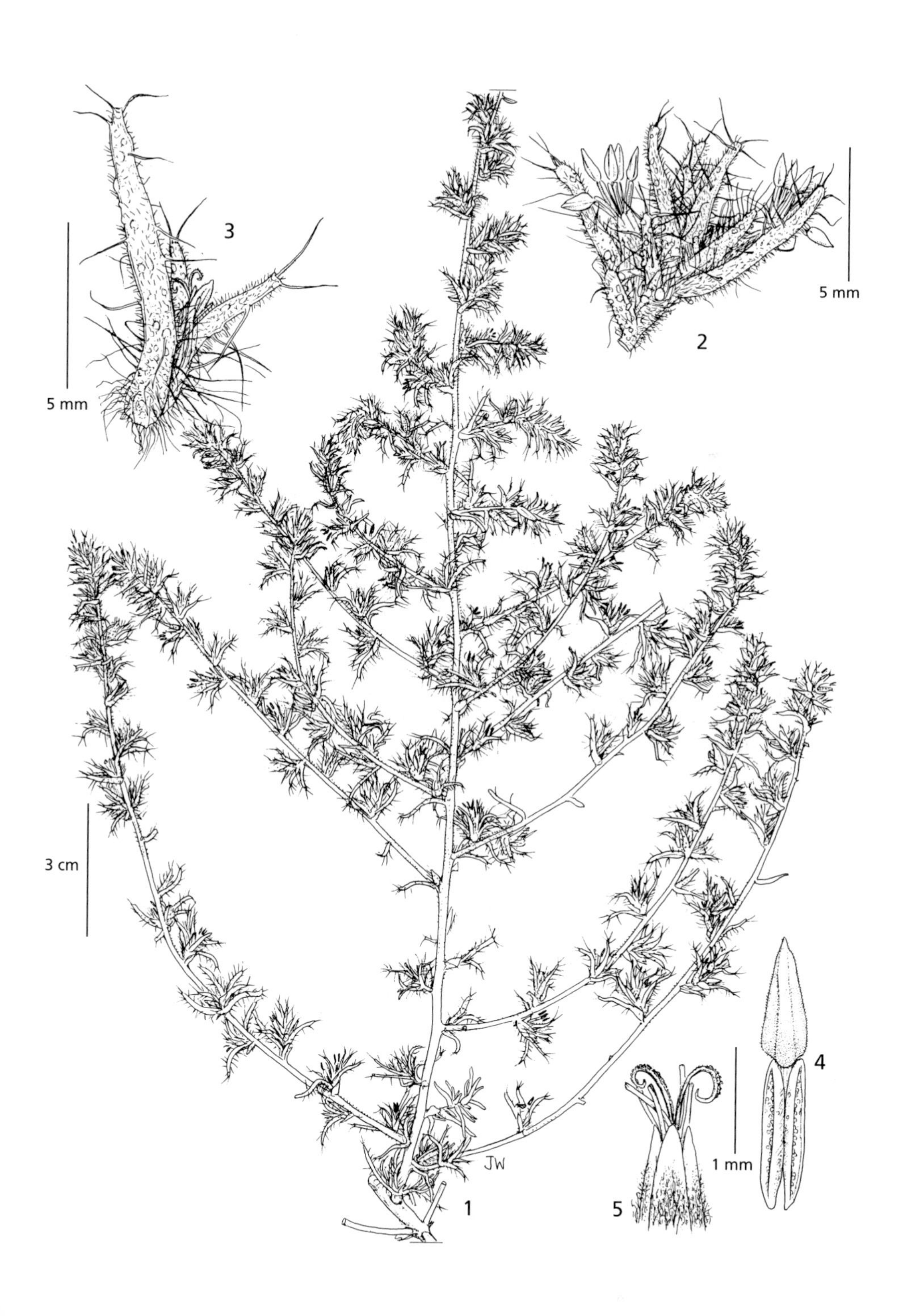

Fig. 89. **Halocharis sulphurea.** 1, habit; 2, group of flowers; 3, flower; 4, anther with appendage; 5, style detail. 1 from *Barkley* 2422; 2–5 from *Rawi* 20693. © Drawn by Juliet Beentje.

1. **Halocharis sulphurea** (*Moq.*) *Moq.* in DC., Prodr. 13 (2): 201 (1849); Rech.f., Fl. Lowland Iraq: 209 (1964); Hedge in Fl. Iranica 172: 331 (1997); Boulos in Fl. Arab. Penins. Socotra 1: 283 (1996); Freitag et al. in Fl. Pak. 204: 198 (2001).

Halimocnemis sulphurea Moq., Chenopod. Monogr. 152 (1840).

Annual to 50 cm, much-branched at base, prostrate to erect, with a copious indumentum of brownish spreading hairs interspersed with shorter hairs. Lower leaves to 30 mm, caducous; other ones 8–15 × 1–2 mm, linear, sessile, often in clusters, sparsely long-hairy throughout and with a few terminal bristles. Flowers arising from leaf axils, clusters with variable numbers of flowers. Perianth segments hyaline, 4 × 1 mm, with hairs on abaxial surface, glabrous within. Filaments usually papillose; anthers with 1–2 mm long vesicle, with terminal appendages that are bright yellow and borne on a short stalk. Style with 2 spreading arms, ± flattened. Fig. 89: 1–5.

HAB. Sandy, clayey and gravelly ground, often in saline areas; alt. s.l.–640 m; fl. Apr.–Jun., fr. Jun.–Nov. DISTRIB. Widespread in semi-desert and foothill regions of Iraq: **FKI**: nr Tuz Khurmatu, *Guest* 666! **FPF**:10 km E of Mandali, *Rawi* 20693!; 5 km W of Mandali, *Rechinger* 9606 (W!); Tursaq, *Barkley* 2422!, 2426!; Qara Tu on Iranian border, *Rawi* 5804!; Jabal Hamrin, *Haines* 1453 (E!); 25 km E of Badra, *Rawi* 20785! **DWD**: Habbaniya, *Rawi & Alizzi* 34458!; 160 km NW of Ramadi, *Rawi* 20895!; Between Karbala and Shithatha, *Rawi* 19982!; Habbaniya, *Fawzi, Hazim & H. Hamid* 38968!; about 50 km NE of Ramadi, *Guest, Rawi & Nuri* 13564; id. 13553! **DSD**: 59 km from Aidha and Ansab, *Fawzi, Hazim & H. Hamid* 38832!; 80 km W of Shabicha, *Chakravarty, Rawi, Khatib & Tikriti* 30075! **LEA**: Fakka to Laqlaq valley on Iranian border, *Rawi* 25827!; Jabal Hamrin nr Shahraban (Muqdadiya), *Haines* W.1453!; Fakka to Msaieda, *Rawi* 14987A! **LCA**: nr Baghdad, *Lazar* 535A! *Ludlow* 214 (BM!); 30 km W of Darraji between Nasiriya and Samawa, *Rawi & Alkas* 16237, 16238!; nr Faluja, *Haines* s.n. (E!). **LSM**: Jaffal, nr Garma, *Rawi & Shahwani* 26946!, 26950A!; 20–30 km N of Amara, *Hazim & Nuri* 30617A!; 20–30 km N of Amara, *Hazim & Nuri* 30617!

Vernacular name KHIDRAF (*Rawi* 14708, nr Rutba,).

Iran, Afghanistan, Pakistan.

2. **Halocharis brachyura** Eig in Palest. J. Bot., Jerusalem ser. 3: 137 (1945); Rech.f., Fl. Lowland Iraq: 209 (1964). —Lectotype: S Iraq [LBA], near the banks of Chankula river, 2 km sideward of the main Baghdad-Basra way, 6.4.1933, *A. Eig & M. Zohary* 896 (HUJ-20769) selected by Sukhorukov in Chenop. Carp. & Syst. p. 349 (2014).

?*Halocharis noaeana* Iljin, Bot. Mat. Gerb. Bot. Inst. Akad. Nauk SSSR 11: 78 (1949). — Type: Iraq, [LEA] Kut [al-Imara], ad ripam fl. Tigris, V. 1861, *Nöe* 1012 (LE!).

Plant similar to *H. sulphurea*, but not hispid. Anthers with thecae ± 1.5 mm and with relatively short vesicle (0.6-0.7 mm).

HAB. River banks, clayey deserts; as ruderal; alt. 0–200 m. fl. Apr.–Jun., fr. Jun.–Aug.
DISTRIB. ?**LBA**: 94 and 133 km SE Baghdad, *Eig & Zohary*" (HUJ!); **LBA**: 50 km S of Basra, *Rawi* 25917!; nr Basra, *staff of Dep. Botany* 15843 (W!), probably belong to this taxon.

Known only from Iraq; a poorly known taxon that requires further investigation.

20. **CLIMACOPTERA** Botsch.

in Sborn. Rabot Akad. Sukachev: 111 (1956)

Previously included in *Salsola* according to Fl. Iranica (1997). The molecular results confirm the distinct generic status of *Climacoptera*.

Annuals, glabrous or hairy (but often glabrescent); indumentum of simple flexouse hairs. Leaves clearly decurrent, terete, alternate, fleshy, without a mucro. Flowers solitary, supported by bract and two bracteoles. Perianth of 5 segment, glabrous or pubescent, with wing-like projections at fruiting stage. Stamens 5, with spongy, white, yellow or red appendages. Stylodia 2, free or diversely connate to the style. Fruit dry. Seed with horizontal embryo; perisperm lost.

The exact species number is not defined, but it can comprise ca. 70 species distributed in arid regions of Eurasia, predominantly in Irano-Turan. Some features like concrescence of the stylodia, pubescence of the plant and its parts play important role in the diagnostics. The genus *Pyankovia* Akhani et Roalson

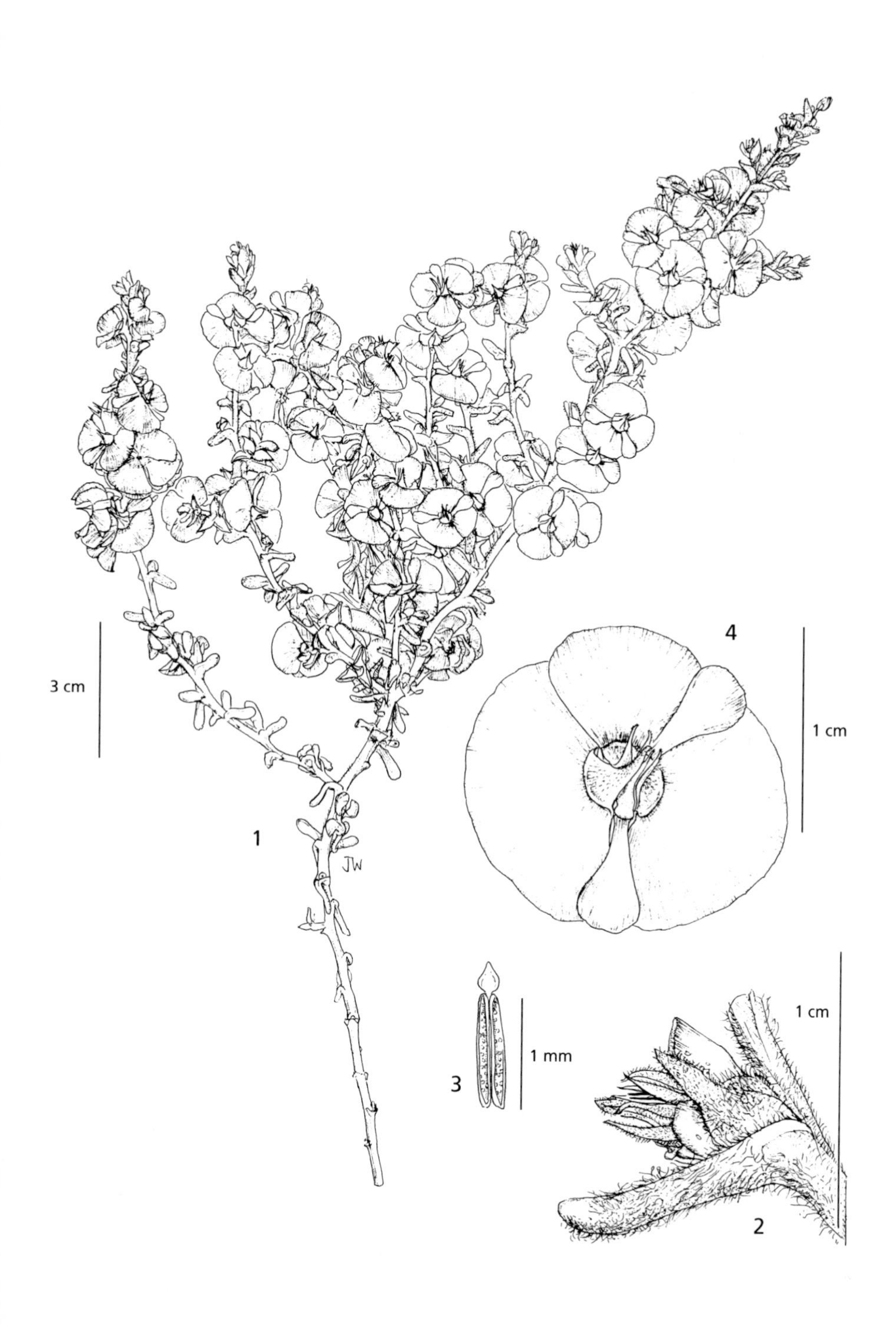

Fig. 90. **Climacoptera iraqensis**. 1, habit; 2, group of flowers; 3, anther with appendage; 4, fruit. 1, 4 from *Rawi & Serkahia* 16301; 2–3 from *Guest, Rawi & Nuri* 13548. © Drawn by Juliet Beentje.

in Int. J. Pl. Sci. 168(6): 949 (2007) with one representative *P. brachiata* (Pall.) Akhani & Roalson (*Salsola brachiata* Pall.=*Climacoptera brachiata* (Pall.) Botsch.) is separated from *Climacoptera* based on molecular phylogeny with regard of some morphological characters (opposite leaves; presence of both flexuose and straight hairs with conical base; vertical embryo position).

Climacoptera (from Gr. κλιμαζ, *klimax*, climax, and πτερό, *ptero*, wing)

Perianth to 6 mm long, pubescent, wings at fruiting perianth 15–20 mm in diameter; anther appendages linear, divided to 1/4 .... 1. *C. iraqensis*
Perianth to 5.5 mm long, glabrous, ca. 12 mm diameter (including wings); anther appendages elliptical, divided to 1/3........................ 2. *C. khalisica*

1. **Climacoptera iraqensis** *Botsch.* in Nov. Sist. Vyssh. Rast. 19: 76 (1982).

Salsola crassa auctt. non Bieb. in Mem. Soc. Nat. Mosc. 1: 137 (1806); Rech.f., Fl. Lowland Iraq: 210 (1964). —Type: Iraq: 5 km NE of Shithatha near the salt riverlet at the mouth to Bahr al Milh, wet salt soil, 15.XI.1973, *Weinert & Mouawi* 14692 (holo BUH; iso LE!, K!).

Annual, branched from the base, to 45 cm tall. Branches alternate, densely clad with flexuous patent hairs. Cauline leaves to 20 × 2 mm. Bracteoles to 5 × 3 mm. Perianth (fruit wing) pubescent, pinkish white or rose-red, to 6 mm long, 15–20 mm diameter including wings. Anthers to 2 × 0.5 mm, with whitish appendages. Style and stigma scarcely exserted from perianth, ± 3 mm. Fig. 90: 1–4.

HAB. Low lying saline marsh land, salty soil on surface, calcareous underneath, salty loamy soil, waste land liable to flooding; alt. 0–50 m; fr. Nov.
DISTRIB. Scattered in deserts and plains of Iraq: **DWD**: 7 km SW of Shithatha, *Rawi* 26903!; nr Shithatha, *Rawi & (Joseph) Serkahia* 16301! 16701! *Guest, Rawi & Rechinger* 16171!; *Weinert & Mousawi* 14692 (type of *C. iraqensis*). **DSD**: Negev to Ruhba, *Haines* W. 2122!; 18 km W by S of Samawa, *Guest, Rawi & Rechinger* 16016!, *Rechinger* 8166A!; 50 km NE of Ramadi, *Guest, Rawi & Nuri* 13548! 13556! 13570! **LCA**: Abu Ghraib, *Shehab* 41391! **LSM**: Qurna, *Graham* 128 (BM!). **LBA**: Basra, *Haines* 1886 (E!).

Syria, Iran.

2. **Climacoptera khalisica** *Botsch.* in Nov. Sist. Vyssh. Rast. 19: 78 (1982). —Type: Iraq, Daltawa (Khalis), salty area liable to flood now dry, 20.X.1960, *Hadač, Haines & Waleed* 1861 (K!).

Salsola crassa auctt.: Rech.f., Fl. Lowland Iraq: 201 (1964)

Annual, branched from base, to 35 cm tall. Branches and leaves alternate, sparsely clad with patent wavy hairs and glabrescent. Cauline leaves to 15 × 1.5 mm. Bracteoles swollen at base, exceeding the perianth, clothed with dense crispate hairs. Perianth to 5.5 mm long, glabrous, ± 12 mm diameter (including wings). Anthers linear, 2 × 0.5 mm, with a whitish appendage. Style and stigmas scarcely exserted from perianth, 3 mm.

HAB. Salty areas; alt. to 200 m; fl. Sep., Oct.
DISTRIB. Very rare, known only from the type: **LCA**: Daltawa (Khalis), *Hadač, Haines & Waleed* 1861!

Endemic.

21. HALIMOCNEMIS C.A. Mey.

in Ledeb., Fl. Altaica 1: 381 (1829)

Halanthium K. Koch, Linnaea 17: 313 (1844); *Halotis* Bunge, Anabas. Rev.: 73 (1862); *Gamanthus* Bunge, Anabas. Rev.: 76 (1862)

Annuals, very branched from the base, with erect, ascending or prostrate main stem, villose. Stems and leaves with two types of simple hairs: shorter straight hairs with no broadened base, and longer (to 5 mm) hairs with conical podium. Leaves alternate or rarely opposite, terete or semiterete, rarely flattened, with the conical persistent mucro or obtuse. Flowers hermaphrodite, solitary, supported by bract and 2 bracteoles; foliar parts of two neighbouring flowers can sometimes be fused to each other forming one diaspore consisting of two fruits. Perianth 4–5-merous, acute and slightly darker at the tip, indurated almost to the top, smooth, tuberculate or winged. Stamens 4–5; anthers with a spongy coloured vesicle that exceed the perianth. Stylodia 2. Seed embryo vertical, spirally coiled; perisperm absent.

About 20 species in the Irano-Turanian region.

Halimocnemis, from Gr.ἅλς, *hals*, salt and κνημη, *kneme*, knee [almost same as *Halocnemum*, though in this case 'cnemum' was defined as 'sheath'.

Leaves numerous, densely arranged, to 30(–35) mm long and 2 mm thick; plant mostly forming tumble-weed habit. . . . . 1. *H. pilifera*
Leaves loosely arranged and distant to each other, to 5 cm long and 2–5 mm thick; plant without tumble-weed habit . . . . 2. *H. purpurea*

1. **Halimocnemis pilifera** *Moq.* in Hist. & Mém. Acad. Roy. Sci. Toulouse, 5 (1): 181 (1839); Hedge in Fl. Iranica 172: 338 (1997); Sukhorukov & Akopyan, Comp. Chenop. Cauc.: 67 (2013).

Halimocnemis pilosa Moq., Chenop. Monogr. Enum. 152 (1840); Boiss., Fl. Orient. 4: 976 (1879).
Halimocnemis gibbosa Wol. ex Stapf, Denkschr. Akad. Wien 51: 277 (1886);
Halotis pilosa (Moq.) Iljin, Fl. USSR 6: 339 (1936); Rech.f., Fl. Lowland Iraq: 210 (1964); Zohary, Fl. Palaest. 1: 178 (1966);
Halotis pilifera (Moq.) Botsch., Nov. Sist. Vyssh. Rast. 8: 262 (1971).

Greyish annual to 20 cm, divaricately branched forming tumble-weed habit, indumentum conspicuous, partially persistent at the fruiting stage; lower leaves to 3–3.5 cm, persistent, at the top with conical yellowish mucro; perianth longer than bracteoles, 8–10 mm, pubescent, brownish or yellowish with white margins; anther 2–3 mm, appendage yellowish, 2–4 mm; fruit ± 3 mm. Fig. 91: 1–2.

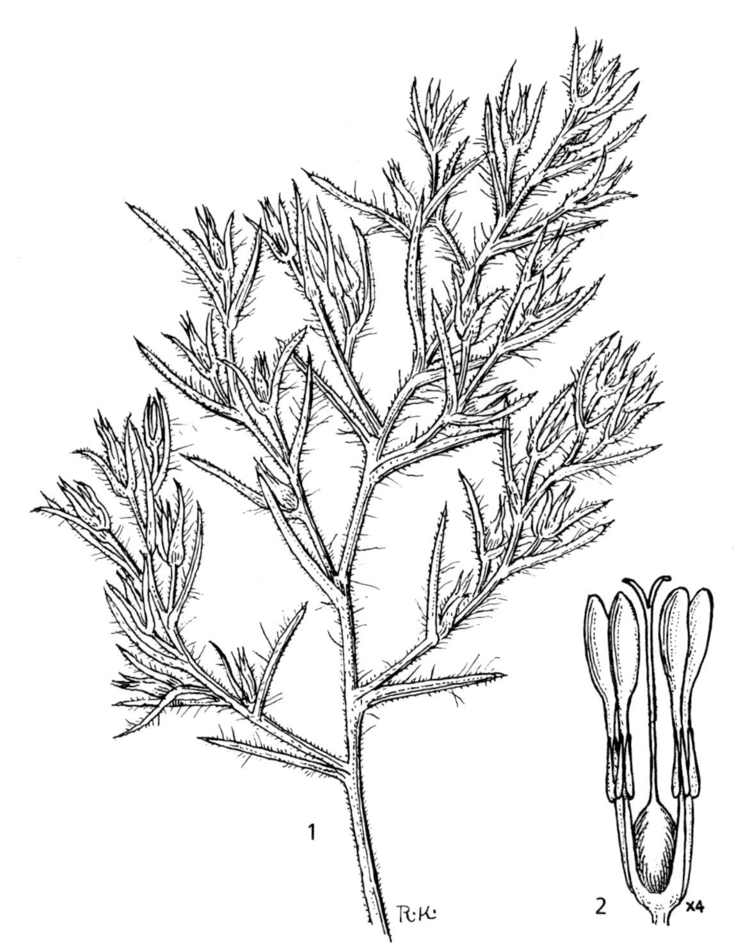

Fig. 91. **Halimocnemis pilifera**. 1, habit × 1; 2, stamens (with appendages) and carpel × 4. Reproduced with permission from Zohary, Flora Palaestina 1: Plates, f. 262. 1966. Drawn by Ruth Koppel. © The Israel Academy of Sciences and Humanities.

HAB. Extreme deserts.
DISTRIB. **DWD**: 30 km N of H3, *Serkahia & Tikrity* 15129! Syrian Desert, 70 km of Rutba, *Eig, Feinbrun & Zohary* s.n. (HUJ!).

Caucasus, Iran, Central Asia (Turkmenistan, Uzbekistan, Tajikistan), Palestine, Syria.

2. **Halimocnemis purpurea** *Moq.* in Hist. & Mem. Acad. Roy. Toulouse 5: 181 (1839); Akhani in Sendtnera 3: 8 (1996).

Physogeton acanthophyllus Jaub. & Spach, Ill. Pl. Orient. 2: 48 (1846).
Halanthium purpureum (Moq.) Bunge in Mém. Ac. Imp. Sci. Pétersb. 7, ser. 4, 11: 85 (1862).
Halotis pedunculata Assadi in Iran. J. Bot. 5 (2): 60 (1993).

Annual up to 50 cm, branched from the base, pubescent but mostly glabrescent at fruiting stage; simple hairs of two types: small hairs (to 0.5 mm) without prominent conical base, and scattered longer multicellular hairs to 2mm with conical base; leaves 20–50 × 2–5 mm; bracts shortened, to 20 mm, semiamplexicaule. Flowers solitary; perianth segments hairy, 8–10 mm, hardened in lower half or more, at fruiting stage smooth or with short tubercle; anthers with long thecae ca. 2 mm, divided to the top; anther vesicles purplish, obovate, 5–8 mm.

HAB. River sides, gypsium soils, disturbed areas; fl. May–Jun; fr. Jul.–Aug.
DISTRIB. **LCA**: Anaiza, *Barkley* 2458! 71 km E of Baquba, *Agnew & Barkley* 2376! **LSM**: 75 km N of Amara, *Hazim & Nuri* 30628!

Iran.

SUBFAMILY SALSOLOIDEAE: TRIBE SALSOLEAE

22. **HALOTHAMNUS** Jaub. & Spach

Ill. Pl. Or. 2: 50 (1845) emend. Botschantzev, Not. Syst. (Leningrad) 18: 146 (1981)
Aellenia Ulbr. in Engler & Prantl., Natürl. Pflanzenfam. ed. 2, 16c: 567 (1934)

Ref.: G. Kothe-Heinrich: Revision der Gattung *Halothamnus* (Chenopodiaceae). Bibl. Bot. 143: 1–176 (1993)

Annual herbs or small shrubs, glabrous or hispidulous, often with unpleasant smell. Branches grey, new growth glaucous. Leaves alternate, entire, semiterete (in ours), ± succulent; axils with some crisped hairs. Bracts and bracteoles ± scale-like except for lowermost bracts which resemble leaves, with membranous margins, bracteoles shorter than perianth. Flowers hermaphrodite. Perianth 5, free, erect, slightly irregular (3 wider than the other 2). Stamens 5, epipetalous; filaments ribbon-like. Anthers oblong to narrowly sagittate. Hypogynous disc glabrous, glandular on upper surface, fleshy, becoming membranous. Stigmas 2; style thickening in fruit. Fruiting perianth with 5 horizontal membranous wings with radiating veins, tubular below and truncate at base; utricle enclosed by perianth, flattened; pericarp membranous. Seeds with horizontal coiled embryo; perisperm absent.

A genus of about 20 species distributed from SW Asia to China and S to Somalia. Most species are halophytes, and some are important for grazing animals while others contain alkaloids. Their characteristic feature is the indurated fruiting perianth.

Halothamnus (from Gr. ἅλς, *als*, salt and θάμνος, *thamnos*, bush)

1. Annual but with basally stiff stem forming tumble-weed habit; branches mostly hispidulous; bracteoles orbicular, exceeding perianth . . . . 1. *H. hierochunticus*
   Shrublet or shrubs, glabrous or hispidulous, branched from the base but not forming tumble-weed habit; bracteoles equalling or shorter than perianth 2
2. Plants grabrous or hispidulous; leaves caducous, ± 2 mm wide at base; upper vegetative leaves to 5 mm . . . . 2. *H. iraqensis*
   Plant glabrous and glaucous; leaves (at least some) persistent, 3–5 mm wide at base. . . . . 3. *H. lancifolius*

1. **Halothamnus hierochunticus** (*Bornm.*) *Botsch.* in Not. Syst. (Leningrad) 18: 156 (1981).

Salsola hierochuntica Bornm. in Beih. Bot. Centralbl. 29 (2): 13 (1912).
Aellenia hierochuntica (Bornm.) Aellen in Verh. Naturf. Ges. Basel 61: 180 (1950); Rech.f., Fl. Lowland Iraq: 203 (1964).

Annual up to 50 cm, sparsely pubescent or glabrous, greyish; stems much branched forming tumble-weed habit. Leaves to 30 × 3 mm (except the lowermost, caducous leaves to 8 cm), semiterete. Bract and bracteoles 3–6 mm, orbicular. Flowers solitary along branches, forming leafy spikes. Fruiting perianth (0.6–)1–1.2(–1.5) cm in diameter (including wings); lobes ovate, obtuse, connivent, with orbicular-obovate wings.

HAB. Deserts, mostly on loess soils, or weed in cultivated areas.
DISTRIB. **MSU**: nr Kirkuk, *Eig, Feinbrun & Zohary* s.n. (HUJ!, LE!). **FUJ**: Hatre, Mosul liwa, *Barkley & Haddad* 9167 (W!); *Guest* 3536 (LE!); *Beije, Barkley & Mohamed* 3354 (LE!); Jabal Hamrin, *Barkley* 3849 (LE!). **FPF**: Mandali, *Haddad* & *Agnew* 6745 (LE!). **LCA**: nr Baghdad, *Lazar* 441 (F, after Rilke, 1999).

Turkey, Iran, Eastern Mediterranean.

2. **Halothamnus iraqensis** *Botsch.* in Not. Syst. Vyssh. Rast. (Leningrad) 18: 152 (1981); Boulos in Fl. Arab. Penins. & Socotra 1: 279 (1996); Kothe-Heinrich in Fl. Iranica 172: 185 (1997).

Aellenia subaphylla auctt. — Type: Iraq: [DWD] Southern Jazira (about 50 km NE of Ramadi), dry saline depression, alt. c. 40 m, 16.XI.1954, *E. Guest, A. al-Rawi & A. Nuri* 13547 (iso LE!).

Small shrub or shrublet to 60 cm tall with stiff woody stems branching copiously from base, lateral branches 10–20 cm; bark of older branches splitting longitudinally. Spring leaves linear, 10–18 × 0.6–1.2 mm, caducous, summer leaves and bracts triangular-ovate, succulent, 3–5 × 2–3 mm, mucronate. Bracteoles clasping perianth, broadly ovate, 2–4 × 2–3 mm, weakly keeled, with narrow membranous margins. Perianth 4–6.5 mm long; tepals pale yellow. Filaments arising from rim of disk, ± 0.75 mm wide; anthers 3–4 mm. Perianth at fruiting stage pale pinkish-yellow, with irregular wings of 11–15 mm in diameter. Fig. 92: 1–3.

With its long narrow spring leaves (rarely seen on herbarium specimens) it can be confused with *H. glaucus* and *H. subaphyllus*, which have not so far been recorded from Iraq.

Two varieties in Iraq; both occupy similar habitats and geographical areas, but var. *hispidulus* has a slightly more southerly distribution; both varieties extend into Saudi Arabia.

Leaves and stem glabrous; bracteoles of upper flowers without membranous apex . . . . a. var. *iraqensis*
Leaves and stem hispidulous; bracteoles of upper flowers with narrow membranous apex . . . . b. var. *hispidulus*

a. var. **iraqensis**: Mandaville, Fl. Eastern Saudi Arabia 44 (1990).

HAB. Sandy desert, saline soils, dunes, 'haswa', gypsiferous sandstone; alt. 0–180 m; fl. Aug.–Nov., fr. Oct.–Dec.
DISTRIB. Widespread in the Desert and Dry Steppe regions of Iraq: **MSU**: Sulaimaniya, *Fawzi* 39531! **FPF**: Kuraiz al-Malih, about 15 km ENE of Jaliba, station, *Guest, Rawi & Rechinger* 16140!, *Rechinger* 8255! **DLJ**: Nr Lake Thirthar, *Barkley & Abbas* 33197!; 3 km SW of Lake Thirthar, *Barkley, Hikmat Abbas & Inman* 1699A!; **DLJ/LCA**: 70 km NW of Falluja nr Thirthar, *Chakravarty & Rawi* 30418! **DWD**: about 30 km S of Ramadi, *Guest, Rawi & Rechinger* 16199!; 50 km NE of Ramadi, *Guest, Rawi & Nuri* 13547 (type of var. *iraqensis*); id., 13571!; by lake Habbaniya, *Guest* 3537!; above Ukhaidhir Valley, Kerbela, *Grant & Barkley* 8576!, *Barkley, Haddad & Rechinger* 6238!; between Shithatha and Rahhalyah, *Rawi & Serkahia* 16312!; **DSD**: 40 km SW of Ur towards Busaiya, *Rechinger* 8181A!; about 40 km S by W of Ur, *Guest, Rawi & Rechinger* 16063!; *Rechinger* 8145!; 30 km SSW of Tal Kalm (nr Ur), *Guest & Ibrahim Mahallal* 15211E!; 40 km W of Karbala, *Rawi & Nuri* 16291, *Rawi* 26894!; 2 km S of Khadar al-Mai, *Guest, Rawi & Rechinger* 16117!; Kuraz al-Malih (about 15 km ENE of Jaliba station), *Guest, Rawi & Rechinger* 16140 (*Rechinger* 8255)!; Khadhar al-Mai, *Alizzi & Omar* 35871!; 18 km W by S of Samawa, Guest, *Rawi & Rechinger* 16018! **LCA**: nr Sharqiya, *Haddad, Safat & Barkley* 3730!; between Nasiriya and Diwaniya, *Husham* (*Alizzi*) 29636!; 30 km W of Darraji, *Rawi & Alkas* 16246!, 16255! **LSM**: 25 km N of Amara, *Guest, Rawi & Rechinger* 17429!

Vernacular name BŪ'AIDH (Habbaniya, *Guest* 3537), SHINAN (nr Shithatha, *Rawi & Serkahia* 16312), 'AIWA (*Rawi & Rechinger* 16117 nr Khidr al-Mai) 'AIWA (nr Busaiya, *Guest, Rawi & Rechinger* 16107), RIMITH (*Rawi* 12543, Busaiyah), RIMTH ABU RAIDA (= WARDA), Jazira (*Guest & Rawi*

15974); an important fodder plant, but known to contain alkaloids.

Syria, Saudi Arabia, Kuwait.

var. **hispidulus** *Botsch.*, op. cit. 153 (1981). —Type: Iraq: [DSD] southern desert, Al-Batin, (about 15 km of Khadhar al-Mai), alt. ± 160 m, 9.XI.1956, *E. Guest, A. al-Rawi & K. Rechinger* 16125 (iso, LE!).

HAB. Sandy and gravelly desert, silty soil prone to flooding, dry hills; alt. 0–165 m; fl. Aug.–Nov., fr. Oct.–Dec. (–May).

DISTRIB. Widespread in the Deserts and Alluvial Plains of Iraq: **FPF**: Jabal Hamrin, *Rechinger* 8080a! **DLJ**: Thirthar, 40 km NNE of Ramadi, *Gillett* 9904!; Nr lake Thirthar, Ramadi, *Barkley & Hikmat Abbas* 3618!; nr Balad, *Baltaxe* 13474, 13474A; *Robertson* 13473!; S Jezira, by ancient Ishaqi canal, *R. Baltaxe* 13471! **DWD**: near the fork in the Falluja-Ramadi highway leading to Lake Habbaniya, *Barkley, Palmatier & Hikmat Abbas* 156!; **DSD**: Al-Batin, about 15 km E of Khadhar al-Mai, *Guest, Rawi & Rechinger* 16125! (type of var. *hispidula* Botsch.); Busaiyah, *Rawi* 12543!; 55 km ESE of Busaiya, *Guest, Rawi & Rechinger*

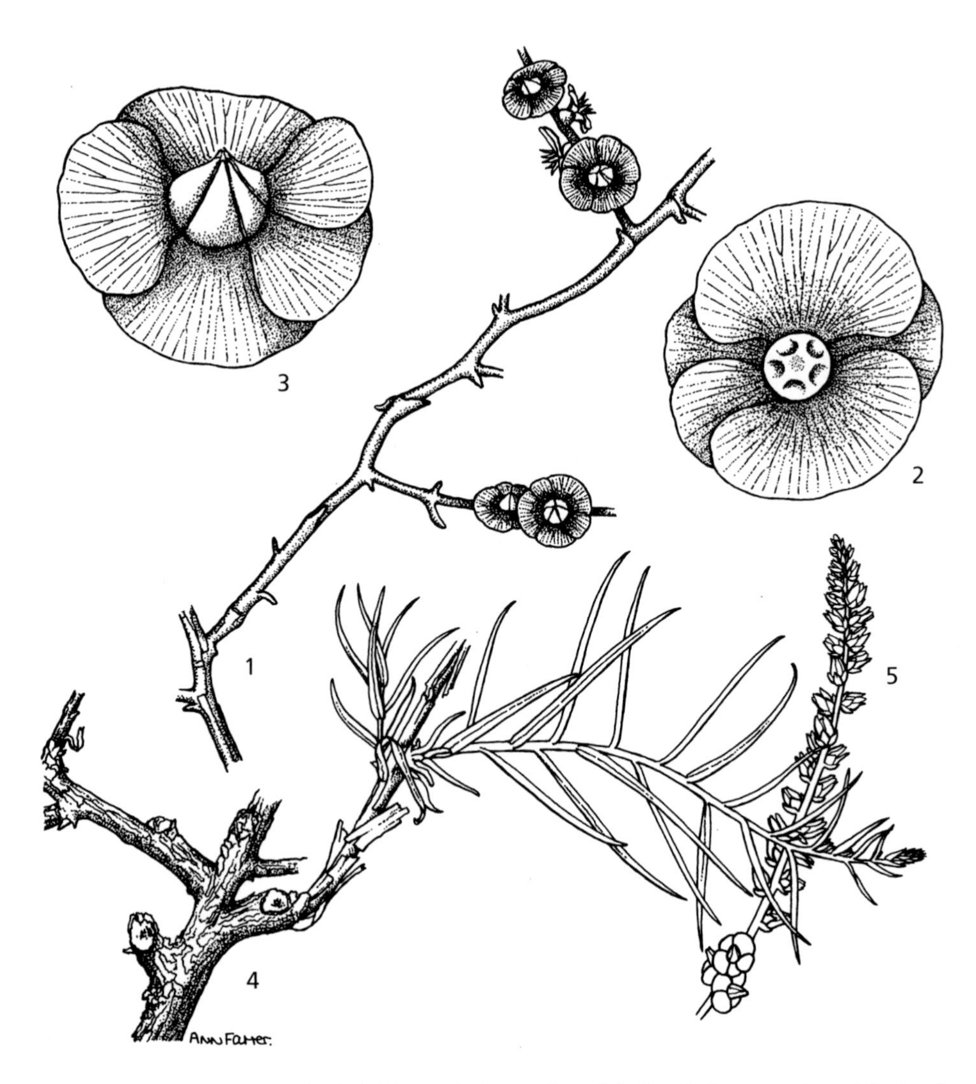

Fig. 92. **Halothamnus iraqensis**. 1, habit × 1; 2, lower view of fruit × 3; 3, upper view of fruit × 3. **Halothamnus lancifolius**. 4, habit × 1; 2, flowering and fruiting branch × 1. Reproduced with permission from Flora of Arabia 1: fig. 53. 1996. Drawn by Ann Farrer.

16107! *Rechinger* 8234 (B!); Khadhar al-Mai, *Alizzi & Omar* 35875!; **LEA**: nr Muqdadiya, *Guest* 15800!; **LCA**: Jazira (55 km N by W of Falluja), *Guest, Rawi & Rechinger* 14529!; Jazira (about 44 km NW by N of Falluja), *Guest & Rawi* 15974!; 15 km W of Falluja, *Barkley, Palmatier & Hikmat Abbas* 148!; **LSM**: Jufail nr Al-Garmah, *Rawi & Shahwani* 26941!

Endemic.

3. **Halothamnus lancifolius** *(Boiss.) Kothe-Heinrich* in Bibl. Bot. 143: 88 (1993); Boulos in Fl. Arab. Penins. Socotra 1: 279 (1996).

Caroxylon lancifolium Boiss., Diagn. Pl. Orient., ser. 1, 12: 98 (1853).
Salsola lancifolia (Boiss.) Boiss., Fl. Orient. 4: 958 (1879); Rech.f., Fl. Lowland Iraq: 201 (1964).

Subshrub to 60 cm, glabrous, glaucous; stem branching mostly in upper part; lower leaves to 40 mm long, succulent, slightly decurrent, triangular with broadened base to 5 mm; the upper leaves shorter, to 15 mm. Bracts longer than perianth 4–5 mm at anthesis, with ovate or triangular segments, slightly concrescent at fruiting stage, with wings originated below the middle, 10–15 mm across. Fig. 92: 4–5.

HAB. Deserts, 0–500 m; fl. Apr.–Jun.; fr. Jun.–Aug.
DISTRIB. **DWD**: Rutba, *Hadač* 4425 (G!); 130 km W of Rutba, *Eig* s.n. (HUJ!); between Ramadi & Rutba, *Rechinger* 12634 (B!); 12635 (B!, W!, E!); 2 km W of Rutba, *Rechinger* 15920 (W!). **DSD**: 514 km W of Baghdad, southern desert, *Eig & Zohary* s.n. (HUJ!).

S Iran, Arabia, Eastern Mediterranean, NE Africa.

EXCLUDED SPECIES

Halothamnus auriculus (*Salsola auricula, Aellenia auricula*): Blakelock in Kew Bull. 12, 3: 487 (1957), probably does not occur in Iraq; no specimens seen.

Halothamnus glaucus (*Salsola glauca, Aellenia glauca*): Anthony J., Plants from Mesopotamia (1935); Blakelock in Kew Bull. 12, 3: 490 (1957), Rech.f., Fl. Lowland Iraq: 203 (1964). The specimen cited in Blakelock (1957) belongs to *Halothamnus iraqensis.*

Halothamnus subaphyllus (*Salsola subaphylla, Aellenia subaphylla*), Blakelock in Kew Bull. 12, 3: 493 (1957), Rech.f., Fl. Lowland Iraq: 204 (1964). No records from Iraq, all specimens belong to *H. iraqensis.*

23. **KALI** Mill.

in Gard. Dict. Abr., ed. 4, 2: 715 (1754)

Ref.: S. Rilke, "Revision der Sektion *Salsola* s.l. der Gattung *Salsola*". Bibl. Bot. 149: 1–189 (1999).

Annuals or subshrubs, glabrous or with papillae. Leaves mostly alternate, or lower leaves arranged opposite, semi-terete or terete, stuff or (lower leaves) fleshy, with a persistent mucro to 3.5–4 mm. Bracts longer than bracteoles or equal in size. Flowers axillary, solitary or 2–3 in clusters. Perianth segments 5, hyaline or membranous, in the fruiting stage sometimes hardened, mostly with tubercules or equal or uinequal wings. Stamens 5, without prominent appendages at the tip of the anthers. Stigmas 2. Fruit dry; seed with horizontal embryo; perisperm lost.

About 12 species in Eurasia and Australia, some species are alien noxious weeds in North America. Previously included in the genus *Salsola.*

Kali, from medieval Lat. *kalium,* Potassium.

Kali tragus (*L.*) *Scop.*, Fl. Carniol., ed. 2, 1: 175 (1772).

Salsola tragus Cent. Pl. 2: 13 (1756); Rilke in Bibl. Bot. 149: 111 (1999); Freitag et al. in Fl. Pak. 204: 148 (2001); Zhu, Mosyakin & Clemants in Fl. China 5: 411 (2003).
Salsola ruthenica Iljin, Sorn. Rast. SSSR 2: 137 (1934); Rech.f., Fl. Lowland Iraq: 199 (1964).

Annual, 20–100 cm, very branched forming tumble-weed habit at fruiting stage; glabrous or rarely papillate; lower leaves to 80 mm, filiform or linear, fleshy and recurved; upper leaves, bracts and bracteoles stuff, subulate, to 40 mm, semi-terete; all mucronate with a yellowish mucro to 4–5 mm. Flowers of two kinds: two-flowered arranged in stiff and easily caducous clusters in the axils of lower leaves supported by floral leaves, and flowers grouped in the terminal lax inflorescence. Perianth of 5 free hyaline segments that is smooth or short-tuberculate in the flowers agglomerated in clusters, and winged at the fruiting stage onto flowers in the inflorescence. These wings 7–12 mm across, white or yellowish, sometimes pinkish. Fig. 93: 1–2.

HAB. Deserts, riversides, sandy places, disturbed areas; alt. ; fl. Aug.–Sep.; fr. Sep.–Oct.
DISTRIB. Scattered but sometimes common in the deserts, plains and lowlands of Iraq: **FPF**: Jabal Hamrin, *Gillett* 12497!, *Sutherland* 477 (BM!). **DWD**: Haditha, *Rawi, Khatib & Alizzi* 32938!; Habbaniya, *Rechinger* 8305 (BM!, LE!). **LCA**: Nasiriya, *Graham* 280 (BM!); Baghdad, *Haines* 71 (E!); Baghdad, *Rechinger, Haines & Khudairi* 25 (E!). **LSM**: Amara, *Evans* s.n. (E!).

Widely distributed in the deserts and steppes of the Irano-Turanian region and N Africa; alien in North America.

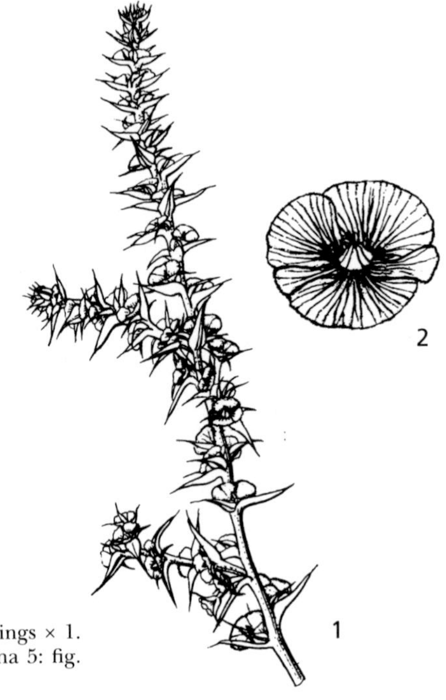

Fig. 93. **Kali tragus**. 1, habit × 1; 2, perianth with wings × 1. Reproduced with permission from Flora of China 5: fig. 333. Drawn by Liu Chunrong.

24. **SEIDLITZIA** Bunge ex Boiss.

Fl. Orient. 4: 950 (1879)

Ref.: Bochantsev, V.P., Generis *Seidlitzia* Bunge species oblivia. Nov. Syst. Vyssh. Rast. (Leningrad), 7: 145–146 (1970).
Iljin, M.M., De genere *Seidlitzia* Bunge notae criticae. Nov. Syst. Vyssh. Nizsh. Rast. (Leningrad), 16: 86–93 (1954).

Shrubs or annual herbs, glabrous or with crimped hairs in leaf axils. Stems whitish, much-branched, with opposite leaves. Leaves succulent, terete, sessile, obtuse. Flowers hermaphrodite or unisexual, in clusters of 3–5 in leaf axils or solitary. Perianth segments 5, obtuse, much expanded in fruit to form wings. Stamens 5, epipetalous; anthers exappendiculate. Styles 2, linear. Fruit horizontal, pericarp smooth. Seeds lacking perisperm; embryo spiral.

Three or four species distributed in the Irano-Turanian region; two species in Iraq.

Seidlitzia (named in honour of Count N.K. von Seidlitz, the noted explorer of the Caucasus (1831) or perhaps his son (Shishkin 1936). Recently, the genus is merged with *Salsola* (Akhani & al., 2007). However, the morphological characters of both genera are quite different, and we still accept the limit of *Seidlitzia* before extended molecular investigations.

Shrubs; cyme mostly with 3 flowers . . . . 1. *S. rosmarinus*
Annuals; flowers at least lower ones solitary . . . . 2. *S. cinerea*

1. **Seidlitzia rosmarinus** *Ehrenb. ex Boiss.*, Fl. Orient. 4: 951 (1879). Shishkin in Fl. S.S.S.R. 6: 275 (1936); Rech.f., Fl. Lowland Iraq: 197 (1964); Zohary, Fl. Palaest. 1: 167 (1966); Hedge in Fl. Iranica 172: 290 (1997); Boulos in Fl. Arab. Penins. Socotra 1: 265 (1996); Mozaffarian, Trees & Shrubs of Iran: 173 (2005); Abdel Bary, Flora of Qatar 1: 123 (2012).

Much-branched shrub up to 1.5–2 m, stems ± whitish, with greyish flaking bark. Leaves opposite, sessile, fleshy, linear, 12–30 × 2–3 mm, deciduous. Flowers arising in groups of 3 in leaf axils, subsessile. Perianth segments partly united, pinkish at anthesis, developing in fruit into 5 unequal spreading reniform wings, purplish or later brownish, 8–12 mm in diameter. Fig. 94: 1–3.

HAB. Sandy, gypsaceous or stony desert, alluvial flats with saline or calcareous substrata, alt. 0–100 m; fl. Aug.–Oct.; fr. Nov.–Dec.

DISTRIB. Widespread across the Deserts and Alluvial Plains of Iraq: **FPF**: 20 km S of Badra, *Rechinger* 12712!; **DLJ**: Thirthar, 40 km NNW of Ramadi, *Gillett* 9901! **DWD**: 50 km NW of Karbala towards Shithatha, *Rawi & Serkahia* 16297!; 40 km W of Karbala, *Rawi* 26892!; 52 km W of Karbala towards Shithatha, *Rawi* 26898, 26901!; 25 km S by W of Ramadi, *Guest, Rawi & Rechinger* 15942!; Shithatha, 42 km W of Karbala, *Guest, Rawi & Rechinger* 16172!; 23 km NW of Rahhaliya, *Rawi* 26932!; 50 km NW of Karbala towards Shithatha, *Rawi & Serkahia* 16296!; Bahr al-Milh, 30 km W by S of Karbala, *Guest, Rawi & Rechinger* 16160!; Shithatha, *Agnew & Haines* W.2117!; between Shithatha and Zakhalyah, *Rawi & Serkahia* 16311!; 30 km NW of Karbala, *Rawi & Chakravarty* 32552!; nr Shithatha, *Rawi* 19952, 19965! **DSD**: 23 km W of Samawa, *Guest & Mahallal* 15208!; Kuraz al-Mali 15 km ENE of Jaliba station, *Guest,*

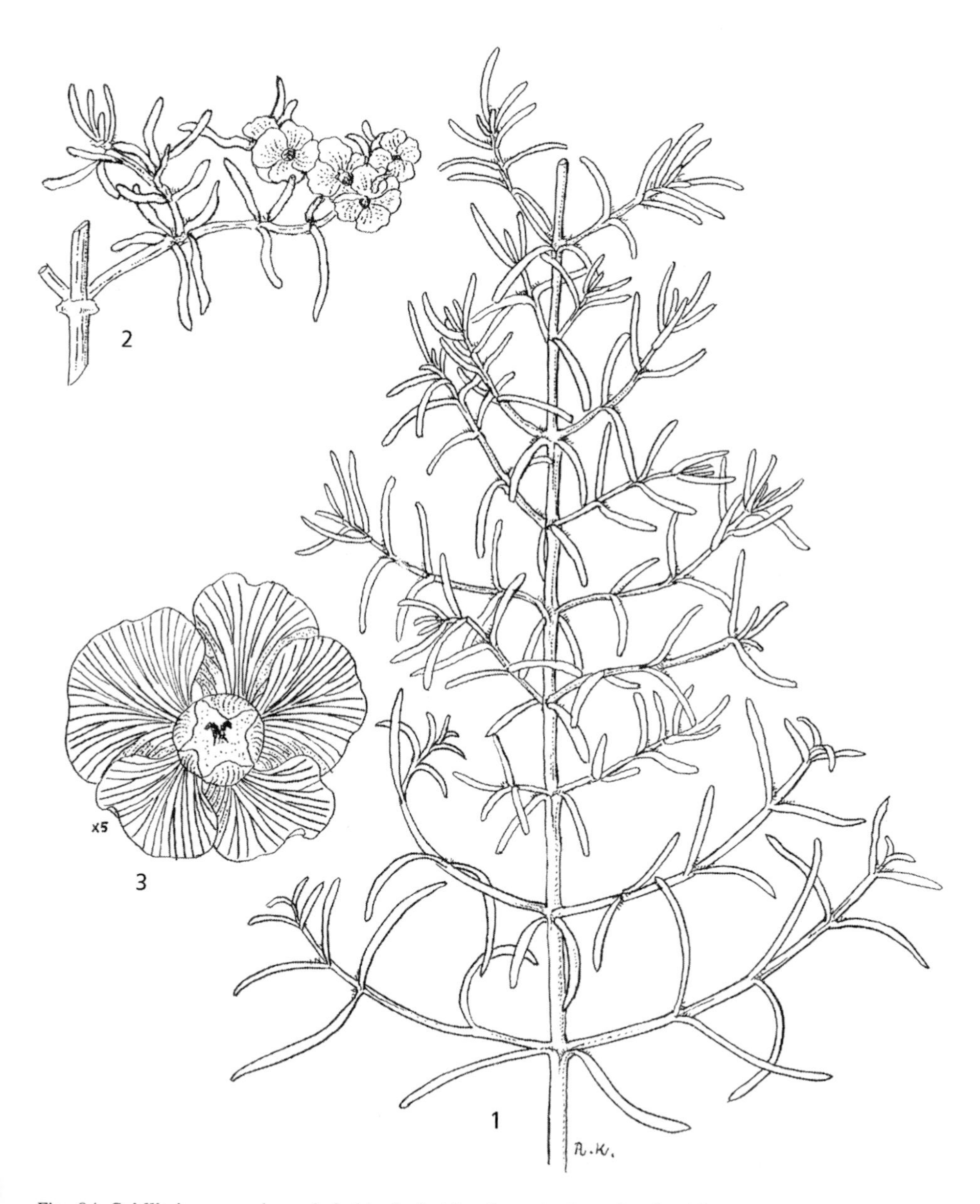

Fig. 94. **Seidlitzia rosmarinus**. 1, habit; 2, fruiting branch; 3, perianth with wings. Reproduced with permission from Zohary, Flora Palaestina 1: Plates, f. 243. 1966. Drawn by Ruth Koppel. © The Israel Academy of Sciences and Humanities.

Rawi & Rechinger 16137!; 38 km SE by S of Zubair, *Guest, Rawi & Rechinger* 16877!; about 20 km SSE of Ur, *Guest, Rawi & Rechinger* 16033!; S of Samawa, *Rechinger* 8163 (E!); Salaibiyat al Hamr, about 35 km S by W of Ur, *Guest, Rawi & Rechinger* 16044B!,16045!, 16046! **LEA**: N of Amara, nr Wadi al-Taiyib, *Guest, Rawi & Rechinger* 17464!; Fakka, 75 km NE of Amara, *Rawi & Alizzi* 32411! **LCA**: 30 km W of Darraji, between Nasiriya and Samarra, *Rawi & Alkas* 16239!; 30 km W of Darraji, nr Al-Khedher, *Rawi & Alkas* 16249!; near lake in Diwaniya region, *Barkley, Safat & Haddad* 3739! **LSM**: shore of Hor Hammah nr Djelibah, *Rechinger* 8256! **LBA**: Basra, inter Ur & Al Busaya, *Rechinger* 8185 (E!).

SHNAN (*Rawi & Alkas* 16249, nr Darraji), AWAISJA (*Gillett* 9901, Thirthar).

Widely distributed in Irano-Turan and Eastern Mediterranean region. Egypt and SW Asia eastwards to Afghanistan, S Uzbekistan and Tajikistan.

2. **Seidlitzia cinerea** (*Moq.*) *Bunge ex Botsch.* in Nov. Sist. Vyssh. Rast. 7: 145 (1971).

Halogeton cinereus Moq., Chenop. Monogr. Enum.: 159 (1840).

Annual, glabrous (esxcept leaf axils), up to 50 cm. Stem whitish, with ascending branches. Leaves glaucous, fleshy, mostly straight, obtuse, to 20 mm. Bract about two times longer than bracteoles. Flowers solitary in leaf axils (the upper flowers can be arranged in 2–3). Perianth lobes oblong-oval, veined, winged at fruiting stage; wings mauve turning brown, 6–11 mm across. Anthers ± 1 mm. Fruit 1.5–2 mm in diameter.

HAB. Sandy or clayey deserts. Very rare in our area.
DISTRIB. **MSU**: Sulaymaniya, *Plitman* 2913 (HUJ!).

Iran, Caucasus, Eastern Mediteranean (rare).

25. **NOAEA** Moq.

in DC., Prodr. 13 (2): 207 (1849)

Noea Boiss. & Bal. in Boiss., Diagn. ser. 2, 4: 76 (1859)

Annual herbs or subshrubs, glabrous, papillate or with crimped hairs in leaf axils. Stems much-branched from the base, often intricately so, tips often spiny. Leaves not succulent, alternate, linear or filiform, sessile, margin ± scarious at base. Flowers hermaphrodite, arising singly or in clusters in axils of leaves; bracts 2, similar to leaves. Tepals 5, membranous, developing in fruit into spreading wings. Stamens 5; anthers ± sagittate, with terminal appendage. Ovary erect, with 2 conspicuous styles. Seed vertical, compressed; perisperm absent; embryo vertical, spiral.

A genus of about five species distributed across SW & C Asia; the limits of species are poorly defined and the subshrubby species might be seen as elements of a polymorphic complex that needs further investigations.

Noaea (following Fl. SSSR vol. 6) commemorates Frank, Vicomte de Noë, born Friedrich Wilhelm Noë (1798–1858), German botanist and pharmacist at Fiume (now Rijeka) who became the director of a Botanic Garden in Istanbul (Nehmé 2000). He was a major contributor of specimens from Turkey and Iraq to Boissier's herbarium in Geneva.)

1. Branches with no spines .... 3. *N. tournefortii*
   Branches terminating in a spine.. 2
2. Plant glabrous; inflorescence consisting more than 15 flowers .... 1. *N. mucronata*
   Plant papillate; inflorescence up to 15 flowers, very lax, internodes 5–20 mm in lower and median part .... 2. *N. kurdica*

1. **Noaea mucronata** (*Forssk.*) *Aschers. & Schweinf.* in Mém. Inst. Egypte 2: 131 (1887); Rech.f., Fl. Lowland Iraq: 206 (1964); Zohary, Fl. Palaest. 1: 175 (1966); Aellen in Fl. Turk. 2: 335 (1967); Hedge in Fl. Iranica 172: 294 (1997); Boulos in Fl. Arab. Penins. Socotra 1: 281 (1996); Sukhorukov & Akopyan, Comp. Chenop. Cauc.: 60 (2013).

Salsola mucronata Forssk., Fl. Aegypt.-Arab. 56 (1775); *Anabasis spinosissima* L.f., Suppl. 173 (1781).

Much-branched spiny subshrub. Stems rigid, terete, branches erecto-patent below to slightly recurved in upper part, ending in spines. Leaves 10–25 × 0.5–1 mm, scarcely fleshy, terete, filiform except at base. Flowers arising from axils along almost whole length of stem,

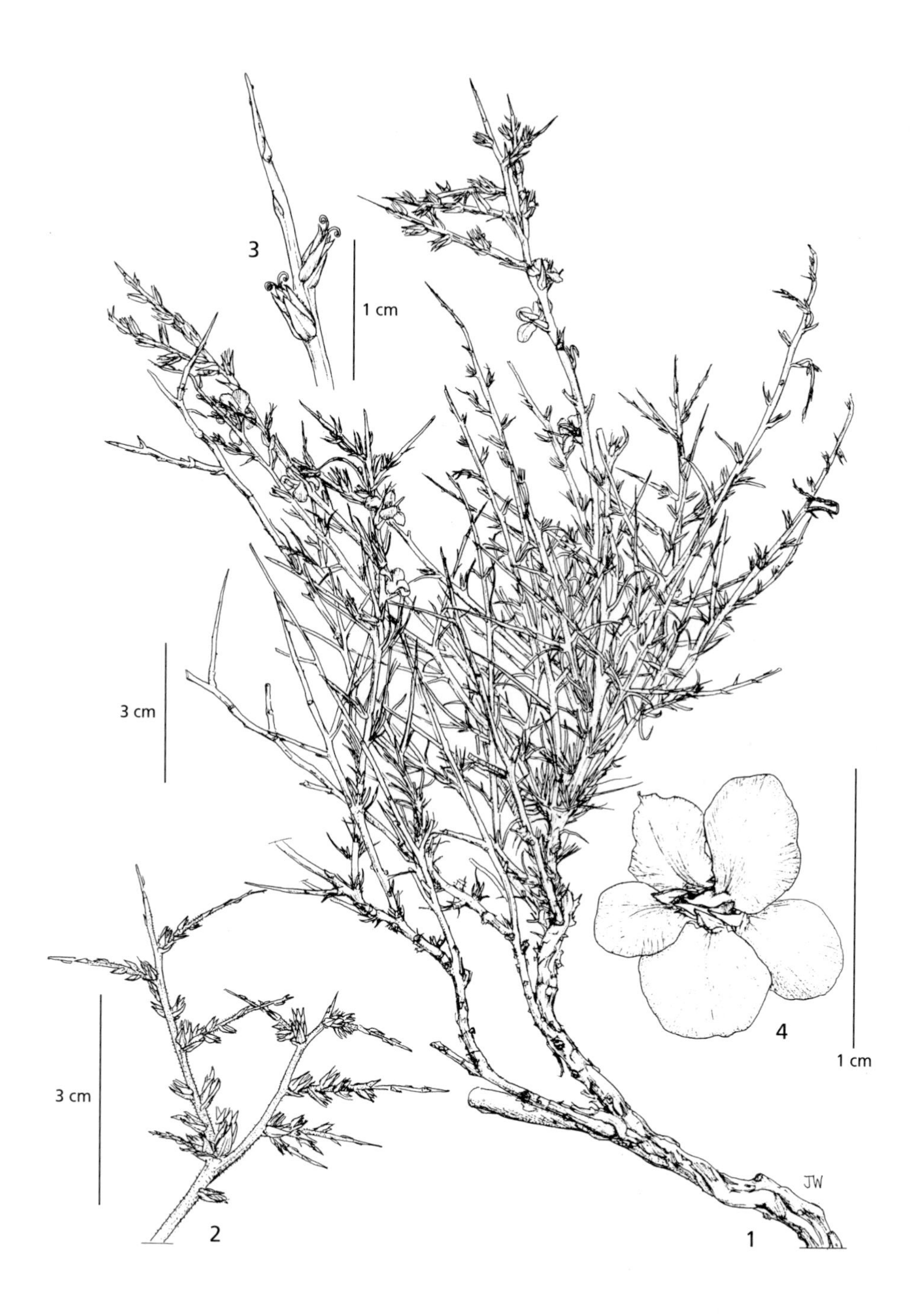

Fig. 95. **Noaea mucronata.** 1, habit; 2, portion of stem; 3, flowers; 4, fruit showing oerianth with wings. 1 from *Lambert & Thorp* 527; 2, 3 from *Mooney* 4613; 4 from *Maitland* s.n. © Drawn by Juliet Beentje.

solitary or in small clusters, subtended by 2 leaflike bracts which barely enclose the flowers. Perianth ± 4 mm, membranous, ovate, apiculate. Anthers ± 2 mm, with a small triangular apical projection. Wings of fruiting perianth to 10 mm across, membranous, pinkish or white, often turning brown. Pericarp slightly fleshy, green or pinkish in upper fruit part. Fig. 95: 1–4.

HAB. Gravelly and sandy places, sandstone hills, roadsides, limestone crags; alt. (100–)700–2500 m; fl. Aug.–Oct.; fr. Oct.–Nov.
DISTRIB. Widespread in the hills and northern desert zones of Iraq: **MRO**: Mt. Qandil above Bisht Ashan, *Rechinger* 11214!; nr Rayat, *Zohary & Amdursky* s.n. (HUJ!); Warshanka to Magar range, *Rawi & Serhang* 24333!; Haji Omran, *Rawi & Serhang* 24278, *Guest & Alizzi* 15890! **MSU**: Sulaimaniya, *Fawzi* 39546! **MJS**: Jabal Sinjar, *Alizzi & Omar* 35327! **FUJ**: Hadhr, *Guest* 3553!; Bara, *Handel-Mazzetti* 1567 (WU!). **FAR**: Makhmur, *Robertson* 211!, *Hunting Aero Surveys* 258! **FKI**: Qara Anjir, nr Kirkuk, *Karim* 39385!; *Uvarov* 25 (HUJ!); Ghurfa plain nr Injana, *Guest* 4007!; nr Tauq bridge, *Guest* 15541!; **FKI/DLJ**: Al-Fatha, Jabal Hamrin 25 km N of Baiji, *Khatib & Tikriti* 29716!; nr Anaisa, *Agnew & Barkley* 2446!; **FPF**: Anaisa nr Tursaq, *Haines* W.1184!; Mandali, Koma Sang, *Guest* 795!; Jabal Hamrin, *Haines* s.n. (E!); Inaiza, *Haines* 1194 (E!); **DGA**: 16 km SW of Samarra, *Barkley* 3822! **DWD**: 40 km E of Rutba, *Omar, Qaisi & Hamad* 43965!; Wadi Fhaimi between Haditha and Ana, *Rawi & Gillett* 6936! **LEA**: Jabal Hamrin between Shahraban (Muqdadiya) and Jalaula, *Rechinger* 8073!

SIRR (*Guest* 3553, Hatra). Eaten by sheep and camels.

N Africa, SE Europe, SW Asia eastwards to Afghanistan.

2. **Noaea kurdica** *Eig* in Palest. J. Bot., Jerusalem ser., 3: 134 (1945). —Type: Iraq [MSU], Iraquian Kurdistan, Suleimani district, [Qara-Dagh] Pir i Mukurum Dagh, rocks near the snow, 19.9.1933, *A. Eig & Sh. Duvdevani* 970 (holo, HUJ-28267!).

Subshrub to 30–40 cm high, similar to *N. mucronata*, but shortly papillate; inflorescences very loose, consists of maximum of 15 flowers; bracts equal with bracteoles, sometimes gibbous at base; perianth brownish. Fruit short papillate.

HAB. Gravelly slopes; alt. 1000–2500 m.
DISTRIB. **MSU**: Pir i Mukurum Dagh, *Eig & Duvdevani* 970 (type) (HUJ!). **FPF**: Kan-i Sakht, *Zohary & Duvdevani* s.n. (HUJ!).

Eig (1945) did not provide any differences between *N. kurdica* and *N. mucronata* in his diagnosis. This is a poorly known species that requires further investigation. Now considered to be endemic to Iraq, but expected to be found in Turkey, W Iran and Syria.

3. **Noaea tournefortii** (*Jaub. & Spach*) *Moq.* in DC., Prodr. 13 (2): 208 (1849); Sukhorukov & Akopyan, Comp. Chenop. Cauc.: 61 (2013).

Anabasis tournefortii Jaub. & Spach, Ill. Pl. Or. 2: 43 (1844–46).
Noanea mucronata subsp. *tournefortii* (Spach) Aellen in Mitt. Basl. Bot. Ges. 1 (1): 13 (1953); Aellen in Fl. Turk. 2: 336 (1967).

Subshrub with no spines. Stems rigid, terete, branches erecto-patent below to spreading, primary branches not or little branched. Leaves 15–30(–40) × 0.5–1.0 mm, scarcely fleshy, terete, filiform except at base. Flowers arising from axils along upper of stem, solitary or in small clusters, subtended by 2 leaflike bracts which distinctly exceed the flowers. Tepals membranous, ovate, apiculate. Anthers ± 2 mm, with a small triangular apical projection. Wings of fruiting perianth pinkish red, accrescent to 5 × 4 mm, membranous.

HAB. Mountain sides, dry stony roadsides, rocky slopes; alt. 1000–1820 m; fl. Jul.–Aug., fr. Oct.
DISTRIB. Mainly in the thorn-cushion zone in montane regions of Iraq, but also extending into the *Quercus/Pistacia* forest zone: **MAM**: Mt. Gara, *Kotschy* 345 (type)!; Jabal Khantur NE of Zakho, *Rawi* 23460!; Sarsang, *Haines* W.432!; Saer Amadiya, *Agnew & Haines* s.n. (E!); **MRO**: 10 km S. of Haji Umran, *Rawi & Serhang* 24270!; Kholan, *Rawi* 13804!; Qandil range, Baisar, *Rawi & Serhang* 24171!; Qandil range NE of Rania, *Rawi* 26820! *Rechinger* 11214 (E!); Mt. Baski Hawaran, *Rawi & Serhang* 23964!; Algurd Dagh, Lalan, *Gillett* 12468! **MSU**: Sulaimaniya, *Haines* W.1315!; Pira Magrun, *Haines* s.n. (E!).

GA SURIK (possibly a contraction of GIAH SURIK), *Rawi* 13804 (Kholan).

Iran, Turkey, South Caucasus.

26. **GIRGENSOHNIA** Bunge ex Fenzl

in Mem. Sav. Etr. Petersb. 7: 478 (1847).

Ref. A.P. Sukhorukov, Notes on the taxonomy of *Girgensohnia* (Chenopodiaceae/Amaranthaceae). Edinb. J. Bot. 64 (3): 317–330 (2007).

Annual herbs, usually branched from the base, glabrous or covered with short papilliform trichomes. Stem epidermis 1-layered. Leaves opposite, linear-subulate, with a small apical mucro, stiff. Flowers axillary, solitary, sessile, with 2 bracteoles. Perianth consisting of 5 whitish membranous tepals, persistent, in fruit forming a cone-shaped structure, with two or three developing wing-shaped projections; less often the projections absent or at least not recorded (Girgensohnia minima). Stamens 5 with short filaments; anthers 0.4–1.0 mm long, usually long persisting in upper part of the perianth; connective with a short terminal extension, up to 0.2 mm long. Stigma capitate or bilobed, sessile or on a very short (less than 0.2 mm) style. Fruit ovoid; pericarp composed of 3 to 5 cell layers, papillate in upper part of the fruit; seed coat 2-layered; embryo vertical, with the radicle oriented upward (terminal or subterminal); perisperm inconspicuous.

Five species in SW and C Asia.

Girgensohnia (in honour of the Estonian botanist and grassland expert Gustav Karl Girgensohn (1786–1872)).

Girgensohnia oppositiflora (*Pall.*) *Fenzl* in Ledeb., Fl. Ross. 3: 835 (1851); Iljin in Fl. S.S.S.R. 6: 277 (1936); Rech.f., Fl. Lowland Iraq: 206 (1964); Hedge in Fl. Iranica 172: 297 (1997); Assadi in Fl. Iran 38: 416 (2001); Freitag et al. in Fl. Pak. 204: 185 (2001); Zhu, Mosyakin & Clemants in Fl. China 5: 399 (2003); Sukhorukov in Edinb. J. Bot. 64, 3: 325 (2007); Sukhorukov & Akopyan, Comp. Chenop. Cauc.: 61 (2013).

Salsola oppositiflora Pall., Reise Russl. Reichs 2: 735 (1773).
Chenopodium oppositifolium (Pall.) L.f., Suppl. 172 (1781).
Noaea oppositiflora (Pall.) Moq. in DC., Prodr. 13 (2): 31 (1849).

Annual, (10–15)20–40(–50) cm, branched mainly in its basal part, with a prominent main stem and arched ascending or spreading branches. Lower leaves up to 15 mm, less often up to 3.0–3.5 cm. Tepals 2.8–3.5 mm in flower, 3.3–4.0 mm in fruit. Wings broadly ovate or rhombic, the margin entire, toothed or erose; in fruit whitish to brownish, located 1.0–1.5 mm below the apex of the tepals. Anthers 0.4–0.6 mm long. Fruits 2.0–2.5(–2.7) mm long. Fig. 96: 1–4.

HAB. Disturbed desert ground, sandy roadsides, clay soil; alt. 0–500 m; fl. Jul.–Sep., fr. Oct.–Nov.
DISTRIB. Occasional and scattered in northern desert and foothills of Iraq: **FUJ**: Mosul, *Handel-Mazzetti*. 3118 (WU!). **DLJ**: 12 km S of Baiji, *E. & F. Barkley* 3680!; Tikrit, *Agnew* W.2126! **DWD**: 20 km S of Rutba, *Dabbagh, Taher & Taufiq* 41783!; 145 km W of Ramadi, *Rawi* 26837; between Ramadi & Rutba, *Rechinger* 12671! (W). **DSD**: 15 km SW of Samawa, *Al-Ani & Barkley* 8963 (W!).

Iran, Syria, Palestine, Afghanistan, Pakistan, Tajikistan, Kyrghyzstan, Kazakhstan, Uzbekistan, Turkmenistan, and China (Xinjiang).

27. **ANABASIS** L.

Sp. Pl.: 223 (1753); Gen. Pl. ed. 5: 104 (1754)

Ref.: Bunge, A. von *Anabasis* in "Anabasearum Revisio", Mem. Acad. Imp. Sci. St. Petersb. 7, ser. 4, 11: 34–47 (1862).
Sukhorukov A.P. Fruit anatomy of the genus *Anabasis* (Salsoloideae, Chenopodiaceae), Austr. Syst. Bot. 21, 6: 431–442 (2008).

Fleshy glabrous shrubs or shrublets, rarely annuals with tufts of hairs at nodes. Stems prominently articulated, branches arising opposite each other. Leaves opposite, well-developed or reduced to scales. Flowers solitary, hermaphrodite and female, subtended by bract and 2 bracteoles, rarely ebracteolate, sunk into axils of upper leaves forming dense spicate inflorescences. Perianth segments 5, ± equal, membranous, divided to near base, at fruiting stage winged or not. Stigmas 2; fruits more or less fleshy in upper part, red, orange or yellow coloured; pericarp smooth or papillate; seeds with vertical embryo; perisperm absent; embryo spiral.

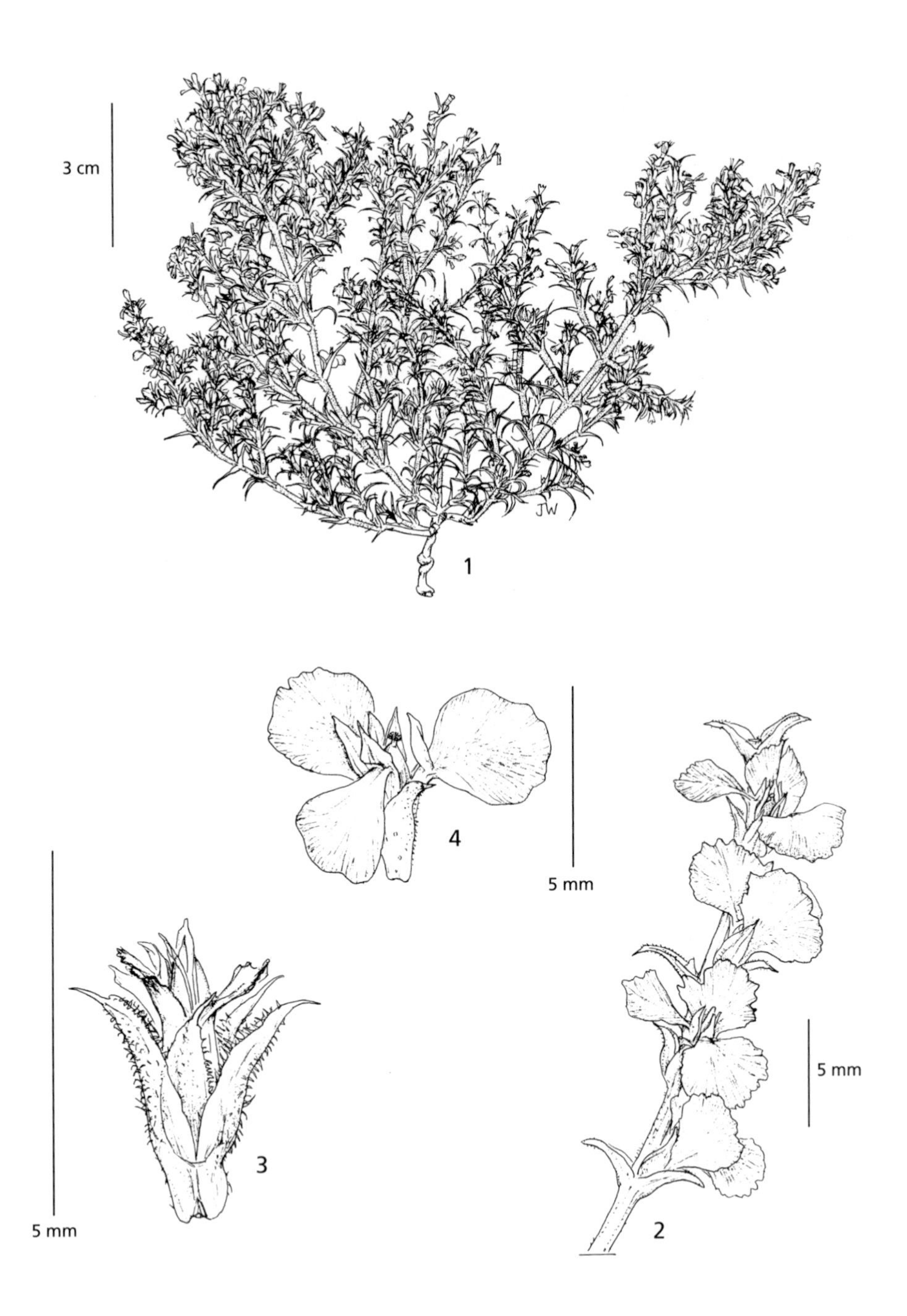

Fig. 96. **Girgensohnia oppositiflora.** 1, habit; 2, portion of flowerng branch; 3, flowe with bracts; 4, fruit with perianth showing the 3 developed wings. 1–2 from *Dabbagh* 41783; 3–4 from *Barkley* 3680. © Drawn by Juliet Beentje.

Anabasis (from the Gr. ἀνά, *ana*, "up" + βασις, *basis*, "step"); according to Fl. SSSR vol. 6, the name Ἀνάβασις was first recorded by Pliny.

1. Leaves well-developed, 5–15(–20) mm; fruit 1.5–2 mm 2
   Leaves scale-like, to 0.5 mm; fruit 2.5–4 mm 3
2. Annual; leaves with caducous mucro .... 4. *A. annua*
   Subshrub; leaves obtuse, rarely with mucro .... 3. *A. setifera*
3. Annual stems white; hairs in the bract's axils of equal length as perianth .... 2. *A. lachnantha*
   Annual stems greyish or green; hairs shorter than perianth .... 1. *A. articulata*

1. **Anabasis articulata** *(Forssk.) Moq. in DC.*, Prodr. 13 (2): 212 (1849); Rech.f., Fl. Lowland Iraq: 207 (1964); Zohary, Fl. Palaest. 1: 177 (1966); Boulos in Fl. Arab. Penins. Socotra 1: 282 (1996); Assadi, Fl. Iran 38: 439 (2001).

Salsola articulata Forssk., Fl. Aegypt. Arab. 55 (1775).

Shrub to 80 cm; stems to 5 mm in diameter, brittle, much-branched, angled or ridged when dry; annual shoots often yellowish or brown in lower part, younger parts greyish to green, often with small lateral buds at nodes. Leaves obsolete, to 0.8 mm. Flowers solitary, arising from lanate tufts at axils of upper leaves; perianth whitish, hyaline, shorter than stamens; at fruiting stage with pink or mauve wings of unequal size. Fruits 3–3.5 mm, with tiny papillae, reddish.

HAB. Stony and sandy desert; alt. 0–700 m; fl. & fr. Aug.–Oct.

DISTRIB. Occasional and scattered in the Southern Desert zone of Iraq: **DWD**: Ik'ara (Qara?) district, 100 km N of Rutba, *Serkahia* 16480!; Qara, N of Rutba, *Tikriti & Hazim* 29713! **DSD**: Busayia, *Eig & Zohary* 18 (HUJ!).

?UDHŪ (*Serekahia* 16480, Ik'ara district); "said to be poisonous to livestock".

Anabasis al-rawii Aellen ex L.M. El-Hakeem & Weinert, Wiss. Zeitung Martin-Luther-Univ. Halle-Wittenberg, Math.-Nat. Reihe 25 (4): 65 (1976), nom. illegit., without herbarium acronym, is closely related to *A. articulata* and the differences are still not clear. No original specimens are in the herbaria visited, and there is no information about the maintenance of the authentic sheets in Iraqi Agricultural Station.

Spain, North Africa, Eastern Mediterranean.

2. **Anabasis lachnantha** *Aellen & Rech. f.*, Anz. Österrr. Akad. Wiss., Math.-Nat. Kl., 1961 (2): 26 (1961). —Type: Iraq: [DSD], district Basra, Southern Desert, about 28 km SW Samawa, in arenosis lapidosis, 15 m, 5.XI.1956, *K.H. Rechinger* 8159 (holo W!; iso B! G! K!); Rech.f., Fl. Lowland Iraq: 207 (1964); Hedge in Fl. Iranica 172: 314 (1997); Boulos in Fl. Arab. Penins. Socotra 1: 282 (1996); Freitag et al. in Fl. Pak. 204: 188 (2001).

Small shrub to 30 cm. Stems much branched, annual shoots yellowish. Leaves minute, to 1 mm, rounded, divergent from the stem. Flowers solitary, in short spikes at end of branches, arisng in axils with white tufts of hairs; perianth whitish, at fruiting stage with pinkish wings, ± 8 mm (incl. wings); wings ovate-orbicular sinuate.

HAB. Sandy and gravelly soils; alt.; fl. Aug.–Oct.; fr. Oct.–Dec.

DISTRIB. Found in the western and southern deserts in Iraq: **DSD**: nr Al-Aidaha c. 110 km SW of Salman, c. 35 km NW of Jumaima, *Guest, Rawi & Rechinger* 19128 !; 9399 *Rechinger* ! 28 km S.W. of Samawa, *Rechinger* 8159 (type of *A. lachnantha*) !; Khidr al Mai, *Rechinger* 8238 (B!, E!); 10 km S. of Khadhar al-Mai, *Guest, Rawi & Rechinger* 16119 ! *Haines* s.n. (E!); inter Ur & Al Busaya, *Rechinger* 8186b (W!); N. of Khadhar al-Mai, *Sarkhaya* 16475 !; Al-Batin on Kuwait border c. 100 km SE of Busaiya, *Guest & Mahallal* 15250 ! Busaiya, *Rawi & Alizzi* 32432 ! Salaibiyat al-Hamir, *Guest, Rawi & Rechinger* 16055 ! **DWD**: Between Shithatha and Rahhaliyah, *Rawi & Alkas* 16322 ! 16327! ; Shithatha, *Agnew & Haines* s.n. (E!).

A rather under-collected species, possibly because it does not readily flower, and which has not been found in fruit. Its characteristic facies is best appreciated from a photograph; see *Fl. Iranica* 172: t. 179 (1997).

Vernacular name RIMITH (*Rawi & Alkas*, Shithatha-Rahhaliyah), IJRŪM (*Guest et al.* 16119 nr Khadr al-Mai), 'AIRUM (*Guest & Mahallal* 15250, Al-Batin).

Kuwait, Saudi Arabia, S Iran.

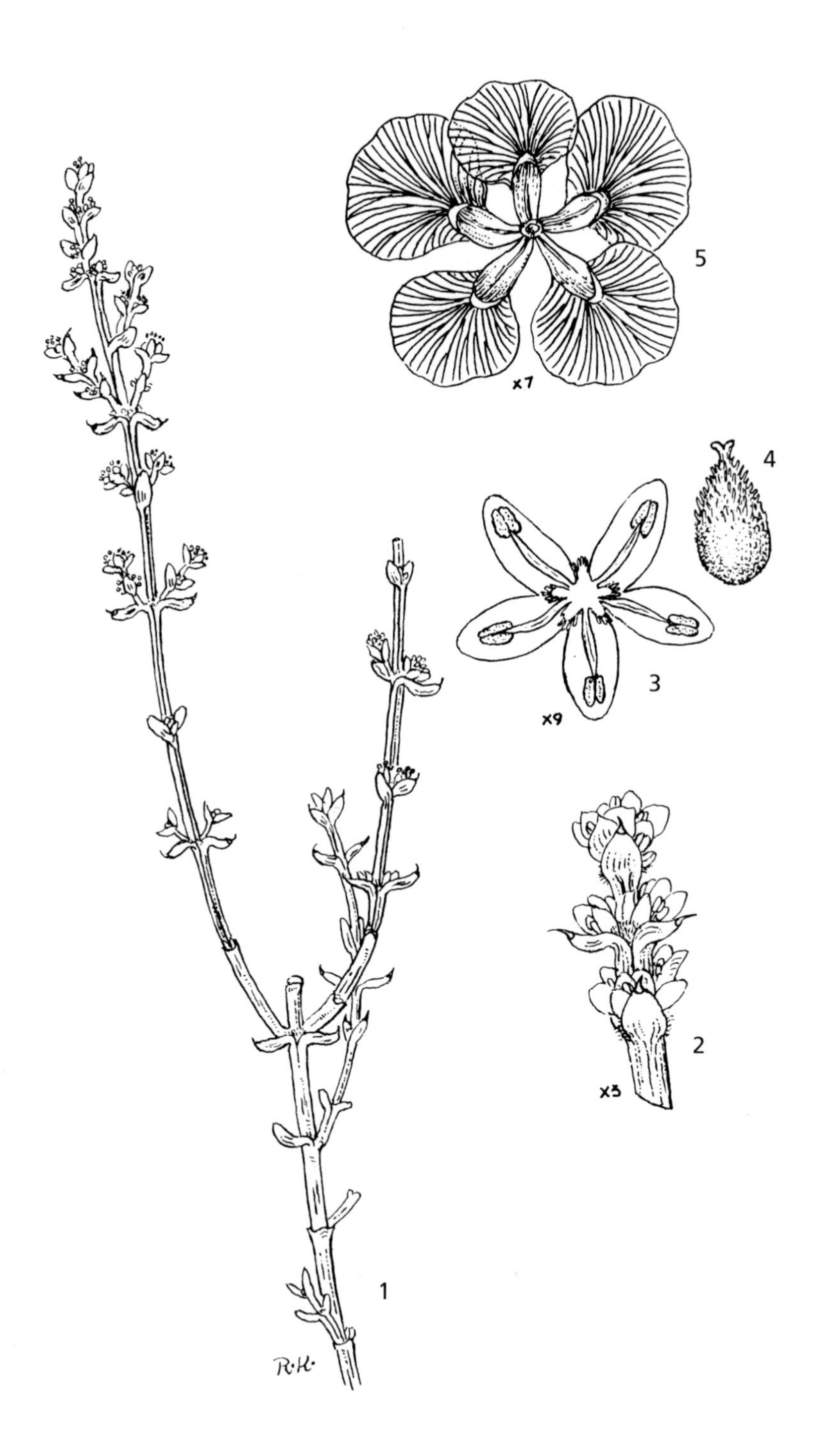

Fig. 97. **Anabasis setifera**. 1, habit × 1; 2, part of flowering stem × 3; 3, flower × 9; 4, ovary, enlarged 5, fruiting perianth × 7. Reproduced with permission from Zohary, Flora Palaestina 1: Plates, f. 261. 1966. Drawn by Ruth Koppel. © The Israel Academy of Sciences and Humanities.

3. **Anabasis setifera** *Moq.*, Chenop. Monogr. Enum.: 164 (1840); Rech.f., Fl. Lowland Iraq: 208 (1964); Hedge in Fl. Iranica 172: 307 (1997). Boulos in Fl. Arab. Penins. Socotra 1: 282 (1996);

Seidlitzia lanigera Post, Fl. Syria: 689 (1896).
Salsola setifera (Moq.) Akhani, Int. J. Pl. Sci. 168(6): 946 (2007) nom. illegit. non Lagasca 1816.

Subshrub to 60 cm, with unpleasant smell. Stem branched, annual shoots with 4 conspicous furrows; leaves 5–15 (20) × 2–5 mm, opposite, semiterete, fleshy, obtuse or rarely with caducous mucro. Flowers 3–7 in clusters. Perianth hyaline, winged at fruiting stage; wings 4–6 mm in diameter, membranous, orbicular or reniform, white, yellowish or (rarely) pinkish. Fruit 1.5–2 mm, pericarp not or slightly fleshy, reddish, throughtoutly papillate. Fl. Aug.–Oct., fr. Oct.–Dec. Fig. 97: 1–5.

HAB. Extreme deserts. Common in desert parts of Iraq, especially in S and SE.
DISTRIB. **DSD**: 28 km W of Samawa, *Rechinger* 8157! ; 5 km of Maaniyah, *Hazim* 32501! 2 km S of Khadhr al-Mai, *Guest, Rawi & Rechinger* 16116! *Rechinger* 8239 (BM! W!, WU!); 60–80 km W of Shabicha, *Chakravarty, Khatib, Rawi & Tikrity* 30044! 30020! 180 km W of Basra, *Rawi & Tikrity* 24985! **DWD**: 10 km N of Rahhaliya, *Rechinger* 8356 (E!, W!); Bahr al-Milh, *Rechinger* 8347 (E!); W of Karbala, *Rechinger* 8313 (W!); Shithatha, *Rawi* 26915! between Shithatha & Hindiya, *Rechinger* 140 (E!, W!); Rutba, *Haines* s.n. (E!); **LCA**: 37 km SW of Nasiriya, *Barkley & Abbas* 8886 (W!).

S Iran, Pakistan, Eastern Mediterranean, Arabia, Egypt, Sudan.

4. **Anabasis annua** *Bunge* in Mém. Acad. Imp. Sci. St.-Pétersb., sér. 7, 4 (11): 46 (1862). Lectotype inter Herat & Tebes, X-XI.1858, herb. Bungeanum (K-000899493, lower left-handed specimen!) selected by Sukhorukov, Chenop. Carp. & Syst.: 349 (2014).

Anabasis lutea Moq. in DC., Prodromus 13 (2): 215 (1849).
Anabasis micradena Iljin in Fl. SSSR 6: 878 (1936).

In all characters similar to *A. setifera*, but the plants annual to 30 cm tall, leaves with caducous mucro.

HAB. clayey deserts. 0–200 m.
DISTRIB. **DWD**: Rutba, *Eig, Feinbrun & Zohary* s.n. (HUJ!). One of the southernmost records of this species. Further findings are possible in the desert zone of Iraq.

Central Asia (S Kazakhstan, Uzbekistan, Turkmenistan), Iran, Afghanistan.

28. **HALOXYLON** Bunge

ex Fenzl in Ledeb., Fl. Ross. 3: 819 (1851).

Shrubs or trees, with ± fleshy scale-like leaves, glabrous except for tufts of hairs in axils. Stems articulated, bark often whitish, with 2-layered epidermis. Leaves opposite, rudimentary (in ours). Flowers borne in axils of subtending bracteoles, hermaphrodite or male, forming ± lax spikes or broad panicles. Perianth segments 5, somewhat chartaceous, in fruit developing into wide-spreading wings. Stamens 5, filaments united at base into a hypogynous disk, alternating with ovoid papillose staminodes. Ovary 2(–3)-styled. Seeds horizontal; albumen absent; embryo spiral.

Haloxylon (from Gr. ἅλς, *hals*, salt and ξύλον, *xylon*, tree or shrub)

Haloxylon persicum *Bunge in Boiss. & Buhse,* Nouv. Mém. Soc. Imp. Nat. Mosc. 12: 189 (1860); Zohary, Fl. Palaest. 1: 166 (1966); Boulos in Fl. Arab. Penins. Socotra 1: 266 (1996); Hedge in Fl. Iranica 172: 318 (1997); Freitag et al. in Fl. Pak. 204: 189 (2001); Norton et al. Illustr. Checklist Fl. Qatar: 29 (2009).

Haloxylon ammodendron auctt., non (C.A. Mey.) Bunge; Rech.f., Fl. Lowland Iraq: 183 (1964).

Shrub or tree to 5 m, very branched from the base; older branched whitish, annual stems green, leaves scale-like, mucronate, to 3 mm; flowers solitary in bract axils, perianth at fruiting stage with white or yellowish wings 8–10 mm across; fruit ± 1.5 mm. Fig. 98: 1–3.

HAB. Sandy desert; alt. ± 50 m; fl. & fr. ?Nov.
DISTRIB. Only found in the desert region: **DLJ**: *Gillett* 9902! (BAG); *Guest, Rawi & Nuri* 13528 (BAG); *Guest & Rawi* 14179 (BAG). **DSD**: *Guest, Rawi & Mahallal* 15210 (BAG). **LBA/DSD**: Basra, inter Ur & Al Busaya, *Rechinger* 8188 (E!).

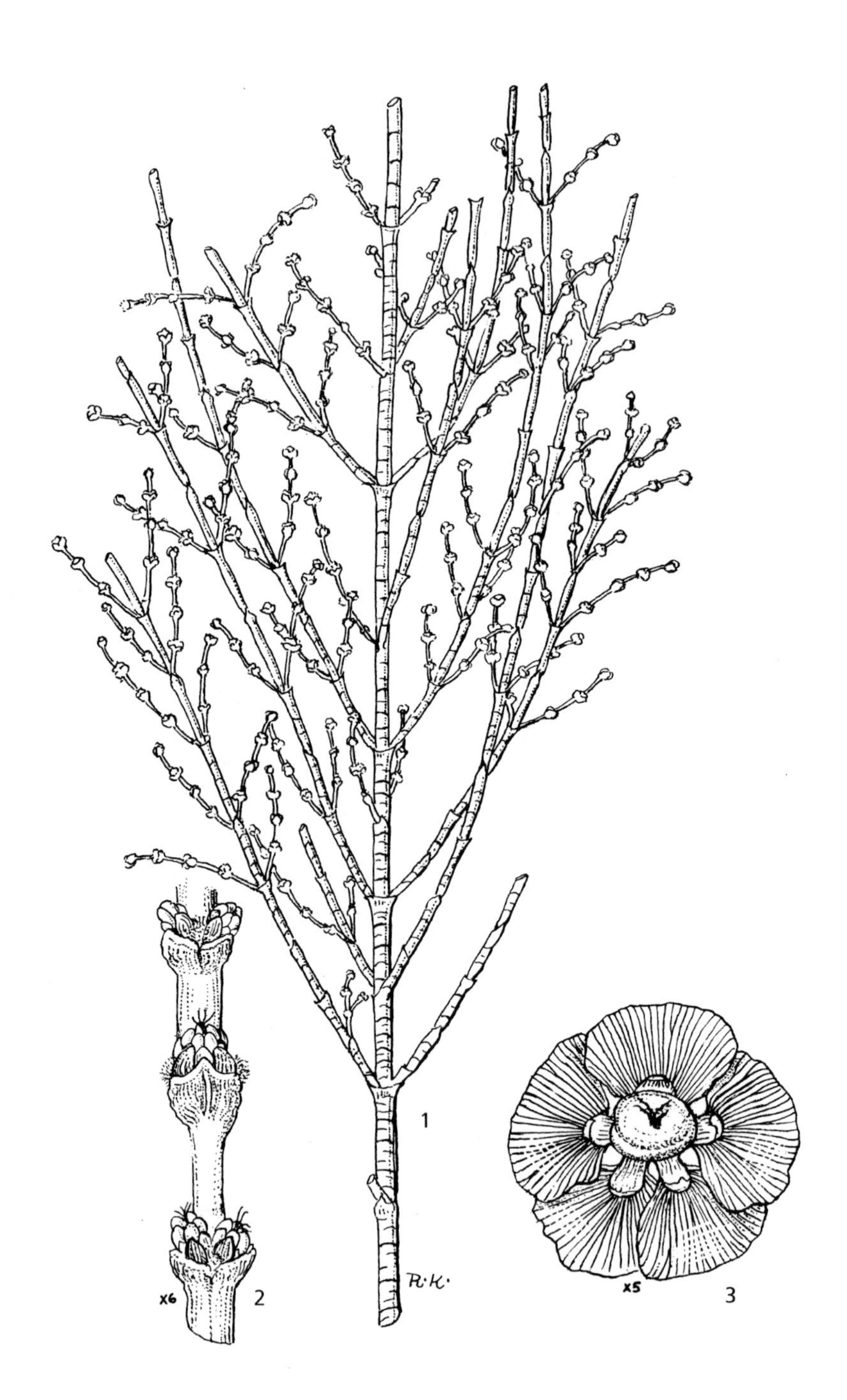

Fig. 98. **Haloxylon persicum**. 1, habit × 1; part of flowering stem × 6; 3, fruit with perianth wings × 5. Reproduced with permission from Zohary, Flora Palaestina 1: Plates, f. 242. 1966. Drawn by Ruth Koppel. © The Israel Academy of Sciences and Humanities.

GHADHA (*Gillet* 9902), GHADA (*Guest, Rawi & Nuri* 13528)

Eastern Mediterraean, Egypt, Iran, C Asia, Afghanistan, Pakistan, Arabian Peninsula.

29. **HAMMADA** Iljin

in Bot. Zhurn. 33 (6): 582 (1948).

Shrubs or shrublets; branches and leaves opposite, leaves subulate or scale-like, tufted in the axils, in other parts glabrous or papillate; flowers solitary, bisexual, supported by bract and 2 bracteoles; perianth of 5 free hyaline segments that develop wings at the fruiting stage; anthers not or very slightly appendiculate; staminodes (5) present, papillose. Fruit dry, white or slightly reddish coloured, included in perianth; seeds with horizontal coiled embryo and with no perisperm.

This genus is often included in *Haloxylon*; however, the recent molecular phylogeny (Kadereit & al., in prep.) suggest the generic rank of *Hammada.* (Gudrun Kadereit, pers. comm.)

About 10 species distributed in North Africa, Spain, Eastern Mediterranean and the Irano-Turanian region; three species in Iraq.

Hammada, from Ar., *hamada*, a type of rocky desert.

1. Plant blackening on drying; spikes short, 2–3 cm; wings of fruiting perianth 4–6 mm across. . . . . 1. *H. articulata*
   Plant not blackening on drying; wings of fruiting perianth 6–10 mm 2
2. Stems divaricate; stem lignification more than 10–15 cm. . . . . .2. *H. salicornica*
   Stems not divaricate; stem lignification to 10–15 cm .3. *H. eigii*

1. **Hammada articulata** *(Moq.) O. Bolòs & Vigo*, Butl. Inst. Catalana Hist. Nat., sec. bot. 38(1): 89 (1974).

Salsola articulata Cav., Icon. 3: 43 (1794) non Forssk. (1775).
Caroxylon articulatum Moq. in DC., Prodr. 13(2): 175 (1849).
Haloxylon articulatum (Cav.) Bunge, Mem. Sav. Etr. Petersb. 7: 469 (1851), nom. illegit.; Rech.f., Fl. Lowland Iraq: 205 (1964).
Haloxylon scoparium Pomel, Nouv. Mat. Fl. Atlant. 335 (1875).
Hammada scoparia (Pomel) Iljin in Bot. Zhurn. 3: 583 (1948); Assadi in Fl. Iran 38: 454 (2001).

Shrub 25–45 cm, glabrous; stems erect, woody, much-branched, new growth fleshy, grey-green and darkening with age; leaves rudimentary, reduced to scales. Inflorescence paniculate, consisting of spikes 2–3 cm. Flowers subtended by 2 bracteoles; perianth segments 5, scarious-margined, much expanded in fruit to form a disk 4–6 mm in diameter. Stamens 5, alternating with 5 subglobose staminodes, papillose on margins. Stigmas 2–3, short. Fruiting perianth with suborbicular wings, pale brown or greenish, 4–6 mm in diameter. Fig. 99: 1–13.

Hab. Dry steppe, sand dunes, sandstone plateau & slopes, gypsum desert, clay soil, sandy and stony places, volcanic rocky slopes, 0–640 m. alt. ; fl. ; fr.
Distrib. Common and widespread in the dried and more desert areas of Iraq: **fpf**: 16 km S of Zurbatiya, *Al-Kaisi & Yahya* 45273!; Jabal al-Muwaila, nr Kuwait (E of Jabal Hamrin), c. 70 km N of Amara, *Guest, Rawi & Rechinger* 17635!; 5 km S of Badra, *Zakirov & Shermatov* 472 (MW!). **dlj**: 60 km N of Rawa, *Hamid* 391353!; Qsaiah nr Rawa, *Rawi* 9916!, 9917!; 25 km NNE of Rawa, *Rawi & Gillett* 7098!; Rawa, Hamid 39118!; 70 km NW of Falluja nr Thirthar, *Chakravarty & Rawi* 30411! **dwd/dlj**: nr Haditha, gypsum desert, *Guest* 3519!; Umm al-Maten (74 km NW on road from Baiji to Haditha), *Chakravarty, Khatib & Alizzi* 31891! **dwd**: 50 km from Rutba to Naqaib, *Sarkaya & Tikriti* 15126!; 260 km NW of Rutba on road to Ramadi, *Rawi* 20966!; Assufi, 80 km N of Rutba, *Rawi & Nuri* 26865!; 25 km W of Rutba, *Rawi* 14695!; Wadi Musaad al Kutbai, 30 km S of Rutba, *Tikriti & Hazim* 29708!; 50 km N of Rutba, *Rawi* 26854!; 270 km W of Ramadi, *Rawi* 26852!; Imhewer (?) to Rutba, Aftan, *Rawi* 23734!; 125 km NE of Rutba, *Chakravarty, Rawi, Khatib & Alizzi* 31568!; Ubailah, 7 km N of Rutba, *Dabbagh, Taher & Taufiq* 41780!; 18 km S of Rutba, *Rawi* 14627!; Nukhaib, *Rawi* 13868!; 75 km W of Rutba, Wadi al-Walag, *Rawi* 14713!; Wadi Hauran, Eig & Zohary 892 (HUJ!). **dsd**: N of Al-Salman, *anon.* 20796!; Jabal Sanam (Basra-Kuwait border), *Rawi & Gillett* 6150!, *Guest, Rawi & H.E. Schwan* 14382!; 50 km S of Ansab to Samah, *Fawzi, Hazim & H. Hamid* 38840!; Malha well (Sur al-Sana'a), 39 km from Khadhr al Mai, *Nuri, Hamid & Kadhim* 40394!; Jabal Sanam, 30 km S of Zubair, *Guest, Rawi & Rechinger* 16987!; 59 km from Aidaha and Ansah, *Fawzi, Hazim & Hamid* 38828!; 10 km W of Karbala, *Rawi & Gillett* 6394!; 20 km SE

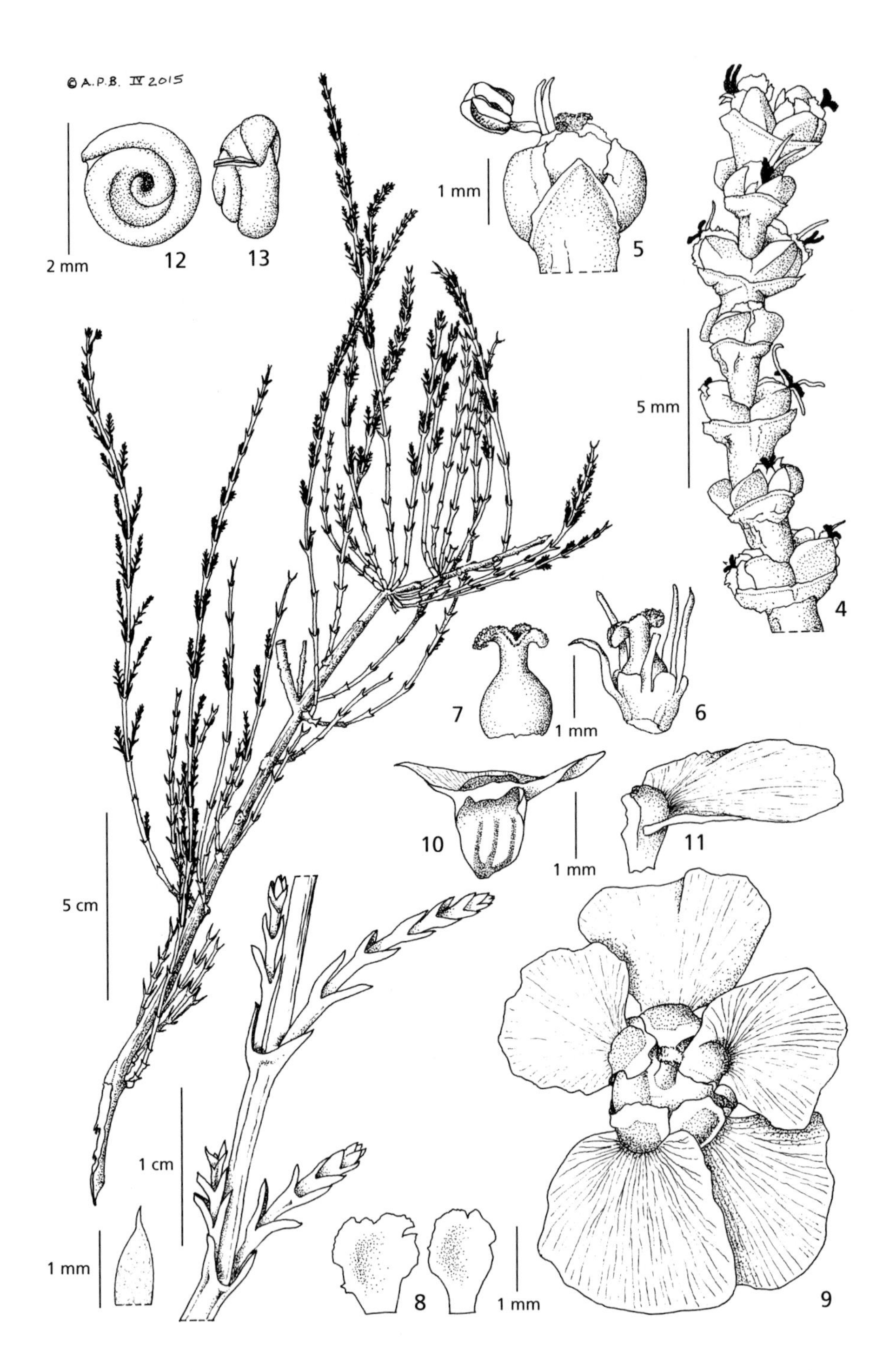

Fig. 99. **Hammada articulata**. 1, habit, vegetative; 2, detail of vegetative shoot; 3, leaf tip; 4, distal portion of inflorescence; 5, flower showing bract and 2 bracteoles; 6, flower lateral view; 7, carpel; 8, two perianth segments; 9, perianth with wings; 10, inner face of winged perianth segment; 12, seed dorsal view; 13, seed side view. 1–3 from *Rawi* 13077; 4–13 from *Tikriti & Hazim* 29708. © Drawn by A.P. Brown.

of Busaiya, *Rawi* 25632!; WNW of Jarishan, about 85 km SW of Basra, *Guest, Rawi & Rechinger* 17090A!; 80 km ESE of Busaiya, *Guest & Rawi* 14272!; 30 km W by S of Safwan, *Guest, Rawi & Rechinger* 16999!; nr Safai al-Maghif (NE of Ghazlani), 100 km WSW of Basra, *Guest, Rawi & Rechinger* 17279!; 65 km SW of Basra, Al-Batin, *Guest, Rawi & Rechinger* 17047!; Nukhaila, 40 km W by N of Basra, *Guest, Rawi & Rechinger* 17347!; Umm Qasr bridge, Al-Ani 39917! *Qatib & Alizzi* 32661!; Al-Ichrishi, 35 km E by N of Busaiya, *Guest & Rawi* 14189!; 75 km ESE of Busaiya, *Guest & Mahallal* 15232A!; Abu Ghvar, 60 km S of Ur, *Guest & Mahallal* 15298!; 46 km WNW of Ansab, on Saudi border, *Guest, Rawi & Rechinger* 19048! **LEA**: nr Wadi al-Tib police post nr Kuwait (*c.* 70 km N of Amara), *Guest, Rawi & Rechinger* 17531! **LCA**: nr Falluja, *Chakravarty & Tikriti* 32458! 32459!; Al-Sudur, *Botany staff* 47174! **LSM**: 90 km SE of Basra, *Chakravarty, Rawi, Khatib & Tikriti* 29940!

Used for fodder (*Sarkaya & Tikrity* 15126). "One of the best sheep feed plants, especially in summer; now seriously depleted by overgrazing" (*Rawi & Guest* 7098).

Colloquial name YITNA (*Rawi* 9916), YATHNAH (*Guest* 3519), ITHNE (*Rawi & Gillett* 7098); RIMTH (*Guest & Rawi* 14189, *Guest & Mahallal* 15298, Anon. 20796), NETOON (*Rawi* 14627).

S Mediterranean area from Spain to Egypt, Syria, Palestine.

2. **Hammada salicornica** *(Moq.) Iljin*, Bot. Zhurn. 33: 583 (1948); Zohary, Fl. Palaest. 1: 164 (1966); Assadi in Fl. Iran 38: 451 (2001); Mozaffarian, Trees & Shrubs of Iran: 160 (2005).

Caroxylon salicornicum Moq. in DC., Prodr. 13 (2): 174 (1849).
Haloxylon salicornicum (*Moq.*) *Bunge ex Boiss.*, Boiss., Fl. Orient. 4: 949 (1879); Rech.f., Fl. Lowland Iraq: 205 (1964); Boulos in Fl. Arab. Penins. Socotra 1: 266 (1996); Hedge in Fl. Iranica 172: 319 (1997); Freitag et al. in Fl. Pak. 204: 194 (2001).
H. elegans (Bunge) Botsch., Nov. Syst. Pl. Vasc. 1: 362 (1964).

Much-branched shrublet, to 1 m tall, usually smaller; stems woody at base, stramineous to whitish; branches grey-green, fleshy, ± 2 mm in diameter, ascending, remaining whitish with age. Leaves reduced to scales, membranous at margins, conspicouosly lanate on adaxial surface. Inflorescence of short dense spikes 3–6 cm. Stamens 5, alternating with fimbriate staminodes. Styles 2(–3), very short. Fruiting perianth 6–8 mm diameter, segments winged, pinkish or greenish with a pale lemon yellow centre.

HAB. Saline subdesert, sandy, silty and gravelly soil; alt. s.l.–570 m; fl. Sep.–Nov.; fr. Oct.–Jan.
DISTRIB. Widespread in the Western and Southern Desert regions, occasional elsewhere in Iraq: **FPF**: Shaikh Sa'ad, *Rawi* 12532! 12535!; Shaikh Sa'ad nr Mandali, *Rawi* 12536!; Qaraghan (Jalaula), *Dummock* (comm. Rogers) 0394!; Al-Sudur, *Omar & Thamer* 47124!; Muwayyib, *Guest* 13148! **DLJ**: Thirthar, 40 km NNE of Ramadi, *Gillett* 9903!; Thirthar lake, *c.* 120 km N of Baghdad, *Mosharraf Hossain* s.n.!; *Barkley & al-Ani* 3620 (W!). **DWD**: Gara 170 km N of Rutba, *Salman* 34847!; 40 km NW of Karbala to Shithatha, *Rawi & Serkahia* 16293!; 20 km W of Karbala, *Rawi* 26887!; Rutba, *Sakin* 34473!; by lake Habbaniya, *Guest* 3540!; beyond Falluja on road S of Lake Habbaniya to sluice gate, *Haines* W.788!; Swab, NW of Rutba, *Rawi* 26872!; 180 km NW of Ramadi, *Rawi* 26849!; 27 km W of Ramadi, *Eig & Feinbrun* s.n. (HUJ!); between Shithatha and Rahhaliya, *Rawi & Alkas* 16314!; by Habbaniya lake, *Guest* 3540!; 20 km NW of Karbala, *Rawi & Alkas* 16277!; *c.* 12 km W by S of Karbala, *Rawi & Rechinger* 16152!; between Shithatha and Rahhahiya, *Rawi & Alkas* 16315!; Qa'ara 90 km N of Rutba, *Khatib & Tikrity* 32198!, 32199!; 13 km W of Samawa, *Guest & Mahallal* 15204A!; Rutba, *Salman* 34821!, 34822! **DSD**: nr Zubair, *Abdul Wahab Mustafa* 3154!; 80 km NW of Nukhaila, *Rawi* 23706!; 100 km NE of Safwan, *Rawi & Haddad* 25989!; Swab NW of Rutba, *Rawi* 26870!; Butain 60 km SW of Zubair, *Alizzi & Omar* 35096!; 60 km SW of Zubair, *Rawi & Alizzi* 32428!; Faithat-um-Rbaian, Almaniya, 95 km W of Shabicha, *Hazim* 30692!; 24 km W of Ma'niya, *Hazim* 32468!; Busaiyah, *Rawi* 12542!; 40 km W of Karbala, *Rawi* 26893!; 30 km W of Ma'niya, *Hazim* 32526!; 25 km W of Ma'niya, *Hazim* 32472!; S of Ur, *Eig & Zohary* s.n. (HUJ!); S of Al Busaiya, 15 km NNW of neutral territory, *Rechinger* 8200!; 3 km NW of Abu Sukhair, *Rawi & Alkas* 16269!; Khidr al Mai, *Rechinger* 8241!; 20 km SSE of Ur, *Guest, Rawi & Rechinger* 16035!, *Rechinger* 8175!; 12 km WSW of Samawa, *Guest, Rawi & Rechinger* 16002!, *Rechinger* 8263! **LEA**: between Muqdadiya & Sadia, *Rechinger & Khudairi* 26 (W!). **LBA**: nr Basra, *Vesey-Fitzgerald* s.n. (HUJ!); Khidr Al-Milkh, *Rechinger* 8241 (E!).

'ARAD, from a Samawa townsman, SHINAN (fodder for camels) *Rawi* 12536, *Guest* 13148), RIMTH (fodder plant) (*Rawi* 12542, *Guest* 3540) or REMITH (*Rawi & Alkas* 16177) or THILAIDH (*Guest* 3540). One label states "poisonous to livestock ... and man" (*Khatib & Tikriti* 32199, 32198), which in view of the overwhelming evidence from other specimens is unlikely.

One of the principal fodder plants eaten by camels, both green and dry. Used also for fuel (*Mustafa* 3154). One specimen has been collected together with its parasite, *Cistanche* (*Guest & Mahallal* 15204).

Iran, Palestine, Egypt, Arabia, Afghanistan, Pakistan.

3. **Hammada eigii** Iljin in Novit. Syst. Pl. Vasc. 1964: 72 (1964); Zohary, Fl. Palaest. 1: 166 (1966).

Haloxylon ramosissimum Benth. & Hook. f., Gen. Pl. 3: 70 (1880) nomen.
Haloxylon articulatum Bunge ssp. *ramosissimum* (Benth. & Hook. f.) Eig, Palest. J. Bot., Jerusalem ser., 3: 128 (1945).
Hammada ramosissima (Benth. & Hook. f.) Iljin, Bot. Zhurn. 33: 583 (1948) comb. illegit.

Shrublet up to 60 cm, stem lignified basally (to 10–15 cm); annual sheets glaucous, glabrous or minutely papillose; leaves to 1.5–2 mm, scale-like; inflorescence up to 20(–30) cm; flowers solitary supported by bract and 2 bracteoles; perianth of 5 hyaline segments that develop wings 7–10 mm; anthers 1 mm; staminodes papillate.

HAB. Deserts; probably rare.
DISTRIB. **DWD/DLJ**: Haditha, *Guest* 3519 (HUJ!); nr Syrian border, *Schwabe* s.n. (B!). **LCA**: near Baghdad, *Lazar* 352 (G!).

Eastern Mediterranean (Palestine, Syria).

30. **CORNULACA** Del.

Fl. d'Egypte: 206 (1813)

Phylloxys Moq. in DC., Prodr. 13 (2): 218 (1849)

Ref.: P. Aellen in Verh. Naturforsch. Ges. Basel 61: 158–166 (1950)

Annual herbs or small shrubs, glabrous, smooth or scabrid with gland-like papillae. Leaves sessile, alternate, tumid or more slender, each aristate with a stout to slender spinous tip, ± amplexicaul. Flowers hermaphrodite, in 1–5-flowered axillary clusters, the clusters normally subtended by a leaf-like bract and 2 similar or shorter bracteoles. Perianth segments 5, at first hyaline, spathulate and free to base, finally indurate at base and ± fused, with 1 or 2 segments developing a strong, sharp or blunt dorsal spine. Stamens 5, filaments (and all floral parts) very short at time of dehiscence of anthers, elongating as fruit develops, shortly to considerably monadelphous, with alternating papillose pulvinate appendages; connectives not appendiculate. Ovary ovoid; style with 2 filiform stigmas. Fruit ovoid, with a firm or delicate pericarp; seeds vertical; perisperm absent, embryo spirally coiled.

Eight species in SW & C Asia and N Africa; three species in Iraq.

Cornulaca (from Lat. *cornu*, horn); the name was employed by Dioscorides, according to Iljin (1936), probably for a species of *Salsola*.

1. Upper leaves forming very long spines (2–4 cm) ....3. *C. setifera*
   Upper leaves more shortly spinous, spine not exceeding 1 cm 2
2. Plant shrubby; fully developed filaments monadelphous near base only, tube much shorter than the free tips......................2. *C. monacantha*
   Plant annual or short-lived perennial; fully developed filaments monadelphous into a tube considerably longer than the free tips...... 1. *C. aucheri*

1. **Cornulaca aucheri** *Moq.*, Chenop. Monogr. Enum.: 163 (1840); Rech.f., Fl. Lowland Iraq: 211 (1964); Boulos in Fl. Arab. Penins. Socotra 1: 260 (1996); Hedge in Fl. Iranica 172: 353 (1997); Assadi in Fl. Iran 38: 398 (2001); Freitag et al. in Fl. Pak. 204: 200 (2001); Ghazanfar, Fl. Oman 1: 45 (2003).

Cornulaca leucacantha Charif & Aellen in Verh. Naturf. Ges. Basel 61: 161 (1950); Rech.f., Fl. Lowland Iraq: 211 (1964).

Annual herb or short lived perennial, rather slender, 6–25 cm or sometimes of cushion habit, with several unbranched oftern papillate stems from base or with 1-few stems with long ascending branches in lower ½ and shorter branches above, smooth or papillose-scabrid. Stems and branches green to whitish, angular when young, ± terete later. Leaves 5–10 mm, straight or recurved, acicular or subulate and white-aristate above, rather abruptly expanded to broad hyaline margins in lower $^{1}/_{3}$–$^{1}/_{4}$, with few to numerous long white flexuose hairs in axils. Flowers in clusters of 2–3, surrounded by dense tufts of short white hairs and subtended by subulate-aristate bracts to ± 5 mm, bracteoles similar to bracts or shorter. Perianth segments 1.5–2 mm, spathulate, obtuse and erose-dentate at apex.

Fig. 100. **Cornulaca aucheri**. 1, habit adult plant × 1; 2, leaves × 2; 3, juvenile plant × 1½ ; 4, leaves juvenile plant × 2. **Cornulaca setifera**. 5, portion of branch × 1; 6, leaves × 2. Reproduced with permission from Flora of Arabia 1: fig. 48. 1996. Drawn by SED.

Filaments when fully developed fused into a distinct tube 1.5 mm, the free tips 0.5 mm with yellowish papillose appendages between; anthers oblong, ± 0.75 mm. Ovary ovoid, with a distinct stylar column 1.5 mm; stigmas filiform, ± 1.25 mm. Fig. 100: 1–4.

HAB. Alluvial and *haswa* plains, gravel ridges, gypsiferous pebbly sandstone, sandy gravel; alt. 0–100 m; fl. Oct.–Nov.
DISTRIB. Scattered and rather local in the Desert and Semi-desert areas of Iraq: **MSU**: Sulaimaniya, *Famzi* 39538! **FKI**: Kirkuk, *Eig, Feinbrun & Zohary* s.n. (HUJ!). **FPF**: Al-Sudur, *anon.* 47170!; Jabal Hamrin, *Rawi* 3401!, *Sutherland* 477 (BM!), *Rechinger & Khudairi* 31 (BM!). **LEA/FPF**, Jabal Hamrin betwen Muqdadiya and Jalaula, *Rechinger* 8084! **DLJ**: nr Balad, *Robertson* 13470!; Sumaicha, *Haines* s.n. (E!), *Handel-Mazzetti.* 3121! (WU!); *Rechinger* 8085 (W!); W of Baiji, *Anders* 1939 (W!). **DWD**: about 50 km NE of Ramadi, *Guest, Rawi & Nuri* 13576!; by L. Habbaniya, *Guest* 3533! **DSD**: Busaiya, *Fawzi, Hazim & Hamid* 38943! **LCA**: Lalefighah (?Harthiya) estate, *Guest* 927!; 25 km W of Abu Ghraib on Falluja road, *Khatib* 30489!; 43 km S of Baghdad, *Gillett* 9897! about 20 km E of Falluja, *Guest, Rawi & Nuri* 13506!; nr Rustamiya, *Guest* 878 (HUJ!); Iskanderiya, *Rechinger* 8270 (E!).

Arabia, Iran, Afghanistan, Pakistan.

2. **Cornulaca monacantha** *Del.*, Fl. d'Egypte: 206, t. 22 (1813); Handel-Mazzetti in Ann. Naturh. Mus. Wien 26: 144 (1912); Rech.f., Fl. Lowland Iraq: 210 (1964); Boulos in Fl. Arab. Penins. Socotra 1: 260 (1996); Hedge in Fl. Iranica 172: 352 (1997); Assadi in Fl. Iran 38: 394 (2001); Freitag et al. in Fl. Pak. 204: 200 (2001); Ghazanfar, Fl. Oman 1: 46 (2003); Mozaffarian, Trees & Shrubs of Iran: 160 (2005).

[*C. aucheri* (non Moq.) Bunge. in Mem. Acad. Imp. Sci. Petersb. ser. 7, 4: 11 (87 (1862) p.p. et Boiss., Fl. Orient. 4: 983 (1879) p.p.]

Low shrub, (5–)15–50 cm, intricately branched with branches frequently elongate, smooth, green or glaucous when young, when older whitish with the cortex cracking circumferentially and peeling, inner cortex brownish. Leaves (3-)5–7(–10) mm, usually ± recurved, triangular-subulate and white- or stramineous-aristate above, gradually expanded with hyaline margins below, with tufts of short white hairs in axils and sometimes around the leaf base. Flowers in clusters of mostly 2–5(–10), clusters rarely congested and often quite well separated, surrounded by dense tufts of short white hairs and subtended by triangular-subulate, aristate bracts to 8 mm and usually shorter, straighter bracteoles from ½ length of to subequalling bracts. Perianth segments 2–3 mm, narrowly spathulate, apex subacute, claw scarcely exserted beyond tuft of hairs; fruiting perianth pyriform, stramineous, angled, 1 or 2 perianth segments bear spines arising from the base, longer than segments. Filaments (1.25–)1.5–3 mm, with yellowish papillose appendages between bases of the long, free tips; anthers oblong, 0.75–1.5 mm. Ovary ovoid; style 0.75–1.5(–2) mm; stigmas filiform, 0.75–2 mm. Fruit ovoid, 1.5–1.75 mm; seeds compressed, roundish, yellow, 1.25–1.5 mm.

HAB. Sand hills, sandy and 'haswa' plains, gypsaceous subdesert; alt. 0–200 m.; fl. & fr. Aug.–Nov.
DISTRIB. Common in the Desert region, occasional in the Alluvial Plain: **DLJ**: S edge of L. Tharthar, *Chakravarty & Rawi* 30438! **DWD**: by Habbaniya lake, *Guest* 3535!; Shithatha, *Agnew & Haines* s.n. (E!). **DSD**: nr Salaibiyat at al-Hamir, about 35 km S by W of Ur, *Guest, Rawi & Rechinger* 16056, 16057A!; 68 km from Busaiya to Zufiad (?Muthanna), *Serkahia & Tikriti* 15117!; 20 km SSE of Ur, *Guest, Rawi & Rechinger* 16034!; Al-Batin, *c.* 30 km SSW of Jarishan, *Guest, Rawi & Rechinger* 16132, 16133!; 100 km NW of Safwan, *Rawi & Haddad* 25990!; *c.* 45 km SE by E of Busaiya, *Guest & Rawi* 14259!; Najaf to Karbala, *Gillett* 9967!; 2 km S of Khadhar al-Mai, *Guest, Rawi & Rechinger* 16118!; 14 km ESE of Khadhr al-Mai, *c.* 100 km SE of Busaya, *Guest & Ibrahim* 15248!; 12 km WSW of Samawa, *Guest, Rawi & Rechinger* 16007!; Faidhat al-Rhasmin, *Karim, Nasir, H. Hamid & Kadhim* 40276!; Umm Qasr port, *Husain al-Ani* 39927!; Kuwaibda, *Eig & Zohary* s.n. (HUJ!); *c.* 50 km ESE of Busaiya, *Guest, Rawi & Rechinger* 16105!; between Abu Ajaj and Tal al-Lahm, *Mahallal* 16471! **LCA**: 30 km N of Darraji, *Rawi & Serkahia* 16254!; Nasiriya to Dewaniyia, *Alizzi* 29635!

Vernacular names: CHIBCHĀB (nr Khidr al-Mai, *Guest, Rawi & Rechinger* 16118) and SHA'RĀN or 'UBAIRAH (nr L. Habbaniya, *Guest* 3535). A note on *Guest, Rawi & Rechinger* 16105 reads "CHIBCHĀB is spiny and hurts man and beast, but SHA'RĀN is not spiny when green and can be eaten by animals". That would seem to suggest that the latter name applies to *C. aucheri* or *C. leucantha*, if recognised as species distinct from *C. monacantha*. The same name is applied to *Agathophora alopecuroides* (q.v.) which implies that the name applies more to the young state of several species than to a single species.

N Africa, Arabia, Iran, Afghanistan, Pakistan.

3. **Cornulaca setifera** (*DC.*) *Moq.* in DC., Prodr. 13 (2): 218 (1849); Rech.f., Fl. Lowland Iraq: 211 (1964); Boulos in Fl. Arab. Penins. & Socotra 1: 260 (1996)

Astragalus setiferus DC., Prodr. 2: 296 (1825).
Cornulaca ?tragacanthoides Moq., Chenop. Enum.: 168 (1840).

Low shrub 12–20 cm; stems several from base, not or sparingly branched above, smooth. Stem and branches whitish or stramineous, angular or striate when young but terete with age. Leaves (1.3–)2–4 cm, straight, subulate, strong, stramineous with a paler arista at tip, abruptly expanded at base with pale membranous margins, with numerous long white flexuose hairs up to 1.5 mm in leaf axils. Flowers in clusters of mostly 2–3, surrounded by tufts of short white hairs and subtended by subulate-aristate bracts mostly 7–9 mm, bracteoles similar or shorter. Perianth segments ± 2 mm, spathulate and erose-dentate at apex. Filaments fused into a distinct tube ± 1.5 mm, free tips ± 1 mm with yellowish papillose appendages between; anthers ovoid, ± 0.75 mm. Ovary ovoid with a stylar column ± 1.5 mm; stigmas filiform, ± 0.5 mm. Mature fruit not seen. Fig. 100: 5–6.

There are three records to be considered for this plant. The first "inter Mejadin (Meyadin) et Salhiye" (Salahiya, now Kifri), *Handel-Mazzetti* 622) is certainly from Syria. This gathering is in WU! The type specimen, collected by Olivier, is vaguely localised as from between Aleppo and Baghdad, and may also be of Syrian origin. The third, cited by Aellen (Verh. naturf. Ges. Basel 61: 164 (1950)) is in G and has been seen. It bears the details "Zwischen Bséra und Folén, Dr Herzfeld, 9 Dec. 1907, comm. Schweinfurth". Where these places are is not clear, but although Aellen places them in Iraq this is very unlikely; he may have assumed "Bséra" to be Basra, but there seems no justification for such an opinion.

Arabia (Saudi Arabia), Syria (Syrian Desert).

31. AGATHOPHORA (Fenzl) Bunge

Mém. Acad. Imp. Sci. Pétersb. ser. 7, 4 (11): 19, 92 (1862).

Ref.: A. von Bunge, Anabasearum Revisio, Mém. Acad. Imp. Sci. Pétersb. ser. 7, 4 (11): 92–94 (1862)
V. Bochantsev, Rod *Agathophora* (Fenzl) Bunge, Bot. Zhurn. 62 (10): 1447–1452 (1977)

Divaricately branched shrubs, glabrous or papillose, with crimped hairs in axils of leaves; scattered glandular hairs can be present on stem and leaves. Stems basally woody, stiff; bark of older wood white, with longitudinal striations. Leaves alternate, fleshy, terete, prominently setose at apex. Flowers hermaphrodite or male, arising in leaf axils in clusters surrounded by tufts of hair. Perianth segments 5, free, membranous; whitish; three of them develop wings near the top, sometimes other segments can bear smaller outgrowths. Stamens 5, with terminal appendages. Disk with papillose lobes. Styles 2, subulate. Fruit dry, enclosed by persistent perianth. Seeds vertical, lacking perisperm; embryo spiral.

The number of the species is still not precisely identified (1–several). Some morphological characters are considered as variations of *A. alopecuroides*. Here, we follow V.P. Bochantsev (l.c.).

Agathophora (from Gr. αγαθος, *agathos*, gift and φέϱν, *pherei*, to bear).

1. **Agathophora iraqensis** *Botsch.* in Bot. Zhurn. 62 (10): 1451 (1977).–Type: Iraq [DWD], nr Shithatha (about 8 km E), about 38 km W of Karbala, ± 40 m, 20.XI.1956, *E. Guest, A. Al-Rawi & K.H. Rechinger* 16168 (holo, K!).

Agathophora alopecuroides auct. non (Delile) Fenzl ex Bunge.
Halogeton alopecuroides Moq., Chenop. Monogr. Enum. 161 (1840); Rech.f., Fl. Lowland Iraq: 212 (1964).

Shrub to 45 cm, intricately branched. Stems whitish, densely papillose at least in younger parts. Leaves 5–10 mm with caducous bristle; axils of leaves densely woolly. Flowers in clusters of 3–7 arising from upper leaf axils. Perianth segments ± ovate, hyaline, 3(–4) of the segments developing into spreading wings in their upper third. Anthers 1 mm with a tiny terminal appendage. Fig. 101: 1–8.

HAB. Sandy & gravelly desert soils, dunes, silty plains; alt. 25–410 m; fl. & fr. Nov.–Mar.
DISTRIB. Common in the Desert regions of Iraq: **DWD**: between Shithatha and Karbala, *Rawi* 19979!; 8 km E of Shithatha, *Guest, Al-Rawi & Rechinger* 16168! (type of *A. iraqensis*); Shithatha, *Agnew & Haines* s.n. (E!); 25 km S by W of Ramadi, *Guest, Rawi & Rechinger* 15938!; 35 km W of Nukhaib, *Rawi* 14755!;

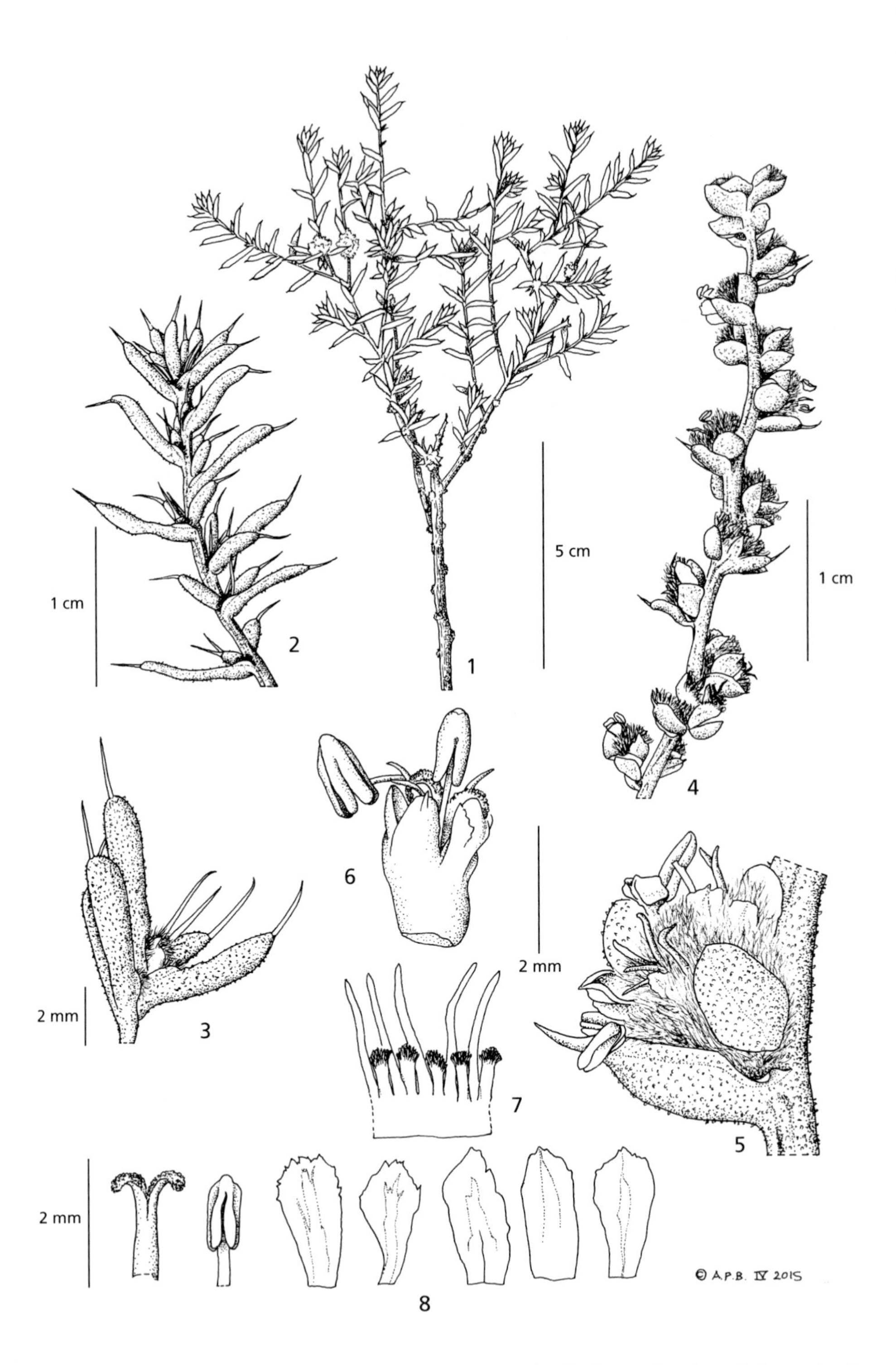

Fig. 101. **Agathophora iraqensis**. 1, habit vegetative shoot; 2, detail of vegetative shoot; 3, leaves detail; 4, distal portion of inflorescence; 5, inflorescence; 6, single flower; 7, filaments and inter filament lobes; 8, style, stamen and perianth segments. 1–3 from *Dickson* 456; 4–8 from *Collenette* 6784. © Drawn by A.P. Brown.

5 km S of Shanana, *Rawi* 14812! **DSD**: nr Khadhr al Mai, *Rechinger* 141!, 8236!, *Guest, Rawi & Rechinger* 16114!, *anon.* 16450!, *Alizzi & Omar* 35873!; 25 km W of Ma'niya, *Hazim* 32471!; about 40 km S by W of Ur, nr Salaibiyat al-Hamid, *Guest, Rawi & Rechinger* 16062!; Khadhr al-Mai, *Sarkahia* 16453, *Rawi, Khatib & Tikriti* 29125!; Shabicha to Salman, *Sarkahia & Tikrity* 16466!; 8 km N by W of Aidaha, about 95 km S of Salman, *Guest, Rawi & Rechinger* 19177!; 180 km W of Basra, *Rawi & Tikriti* 24986!; Shaib al Batin, about 100 km SE of Busaiya, *Rawi, Khatib & Tikriti* 29131!, *Guest & Mahalhal* 15251!, *Rawi, Khatib & Tikriti* 29134!; nr Al-Aidaha, *Guest, Rawi & Rechinger* 19129!

The general distribution of *A. iraqensis* is not traced so far. After Bochantsev (1977), *Agathophora alopecuroides* may have small distribution in Egypt and is distinguished by the very broad base of the bract's. The close related *A. postii* (Eig) Botsch. described from Syria (El Jebah to el Beida, 1890, *Post* 1035, HUJ-28250!) is with mostly glabrous leaves with no mucro and the 3 perianth segments bear the wing near their middle part.

Several vernacular names are recorded: SHA'RĀN, *Guest et al.* 15251 (al-Batin), *Guest et al.* 16114 (nr Khidr al-Mai), SHIRĀN (*Rawi* 14755, Nukhaib), DUMRĀN (*Sarkahia & Tikrity*, Shabicha to Salman), DHUMRĀN (Khadr al-Mai, *anon.* 16450).

So far known only from Iraq.

32. TRAGANUM Del.

Fl. Egypt. 204 (1813)

Small shrubs to 100 cm. Stem divaricately branched, with wooly tufts of hairs in the leaf axils, in other parts glabrous, rarely stem and leaves papillate. Leaves alternate, fleshy, semiterete or flattened, up to 5 mm, cuspidate with a mucro to 0.5 mm or obtuse. Flowers supported by a bract and 2 bracteoles, small, bisexual. Perianth with 5 membranous lobes; tubus (lower part of the perianth) indurated and thickened in fruit, furnished with small, somewhat horn-shaped, hard protuberances. Stamens 5. Style 2-partite into 2 subulate stigmas. Fruit included in the woody perianth, more or less globular, somewhat depressed, with membranous pericarp tightly adherent to the tubus. Seeds horizontal.

Two or three species distributed in North Africa, Arabia and the eastern Mediterranean, with irradiations into Macaronesia and Iraq; one species in Iraq.

1. **Traganum nudatum** Delile, Fl. Aeg. Ill. 60 (1813); Descr. Egypte, Hist. Nat. 204 (1813); Boiss., Fl.Orient. 44: 946 (1879); Rech.f., Fl. Lowland Iraq: 196 (1964); Zohary, Fl. Palaest. 1: 162 (1966); Boulos in Fl. Arab. Penins. Socotra 1: 259 (1996); Boulos, Fl. Egypt 1: 116 (1999).

Shrublet to 50 cm. Stem divaricately branched, with wooly tufts of hairs in the leaf axils, in other parts glabrous, rarely stem and leaves papillate. Leaves alternate, fleshy, semiterete, the lower (spring) leaves to 10(–15) mm, other ones up to 5 mm, short cuspidate or obtuse. Perianth glabrous, membraneous; tubus indurated, without appendages or with small protuberances.

HAB. Deserts. Probably rare; alt. ; fl. Mar.–Apr.; fr. May–Jun.
DISTRIB. **DWD**: Shithatha, *Rawi & Gillett* 6475! **DSD**: Busaiya, *Rawi* 26025! *Rechinger* 8183 (BM!); between Ur & Al Busaya, *Rechinger* 8181B (W!); Khidr al Mai, *Rawi, Khatib & Tikrity* 29115!; 15 km SE of Ashuriya, *Rechinger* 13612 (W!); Ukhaidhir, *Grant & Barkley* 8571! **LCA**: 30 km W of Darraji, *Rawi* 16243 (LE!); nr Sharrikiya, *Haddad, Safat & Barkley* 3731A (W!).

EXCLUDED SPECIES

Camphorosma sp.: The material at Kew identified as *Camphorosma* sp. (DWD, 70 km N from Rutba, *Rawi & Nuri* 27146) is referable to *Caroxylon vermiculatum.*

Arthrocnemum glaucum Del. was cited in Rech.f., Fl. Lowland Iraq: 192 (1964). The specimen seen belongs to *Halocnemum strobilaceum.* However, the presence of *Arthrocnemum macrostachyum* (Moric.) Bunge is predictable due to its presence in neighboring countries (e.g. Eastern Mediterranean, Arabia).

Petrosimonia brachiata (Pall.) Bunge was noted for Iraq by Zohary (1950), and he was followed by Rawi (1964) and Aellen in Rech.f., Fl. Lowland Iraq: 208 (1964). Handel-Mazzetti's collection of the species from Lake al-Chattuniye ("El-Chattunije, inter fluvium Chabur et montes Dschebel Sindschar": *Handel-Mazzetti* 1643, see also the specimen in WU!) is the source of this record, and as has been pointed out by Blakelock (1958: 485) the locality is in Syria. The species may yet be discovered in N Iraq, though the probability is not perhaps a high one. It is an annual with leaves densely clothed with closely appressed, bifurcate hairs; the perianth segments are free and unchanged in fruit, bearing neither wings nor tubercles.

Salsola gossypina (=*Kaviria gossypina*): Blakelock in Kew. Bull. 12, 3: 490 (1957): no specimens seen. The area of this taxon is N & C Iran, Turkmenistan, Tajikistan and Afghanistan.

Salsola longifolia: Blakelock in Kew. Bull. 12, 3: 493 (1957): specimen cited (no 15208) belongs to *Seidlitzia rosmarinus.*

96. **AMARANTHACEAE** Jussieu

Gen. Pl. 87 (1789); H. Schinz in Engler & Prantl, Pflanzenfam. ed. 2, 16c: 7–85 (1934)

P. Aellen[17]

Annual or perennial herbs or subshrubs, rarely lianes. Leaves simple, alternate or opposite, exstipulate, entire or almost so. Inflorescence a dense head, loose or spike-like thyrse, spike, raceme or panicle, basically cymose, bracteate; bracts hyaline to white or coloured, subtending one or more flowers. Flowers hermaphrodite or unisexual (plants dioecious or monoecious), actinomorphic, usually bibracteolate, frequently in ultimate 3-flowered cymules (triads); lateral flowers of such cymules sometimes (not in Iraq) modified into scales, spines or hooks. Perianth uniseriate, membranous to firm and often finally ± indurate, usually falling with the ripe fruit included, tepals free or somewhat fused below, frequently ± pilose or lanate, green to white or variously coloured. Stamens as many as and opposite to the petals, rarely fewer filaments free or fused into a cup below, sometimes almost completely fused and 5-toothed at apex with entire or deeply lobed teeth, some occasionally anantherous, alternating with variously shaped pseudostaminodes or not; anthers unilocular (bilocellate) or bilocular (4-locellate). Ovary superior, unilocular; ovules 1-many, erect to pendulous, placentation basal; style very short to long and slender; stigmas capitate to long and filiform. Fruit an irregularly rupturing, circumscissile or indehiscent capsue (utricle), rarely a berry or crustaceous, usually with thin, transparent walls. Seeds round to lenticular or ovoid, embryo curved or circular, surrounding the ± endosperm.

A large and predominantly tropical family of some 65 genera and over 1000 species, including some cosmopolitan 'weeds' and a large number of xerophytic plants.

In the APG III classification (2009) Amaranthaceae include the plants formerly treated in the family Chenopodiaceae (in this Flora treated as a separate family). The monophyly of the broadly defined Amaranthaceae has been strongly supported by both morphological and phylogenetic analyses. The main differences between Amaranthaceae (sens. strict.) and Chenopodiaceae (sens. strict.) are the membraneous petals and stamens often united in a ring structure in Amaranthaceae.

Kai Müller, K. & Borsch, T. (2005). Phylogenetics of Amaranthaceae using matK/trnK sequence data – evidence from parsimony, likelihood and Bayesian approaches. Ann. Miss. Bot. Gard. 92: 66–102.

Kadereit, G., Borsch, T., Weising, K. & Freitag, H. (2003). Phylogeny of Amaranthaceae and Chenopodiaceae and the evolution of C4 photosynthesis. Int. J. Pl. Sciences 164 (6): 959–986.

1. Leaves opposite 2
   Leaves alternate 3
2. Bracteoles with a projecting dorsal keel along the midrib . . . . . . 5. *Gomphrena*
   Bracteoles with no dorsal keel along the midrib . . . . . . 4. *Alternanthera*
3. Flowers white, woolly, in cylindrical spikes; dioecious . . . . . . 3. *Aerva*
   Flowers never white-woolly; monoecious or hermaphrodite 4
4. Flowers monoecious, the males scattered among the females or situated at the tip of the inflorescence . . . . . . 2. *Amaranthus*
   Flowers hermaphrodite . . . . . . 1. *Celosia*

[17] Letter in the files from P. Aellen to C.C. Townsend dated 20 June 1968 ... " I come now to the end of working out the Chenopodiaceae and Amaranthaceae for your Flora of Iraq".
Updated by Shahina A. Ghazanfar

1. **CELOSIA** L.

Sp. Pl.: 205 (1753); Gen. Pl. ed. 5: 96 (1754)

Annual or perennial herb with alternate leaves. Flowers hermaphrodite, bibracteolate, in bracteate spikes or thyrses. Perianth segments 5, free. Stamens 5, deltoid below and filiform above, or (not in Iraq) ± swollen laterally or toothed on each side of the anthers. Ovary with few to many ovules (rarely 1–2, and not in Iraq), stigmas 2–3, capsule circumscissile.

About 50 species in the warmer regions of the Old and New Worlds, especially tropical Africa; one species in Iraq.

Celosia argentata *L.*, Sp. Pl.: 205 (1753); Boiss., Fl. Orient. 4: 967 (1875); Zohary, Fl. Pal. ed. 2, 2: 456 (1933); Guest in Kew Bull. 27: 20 (1933); Fl. SSSR 6: 356 (1936); Husain & Kasim, Cult. Pl. Iraq: 99 (1975).

The wild form of this plant, with dense silvery or pinkish spikes terminal on stem and branches, is a weed of more tropical regions than Iraq. In Iraq the species is only represented by the selected cutivated form:

f. **cristata** (*L.*) *Schinz* in Pflanzenfam. ed. 2, 16c: 29 (1934)

C. argentata var. *cristata* (L.) O.Kuntze, Rev. Gen. Pl. 2: 541 (1891), incl. cvs. *childsii*, *plumosa* & c.
[*C. cristata* (non L.) Dinsm., Fl. Pal. ed. 2, 2: 456 (1933); Guest in Dep. Agr. Iraq Bull. 27: 20 (1933); Rawi & Chakravarty in Dep. Agr. Iraq Tech. Bull. 15: 24 (1964); Mouterde, Nouv. Fl. Syr. 1: 439 (1966); Aellen in Fl. Iranica 91: 16 (1972).

This is the well-known garden Cockscomb, in which the upper part or almost the whole of the inflorescence produces long and plumose to short, squat, fan-shaped shoots with narrow bracts and bracteoles. It is widely grown as a decorative in the warmer regions of the world, including Iraq. An account of its cytology is given by Khoshoo & Pal, "The probable origin and relationships of the garden cockscomb", J. Linn. Soc. Bot. 66: 127–141 (1973).

HAB. & DISTRIB. **LCA**: Zafaraniy station, 15/10/1957, *Al-Kas* 20502 (BAG); Ghazi garden Baghdad, 9/10/1947, *Baker Haj* 10082 (BAG); Baghdad, 20/9/1970, cultivated, *Sahira* 324 (BAG); Baghdad, 13/7/72, cultivated, *Ali Abdul Latif* 739 (BAG).

Guest (1931) reported this form as "cultivated in flower gardens" in Iraq and Husain & Kasim (1975) state that it is "cultivated throughout the country".

Cock's-comb (Eng.); 'URF AD-DÎK (Guest, 1933).

2. **AMARANTHUS** L.

Sp. Pl.: 989 (1753); Gen. Pl. ed. 5: 427 (1754)

Ref. Sauer, J. (1967). The grain amaranths and their relatives: a revised taxonomic and geographical study. Ann. Miss. Bot. Gard. 54: 103-137 (1967).

Annual or more rarely perennial herbs, glabrous or furnished with short and gland-like or multicellular hairs. Leaves alternate, long-petiolate, simple and entire or sinuate. Inflorescences basically cymose, bracteate, consisting entirely of dense to lax axillary clusters or the upper clusters leafless and ± approximate to form a lax or dense "spike" or panicle. Flowers dioecious (not in Iraq) or monoecious, bibracteolate; perianth segments (2–)3–5, free or connate at base, membranous, those of female flowers slightly accrescent in fruit. Stamens free, usually similar in number to perianth segments; anthers bilocular (4-locellate); stigmas 2–3. Ovule solitary, erect. Fruit a dry capsule, indehiscent, irregularly rupturing, or commonly dehiscing by a circumscissile lid. Seeds usually black and shining, testa thin; embryo annular, endosperm present.

Amaranthus (from Gr. αμαϱανφος, *amaranthos*, unfading, according to Stearn (1972) who adds in explanation that the flowers of some species retain their colour for a long time like everlastings; Gilbert-Carter (1964) also mentioned that *amaranthus* is the name of a plant in Lat. authors; Amaranth.

About 50 species, chiefly in the warmer temperate and subtropical regions of the world, about 12 species being ± cosmopolitan weeds in the tropics also. A genus of considerable difficulty taxonomically, especially the "grain amaranths" such as *A. caudatus* and *A. hybridus*, which have been cultivated since ancient times. The foliage of several species is eaten as a kind of spinach in numerous parts of the world; nine speces in Iraq.

1. Inflorescence leafy to top, consisting entirely of axillary, cymose clusters 2
   Inflorescence not leafy to tip, a terminal leafless spike or panicle present. 5
2. Tepals of female flowers with slender, divergent, usually colourless awns. . .5. *A. tricolor*
   Tepals of female flowers sharply mucronate 3
3. Bracteoles strongly spinose-acuminate, about twice as long as perianth . . . . . 7. *A. albus*
   Bracteoles not spinose-acuminate, subequalling or shorter than perianth. 4
4. Female flowers with 3 perianth segments; fruit strongly wrinkled, exceeding perianth . . . . 8. *A. graecizans*
   Female flowers with 4-5 perianth segments; fruit smooth, subequalling or shorter than perianth . . . . 9. *A. blitoides*
5. Capsule indehiscent, irregularly rupturing, very strongly muricate; seeds with low, scurfy papillae on the reticulate pattern of the testa . . . . . 6. *A. viridis*
   Capsule dehiscent by a circumscissile lid, wrinkled or not but never strongly muricate; seeds without low papillae 6
6. Terminal spike of inflorescence produced into a long, pendulous "tail"; female perianth segments very broadly obovate or spathulate, distinctly imbricate. . . . . 1. *A. caudatus*
   Terminal spike of inflorescence erect or somewhat nodding only; female perianth segments lanceolate to oblong or narrowly spathulate, not or indistinctly imbricate. 7
7. Female perianth segments spathulate to narrowly oblong-spathulate, obtuse or many emarginate, longer than fruit. 4. *A. retroflexus*
   Female perianth segments lanceolate to oblong, acute, or if obtuse then not longer than fruit. 8
8. Terminal inflorescence alone usually spiciform, axillary inflorescences usually ± rounded clusters – or if axillary spikes present then these with globose clusters at junction with stem; bracteoles and tepals of female flowers terminating in fine, flexuous, hair-like awns. . . . . .5. *A. tricolor*
   Terminal and at least some (usually most) of upper axillary inflorescences spiciform, without dense globose clusters at junction with stem; bracteoles and tepals of female flowers terminating in stout, rigid awns 9
9. Male flowers confined to terminal region of spike or almost so . . . . .3. *A. dubius*
   Male flowers scattered among the females . . . . .2. *A. hybridus*

1. **Amaranthus caudatus** *L.*, Sp. Pl. ed. 1: 990 (1753); Boiss., Fl. Orient. 4: 988 (1879); Zohary, Fl. Pal. ed. 2, 2: 456 (1933); Guest in Dep. Agr. Iraq 27: 6 (1933); Kom., Fl. USSR 6: 360 (1936); Blakelock in Kew Bull. 8: 228 (1953); Rawi in Dep. Agr. Iraq Tech. Bull. 14: 168 (1964); Zohary, Fl. Palaest. 1: 184 (1966); Mouterde, Nouv. Fl. Syr. 1: 440 (1966); Aellen in Fl. Iranica 91: 3 (1972); Tackholm, Stud. Fl. Egypt ed. 2: 131 (1974); Townsend in Fl. Pak. 71: 10 1974); Husain & Kasim, Cult. Pl. Iraq: 99 (1975); Miller in Fl. Arab. Penins. & Socotra 1: 288 (1996); Bao Bojian, Clements & Borsch in Fl.China 5: 418 (2003).

Annual herb, erect, to 1.5 m high but generally shorter, commonly reddish or purplish throughout. Stems rather stout, not or sparingly branched, glabrous or thinly furnished with rather long, multicellular hairs which are increasingly numerous upwards. Leaves glabrous or ± sparingly pilose along margins and lower surface of primary venation, long-petiolate (petiole to 8 cm but not longer than lamina), lamina broadly ovate to rhomboid-ovate or ovate-elliptic, 2.5–15 × 1–8 cm, obtuse to subacute at mucronate tip, shortly cuneate to attenuate below. Flowers in axillary and terminal red or green to yellowish spikes formed of increasingly approximated cymose clusters, terminal inflorescences varying from a single, elongate, tail-like, pendulous spike to 30 cm or more long and ± 1.5 cm wide, to a panicle with the ultimate spike so formed; male and female flowers intermixed throughout the spikes. Bracts and bracteoles deltoid-ovate, pale-membranous, acuminate and with a long, pale or reddish, rigid, erect arista formed by the yellow-green or reddish, stout, excurrent midrib, the longest up to twice as long as the perianth. Perianth segments 5; those of the male flowers oblong-elliptic, 2.5–3.5 mm, acute, aristate; those of the female flowers 1.8–2.5 mm, broadly obovate to spathulate, distinctly imbricate, abruptly narrowed to a blunt or sometimes faintly emarginate, mucronate tip. Stigmas 3, erect or flexuous. Capsule

2–2.5 mm, ovoid-globose, circumscissile, slightly urceolate, lid smooth or furrowed below, abruptly narrowed into a short, thick beak. Seeds compressed, black, almost smooth and shining, or commonly subspherical with a thick yellowish margin and a translucent centre, 1–1.25 mm.

Hab. & Distrib. Reported by Guest (1933) and others to be cultivated as an ornamental in gardens in Iraq, but no specimens have been seen.

Not known in the wild state except as a casual or escape, but widely distributed in most parts of the world as an ornamental and in some regions (e.g. Nepal) also as a grain crop. Postulated by Sauer (l.c.: 127, 1967) to be a cultigen derived from the tropical American *A. quitensis* Kunth. In spite of its probable American origin it has been reported from Europe (including Crimea), Palestine, Egypt, Caucasus, Iran, Pakistan, C Asia (Turkmenistan to Tian Shan), India, Africa (Sudan, Ethiopia) and many other places.

Love-lies-bleeding (Eng.), so called on account of its showy blood-red flowers; 'URF AD-DÎK ("cock's comb") (Ar., *Dinsmore* 1932, *Guest* 1933, a name also current for other species of *Amaranthus, Celosia argentea* and probably certain other plants).

2. **Amaranthus hybridus** *L.*, Sp. Pl. ed. 1: 990 (1753); Zohary, Fl. Palaest. 1: 181 (1966); Mouterde, Nouv. Fl. Syr. 1: 441 (1966); Aellen in Fl. Turk. 2: 1967); Townsend in Fl. Pak. 71: 11 (1974); Miller in Fl. Arab. Penins. & Socotra 1: 288 (1996); Bao Bojian, Clements & Borsch in Fl.China 5: 419 (2003).

Annual herb, erect or less commonly ascending, up to 2(–3) m high in cultivated forms but much less in subspontaneous plants, often reddish-tinted throughout. Stems stout, branched, angular, glabrous or thinly to moderately clad with short or long multicellular hairs (increasingly so above, especially in the inflorescence). Leaves glabrous or thinly pilose on the lower margins and underside of the primary nervation, long-petiolate (petioles up to 15 cm, scarcely exceeding the lamina), lamina broadly lanceolate to rhomboid or ovate, 3–19(–30) × 1.5–8(–12) cm, gradually narrowed to the blunt to subacute mucronulate tip, attenuate or shortly cuneate into the petiole below. Flowers in yellowish, green, reddish or purple axillary and terminal spikes formed of cymose clusters, which are increasingly closely approximate upwards, the terminal inflorescence varying from a single spike to a broad, much-branched panicle up to 45 × 25 cm, ultimate spike often nodding; male and female flowers intermixed throughout the spikes. Bracts and bracteoles deltoid-ovate to deltoid-lanceolate, pale-membranous, acuminate and with a long, pale to reddish-tipped, erect arista formed by the stout, excurrent, yellow or greenish midrib, subequalling to or exceeding the perianth. Perianth segments 5, 1.5–3.5 mm, lanceolate or oblong, acute-aristate or the inner sometimes blunt in the female flowers, only the midrib at most greenish. Stigmas (2–)3, erect flexuous or recurved, ± 1 mm. Capsule subglabrous to ovoid or ovoid-urceolate, 2–3 mm, circumscissile, with a moderately distinct to obsolete beak; lid smooth, longitudinally sulcate, or sometimes rugulose below the beak. Seeds black and shining or pale, compressed, ± 1 mm, almost smooth centrally, faintly reticulate around the margins. Fig. 102: 1.

The species is found in Europe, Syria, Lebanon, Palestine, Sinai, Lower Egypt, Turkey, Caucasus, Iran, Pakistan, Afghanistan, C Asia (Turkmenistan) and many other parts of the world. Several subspecies are recognized of which two occur in Iraq.

subsp. **hybridus**.

Amaranthus hypochondriacus L., Sp. Pl. ed. 1, 991 (1753).
A. chlorostachys Willd., Hist. Amaranth.: 34, t. 10 f. 19 (1790); Boiss., Fl. Orient. 4: 988 (1879); Aellen in Fl. Iranica 91: 4 (1972).

Larger bracteoles of female flowers mostly 1.5–2 × as long as perianth. Style bases and upper part of lid of fruit swollen, so that the fruit has an inflated beak.

Hab. In a sandy clay field nr r. Euphrates; alt. ± 150 m; fl. (& fr.?) Sep.
Distrib. (of subsp.). Apparently rare in Iraq, only found once in the desert region: **dlj**: Rawa, *Alizzi & Omar* 35347!

Distribution (of subspecies): found practically throughout the warmer regions of the world as a spontaneous naturalised weed, frequent also in the temperate regions as a casual. Grown in many areas as a grain crop, and coloured forms are widespread as decoratives.

Green-flowered forms (var. *hybridus*) and forms with inflorescences and frequently vegetative parts also red (var. *erythrostachys* Moq.) occur. The selected cultivated forms are frequently very large with very broad inflorescences compared with the weedy wild forms exemplified by *A. powellii* S.Wats., recognised by Sauer, l.c., as a distinct species.

Red-flowered forms lose their colour in the herbarium, and unfortunately the Iraqi specimen bears no colour note.

subsp. **cruentus** (*L.*) *Thell.* in Fl. Adv. Montpellier: 205 (1912).

Amaranthus cruentus L., Syst. Nat. ed. 10, 2: 1269 (1759); Mouterde, Nouv. Fl. Lib. et Syr. 1: 441 (1966); Aellen in Fl. Iranica 91: 6 (1972); Townsend in Fl. Pak. 71: 12 (1974); Boulos, Fl. Egypt 1: 132 (1999).

A. paniculatus L., Sp. Pl. ed. 2: 1406 (1763); Boiss., Fl. Orient. 4: 989 (1879); Zohary, Fl. Pal. ed. 2, 2: 456 (1933); Vassilczenko in Fl. SSSR 6: 361 (1936).

A. hybridus var. *paniculatus* (L.) Thell. in Fl. Adv. Montpellier: 205 (1912); Rawi, Dep. Agr. Iraq Tech. Bull. 14: 168 (1964); Mouterde, Nouv. Fl. Lib. et Syr. 1: 441 (1966).

Larger bracteoles of female flowers mostly 1–1.5 × as long as perianth. Style bases and upper part of lid of fruit not swollen, so that the fruit has a smooth, narrow beak.

HAB. In mountain valleys, cultivated by stream, in irrigated gardens on the plain; alt. up to 1100 m; fl. Sep.–Dec.

DISTRIB. (of subsp.). Occasional in the forest zone and desert region of Iraq: **MAM**: S of Sharanish, 25 km NE of Zakho, *Rawi, Nuri & Tikriti* 29067; **MAM**: Baikado valley (cult.), 5 km from Kani Mazi, *Al-Kaisi et al.* 43903; **MRO**: Khazna, on S slope of Ser-i Khazni, *Haley* 244. **DSD**: Umm Qasr (cult.), *H. Hamid* 38555! **LCA**: Baghdad (cult.), *Sahira* C. 313!

As with subsp. *hybridus*, green- and red-flowered forms occur, var. *cruentus* being the red form; the green-flowered forms seems to have no name, but since the red forms lose colour in the herbarium and transitional forms occur there seems little profit in bestowing one. As with subsp. *hybridus*, there are numerous decorative cultivars. Nat. Herb. Iraq no. 43903, with its main terminal spike clavate-thickened and congested at the tip, is clearly one of these.

Distribution (of subspecies): probably Central American in origin. Now found throughout the warm-temperate regions of the world as a garden ornamental, and in some regions as a grain crop or potherb. As a spontaneous weed it is however found principally in Asia eastwards from Malaya and in tropical Africa.

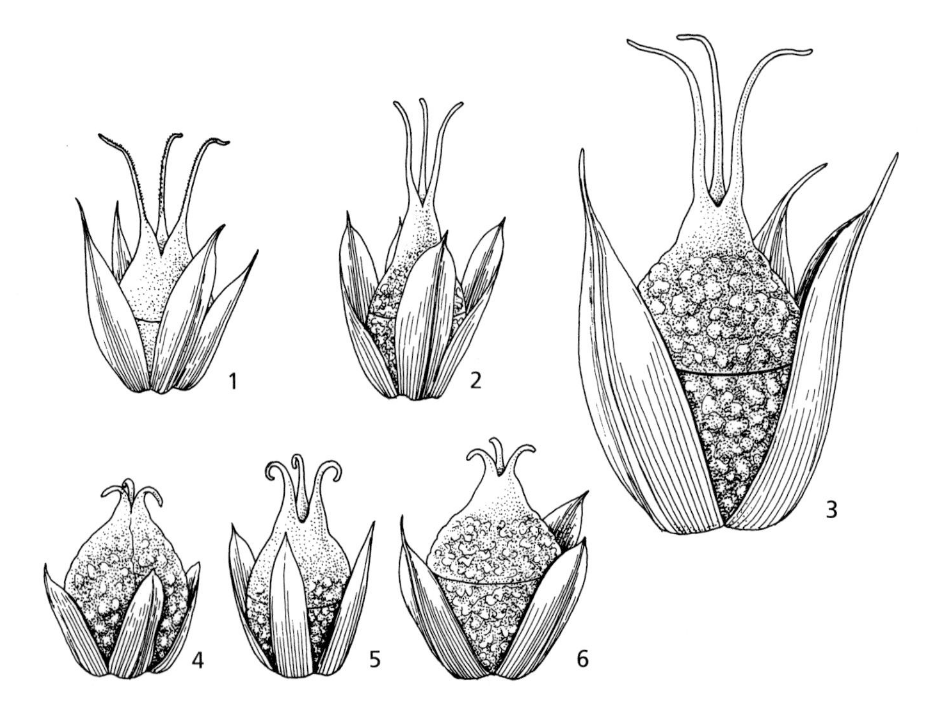

Fig. 102. **Amaranthus** fruits (× 20). 1, **A. hybridus**. 2, **A. dubius**. 3, **A. tricolor**. 4, **A. viridis**. 5, **A. albus**. Reproduced with permission from Flora of Arabia 1: fig. 55. 1996.

3. **Amaranthus dubius** *Mart.*, Pl. hort. Acad. Erlang.: 197 (1814) nomen nudum, ex Thell., Fl. Adv. Montpellier: 203 (1912).

[*A. tristis* (non L.) Moq. in DC., Prodr. 13(2): 260 (1849) et auctt. mult., non L.]

Erect annual herb, mostly to 90 cm (rarely to 1.5 m) tall. Stem rather slender to stout, usually branched, angular, glabrous or increasingly clothed upwards (especially in the inflorescence) with short to rather long, multicellular hairs. Leaves glabrous, or thinly and shortly pilose on lower surface of primary venation, long-petiolate (petioles up to 8.5 cm, sometimes longer than the lamina), lamina ovate or rhomboid-ovate, 1.5–8(–12) × 07–5(–8) cm, blunt or retuse at the tip with a distinct, fine mucro formed by the percurrent nerve, cuneate (usually shortly so) at base. Flowers green, in lower part of plant in axillary clusters 4–19 cm in diameter, towards the ends of the stem and branches the leafless clusters approximate to form simple or (the terminal at least) branched spikes 3–15(–25) × 0.6–0.8(–1.0) cm. Lower clusters of flowers entirely female, spikes generally showing a few male flowers at tips only (rarely in more than the apical 1 cm). Bracts and bracteoles deltoid-ovate, pale-membranous, with an erect awn formed by the excurrent midrib, bracteoles shorter than or subequalling perianth, rarely slightly exceeding it. Perianth segments (4–)5, those of the female flowers 1.5–3 mm, narrowly oblong or spathulate-oblong, obtuse or sometimes (particularly those approaching the male flowers) acute, mucronulate, frequently with a greenish dorsal vitta above, those of the male flowers broadly lanceolate or lanceolate-oblong, generally acuminate, the thin midrib green. Stigmas 3, flexuous or reflexed, ± 1 mm. Capsule ovoid-urceolate, 1.5–2 mm, with a short inflated beak, circumscissile, the lid strongly rugulose below the beak. Seeds ± 1 mm, compressed, black, shining, faintly reticulate. Fig. 102: 2.

HAB. Cultivated in irrigated places on the alluvial plain; alt. not recorded; fl. & fr. Jul.
DISTRIB. Apparently very rare in Iraq, only found once in cultivation: **LCA**: Baghdad (cult.), *Ali Abdul-Latif* C. 741!

Of American origin, now found practically throughout the tropics, frequently grown as a spinach plant in subtropical regions also.

Apart from the key character to separate this species from *A. hybridus* (*A. dubius* agg.), it is easily separable from subsp. *hybridus* by its short bracteoles and from subsp. *cruentus* by the inflated beak of the fruit.

4. **Amaranthus retroflexus** *L.*, Sp. Pl. ed. 1: 991 (1753); Boiss., Fl. Orient. 4: 989 (1879); Hand.-Mazz. in Ann. Naturh. Mus. Wien 26: 144 (1912); Nábělek in Publ. Fac. Sci. Univ. Masaryk 105: 11 (1929); Zohary, Fl. Pal. ed. 2, 2: 456 (1933); Guest in Dep. Agr. Iraq Bull. 27: 6 (1933); Vassilczenko in Fl. SSSR 6: 362 (1936); Blakelock in Kew Bull. 4: 445 (1950); Zohary in Dep. Afr. Iraq Bull. 31: 49 (1950); Aellen in Fl. Lowland Iraq: 214 (1964); Rawi in Dep. Agr. Iraq Tech. Bull. 14: 168 (1964); Mouterde, Nouv. Fl. Lib. et Syr. 1: 442 (1966); Zohary, Fl. Palaest. 1: 182 (1966); Aellen in Fl. Turk. 2: 340 (1967) & in Fl. Iranica 91: 3 (1972); Townsend in Fl. Pak. 71: 12 (1974); Boulos in Fl. Egypt 1: 132 (1999); Bao Bojian, Clements & Borsch in Fl. China 5: 419 (2003).

A. retroflexus var. *delilei* (Richat & Loret) Thell. in Vierteljahrsschr. nat. ges. Zürich 52: 442 (1907).

Annual herb, erect or with ascending branches, (6–)15–80(–100) cm, simple or branched (especially from base to about middle of stem). Stem stout, subterete to angled, densely clothed with multicellular hairs. Leaves hairy along lower surface of primary venation and often the basal margins, long-petiolate (petiole to 6 cm, and in large plants not or rarely equalling the lamina), lamina ovate to rhomboid or oblong-ovate, (1–)5–11 × (0.6–)3–6 cm, obtuse to subacute at the mucronulate tip, shortly cuneate or attenuate into petiole. Flowers in greenish or rarely somewhat pinkish-suffused, stout axillary and terminal spikes, which are usually shortly branched to give a lobed appearance, more rarely with longer branches; terminal inflorescence paniculate, very variable in size, male and female flowers intermixed, the latter generally much more plentiful except sometimes at tips of spikes. Bracts and bracteoles lanceolate-subulate, pale-membranous with a prominent green midrib excurrent into a stiff, colourless arista, longer bracteoles subequalling to twice as long as perianth. Perianth segments 5, those of male flowers 1.5–2.5 mm, lanceolate-oblong, blunt to subacute, those of female flowers 2–3 mm, narrowly oblong-spathulate to spathulate, obtuse or emarginate, ± green-vittate along midrib, which ceases below apex or is excurrent in a short mucro. Stigmas 2–3, patent-flexuous or erect, 1 mm. Capsule subglobose, ± 2 mm, usually shorter than perianth, circumscissile, rugose below lid, beak indistinct. Seeds black and shining, compressed, ± 1 mm, almost smooth centrally, faintly reticulate around margins.

HAB. In the lower mountains on rocky slopes and stony hillsides, on a scree, in shade of *Quercus* by a stream, also as a weed in orchards and gardens, on the plains as a weed in a rice field, along irrigation channels and near canal by a saline seepage pool; alt. 50–1100(–1650) m; fl. & fr. Jul.–Sep.
DISTRIB. Common in the forest zone of Iraq, occasional in the steppe and on the alluvial plain in the desert region: **MAM**: Mar Ya'qub, *Nábělek* 1080; Sharanish, NE of Zakho, *Rawi, Tikriti & Nuri* 29035!; Kani Mazi bridge, *Botany Staff* 43852!; Baidaho, 5 km from Kani Mazi, *id.* 43907!; Zawita, *Guest* 3746!; Sersang, *Haines* W.405! **MRO**: Shaqlawa, weed in garden, *Sahira* 37552!; Rowanduz gorge, *Guest* 2981!; Ari, below Ser Kurawa, *Gillett* 9708!; Haji Umran, *Rawi* 24281!; Pushtashan, NE of Rania, *Rawi & Serhang* 26550!; **MSU**: Penjwin, *Guest* 12950!; **FUJ**: between Sinjar and Tal Afar, *Alizzi & Omar* 35314!; **FKI/DGA**: between Kirkuk and Khalis, *Rawi, Chakravarty, Alizzi & Nuri* 19699!; **LCA**: Harathiya, Baghdad, *Guest* 3237!; Karrada, Baghdad, *Paranjpye* s.n.! Abu Ghraib, *Fawzi Karim & Ali Abdul-Latif* 34635!; **LEA**: Dabuni, nr Aziziya, *Guest* 218!; 30 km SE of Aziziya, on road to Kut, *Alizzi & Omar* 34635!

Native of N America S to Mexico. Introduced into the Old World as a weed, occurring in more temperate regions than most other species of the genus. S and C Europe, N Africa, temperate Asia from Cyprus and Turkey to Iran, Siberia, C Asia, Mongolia, China, Japan. Adventive in Australia, S Africa and probably elsewhere. Generally a weed of cultivation.

5. **Amaranthus tricolor** *L.*, Sp. Pl. ed. 1: 989 (1753); Vassilczenko in Fl. SSSR 6: 363 (1936); Zohary, Fl. Palaest. 1: 184 (1966); Townsend in Fl. Pak. 71: 13 (1974); Husain & Kasim, Cult. Pl. Iraq: 99 (1975); Boulos in Fl. Egypt 1: 132 (1999); Bao Bojian, Clements & Borsch in Fl. China 5: 420 (2003).

Amaranthuis tristis L., Sp. Pl. ed. 1: 989 (1753).
A. melancholinus L., Sp. Pl. ed. 1: 989 (1753); Moq. in DC., Prodr. 13 (2): 262 (1849).
A. gangeticus L., Syst. Veg. ed. 10: 268 (1759); Moq. in DC., op. cit. 260 (1849); Boiss., Fl. Orient. 4: 990 (1879); Blatter in Journ. Ind. Bot. Soc. 11: 40 (1932).
A. tricolor var. *gangeticus* Fiori in Fiori & Paoletti, Fl. Anal. d'Ital. 1: 322 (1896-98); Rawi in Dep. Agr. Iraq Tech. Bull. 14: 168 (1964); Tackholm, Stud. Fl. Egypt 2: 133 (1974).

Annual herb, ascending or erect, attaining 1.25 m or more in cultivation. Stem stout, usually much-branched, it and the branches angular, glabrous or clothed in upper parts with sparse (or denser in the inflorescence), ± crisped hairs. Leaves glabrous, or thinly pilose on lower surface of primary venation, green or variably purplish-infused, very variable in size, long- (up to 8 cm) petiolate, lamina broadly ovate, rhomboid-ovate or broadly elliptic to lanceolate-oblong, emarginate to obtuse or acute at apex, at base shortly cuneate to attenuate, decurrent along petiole. Flowers green to crimson in ± globose clusters 4–25 mm in diameter, all or only the lower axillary and distant, upper sometimes (probably usually in Iraq) without subtending leaves and increasingly approximate to form a thick terminal spike of variable length, male and female flowers intermixed. Bracts and bracteoles broadly or deltoid-ovate, bracteoles subequalling or shorter than perianth, pale-membranous, broadest near base and narrowed upwards to green midrib, which is excurrent to form a long, pale-tipped awn usually at least ½ as long as the basal portion and sometimes equalling it. Perianth segments 3, 3–5 mm, elliptic or oblong-elliptic, narrowed above, pale-membranous, green midrib excurrent into a long, pale-tipped awn; female flowers with perianth segments slightly accrescent in fruit. Stigmas 3, erect or recurved, ± 2 mm. Capsule ovoid-urceolate with a short beak, 2–3 mm, circumscissile, membranous, obscurely wrinkled. Seeds 1–1.5 mm, black or brown, shining, very faintly reticulate, lenticular. Fig. 102: 3.

HAB. Cultivated, or an escape from cultivation; alt. up to 50 m; fl. & fr. May–Jun. & Sep.–Oct.
DISTRIB. Uncommon in the desert region of Iraq on the irrigated alluvial plain. **LCA**: Rustamiya, *Lazar* 429!; Baghdad (cult.), *Sahira* C.323!; C.564! C.566!; Za'faraniya, *Rawi & Shahwani* 25331! **LBA**: Basra (Makina Masus), *Whitehead* 160, 169 in Blatter, l.c.

As stated elsewhere [Fl. W. Pakistan 71: 14 (1974)] I do not find the subspecific divisions of this species given by Aellen (l.c.: 494–496) to be practical, and regard *A. tricolor* as a single polymorphic unit.

Asia from India to China and Japan, south to Indonesia; also in New Guinea and the New Hebrides and small Pacific Island groups such as Fiji. Introduced and/or cultivated elsewhere (including Africa, West Indies etc.).

Although Husain & Kasim (1974) state that this species is "widely" cultivated in Iraq the only specimens we have seen are from Baghdad and Basra.

6. **Amaranthus viridis** *L.*, Sp. Pl. ed. 2: 1405 (1763); Zohary, Fl. Pal. ed. 2, 2: 457 (1933); Anthony in Notes Roy. Bot. Gard. Edinb. 18: 296 (1935); Aellen in Fl. Turk. 2: 341 (1967) & in Fl. Iranica 91: 8 (1972); Townsend in Fl. Pak. 71: 14 (1974); Boulos in Fl. Egypt 1: 135 (1999); Bao Bojian, Clements & Borsch in Fl.China 5: 420 (2003).

A. gracilis Desf., Tabl. Ecole Bot.: 43 (1804), nomen ex Poir., Lam. Encycl. Suppl. 1: 312 (1810); Blakelock in Kew Bull. 5: 445 (1950); Aellen in Fl. Lowland Iraq: 215 (1964); Mouterde, Nouv. Fl. Lib. et Syr. 1: 442 (1966); Zohary, Fl. Palaest. 1: 186 (1966).

Euxolus caudatus (Jacq.) Moq. in DC., Prodr. 13(2): 274 (1849).

Annual herb, erect or more rarely ascending, 10–75(–100) cm. Stem rather slender, sparingly to considerably branched, angular, glabrous or more frequently increasingly hairy upwards (especially in the inflorescence) with short or longer and rather floccose multicellular hairs. Leaves glabrous, or shortly to fairly long-pilose on lower surface of primary and most of venation, long-petiolate (petioles to 10 cm and the longest commonly longer than lamina), lamina deltoid-ovate to rhomboid-oblong, 2–7 × 1.5–5 cm, margins occasionally obviously sinuate, shortly cuneate to subtruncate below, obtuse and narrowly to clearly emarginate at tip, minutely mucronulate. Flowers green in slender axillary or teminal, often paniculate spikes 2.5–12 × 2–5 mm, or in lower part of stem in dense axillary clusters to 7 mm in diameter; male and female flowers intermixed but the latter more numerous. Bracts and bracteoles deltoid-ovate to lanceolate-ovate, whitish-membranous with a very short, pale or reddish awn formed by the excurrent green midrib, bracteoles shorter than perianth (± 1 m). Perianth segments 3, very rarely 4, those of the male flowers oblong-ovate, acute, concave, 1.5 mm, shortly mucronate; those of the female flowers narrowly oblong to narrowly spathulate, finally 1–2 mm, the borders white-membranous, minutely mucronate or not, midrib green and often thickened above. Stigmas 2–3, short, erect or almost so. Capsule subglobose, ± 1 mm, not or slightly exceeding perianth, indehiscent or rupturing irregularly, very strongly rugose throughout. Seeds 1–1.5 mm, round, only slightly compressed, dark brown to black with an often paler thick border, ± shining, reticulate and with shallow scurfy verrucae on the reticulum, the verrucae with the shape of the areolae. Fig. 102: 4 & Fig. 103: 1–2.

HAB. Weed of orchards and date gardens, fields and shady ditches, also on muddy banks of receding river and tidal estuary; alt. up to ± 150 m; fl. & fr. Mar.–May & Aug.–Oct.
DISTRIB. Occasional, locally common, on the alluvial plain in the desert region of Iraq; very rare on the lower margins of the dry steppe zone: **FKI/FPF**: Jabal Hamrin, in Anth., l.c.; **DWD**: Hit, *Guest* 3527!; **LCA**: Baghdad, *Paranjpye* s.n.! comm. Graham, *Gillett* 8401!, *Agnew* 33633!, *Haines* W.13!; banks of r. Tigris nr Baghdad, *Guest* 13466!; Za'faraniya, *Aftan Rawi* 19910! *Sahira & T. Sakin* 34763!, 34776! *Sahira* 39479!; **LSM**: Garm al Bani Sa'd, *Rawi* 16602!; **LBA**: Basra, *Gillett* 10051!, *Rawi* 16574!; bank of Shatt al-'Arab nr Basra, *Haines* W. 878!

Fig. 103. **Amaranthus viridis**. 1, habit × ½; 2, fruit × 15. Reproduced with permission from Flora of Pakistan 71: f. 2, 1974. Drawn by M.Y. Saleem.

Probably the most cosmopolitan species of the entire genus, found throughout the tropics and subtropics and more tolerant of temperate regions than most of its allies. In the Orient region it has been found in Lebanon, Palestine, Jordan, Lower Egypt, Turkey, Pakistan, Afghanistan, N Africa (Libya) and Macaronesia.

HUMATA AL-GHANAM (Ir.-Baghdad, *Gillett* 8401), KHEIYES (Ir.-Basra, *Gillett* 10051 – "not eaten by livestock").

7. **Amaranthus albus** *L.*, Syst. Veg. ed. 10, 2: 1268 (1759); Boiss., Fl. Orient. 3: 990 (1879); Zohary, Fl. Pal. ed. 2, 2: 457 (1933); Vassilczenko in Fl. SSSR 6: 364 (1936); Aellen in Fl. Lowland Iraq: 214 (1964); Rawi in Dep. Agr. Iraq Tech. Bull. 14: 168 1964); Mouterde, Nouv. Fl. Lib. et Syr. 1: 442 (1966); Zohary, Fl. Palaest. 1: 185 (1966); Aellen in Fl. Turk. 2: 343 (1967) & in Fl. Iranica 91: 9 (1972); Boulos, Fl. Egypt 1: 133 (1999); Bao Bojian, Clements & Borsch in Fl. China 5: 420 (2003).

Annual herb, (10–)20–45(–55) cm, erect to prostrate, much-branched from base upwards. Stems tough and wiry, whitish to pale green, angular, glabrous or frequently ± whitish-floccose with multicellular hairs at least in youngest parts. Leaves glabrous, oblong-obovate to spatulate, rounded to truncate or slightly emarginate at the pale-mucronate tip (mucro 0.5–1 mm), long-attenuate below, those of the main stem and branches 1–5(–8.5) × 0.4–1.5(–2.5) cm including the 0.3–1.5(–4) cm petiole, margins frequently minutely crispate, upper leaves rapidly diminishing in size. Flowers in congested, axillary, rather prickly cymes to 2 cm in diameter, cyme-branches spike-like; male flowers scattered or some solitary in axils of uppermost leaves. Bracteoles much exceeding perianth, subulate or lanceolate-subulate, 2–3 mm, pale-margined with margin widest below, sharply pungent with stoutly aristate, excurrent midrib. Perianth segments 3, ± 1 mm; those of male flowes lanceolate-oblong, pale-membranous with a narrow green midrib excurrent in a short pale mucro; those of female flowers narrowly oblong to narrowly spathulate, acute or more commonly obtuse with a distinct sharp mucro, narrow green vitta at midrib often widened above. Stigmas 3, slender, pale, recurving. Capsule 1.5–2 mm, pyriform, circumscissile, greenish, and wrinkled except for the swollen, whitish beak, obviously exceeding the perianth. Seeds shining, black, somewhat compressed, ± 1 mm, faintly reticulate. Fig. 102: 5.

HAB. Mountainsides, weed in gardens and on sandy waste land on a river bank in the steppe, weed in fields, along ditchsides and on silty banks of a receding river on alluvial plain in desert region; alt. up to 750 m; fl. & fr. (Jul.-)Sep.–Oct.(–Nov.).
DISTRIB. Occasional in the lower forest zone and steppe region of Iraq, common on the irrigated alluvial plain in the desert region: **MAM**: Bekher, *Rawi, Nuri & Tikriti* 28958!; **MRO**: Gali Ali Beg, *Omar, Sahira, Karim & Hamid* 38315!; Bustana village nr Koi Sanjaq, *M. Nuri & K. Hamad* 41194!; **MSU**: nr Halabja, *id.* 41212!; **FUJ**: Hamam Alil, *Rawi, Nuri & Takriti* 28948!; **FUJ/FNI**: Mosul, *Anders* 2420; **FNI**: Nineveh, *E. Chapman* 26179!; **LCA**: Sudur, *Omar, Janan, Sahira & Fauzi* 36976!; Tarmiya, *Haines* W.15!; Abu Ghraib, *Alizzi* 34175!; Baghdad, *Guest* 13467!, *Chakravarty* 29685!, *Ali Abdul-Latif* 39499!; Suwaira, *Muhammad Shabul* 35772!; Na'maniya, *Jum'a Brahim* 6188!; 10 km S of Hamza, *Fuad Safwat, F. & E. Barkley* 3722!; **LEA**: 3 km N of Chabbab bridge, about 40 km E by N of Kut, *Alizzi & Rawi* 32392!

Native of N America, now dispersed over much of the warmer N temperate region of the world as an introduced weed. In the Orient region it has been found in Syria, Lebanon, Palestine, Lower Egypt, Caucasus, Iran, Afghanistan and C Asia. According to Vassilczenko (1936) this plant is a suitable component of silage.

8. **Amaranthus graecizans** *L.*, Sp. Pl. ed. 1: 990 (1753); Hand.-Mazz. in Ann. Naturh. Mus. Wien 26: 144 (1912); Nábělek in Publ. Fac. Sci. Univ. Masaryk 105: 12 (1929); Guest in Dep. Agr. Iraq Bull. 27: 6 (1933); Vassilczenko in Fl. SSSR 6: 365 (1936); Zohary in Dep. Agr. Iraq Bull. 31: 49 (1950); Blakelock in Kew Bull. 5: 445 (1950) & 8: 228 (1953); Aellen in Fl. Lowland Iraq: 214 (1964); Rawi in Dep. Agr. Iraq Tech. Bull. 14: 168 (1964); Mouterde, Nouv. Fl. Lib. et Syr. 1: 443 (1966); Zohary, Fl. Palaest. 1: 185 (1966); Aellen in Fl. Turk. 2: 343 (1967) & in Fl. Iranica 91: 10 (1972); Townsend in Fl. Pak. 71: 17 (1974); Miller in Fl. Aran. Penins. & Socotra 1: 291 (1996); Boulos, Fl. Egypt 1: 133 (1999).

[?*A. blitum* (non L.) Blatter in Journ. Ind. Bot. Soc. 11: 40 (1932)]

Annual herb, branched from base and usually also above, erect, decumbent or prostrate, mostly up to 45 cm (rarely to 70 cm). Stem slender to stout, angular, glabrous or thinly to moderately clothed with short to long, often crispate multicellular hairs which increase upwards, especially in inflorescence. Leaves glabrous or sometimes sparingly clothed on the lower surface of the principal veins with very short, gland-like hairs, long-petiolate (petioles 3–4.5 mm, sometimes longer than the lamina), lamina broadly ovate or rhomboid-ovate to narrowly linear-lanceolate, 4–55 × 2–30 mm, acute to obtuse or obscurely retuse at mucronulate tip, cuneate to long-attenuate at base. Flowers all in axillary cymose clusters, male and female intermixed, male commonest in upper clusters. Bracts and bracteoles narrowly lanceolate-oblong, pale-membranous, acuminate and with a pale or reddish arista formed by the excurrent green midrib, bracteoles subequalling or usually shorter than perianth. Perianth segments 3, all 1.5–2 mm; those of male flowers lanceolate-oblong, cuspidate, pale-membranous with a narrow green midrib excurrent in a short, pale arista; those of female flowers lanceolate-oblong to linear-oblong, gradually to abruptly narrowed to mucro, midrib often bordered by a green vitta above and apparently thickened, margins pale whitish to greenish. Stigmas 3, slender, usually pale, flexuous, ± 0.5 mm. Capsule subglobose to shortly ovoid, 2–2.25 mm, usually strongly wrinkled throughout with a very short, smooth beak, exceeding the perianth, circumscissile or sometimes not. Seeds shining, compressed, black, 1–1.25 mm, faintly reticulate. Fig. 102: 6.

HAB. & DISTRIB. of species: as for subsp. *graecizans* (below).

subsp. **graecizans**.

A. angustifolius Lam., Encycl. 1: 115 (1783), nom. illegit. *A. blitum* L. var. *graecizans* (L.) Moq. in DC., Prodr. 13 (2): 363 (1849).

A. silvestris Vill. var. *graecizans* (L.) Boiss., Fl. Orient. 4: 990 (1879) sub "sylvestris" (sphalm.); Zohary, Fl. Pal. ed. 2, 2: 457 (1933); Rawi in Dep. Agr. Iraq Bull. 14: 168 (1964); Mouterde, Nouv. Fl. Lib. et Syr. 1: 443 (1966); Zohary, Fl. Palaest. 1: 185 (1966).

A. silvestris (non Vill.) Anth. in Notes Roy. Bot. Gard. Edinb. 18: 296 (1935). *A. angustifolius* L. subsp. *graecizans* (L.) Thell. in Asch. & Graebn., Syn. Mitteleur. Fl. 5: 307 (1914); Mouterde, Nouv. Fl. Lib. et Syr. 1: 443 (1966); Blakelock in Kew Bull. 5: 445 (1956).

A. graecizans var. *pachytepalus* Aellen in Fl. Iranica 91: 12 (1972).

HAB. Weed in gardens and orchards and among crops in cotton fields, rice fields etc., on banks of receding rivers and by roadsides; alt. 250–300(–900) m; fl. & fr. (Jan.-)Jul.–Aug. & sometimes also Sep.–Nov.

DISTRIB. Very common on the alluvial plain in the desert region of Iraq; occasional in the steppe region. ?**MSU**: Susa, *F. Karim* 39265!; **FUJ/FNI**: banks of r. Tigris at Mosul, *Nábělek* 1083; **FNI**: Nineveh, *Bornm.* 1769!; *E. Chapman* 25391!; Tal Kaif, *Muhammad Najib* 5164!; Ba'shiqa, *J.F. Tikriti* 16387!; **DWD**: Hit, *Guest* 3528!; **DLJ**: Rawa, *Alizzi & Omar* 35342!; **DSD**: Um Qasr, *H. Hamid* 38553!; **LCA**: nr Baghdad, *Haussknecht* s.n.! *Paranjpye* s.n.! comm. Graham, *Gillett* 9893! *Haines* s.n.; Qasr Naqib, nr Baghdad, *Handel-Mazzetti* 439; Abu Ghraib, *Gillett* 12506! *anon.* 21539!; Rustamiya, *Guest* 168!; Zafaraniya, *Gillett* 5871! *Sahira & T. Sakin* 34766!; Babylon, *Nábělek* 1088; ? "Babylonia", *Noë* s.n. in Fl. Orient.; Suwaira, *Alizzi & Omar* 34656!; Shatra, *Omar & Karim* 36761!; ? **LEA**: Diyala, *Sutherland* 387! 388!; 30 km SE of Baghdad along the Kut highway, *Barkley & Majid Brahim* 4018!; Aziziya, *Guest* 3475!; **LSM**: by Musharrah Canal, Amara, *Evans* M/302!; Nasiriya, *Abdul-Razzaq Barbuti* in Guest 25381!; Garma district, *Rawi* 16611!; Garmat Bani Sa'd, *Rawi* 16566!; 40 km E of Amara, *Alizzi & Omar* 34733!; Musaida, *Rawi* 16524!; **LBA**: Basra, *Whitehead* 121, in Blatter, l.c.

This, the typical subspecies, is the characteristic form of *A. graecizans* not only in Iraq but throughout SW Asia, also occurring in NW India and most parts of Africa, and in other regions chiefly as a casual.

'URF AD-DÎK ("cock's comb", Ir.-Hit, *Guest* 3475; Ir.-Baghdad, *Gillett* 5871); SARMAQ (Ir.-Aziziya, *Guest* 3475 – "cooked and eaten by people as a vegetable"; Ir.-Nasiriya, *Abdul-Razzaq al-Barbuti* in *Guest* 2538 – "eaten by all animals").

subsp. **silvestris** (*Vill.*) *Brenan* in Watsonia 4: 273 (1961); Aellen in Fl. Iranica 91: 11 (1972). Fig. 103: 5.

A. silvestris Vill., Cat. Pl. Jard. Strasb.: 111 (1807).

A. graecizans var. *sylvestris* (Vill.) Asch. & Schweinf., Beitr. Fl. Aethiop.: 176 (1867); Aellen in Fl. Iranica 91: 11 (1972).

Leaf blade (especially of the larger leaves of the main stem) broadly to rhomboid-ovate or elliptic-ovate, less then 2.5 × as long as broad. Perianth segments very shortly (sometimes scarcely) mucronate.

Distribution (of subspecies): Not yet recorded from Iraq, Anthony's record of "*A. sylvestris* Desf." from Amara being based on misidentifications of subsp. *graecizans*. The present plant may well occur as a weed in the hill country of Iraq.

Occurs in the Old World from the warmer parts of Europe to the cooler regions of western Asia (Caucasus, N Iran etc.) and NW India; also in (chiefly eastern) tropical Africa.

The name *Amaranthus blitum* L. is correctly applied to the plant now known as *A. lividus* L. It has, however, been widely misapplied as *A. graecizans* and should never be used, being rejected under Article 65 of the International Code of Botanical Nomenclature as a long-standing source of confusion. Whitehead's specimen so named by Blatter has not been seen, and so its identity is uncertain – but in all probability it is the typical form of the present species.

9. **Amaranthus blitoides** *S. Wats.*, Proc. Am. Acad. 12: 273 (1877); Vassilczenko in Fl. SSSR 6: 363 (1936); Rawi in Dep. Agr. Iraq Tech. Bull. 14: 168 (1964); Mouterde, Nouv. Fl. Lib. et Syr. 1: 442 (1966); Zohary, Fl. Palaest. 1: 184 (1966); Aellen in Fl. Turk. 2: 342 (1967 & in Fl. Iranica 91: 6 (1972); Bao Bojian, Clements & Borsch in Fl.China 5: 420 (2003).

Annual herb, (10–)20–50(–75) cm, decumbent or more usually prostrate with the tips of the branches ascending, much-branched from base upwards. Stems tough and wiry, whitish to pale green, angular, glabrous or with short whitish papillate hairs in upper parts. Leaves glabrous, oblong to oblong-obovate or spatulate, rounded to subacute at the pale-mucronate tip (mucro to 0.5 mm), attenuate below, those of the main stem and branches 1.5–5(–6.5 × 0.3–1.25(–1.6) cm including the 0.4–1.5(–2.5) cm petiole, with distinct, white, scabridulous margins; upper leaves rapidly diminishing in size. Flowers in congested, axillary cymes, cyme-branches spike-like; male flowers scattered, few. Bracteoles lanceolate-ovate or lanceolate, to ± 2 mm, pale-margined, mucronate but not rigidly aristate and spinescent, not exceeding perianth segments. Perianth segments 4–5, all 2–2.5 mm, those of male flowers lanceolate-oblong, pale-membranous with a narrow green midrib excurrent in a short pale mucro; those of female flowers somewhat unequal, lanceolate-ovate, acuminate or more shortly and abruptly pointed, green-vittate, mucronate. Stigmas 3, slender, pale, recurving. Capsule ± 2 mm, roundish or broadly pyriform, circumscissile, pale green, smooth, equalling or somewhat shorter than perianth, beak short and firm. Seeds shining, black, compresssed, 1–1.5 mm, faintly reticulate.

HAB. On open roadsides near bungalows, in stony places among ruins; alt. 100–600 m; fl. & fr. Sep.–Nov.
DISTRIB. Rare in Iraq, only found comparatively recently in two localities, one in the lower forest zone, the other in the dry steppe zone. **MSU**: Dukan, *Haines* W.2154! **FUJ**: Hadhr, *Alizzi & Omar* 35257!

Native of N America, now widespread in the Old World as a weed. It is naturalized in C and S Europe and occurs in our area in Syria, Lebanon, Iran and C Asia as well as N Africa and elsewhere.

3. **AERVA** Forssk. nomen. conserv.

Fl. Aegypt.-Arab.: 170 (1775)

Ouret Adans., Fam. Pl. 2: 268, 586 (1763)

Perennial herbs (sometimes flowering in first year), prostrate to erect or scandent. Leaves and branches opposite or alternate, leaves entire. Flowers hermaphrodite or dioecious, sometimes probably polygamous, bibracteolate, in axillary and terminal sessile or pedunculate bracteate spikes, with one flower in axil of each bract. Perianth segments 5, oval or lanceolate-oblong, membranous-margined with a thin to wider green centre, perianth deciduous with fruit but bracts and bracteoles persistent. Stamens 5, shortly monadelphous at base, alternative with subulate or rarely narrowly oblong and truncate or emarginate pseudostaminodes; anthers bilocular (4-locellate). Ovary with a single pendulous ovule; styles very short to slender and distinct; stigmas 2, short to long and filiform (sometimes solitary and capitate, flowers then probably functionally male). Utricle thin-walled, bursting irregularly. Seeds compressed-reniform, firm, black.

About 10 species in the tropics, chiefly centred on Africa; one species recorded from "Babylonia" some 130–140 years ago but not found since.

Aerva javanica (*Burm.f.*) *Juss. ex Schult.*, Syst. Veg. ed. 15, 5: 565 (1819); Moq. in DC., Prodr. 13 (2): 299 (1849); Boiss., Fl. Orient. 4: 992 (1879); Zohary in Dep. Agr. Iraq Bull. 31: 49 (1956); Rawi, ibid. Tech. Bull. 14: 68 (1964); Aellen in Fl. Lowland Iraq: 215 (1964); Rawi, ibid. Tech. Bull. 14: 118 (1964).

Fig. 104. **Aerva javanica**. 1, habit × ½; 2, female flower × 7. **Alternanthera sessilis**. 3, habit flowering branch × ½; 4, flower with front tepal and stamen removed × 12; 5, fruiting perianth × 10. Reproduced with permission from Flora of Pakistan 71: f. 4 & 7, 1974. Drawn by M.Y. Saleem.

Celosia lanata L., Sp. Pl. ed. 1: 205 (1753).
Iresine javanica Burm.f., Fl. Ind.: 212, t. 65 f. 2 (1768).
I. persica Burm.f., op. cit.: 212 t. 65 f. 1 (1768).
Illecebrum javanicum (Burm.f.) Murr., Syst. Veg. ed 13: 206 (1774).
Aerva tomentosa Forssk., Fl. Aegypt.-Arab.: 170 (1775).
A. persica (Burm.f.) Merrill in Philippine Journ. Sci. 19: 348 (1921); Aellen in Fl. Iranica 91: 13 (1972).

Perennial herb, frequently woody and suffruticose or growing in erect clumps, 0.3–1.5 m, branched from about the base with simple stems or stems with long ascending branches. Stem and branches terete, striate, ± densely whitish- or yellowish-tomentose or pannose, when dense the indumentum often appearing tufted. Leaves alternate, very variable in size and form, from narrowly linear to suborbicular, ± densely whitish- or yellowish-tomentose but usually more thinly so and greener on upper surface, margins plane or ± involute (when strongly so the leaves frequently ± falcately recurved), sessile or with a short and indistinct petiole or the latter rarely to 2 cm in robust plants. Flowers dioecious. Spikes sessile, cylindrical, dense and stout (up to 10 × 1 cm), to slender and interrupted with lateral globose clusters of flowers and with some spikes apparently pedunculate by branch reduction; male plants always with more slender spikes (but plants with slender spikes are not always male); upper part of stem and branches leafless, upper spikes forming terminal panicles; bracts 1–2 mm, broadly deltoid-ovate, hyaline, acute or obtuse with the obscure midrib ceasing below apex, densely lanate throughout or only about the base or apex, persistent; bracteoles similar, also persistent. Female flowers with 2 outer tepals 2–3 mm, oblong-obovate to obovate-spathulate, lanate, acute to obtuse or apiculate at tip, yellowish midrib ceasing well below apex; inner 3 slightly shorter, elliptic-oblong, ± densely lanate, acute, with a narrow green vitta along midrib, which extends for ± $^{2}/_{3}$ the length of each tepal; style slender, distinct, with two filiform flexuous stigmas at least equalling it in length; filaments reduced, anthers absent. Male flowers smaller, outer tepals 1.5–2.25 mm, ovate; filaments delicate, anthers about equalling perianth; ovary small, style very short, stigma rudimentary. Capsule 1–1.5 mm, rotund, compressed. Seeds ± 1 mm, round, slightly compressed, brown or black, shining and smooth or faintly reticulate. Fig. 104: 1–2.

HAB. Not recorded.
DISTRIB. Very rare if indeed it occurs in Iraq. "Babylonia", *Aucher* 3193, *Noë* s.n.

Widespread in the drier parts of the tropics and subtropics of the Old World from Burma, India and Sri Lanka through Pakistan, S Iran, Upper Egypt, Arabia and N Africa to Morocco, S to Cape Verde Is. and Cameroon through Uganda and Tanzania to Madagascar. Introduced into Australia and elsewhere.

4. **ALTERNANTHERA** Forssk.

Fl. Aegypt.-Arab.: 28 (1775)

Annual or perennial herbs, prostrate or erect to floating or scrambling, with entire, opposite leaves. Inflorescences of sessile or pedunculate heads or short spikes, axillary, solitary or clustered, bracteate. Flowers hermaphrodite, solitary in axil of a bract, bibracteolate, bracts persistent but perianth falling with fruit, bracteoles persistent or not. Perianth segments 5, free, equal or unequal, glabrous or clothed with smooth or denticulate hairs. Stamens 2–5, some occasionally without anthers, filaments distinctly monadelphous at base into a cup or tube, alternative with large and dentate or laciniate to very small pseudo-staminodes (rarely these obsolete); anthers unilocular (bilocellate). Style short, stigma capitate. Ovary with a single pendulous ovule. Fruit an indehiscent utricle, thin-walled or sometimes corky, seeds ± lenticular.

A large genus of over 150 species, far the greatest number occurring in the American tropics and subtropics, from which several have spread to become widely distributed weeds.

Alternanthera (from Lat. *alternans*, alternating, *anthera*, anthers, according to Stearn (in Smith, 1972) because alternate anthers in this plant are sterile); Joy weed.

Tepals glabrous or almost so, 1-nerved throughout . . . . 1. *A. sessilis*
Tepals strongly hairy, 3-nerved in lower half. . . . . 2. *A. bettzickiana*

1. **Alternanthera sessilis** (*L.*) *DC.*, Cab. Hort. Monsp.: 77 (1813); Moq. in DC., Prodr. 13 (2): 357 (1849). Aellen in Fl. Lowland Iraq: 216 (1964); Zohary, Fl. Palaest. 1: 189 (1966); Townsend in Fl. Pakk. 71: 41 (1974); Miller in Fl. Aran. Penins. & Socotra 1: 302 (1996); Boulos, Fl. Egypt 1: 142 (1999); Bao Bojian, Clements & Borsch in Fl. China 5: 427 (2003).

Gomphrena sessilis L., Sp. Pl. ed. 1: 225 (1753).
Illecebrum sessile (L.) L., Sp. Pl. ed. 2: 300 (1762).
Alternanthera "achyranth." Forssk., op. cit. lix (1775).
A. repens Gmel., Syst. Nat. ed. 14, 2 (1): 106 (1791).
A. triandra Lam., Encycl. Meth. 1: 95 (17xx); Trimen, Handb. Fl. Ceylon 3: 405 (1895); Gamble, Fl. Madras: 1179 (1925).
Achyranthes alternifolia L.f., Sp. Pl. Suppl.: 159 (1781), quoad pl. zeyl., non L., Mantissa: 50 (1767).

Annual or usually perennial herbs; in drier situations with slender, more solid stems, erect, ± much-branched, to 30 cm; in wetter places ascending or commonly prostrate, with stems 0.1–1 m long, rooting at nodes, ± fistular, with numerous lateral branches; when floating very fistular, stems attaining several metres in length and over 1 cm thick, with long clusters of whitish rootlets at nodes. Stem and branches green or purplish, with a narrow line of whitish hairs down each side of stem and tufts of white hairs in branch and leaf axils, otherwise glabrous, striate, terete. Leaves extremely variable in shape and size, linear-lanceolate to oblong, ovate, or obovate-spathulate, 1–9(–15) × 0.2–2(–3) cm, blunt to shortly acuminate at apex, cuneate to attenuate at base, glabrous or thinly pilose, especially on lower surface of midrib; petiole obsolete or to ± 5 mm. Inflorescence sessile, axillary, solitary or in clusters of up to 5, subglobose or slightly elongate in fruit, ± 5 mm in diameter; bracts scarious, white, deltoid-ovate, mucronate with excurrent pale midrib, glabrous, ± 1 mm; bracteoles similar, also persistent. Tepals ovate-elliptic, equal, 1.5–2.5 mm, acuminate to rather blunt, white, glabrous or almost so, shortly but distinctly mucronate with stout excurrent midrib, margins obscurely denticulate. Stamens 5 (2 filaments anantherous), at anthesis subequalling ovary and style, alternating pseudostaminodes resembling filaments but usually somewhat shorter. Ovary strongly compressed, roundish, style extremely short. Fruit obcordate or cordate-orbicular, 2–2.5 mm, strongly compressed with a narrow, pale, somewhat thickened margin. Seeds discoid, ± 1 mm, brown, shining, faintly reticulate. Fig. 104: 3–5.

HAB. In damp swampy places, marshy areas, low-lying ditches near a muddy tidal river bank etc.; alt. s.l.–10 m; fl. & fr. Dec.–Mar.
DISTRIB. Occasional in the southern marsh district of the desert region of Iraq: **LSM**: nr Dibin (Diba), *Thesiger* 1264!; Chabaish (Kaba'ish), in Hor al-Hammar, *H. Hamid* 38547!; Al-Halfaya (Amara liwa), *Rawi* 16634 (BAG); Hor Al-Shain S of Majaar Al-Kabbeer, 17/1/1977, *Thamer* 46657 (BAG); Hor AL-Hawiza (10 km E of Qurna, 21/3/1977, *Thamer* 46664 (BAG); Hor Al-Hawiza (18 km E of Qalat Salih), 17/5/1977, *Thamer, Wedad & Hana* 46791 (BAG). **LBA**: Basra, *Whitehead* 59 in Blatter, l.c., *Haines* W.876!

A common species, widespread in warmer regions of Old and New Worlds in waste and cultivated ground, especially in damp or wet conditions. The name *A. sessilis* (L.) R.Br. is sometimes cited in error as an earlier combination.

2. **Alternanthera bettzickiana** (*Regel*) *Voss* in Vilm., Blumengart. ed. 3, 1: 869 (1895) – "bettzichiana" sphalm.; Aschers. & Graebn., Syn. Fl. Mitteleur. 5 (1): 365 (1920); Bailey, Manual Cult. Pl. ed. 2: 357 (1949).

Telanthera bettzickiana Regel in Index Sem. Horti Petrop. 1862: 28 (1862) & Gartenflora 11: 178 (1862).

Erect or ascending, bushy perennial herb commonly cultivated as an annual, 5–45 cm, stem and branches villous when young but soon glabrescent. Leaves narrowly or more broadly elliptical to oblanceolate or rhomboid-ovate, acute or acuminate at apex, long-attenuate into an indistinctly demarcated petiole below, often purple-suffused and sometimes variegated. Heads axillary, sessile, usually solitary, globose or ovoid, 5–6 mm in diameter. Tepals white, lanceolate, acute, mucronate with excurrent midrib, prominently 3-nerved below and darker in nerved area, with a line of minutely barbellate white hairs along each side of this area, hairs becoming denser towards base of tepal. Pseudo-staminodes as long as filaments, laciniate at apex.

DISTRIB. **LCA**: Baghdad, 4/12/1947, cultivated in open sunlight, *Gillett* 10143 (BAG); Abu-Ghraib, 12/10/1972, *Fawzi & Ali Abdul Latif* 39511 (BAG); Abu-Chesaf, 25 km S of Al-Kahla, *Thamer* 46661 (BAG).

Husain & Kasim, Cult. Pl. Iraq: 98 (1975) record *Alternanthera amoena, A. chromopetala* and *"A. paranchoides"* in cultivation in C & S Iraq. No *Alternathera "chromopetala"* has ever been described, and this is presumably a gardeners' name. *A. "paranchoides"* should probably be *"paronychioides"* and presumably refers to cv. *paronychioides* of *A. bettzickiana* rather than the botanical species *A. paronychioides* St. Hil. – a weedy species not likely to be cultivated. The *A. amoena* of Husain & Kasim probably also refers to *A. bettzickiana*. These two species, together with *A. versicolor* (Regel) Voss., are not well understood botanically but *A. bettzickiana* is the common plant in cultivation. Said to be a native of tropical S America, it is unknown truly wild, apparently always sterile and propagated by cuttings. It is probably a cultigen of *A. flavogrisea* (Urb.) Urb. (=*A. ficoidea* auctt.). No specimens of any plants of this group have been seen from Iraq.

5. **GOMPHRENA** L.

Sp. Pl.: 224 (1753); Gen. Pl. ed. 5: 105 (1754)

Annual or occasionally perennial herbs with entire, opposite leaves. Inflorescences terminal or axillary, capitate or spicate, solitary or glomerate, often subtended by a pair of sessile leaves. Flowers hermaphrodite, each solitary in axil of a bract, bibracteolate; bracteoles laterally compressed, carinate, often ± winged or cristate along dorsal surface of midrib. Tepals 5, erect, free or almost so, ± lanate dorsally, at least the inner 2 usually ± indurate at base in fruit. Stamens 5, monadelphous, tube shortly 5-dentate with entire to very deeply lobed teeth, with or without alternating pseudostaminodes. Stigmas 2, suberect or ± divergent to very short. Fruit a thin-walled, irregularly rupturing utricle.

A rather large genus of about 90 species, centred on tropical America but also with several Australian representatives; one species in Iraq.

According to Bailey (1939) *Gomphrena* (thought to be a name suggested by Lat. *gromphraena*, the name in Pliny for some kind of Amaranth and supposed to have been derived from the Lat. *grapho*, to write or paint, alluding to the plant's highly coloured or "painted" foliage. Also sometimes known as Batchelor's Buttons though two utterly distinct plants belonging to two different families have the same popular name in English; Globe Amaranth (Eng., Am.). Bedev. (1936?) ANBAR; DAMM AL-ĀSHAQ; ZIRR HABASHI.

Gomphrena globosa *L.*, Sp. Pl. ed. 1: 224 (1753); Zohary, Fl. Pal. ed. 2, 2: 461 (1933); Vassilchenko in Fl. SSSR 6: 369 (1936); Husain & Kasim, Cult. Pl. Iraq: 99 (1975).

Annual herb, decumbent or erect, branched from base and also above, 15–60 cm; stem and branches ± densely clothed with appressed white hairs at least when young. Leaves broadly lanceolate to oblong or elliptic-oblong, narrowed to an ill-defined petiole below, thinly pilose, the pair of leaves subtending the terminal inflorescence sessile or almost so, broadly to subcordate-ovate. Inflorescence sessile above the uppermost pair of leaves, usually solitary, globose or depressed-globose, ± 2 cm in diameter, stramineous to pinkish or deep red; bracteoles strongly laterally compressed, with a broad irregularly dentate crest from apex almost to base of dorsal surface of midrib. Tepals 6–6.5 mm, outer considerably lanate. Staminal tube subequalling perianth; 5 teeth deeply bilobed with obtuse lobes subequalling anthers; pseudostaminodes absent.

HAB. Cultivated in gardens as an ornamental in C Iraq; alt. ± 35 m; fl. & fr. Jun.
DISTRIB. Cultivated in the desert region of Iraq on the irrigated alluvial plain: **LCA**: Baghdad (cult.), *Janan al-Mukhtar* 35771!; *Sahira* C. 311! C.525!; Zafaraniya (cult.) *Al-Kas* 20500 (BAG!).

Widely grown as an ornamental in many parts of the world (including Europe, Palestine, Syria, Lebanon, C Asia, India etc.) and sometimes occurring as an escape. According to Husain & Kasim (1975) it is "cultivated throughout the country" in Iraq; but only specimens from Baghdad have been seen.

Globe Amaranth or Bachelor's Buttons (Eng.), Common Globeamaranth (Am.); WARD AD-DUKMA (Ir., *Janan* 35711, *Sahira* C. 311); among other Arabic colloquial names for this plant are DAMM AL-'ĀSHIQ and ZIRR HABASHI as given by Schweinfurth (1912) and Bedzuian (1936).

97. THELIGONACEAE E.Ulbrich

(Cynocrambaceae)

Pflanzenfam. ed. 2, 16C: 368–378 (1934)

C. C. Townsend[18]

Fleshy annual or perennial herbs. Leaves simple, entire, stipulate, the lower opposite, the upper commonly alternate by suppression of one of the pair. Flowers monoecious, in few-flowered axillary cymes opposite the upper alternate leaves, cymes containing one or both sexes. Petals absent. Male flowers valvately 2–5-partite; stamens up to ± 30, mostly 7–12, filaments filiform, anthers linear, dehiscing by longitudinal slits. Female flowers with a caducous, membranous, strongly asymmetrical tubular perianth, 24-dentate at the mouth. Ovary ovoid, ovule solitary, erect; style simple, becoming basal by the unilateral development of the ovary, exserted. Fruit a subglobose nutlike drupe. Seeds spherical, with a fleshy endosperm, dispersed by ants.

Traditionally, Theligonaceae have been treated as a monogeneric family (as in this Flora), however, recent reliable molecular data place it within a highly derived group of Rubiaceae (along with *Galium*, *Hedyotis* and *Rubia*: see Pl. Syst. Evol. 225: 43–75. 2000).

One to three species in the Mediterranean region, Macaronesia, China and Japan.

1. THELIGONUM L.

Sp. Pl.: 993 (1753); Gen. Pl. ed. 5: 430 (1754)

Cynocrambe Gaertn., Fruct. 1: 362, t. 75 (1791)

Monotypic. Description as for family.

Theligonum cynocrambe *L.*, Sp. Pl. ed. 1: 993 (1753).

Theligonium alsinoideum Lam., Fl. Fr. 2: 198 (1778).
Cynocrambe prostrata Gaertn., Fruct. 1: 362, t. 75 (1791); Boiss., Fl. Orient. 4: 897 (1879); Zohary, Fl. Pal. ed. 2, 2: 426 (1933); Cullen in Fl. Turk. 2: 344 (1967).
Parietaria debilis (non Forst.f.) Standley in Publ. Field Mus. Nat. Hist. publ. 469, Anthrop. Ser., 30 (1): 181 (1940).

Fleshy, flaccid annual with a slender, fibrous root, branched from about the base, (5–) 10–20(–30) cm. Stem and branches weak, glabrous or sparsely shortly stiffly hairy. Leaves rhomboid-ovate or subdeltoid-ovate, very variable in size, rather abruptly narrowed to a long petiole; petioles of lower leaves about as long as lamina, those of upper and floral leaves shorter; margins of lamina stiff papillose-hairy. Stipules membranous, entire or remotely toothed, those of the lower paired leaves connate, those of the upper alternate leaves erect. Cymes in upper leaf axils male, those of lower axils female, with some mixing of the sexes about the middle of the stem. Male flowers botuliform in bud, the (usually 2) perianth segments narrowly elliptic-lingulate, 2–3 mm, recurved and frequently coiling after anthesis. Filaments thread-like, shorter than the linear, finely spirally twisted, 1.5–2 mm anthers. Female flowers minute, tubular perianth 0.75–1 mm, transparent-membranous. Style slightly exserted, papillose above. Drupe subglobose, ± 1.25 mm, with minute oblong whitish verruculae arranged in longitudinal rows. Seed ±1.5 mm, globose, verruculose endosperm backed by the large, curved, shining embryo. Fig. 105: 1–6.

HAB. Rocky mountain slopes, silt ridges; alt. 1000–1600; fl. Mar.–May.
DISTRIB. **MAM**: Qara on road to koi Sanjar, 6/5/1959, *Rawi, Nuri &Alkass* 28040 (BAG); Aqra, 30/5/1948, *Rawi* 11303A (BAG). **MRO:** Shakh-i- Harir (near Harir, Rowanduz), 1/5/56, *Emberger et al.* 15457 (BAG); Ser kupkan village, c.7 km NW of Rania, 10/5/1959, *Rawi, Nuri & Kass* 28514 (BAG). **MSU**: Ahmad Awa (Zalan) 4km E Khurmal, 1/5/1971, *Al-Khyat* 52870 (BAG). **MJS**: Jabel Sinjar, 16/4/1980, *Al Kaisi & Wedad* 52031 (BAG); Karsi,19/4/1980, *Al Kaisi & Wedad* 52251 (BAG); Jabel Sinjar, E slope, 20/5/1980, *Omar, Al Kaisi & Al Khayat* 52573 & 52572 (BAG); Jabel Sinjar, SE slope, 22/4/1981, *Al-Khyat & Wedad* 53360 (BAG); Kursi, Jabel Sinjar, 23/5/1948, *Gillett* 10915 (BAG). **FNI**: Eski Kellek, 23/3/1948, *Rawi & Gillett* 10399 (BAG). **FAR**: Bastura Chai, 27/3/1948, *Rawi & Gillett* 10492 (BAG).

Mediterranean region, Syrian Desert, China, Japan.

---

[18] Royal Botanic Gardens Kew, UK

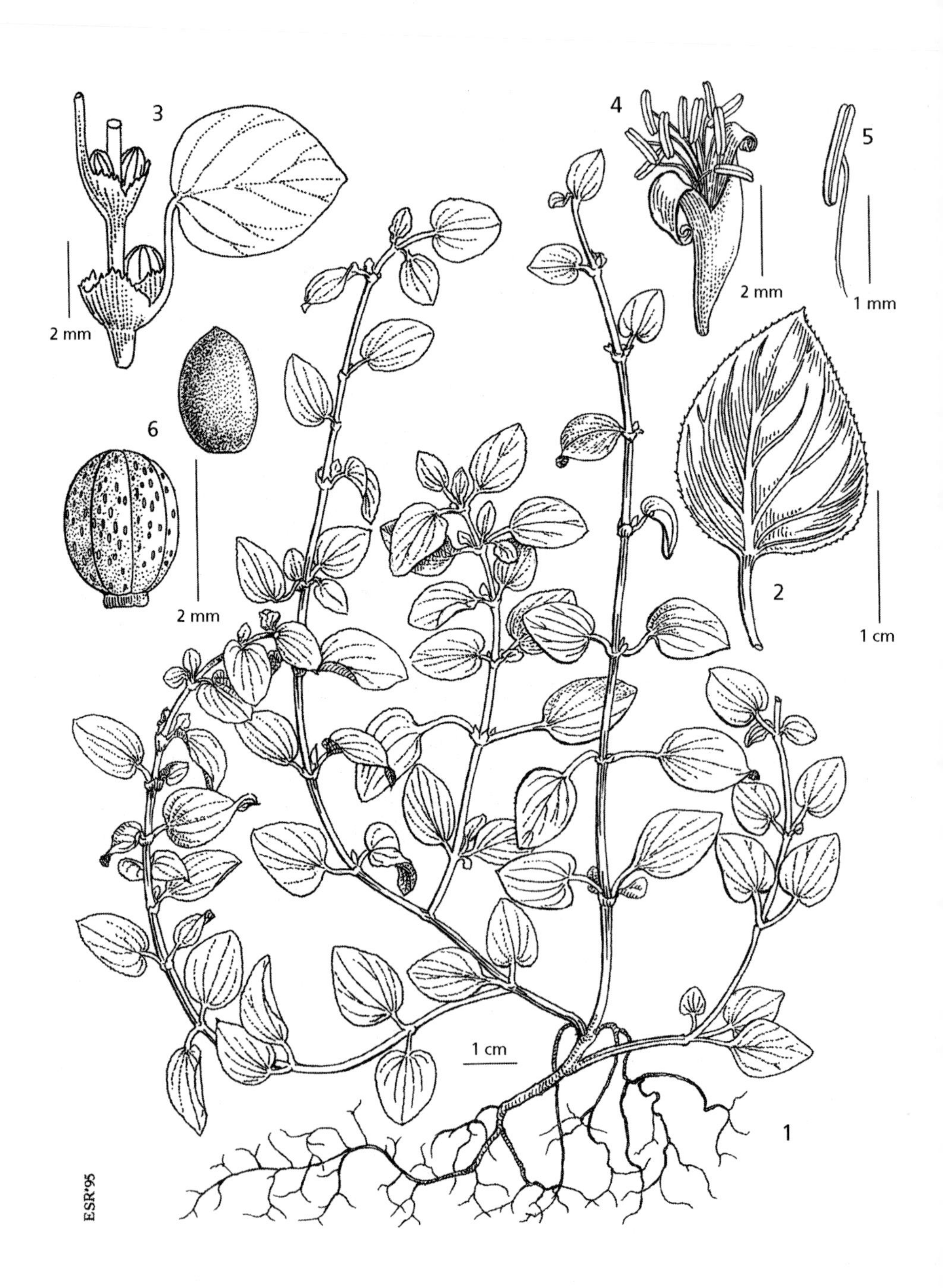

Fig. 105. **Theligonum cynocrambe**. 1, habit; 2, leaf; 3, portion of stem detail; 4, flower; 5, stamen; 6, fryuit and seed. Reproduced with permission from Flora Iberica.

98. SPHENOCLEACEAE

(Lindl.) Mart. ex DC.

H. A. Alizzi

Annual herbs, with the appearance of a small *Phytolacca*, rather fleshy without milky latex. Leaves alternate, simple, entire, exstipulate. Inflorescence a dense spike, terminal. Flowers actinomorphic, hermaphrodite, sessile; bract 1, bracteoles 2. Calyx epigynous, 5-lobed, imbricate, persistent; calyx-tube adnate to the ovary. Corolla epigynous, urceolate-campanulate, 5-lobed to middle, lobes imbricate. Stamens 5, inserted on the corolla-tube, alternating with lobes of corolla; filaments very short, anthers 2-celled, longitudinally dehiscent. Ovary inferior, 2-locular, ovules numerous in each locule, placentation axile; style short; stigma capitate or obscurely 2-lobed. Capsule membranous, circumscissile. Seeds minute, endosperm scanty, embryo straight.

The family has previously been placed in Asterales, especially near Campanulaceae and included by some authorities in that family. In the APG III (2009) classification it is placed in lamiids in the Solanales.

Monotypic; widely distributed in the Old World Tropics; introduced elsewhere.

SPHENOCLEA Gaertn.

Fruct. Sem. Pl. 1: 113, tab. 24/5 (1788), nom. cons.

Pongatium Juss., Gen. 423 (1789); Lam., Tabl. Encycl. 3, 1: 444 (1797).

Characters same as those for the family.

Sphenoclea (from the Gr. σφήνα, wedge, κλειστός, closed).

Sphenoclea zeylanicum Gaertn., Fruct. Sem. Pl. 1: 113, tab. 24 (1788); Federov in Fl. URSS 24: 449 (1957); Rawi in Dep. Agr. Iraq Tech. Bull. 14: 129 (1964); Hepper in Fl. West Trop. Africa, ed. 2, 2: 307, fig. 272 (1963); Shaw in Fl. Trop. East Africa: Sphenocleaceae 1 (1968); Boulos, Fl. Egypt 3: 134 (2002); Deyuan Hong & Turland in Fl. China 19: 504 (2011).

Sphenoclea pongatium DC., Prodr. 7: 548 (1839), nom. illeg.; Boiss., Fl. Orient. 3: 963 (1875).
Pongatium indicum Lam., Tabl. Encycl. 2: 44 (1794), nom. illeg. superfl.

Annual herb, usually occurring in swamps and stagnant water. Stems erect, 40–100 cm tall, branched, glabrous, spongy or fleshy, especially the lower parts. Leaves shortly petiolate, , linear-lanceolate to elliptic-lanceolate, 2.8–7 × 4–8 mm, entire, tapering at both ends, more or less acute at the apex, glabrescent; petiole 1.5–2 mm. Flowers < 2 mm, sessile, densely crowded in spikes; spike terminal, cylindrical with a conical apex, 2–4 cm. Calyx lobes triangular, 1 × 1 mm, glabrous; calyx tube obconical or cup-shaped. Corolla white, campanulate; lobes 5, patent, valvate in bud. Capsule depressed globose, ± 2 mm across, dehiscing by a transverse slit. Seeds numerous, ± 0.5 mm, brownish, scabrous, papillose. Fig. 106: 1–14.

HAB. In wet places, inundated rice fields, ditches etc.; alt. 5–10 m.; fl.: Oct.–Nov.
DISTRIB. Apparently rare in Iraq (though probably more common than indicated); found only twice in the Southern Marsh district of the desert region. **LSM**: 10 km E of Amara, *Rawi & Alizzi* 32409!; Musaida, *Husham Alizzi & Muhammed* 29608.

SAIYĀGH RŪHA (Ir.-Amara, *Alizzi & Muhammad* 29608.)

A common weed in rice fields in many tropical countries. According to Burkill (1935) this plant is eaten with rice in Java, the young plants being consumed whole but only the tips of old plants; it is said to have a bitter taste. Uphof (1968) adds that the leaves are on sale in the market of that country.

Native of tropical Asia (India, Sri Lanka, Myanmar, Malaysia, China, Japan, etc.) but now found throughout the tropics of Africa, Madagascar, Australia and S America; also in the Nile Delta and W Mediterranean district of Egypt, S Iran, C Asia (Pamir Alai), Pakistan and elsewhere.

---

[19] National Harbarium, Baghdad.
Updated by Shahina A. Ghazanfar

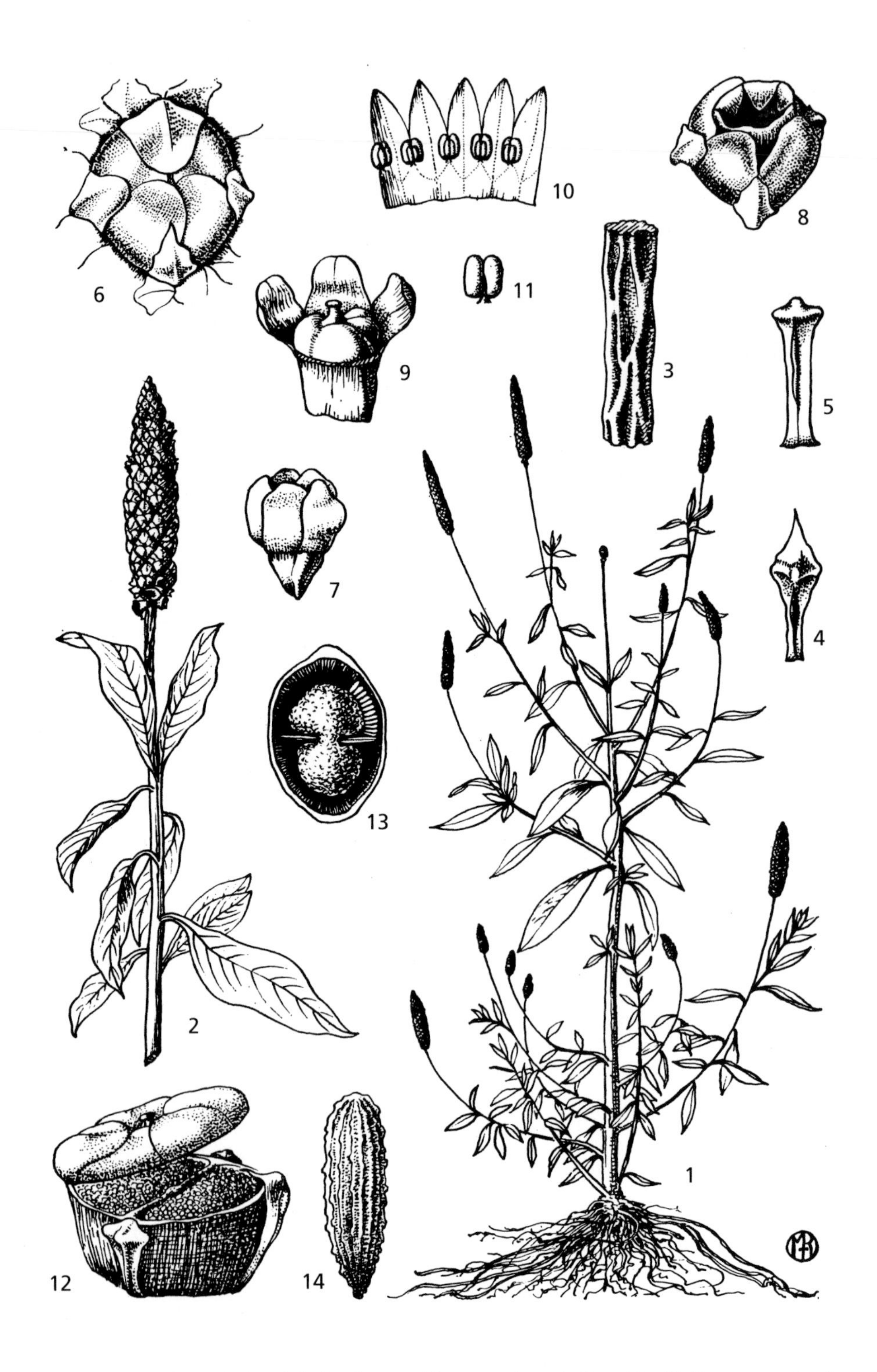

Fig. 106. **Sphenoclea zeylanica.** 1, habit; 2, part of flowering branch; 3, rachis of inflorescence showing scars left by fallen capsules; 4, bract; 5, bracteole; 6, flower bud, apical view; 7, bud opening; 8, flower showing partially opened corolla; 9, ovary and calyx (with two sepals removed); 10, corolla, opened; 11, stamen; 12, capsule, partially dehisced; 13, transverse section of capsule; 14, seed. Reproduced with permission from Flora Tropical East Africa: Sphenocleaceae: fig. 1. 1963.

INDEX TO FAMILIES, GENERA AND SPECIES

Scientific names of accepted species and infraspecific taxa are in **bold**, synonyms and misapplied names are in *italics* and subgeneric and sectional names are in small capitals. Where a name appears on more than one page, the principal page reference is given in **bold**.

Acanthophyllum C. A. Meyer 90
- **acerosum** Sosn. 92
- **bracteatum** Boiss. 90, fig. 36
- **caespitosum** Boiss.94, fig. 37
- *confertiflorum* Rech. f. 92
- **khuzistanicum** Rech. f. 90
- **kurdicum** Boiss. & Hausskn. 92
- *mucronatum* Fenzl non. C.A. Mey. 92
- *tournefortii* Fenzl 92
- **verticillatum** (Willd.) Hand. Mazz. 92

Achyranthes alternifolia L.f. 268

Aellenia Ulbr. 232
- *auricula* (Moq.) Ulbr. 235
- *glauca* (M. Bieb.) Aellen 235
- *hierochuntica* (Bornm.) Aellen 233
- *subaphylla* (C.A. Mey.) Aellen 235

Aerva Forssk. 266
- **javanica** (Burm.f.) Juss. ex Schult. 266, fig. 104
- *persica* (Burm.f.) Merrill 266
- *tomentosa* Forssk. 266

Agathophora Bunge 253
- alopecuroides (Delile) Fenzl ex Bunge 253
- *alopecuroides* auct. non (Delile) Fenzl ex Bunge 253
- **iraqensis** Botsch. 253, fig. 101
- postii (Eig) Botsch. 255

Agriophyllum Bieb. 204
- **minus** Fisch. & C.A. Mey. ex Ledeb. 204, fig. 80

Agrostemma L. 123
- **githago** L. 123

AIZOACEAE 124

Aizoanthemum L. 124
- **hispanicum** (L.) H.E.K. Hartmann 124, fig. 48

Aizoon hispanicum L. 124

Alsinanthus gypsophiloides (L.) Desv. 32
- *serpyllifolius* (L.) Desv. 33

Alsine akinfiewii (Schmalh.) Woronow 38
- *bocconii* Scheele 27
- *brevis* Boiss. 38
- *decipiens* Fenzl 40
- *intermedia* Boiss. 40
- *juniperina* (L.) Wahlenb. 38
- *media* L. 44
- *mesogitana* Boiss. 41
- *meyeri* Boiss. 38
- *montana* sensu Boiss. non L. 40
- *montana* var. *caucasica* (Boiss.) Boiss. 40
- *multinervis* (Boiss.) Bornm. 38
- *pallida* Dumort. 46
- *picta* (Sibth. & Sm.) Bornm. 35
- *recurva* (All.) Wahlenb. 37
- *recurva* (All.) Wahlenb var. *nivalis* Boiss. 37
- *rudbaransis* Stapf 40
- *sclerantha* Fisch. & Mey. 41
- *serpyllifolia* (L.) Crantz 33
- *succulenta* Delile 19
- *tenuifolia* (L.) Crantz 42
- *tenuifolia* var. *genuina* sensu Boiss. 41, 42
- *tenuifolia* var. *subtilis* Fenzl 42
- *wiesneri* Stapf 40

Alsinella serpyllifolia (L.) S.F. Gray 33

Alternathera Forssk. 268
- *"achyranth."* Forssk. 268
- amoena (Regel) Voss. 269
- **bettzickiana** (Regel) Voss 269
  - cv. paronychioides 269
- chromopetala 269
- *ficoidea* auctt. 269
- flavogrisea (Urb.) Urb. 269
- paronychioides St. Hil. 269
- *repens* Gmel. 268
- **sessilis** (L.) DC. 268, fig. 104
- *triandra* Lam. 268
- versicolor (Regel) Voss. 269

AMARANTHACEAE Juss. 256

Amaranthus L. 257
- **albus** L. 263, fig. 102
- *angustifolius* Lam. 265
  - subsp. *graecizans* (L.) Thell. 265
- **blitoides** S. Wats. 265
- blitum L. 265
- *blitum* L. var. *graecizans* (L.) Moq. 265
- *blitum* (non L.) Blatter
- **caudatus** L. 257, **258**
- *chlorostachys* Willd. 259
- *cruentus* L. 259
- **dubius** Mart. ex Thell. 260, fig. 102
- *gangeticus* L. 262
- *gracilis* Desf. 262
- **graecizans** L. 264
  - subsp. **graecizans** 264
  - subsp. **silvestris** (Vill.) Brenan 265
  - var. *pachytepalus* Aellen 265
  - var. *sylvestris* (Vill.) Asch. & Schweinf. 265
- **hybridus** L. 257, **259**, fig. 102
  - subsp. **cruentus** (L.) Thell. 259
  - subsp. **hybridus** 259
  - var. **erythrostachys** Moq. 259
  - var. **hybridus** 259
  - var. *paniculatus* (L.) Thell. 259

hypochondriacus L. 259
lividus L. 265
melancholinus L. 262
paniculatus L. 259
retroflexus L. 261
var. *delilei* (Richat & Loret) Thell. 261
silvestris Vill. 265
var. *graecizans* (L.) Boiss. 265
silvestris (non Vill.) Anth. 265
tricolor L. 262, fig. 102
var. *gangeticus* Fiori 262
tristis L. 262
tristis (non L.) Moq. in DC.
viridis L. 262, figs. 102 & 103
Anabasis L. 241
al-rawii Aellen ex L.M. El-Hakeem & Weinert
annua Bunge 245
articulata (Forssk.) Moq. 243
lachnantha Aellen & Rech.f. 243
lutea Moq. in DC. 245
micradena Iljin 245
setifera Moq. 245, fig. 97
spinosissima L.f. 236
tournefortii Jaub. & Spach 240
Ankyropetalum Fenzl 88
gypsophiloides Fenzl 88, fig. 35
var. *coelesyriacum* (Boiss.) Bark. 88
var. *glandulosum* Bornm. 88
var. *viscosum* Bark. 88
Arenaria L. 31
anomala (Waldst. & Kit.) Skinners 48
balansae Boiss. 32
var. **balansae** 32
var. **glaberrima** (Fenzl) McNeill 32
cucubaloides Smith
diandra Guss. 27
filiformis Labill. 35
globulosa Labill. var. *nana* C.A. Mey. 38
glutinosa Bieb. 55
gypsophiloides L. 32
var. glabra Fenzl 32
var. **gypsophiloides** 32
holosteoides (C. A. Meyer) Edgew. 44
holosteoides Edgew. var. *stellarioides* Williams 44
hybrida Vill. 42
intermedia (Boiss.) Fernald 40
juniperina L. 38
kashmirica Edgew. ex Edgew. & Hook. f. 37
kurdica McNeill 32
leptoclados (Reichenb.) Guss. 34
media L. 25
meyeri (Boiss.) Edgew. & Hook.
mirdamadii Rech. f. 32
neelgherense Wight & Arnott var. *glaberrima* Fenzl 32
neelgherense Wight & Arnott var. *glanduloso-pubescens* Fenzl 32
picta Sibth. & Sm. 35
recurva All. 37
rubra L. var. *marina* L. 25
sabulinea Griseb. ex Fenzl. 34
sabulinea Gris. ex Fenzl var. *brevipes* Bornm. 34
serpyllifolia L. 33, fig. 14
serpyllifolia L. var. *leptoclados* Reichenb. 34
serpyllifolia L. subsp. *leptoclados* (Reichenb.) Nyman 34
spinuliflora Ser. 31
tenuifolia L. 42
verticillata Willd. 92
Arthrocnemum glaucum auct. (non (Del.) Ung.-Sternb.) 207
glaucum (Del.) Ung.-Sternb. 255
macrostachyum (Moric.) Bunge 255
Astragalus setiferus DC. 252
Atraphaxis L. 142
billardieri *Jaub. & Spach* 144, fig. 55
var. *billardieri* 144
var. *tournefortii* (Jaub. & Spach) Cullen 144
tournefortii Jaub. & Spach 144
Atriplex L. 179
aellenii Sukhor. 183
asphaltitis Kasapligil 181
aucheri Moq. 190
autrani Post 184
belangeri (Moq.) Boiss. 189
bracteosa Trautv. 186
brenanii Sukhor. 183
davisii Aellen 184
dimorphostegia Kar. & Kir. 186, fig. 72
flabellum Bunge 189
halimus L. 183
hastata L. 190
heterosperma Bunge 184
holocarpa F. Muell. 189
hortensis L. 183, fig. 71
laevis C.A. Mey. 190
lasiantha Boiss. 184, fig. 72
leucoclada Boiss. 181, fig. 70
leucoclada var. turcomanica (Moq.) Zoh. 183
mesopotamica Eig 184
micrantha C.A. Mey. 184, fig. 71
nitens Schkuhr 190
nitens auctt. non Schkuhr
nummularia Lindl. 181
olivieri Moq. 186
patula L.
prostrata Boucher ex DC. 190
sagittata Borkh. 190
semibaccata R.Br. 190
spongiosa F. Muell. 189
tatarica L. 186, fig. 73
var. *desertorum* Eig 186
var. *tenera* Eig 184
var. *virgata* Boiss. **184**, 186
thunbergiifolia (Boiss. & Noë) Boiss. & Noë 189
transcaspica Bornm. & Sint. 186
turcomanica (Moq.) Boiss.
Axyris ceratoides L. 196

Bassia All. 196
arabica (Boiss.) Maire & Weiller 203
eriophora (Schrad.) Aschers. 197, fig. 77
hyssopifolia (Pall.) O. Kuntze 199, fig. 78
indica (Wight) A.J. Scott 200
joppensis Bornm. & Dinsmore 200
monticola (Boiss.) Kuntze 204
muricata (L.) Aschers. & Schweinf. 199
pilosa (Fisch. & C.A. Mey.) Freitag & G. Kadereit 203

prostrata (L.) A.J. Scott 201, fig. 79
pulverulenta H. Lindb. 219
reuteriana Guerke 200
scoparia (L.) A. J. Scott 201, fig. 79
subsp. *densiflora* (Turcz. ex Moq.) Cirujano & Velayos 201
Belovia Moq. 210
Bergia L.
ammanioides *Roxb.* 3, fig. 1
var. *pentandra* Wight 3
aquatica Roxb. 1
capensis L. 1, fig. 1
pentandra Camb. ex Guill. & Perr. 3
verticillata Willd. 1
Beta L. 168
maritima L. 169
vulgaris L. subsp. lomatogonoides Aellen 169
vulgaris L. subsp. maritima 169
var. **cicla** L. 169
var. *foliosa* Thell. 169
var. **rubra** (L.) Moq. 169
var. **vulgaris** **169**
Bienertia Bunge ex Boiss. 215
cycloptera auct. non Bunge ex Boiss. 215
sinuspersici Akhani 216, fig. 85
Bilderdykia convolvulus (L.) Dumort. 162
Bistorta officinalis Delarbre 149
Blitum L. 192
chenopodioides L. 178
rubrum L. var. *subintegrum* Boiss. 178
virgatum L. 194, fig. 75
subsp. **virgatum** 194
subsp. **montanum** (Uotila) S. Fuentes, Uotila & Borsch 194
Brezia Moq. 210
Bufonia L. 55
calycina Boiss. & Hausskn. 55
kotschyana Boiss.
leptoclada Rech.f. 55, fig. 22
oliveriana Ser. 57

Calligonum L. 146
comosum L'Herit. 147, fig. 57
crinitum Boiss. subsp. arabicum (Sosk.) Sosk.
polygonoides L. 147
subsp. *comosum* (L'Herit.) Soskov 149
tetrapterum Jaub. & Spach 147. fig. 57
Camphorosma 9 255
Caroxylon Thunb. 217
aphyllum (L.f.) Tzvelev
articulatum Moq. 247
azaurenum (Mouterde) T.A. Theodorova 226
boissieri (Botsch.) Freitag 223
canescens (Moq.) Akhani & Roalson
cyclophyllum (Baker) Akhani & Roalson 221
dendroides (Pall.) Tzvelev 223
hispidulum Bunge
imbricatum (Forssk.) Moq. 223
imbricatum (Forssk.) Akhani & Roalson 223
incanescens (C.A. Mey.) Akhani & Roalson 218
inerme (Forssk.) Akhani & Roalson 219, fig. 87
jordanicola 221, fig. 86
lancifolium Boiss. 235
nitrarium (Pall.) Akhani & Roalson *218*
salicornicum Moq. 249
tetrandrum (Forssk.) Akhani & Roalson 224
tomentosum (Moq.) Tzvelev 226
vermiculatum (L.) Akhani & Roalson 219, fig. 86
volkensii (Asch. & Schweinf.) Akhani & Roalson 219
CARYOPHYLLACEAE A.L. de Jussieu 6
subfam. **ALSINOIDEAE** Fenzl 6, 8, **29**
subfam. **CARYOPHYLLOIDEAE** (Juss.) Rabeler & Bittrich 6, 8,
subfam. **PARONYCHIOIDEAE** Meisn. 6, 8
Celosia L. 256
argentata L. 257
f. **cristata** (L.) Schinz 257
cvs. 'childsii', 'plumosa' &c. 257
var. *cristata* (L.) O. Kuntze 257
cristata (non L.) Dinsm. 257
lanata L. 266
Cerastium L. 46
anomalum Waldst. & Kit. ex Willd. 48
argeum Boiss. & Bal. 47
atticum Boiss. & Heldr. 51
balearicum F. Hermann 49
blepharophyllum Fisch. & C. A. Mey. ex Fenzl 51
blepharostomum Fisch. & C.A. Mey. ex Hohen. 51
brachypetalum Pers. 51
subsp. *luridum* (Boiss.) Nyman 51
subsp. **roeseri** (Boiss. & Heldr.) Nyman 51
var. *luridum* Boiss. 51
caespitosum Gil. 48
cerastioides (L.) Britton 47
var. *lalesarense* Bornm. 47
dentatum Möschl 49
dichotomum L. 51, fig. 20
var. **dichotomum** 53
var. **inflatum** (Link) Kandemir 53
subsp. *inflatum* (Link) Cullen 53
dubium (Bastard) Guepin 47
elbrusense Boiss.
fontanum Baumg. 48
subsp. *trivale* (Link) Jalas 48
subsp. **vulgare** (Hartman) Greuter & Burdet 48
fragillimum Boiss. 49, fig. 19
glomeratum Thuill. 49
holosteoides Fries 48
subsp. *triviale* (Link) Möschl 48
inflatum Link 53
intermedium Williams 47
longifolium Willd. 51
luridum (Boiss.) Lonsing 51
pentandrum L. 49
perfoliatum L. 48
purpurascens Adams 48
var. **elbrusense** (Boiss.) Möschl
roeseri Boiss. & Heldr. 51
semidecandrum L. 49
tmoleum Boiss. 49
trigynum Vill. 47
trivale Link 48
viscosum L. 49
vulgare Hartman 48
vulgatum L. 48, **49**
Ceratoides Gagnebin 194

papposa (Pers.) Botsch. et Ikonn. 196
Ceratospermum Pers. 194
papposum Pers. 196
Chenolea arabica Boiss. 203
Chenoleoides arabica (Boiss.) Botsch. 203
CHENOPODIACEAE Vent. 164
subfam. **Betoideae** Ulbr. 168
subfam. **Camphorosmoideae** 196
subfam. **Chenopodioideae** 169
subfam. **Corispermoideae** 204
subfam. **Salicornioideae** 206
subfam. **Salsoloideae** Raf. 217
subfam. **Suaedoideae** Ulbr. 210
tribe **Anserineae** Dumort. 190
tribe **Axyrideae** G. Kadereit & Sukhor. 194
tribe **Caroxyleae** Akhani & Roalson 217
tribe **Chenopodieae** 171
tribe **Dysphanieae** 169
tribe **Salsoleae** 232
Chenopodiastrum S. Fuentes, Uotila et Borsch 176
murale (L.) S. Fuentes, Uotila & Borsch 176, fig. 68
Chenopodina Moq. 210
Chenopodium L. 171
aegyptiacum Hasselq. 214
album L. 174, fig. 67
subsp. *iranicum* Aellen 176
subsp. *novopokrovskyanum* Aellen 176
altissimum L. 213
ambrosioides L. 171
baryosmum Roemer et Schultes 223
blomianum Aellen 174
botryoides Sm. 178
botrys L. 170
chenopodioides (L.) Aellen 178
crassifolium Hornem. 178
ficifolium Sm. 174
subsp. **blomianum** (Aellen) Aellen 174
subsp. ficifolium 174
foliosum (Moench) Aschers. 194
subsp. *montanum* Uotila 194
glaucum L. 179
iranicum (Aellen) Hamdi & Malekloo 176
murale L. 176
novopokrovskyanum (Aellen) Uotila *176*
oppositifolium (Pall.) L.f. 241
opulifolium Schrad. 172
rubrum L. 179
scoparium L. 201
urbicum L. 178
vulvaria L. 172, fig. 66
zerovii Iljin 176
Cherleria juniperina (non D. Don) auctt. 38
Climacoptera Botsch. 228
brachiata (Pall.) Botsch. 230
iraqensis Botsch. 230
khalisica Botsch. 230
Comphrosma pteranthus L. 17
Cornulaca Del. 250
aucheri Moq. 250, fig. 100
aucheri (non Moq.) Bunge 252
leucacantha Charif & Aellen 250
monacantha Del. 252
setifera (DC.) Moq. 252, fig. 100
tragacanthoides Moq. 252
Corrigiola albella Forssk. 9
repens Forssk. 19
Cryophytum nodiflorum (L.) Rothm. 126
Cucubalus aegyptiacus L. 117
spergulifolia Desf. 102
Cynocrambaceae Meisn. 271
Cynocrambe Gaertn. 271
prostrata Gaertn. 271

Dianthus L. 58
anatolicus Boiss. 70
asperulus Boiss. & Huet 70
atomarius Boiss. 69
auraniticus Post 62
barbatus L. 58, **72**
basianicus Boiss. et Hausskn. 68
caryophyllus L. 58, **72**
chinensis L. 58, **72**
crinitus Sm. var. *tomentellus* Boiss. 68
cyri Fisch. & C.A. Mey. 61
fimbriatus Bieb. 65
var. *brachyodontus* 67
var. *macropetalus* Boiss. & Huet 65
floribundus Boiss. 70
hymenolepis Boiss. 69
judaicus Boiss. 62
kotschyanus Boiss. 70
libanotis Labill. 69
liboschitzianus Ser. 62
longivaginatus Rech.f. 65, fig. 25
macranthus Boiss. 68
macrolepis Boiss. 62
masmenaeus Boiss. 70
var. **glabrescens** Boiss. 70
monadelphus Vent. subsp. *judaicus* (Boiss.) Greuter & Burdet 62
multicaulis Boiss. & Huet 62
multipunctatus Ser. 61
var. *gracilior* Boiss. 61
var. *subenervis* Boiss. 61
var. *velutina* Boiss. 61
nassireddini Stapf 67
noeanus Boiss. 70
orientalis Adams 65
subsp. **macropetalus** (Boiss.) Rech.f. 65, fig. 26
subsp. **nassireddini** (Stapf) Rech.f. 67, fig. 27
subsp. **orientalis** **65**, fig. 26
var. *brachyodontus* (Boiss. & Huet Bornm.) Bornm. 67
var. *macropetalus* (Boiss.) Bornm. 65
paniculatus Pau 61
parviflorus Boiss. 70
pendulus Boiss. & Blanche 68
persicus Hausskn. 70, fig. 28
quadrilobus Boiss. 61
robustus Boiss. & Kotschy ex Boiss. 70
siphonocalyx Blakelock 62, fig. 24
striatellus Fenzl 61
strictus Banks & Soland. 59, fig. 23
subsp. **strictus** var. **gracilior** (Boiss.) Eig 59, **61**
subsp. **strictus** var. **strictus** 59, **61**
subsp. **strictus** var. **subenervis** (Boiss.) Eig 59, **61**

subsp. **velutinus** (Boiss.) Greuter & Burdet 61
var. *axilliflorus* (Fenzl) Reeve 59
var. *velutinus* (Boiss.) Eig 61
sulcatus Boiss. 61
transcaucasicus Schischk. 70
zonatus Fenzl 69
var. **hypochlorus** (Boiss. & Heldr.) Reeve 69
Dichodon cerastioides (L.) Reichenb. 47
dubium (Bastard) Ikonn. 48
Dichoglottis linearifolia Fisch. & Mey. 83
Dondia Adans. 210
Dorotheanthus Schwantes 126
gramineus (Haw.) Schwantes 127
Dysphania R. Br. 169
ambrosioides (L.) Mosyakin & Clemants 170, fig. 65
botrys (L.) Mosyakin & Clemants 170, fig. 65

Echinopsilon caspicum Bunge 199
hyssopifolium (Pall.) Moq. 199
lanatum (Maerckl.) Moq. 199
Einadia Raf. 171
ELATINACEAE Dumort. 1
Elatine ammanioides (Roxb. ex Roth) Wight & Arn. 3
luxuriana Del. 1
verticillata (Willd.) Wight et Arnott 1
Emex Campd. 131
spinosa (L.) Campd. 131, fig. 51
Eremogone gypsophiloides (L.) Fenzl 32
Euthalia serpyllifolia (L.) Rupr. 33
Eurotia Adans. 194
Euxolus caudatus (Jacq.) Moq. 262

Fagopyrum Mill. 161
convolvulus (L.) H. Gross 162
esculentum Moench 162, fig. 63
vulgare T. Nees 162
Fallopia (L.) Á. Löve 162
convolvulus (*L.*) Á. Löve 162, fig. 64
vulgare T. Nees 162

Gamanthus Bunge 230
Gasoul nodiflorum (L.) Rothm. 126
Gastrocalyx ampullatus (Boiss.) Schischk. 106
Girgensohnia Bunge ex Fenzl 241
oppositiflora (Pall.) Fenzl 241, fig. 96
Githago segetum Desf. 123
Glinus L. 4
dictamnoides Burm. f. 4
lotoides L. 4, fig. 2
Gomphrena L. 270
globosa L. 270
sessilis L. 268
Gouffeia holosteoides C. A. Meyer 44
Gypsophila L. 78
alpina Habl. 74
anatolica Boiss. & Heldr. 80
antari Post & Beauverd ex Dinsmore 83
arabica Bark. 83
boissieriana Hausskn. & Bornm. 86, fig. 34
capillaris (Forssk.) C. Chr. 83
subsp. **capillaris** 83
subsp. **confusa** Zmarzty 83
caricifolia Boiss. 79
diaphylla Azn. 80
gypsophiloides (Fenzl) Blakelock 88
haussknechtii Boiss. 82
heteropoda Freyn & Sint. 82
subsp. **heteropoda** 82
koeiei Rech. f. 85
libanotica Boiss. 80, fig. 32
nabelekii Schischk.
lignosa Náb. nec Hemsl. & Lace 82
linearifolia (Fisch. & Mey.) Boiss. 83, fig. 33
lipskyi Schischk. 82
lurorum Rech. f. 88
nabelekii Schischk. 82
var. *lipskyi* (Schischk) Bark. 82
nanella Grossh. & Schischk. 82
obconica Bark. 83
pallida Stapf 82
var. *haussknechtii* (Boiss.) Bark. 82
pallidifolia Bark. 80
perfoliata L. 80
var. **anatolica** (Boiss. & Heldr.) Bark. 80
persica Bark. 85
pilosa Hudson 85, fig. 83
pilulifera Boiss. & Heldr. 79
platyphylla Boiss. 86
platyphylla auct. non Boiss. 86
polyclada Fenzl ex Boiss. 85
var. **polyclada** 85
porrigens (L.) Boiss. 85
pulchra Stapf 85
rokejeka Delile 83
ruscifolia Boiss. 80, fig. 32
sphaerocephala Fenzl ex Tchihat. 79
var. **sphaerocephala** 79, fig. 34
stricta Bunge 74
subaphylla Rech. f. 88
szowitsii Fisch. & Mey. var. *glandulosa* Ledeb. 83
trichopoda Boiss. 83
trichotoma Wenderoth 80
var. *anatolica* (Boiss. & Heldr.) Bornm. 80
venusta Fenzl 86
wiedemanni Boiss. 86

Habrosia Fenzl 31
spinuliflora (Ser.) Fenzl 31
Hagenia filiformis Moench 85
Halanthium C. Koch 230
purpureum (Moq.) Bunge 232
Halimocnemis C.A. Mey. 230
gibbosa Wol. ex Stapf 231
pilifera Moq. 231, fig. 91
pilosa Moq. 231
purpurea Moq. 232
sclerosperma (Pall.) C.A. Mey.
sulphurea Moq. 228
Halocharis Moq. 226
brachyura Eig 228
noaeana Iljin 228
sulphurea (Moq.) Moq. 228, fig. 89
Halocnemum Bieb. 207
strobilaceum (Pall.) Bieb. **207**, 255, fig. 82
Halogeton alopecuroides Moq. 253
cinereus Moq. 238
Halopeplis Bunge ex Ung.-Sternb. 206

pygmaea (Pall.) Bunge ex Ung.-Sternb. 206, fig. 81
Halothamnus Jaub. & Spach
glaucus (Bieb.) Botsch. 235
hierochunticus (Bornm.) Botsch. 233
iraqensis Botsch. 233, fig. 92
var. **hispidulus** Botsch. 234
var. **iraqensis** 233
lancifolius (Boiss.) Kothe-Heinrich 235, fig. 92
subaphyllus (C.A. Mey.) Botsch. 233
Halotis Bunge 230
pedunculata Assadi 232
pilifera (Moq.) Botsch. 231
pilosa (Moq.) Iljin 231
Haloxylon Bunge 245
ammodendron auctt., non (C.A. Mey.) Bunge 245
articulatum (Cav.) Bunge 247
ssp. *ramosissimum* (Benth. & Hook. f.) Eig 250
elegans (Bunge) Botsch. 249
persicum Bunge ex Boiss. & Buhse 245, fig. 98
ramosissimum Benth. & Hook. f. 250
salicornicum (Moq.) Bunge ex Boiss. 249
scoparium Pomel 247
Hammada Iljin 247
articulata (Moq.) O. Bolòs & Vigo 247, fig. 99
eigii Iljin 249
ramosissima (Benth. & Hook. f.) Iljin 249
salicornica (Moq.) Iljin 249
scoparia (Pomel) Iljin 247
Herniaria L. 12
arabica Hand.-Mazz. 13
cinerea DC. 12
densiflora Williams 14
fruticosa L. var. *hemistemon* (J. Gay) Barratte 13
glabra L. 12
var. *glaberrima* sensu Chaudhri non Fenzl 12
hemistemon J. Gay 13
hirsuta L. 12, fig. 4
incana Lam. 14
var. *angustifolia* Fenzl 14
lenticulata Forssk. 9
macrocarpa Sm. 14
Holosteum L. 53
glutinosum (Bieb.) Fisch. & C.A. Mey. 55, fig. 21
liniflorum Fisch. & C.A. Mey. 55
polygamum C. Koch 55
umbellatum L. 53, fig. 21
var. *glutinosum* (M. B). Gay 55

Illecebrum arabicum L. 9
cephalotes M. Bieb. 11
javanicum (Burm.f.) Murr. 266
longisetum Bertoloni 9
paronychia L. 9
sessile (L.) L. 268
Iresine javanica Burm.f. 266
persica Burm.f. 266

Kali Mill. 235
tragus (L.) Scop. 235, fig. 93
Kaviria Akhani 224
azaurena (Mouterde) Sukhor. 226
gossypina (Boiss.) Akhani 255
lachnantha (Botsch.) Akhani 224
tomentosa (Moq.) Akhani 226, fig. 88
Kochia Schrad. 196
densiflora (Turcz. ex Moq.) Aellen 201
f. **trichophylla** Schinz & Thell. (within *Kochia*) 201
eriophora Schrad. 197
griffithii Bunge ex Boiss. 200
hyssopifolia Pall. 199
indica Wight 200
latifolia Fresen. 197
monticola Boiss. ex Moq. 204
muricata (L.) Schrad. 199
noeana Boiss. 203
parodii Aellen 201
prostrata (L.) Schrad. 201
scoparia (L.) Schrad. 201
subsp. *densiflora* (Turcz. ex Moq.) Aellen 201
var. *densiflora* Turcz. ex Moq. 201
sieversiana (Pall.) C.A. Mey. 201
suffruticulosa Less. 201
Krascheninnikovia Gueldenst. 194
ceratoides (L.) Gueldenst. 196, fig. 76
subsp. lanata (Pursh) Kuntze 194

Lampranthus violaceus (non (DC.) Schwantes) Blakelock 127
Lepyrodiclis Fenzl 42
cerastioides Kar. & Kır. 44
holosteoides (C. A. Meyer) Fenzl ex Fisch. & Mey. 44, fig. 17
paniculata Stapf 46
stellarioides Schrenk ex Fisch. & C.A. Mey. 44
Loeflingia L. 21
hispanica *L.* 21, fig. 9

Mesembryanthemum L. 126
gramineum Haw. 127
nodiflorum L. 126, fig. 49
Mesostemma kotschyana (Fenzl ex Boiss.) Vved. 46
Minuartia L. 34
akinfiewii (Schmalh.) Woronow 38
decipiens (Fenzl) Bornm. 40
subsp. **decipiens** 40
hamata (Hausskn.) Mattf. 41
hirsuta (Bieb.) Hand.-Mazz. subsp. *oreina* Mattf. 37
hybrida (Vill.) Schischk. 42
intermedia (Boiss.) Hand.-Mazz. 40
juniperina (L.) Maire & Petitmengin 38, fig. 16
kashmirica (Edgew.) Mattf. 37
lineata (C.A. Meyer) Bornm. 37
mesogitana (Boiss.) Hand.-Mazz. 41
subsp. *lydia* (Boiss.) McNeill 41
subsp. **turcomanica** (Schischk.) McNeill 41
var. *turcomanica* (Schischk.) McNeill 41
meyeri (Boiss.) Bornm. 38
montana *L.* 40
subsp. **wiesneri** (Stapf) McNeill 40
multinervis (Boiss.) Bornm. 38

oreina (Mattf.) Schischk. 37
picta (Sibth. & Sm.) Bornm. 35, fig. 15
recurva (All.) Schinz & Thellung 37
subsp. **oreina** (Mattf.) McNeill 37
sclerantha (Fisch. & Mey.) Thellung 40
sublineata Rech. f. 37
subtilis (Fenzl) Hand.-Mazz. 42
tenuifolia (L.) Hiern non Nees ex Mart. 42
subsp. *hybrida* (Vill.) Mattf. 42
turcomanica Schischk. 41
MOLLUGINACEAE Bartl. 3
Mollugo hirta Thunb. 4
lotoides (L.) O. Kuntze 4
tetraphylla L. 17
Morocarpus foliosus Moench 194

Nitrosalsola nitraria (Pall.) Tzvel. 218
Noaea Moq. 238
canescens Moq. 223
kurdica Eig 240
mucronata (Forssk.) Aschers. & Schweinf. 238, fig. 95
subsp. *tournefortii* (Spach) Aellen 240
oppositiflora (Pall.) Moq. 241
tournefortii (Jaub. & Spach) Moq. 240

Obione flabellum (Bunge) Ulbrich 189
thunbergiifolia Boiss. & Noë 189
Oxybasis Kar. & Kir. 178
chenopodioides (L.) Fuentes, Uotila & Borsch 178, fig. 69
glauca (L.) Fuentes & Borsch 179
minutiflora Kar. & Kir. 178
rubra (L.) Fuentes & Borsch 179
urbica (L.) Fuentes, Uotila & Borsch 178
Oxyria Hill 140
digyna (L.) Hill 140, fig. 53
elatior R.Br. 140

Panderia Fisch. & C.A. Mey. 196
iraqensis Sukhor. (ined.) 203
pilosa Fisch. & C.A. Mey. 203
turkestanica Iljin 203
Parietaria debilis (non Forst.f.) Standley 271
Paronychia Miller 8
arabica (L.) DC. 9, fig. 3
subsp. *breviseta* (Aschers. et Schweinf.) Chaudhri var. *breviseta* 9
argentea Lam. 9
capitata (L.) Lam. var. *pubescens* Fenzl 11
desertorum Boiss. 9
glomerata Moench. 9
hispanica DC. 9
kurdica Boiss. 11
var. *imbricata* Chaudhri 11
longiseta (Bertoloni) Webb & Berthelot 9
mesopotamica Chaudhri 11
nitida Gaertn. 9
sclerocephala Decne. 14
Pentodon barbatus Ehrenb. 204
Persicaria (L.) Mill.149
amphibia (L.) Delarbre *151*
bistorta (L.) Samp. 149, fig. 58
subsp. **bistorta 149**
hydropiper (L.) Delarbre 154, fig. 59
lapathifolia (L.) Delabre 152
maculata (Rafin.) A. Löve & D. Löve 151
maculosa Gray 151, fig. 59
orientalis (L.) Spach 151
salicifolia (Brouss. ex Willd.) Assenov 152
Petrorhagia (Ser.) Link 72
alpina (*Habl.*) Ball & Heywood 74
cretica (*L.*) Ball & Heywood 74, fig. 29
macra (Boiss. & Hausskn.) Ball & Heywood 72
wheeler-hainesii Rech. f. 74
Petrosimonia brachiata (Pall.) Bunge 255
Phylloxys Moq. 250
Physogeton acanthophyllus Jaub. & Spach 232
Pleconax conoidea (L.) Sourkova 122
Polycarpaea Lam. 19
fragilis Del. 19
repens (Forssk.) Aschers. & Schweinf. 19
robbairea (Kuntze) Greuter & Burdet 21, fig. 8
Polycarpon Loefl. ex L. 17
arabicum Boiss. 19
robbairea Kuntze 21
succulentum (Del.) J. Gay 19
tetraphyllum (L.) L. 17, fig. 7
POLYGONACEAE 130
Polygonum L. 154
alpestre C.A. Mey. 159
ammanioides Jaub. & Spach 159
amphibium L. 151
arenarium Waldst. & Kit. 160
arenastrum (Boreau) auctt. 157
argyrocoleum Steud. ex Kunze 155, fig. 60
aviculare L. 157
alpestre C.A. Mey. 159
bellardii (non All.) auctt. 155
bistorta L. 149
var. **angustifolia** (Meisn.)
bistortoides (non Pursh) Boiss. 149
chlorocoleum Steud. ex Boiss. 155
cognatum Meisn. 159
var. *alpestre* (C.A. Mey.) Meisn. 159
var. *ammanioides* (Jaub. & Spach) Meisn.
convolvulus L. 162
corrigioloides Jaub. & Spach 157
equisetiforme (non Sm.) auctt. 155
fagopyrum L. 162
hydropiper L. 154
lapathifolium L. 152
luzuloides Jaub. & Spach 159
mucronatum Royle ex Bab. 160
nodosum Pers. 152
noeanum Boiss. 155
obtusatum Steud. ex Meisn. 152
olivieri Jaub. & Spach 158
paronychioides C.A. Mey. ex Hohen. 160, fig. 62
patulum Bieb. 155, fig. 60
persicaria L.
polycnemoides Jaub. & Spach 158, fig. 61
salicifolium Brouss. ex Willd. 152
setosum Jacq. 159
subsp. *luzuloides* (Jaub. & Spach) Leblebici 159
PORTULACACEAE Engler 128
Portulaca L.
grandiflora Hook. 130

oleracea L. 128, fig. 50
Portulacaria afra Jacq. 128
Pseudosaponaria pilosa (Huds.) Ikon. 85
Pteranthus Forssk. 16
dichotomus Forssk. 17. fig. 6
echinatus Desf. 17
trigynus Caball. 17
Pteropyrum Jaub. & Spach 144
aucheri Jaub. & Spach 144, fig. 56
ericoides Boiss. 144
gracile Boiss. 146
noeanum Boiss. ex Meisner 146
olivieri Jaub. & Spach 146, fig. 56
var. *gracile* Boiss. 146
Pongatium Juss. 273
indicum Lam. 273
Pyankovia Akhani & Roalson 228
brachiata (Pall.) Akhani & Roalson 230

Queria hispanica L. 40

Rhagodia R. Br. 171
Rheum L. 140
ribes L. 142, fig. 54
palaestinum Feinbr. 142
Robbairea confusa Maire 21
delileana Milne Redhead 21
major (Asch. & Schwinf.) Botsch. 21
Rokejeka capillaris Forssk. 83
Rumex L. 132
angustifolius Campd. 136
subsp. **angustifolius** 136
× **baqubensis** Rech. f. 138
conglomeratus Murray 137, fig. 51
conglomeratus × *dentatus* subsp. *mesopotamicus*
crispus L. 136
cyprius Murb. 134, fig. 51
subsp. **horizontalis** (C. Koch) Rech. f.
dentatus L. 138, fig. 51
subsp. **mesopotamicus** Rech.f. 138
denticulatus (non Campd.) C. Koch 138
dictyocarpus Boiss. & Buhse 138
digynus L. 140
halepensis Miller 138
horizontalis C. Koch 133
patientia L. 140
var. *kurdica* Boiss. 136
ponticus E.H.L. Krause **136**, 140
pulcher L. 137
subsp. **anodontus** (Hausskn.) Rech. f.) 137
subsp. *divaricatus* (L.) Murb. 137
subsp. **woodsii** (De Not.) Arcangeli 137
syriacus Meisn. 138
tuberosus L. 133
subsp. **horizontalis** (C. Koch) Rech. f. 133
subsp. **turcomanicus** (Rech. f.) Rech. f. 133
var. *turcomanicus* Rech. f. 133
vesicarius L. 134, fig. 52
Sagina L. 57
apetala Arduino 58
ciliata Fries in Liljebl. 58
linnaei Presl 57
procumbens L. var. *apetala* (Ardunio) Huds. 58
reuteri Boiss. 58
saginoides (L.) Karst. 57
Salicornia L. 208
europaea (non L.) Rech. f. 210
herbacea (non L.) Rech. f. 210
perennans Willd. 210, fig. 83
persica Akhani 210
pygmaea Pall. 206
strobilacea Pall. 207
Salsola L.
altissima (L.) L. 213
articulata Forssk. 243
articulata Cav. non Forssk. 247
aurantiaca Bunge ex Boiss. 226
auricula Moq. 235
azaurena Mouterde 226
baryosma (Roem. et Schultes) Dandy 223
boissieri Botsch. 223
brachiata Pall. 230
canescens (Moq.) Boiss. 223
canescens sensu Aellen 224
chrysoleuca Aellen 223
crassa Bieb.
crassa (non Bieb.) Rech. f. 230
cyclophylla Baker 221
dendroides Pall. 223
subsp. *trichantha* Botsch. 223
glauca Bieb. 235
gossypina Bunge ex Boiss. 255
hierochuntica Bornm. 233
hispidula (Bunge) Boiss.
hyssopifolia Pall. 199
imbricata Forssk. 223
incanescens C.A. Mey. 218
inermis Forssk. 219
jordanicola Eig 221
lachnantha (Botsch.) Botsch. 224
lancifolia Boiss. *235*
longifolia Forssk. 255
monobractea Forssk. 199
mucronata Forssk. 238
muricata L. 199
nitraria Pall. 218
oppositiflora Pall. 241
palaestinica Botsch. 219
prostrata L. 201
pseudonitraria Aellen 218
ruthenica Iljin 235
setifera (Moq.) Akhani 245
spinosa Pabot (non Lam.) 226
spissa M. Bieb. 218
subaphylla C.A. Mey. 235
tetrandra Forsk.
tomentosa (Moq.) Spach 226
subsp. *lachnantha* Botsch. 224
tragus L. 235
trichantha (Botsch.) Botsch. 223
vermiculata L. 219
var. *villosa* (Del. ex Roem. & Schult.) Moq. 219
villosa Del. ex Roem. & Schult.
villosa (non Del.) Roem. & Schult. 219
volkensii Asch. & Schweinf. 219
Saponaria L. 76
boissieriana (Hausskn. & Bornm.,) Preobr. ex Popov 86
cretica L. 74

porringens (Gouan) L. 85
suffruticosa Nábělek 76
tridentata Boiss. 76
vaccaria L. 94
viscosa C. A. Mey. 76, fig. 31
Schanginia C.A. Mey.
aegyptiaca (Hasselq.) Aellen 214
altissima (L.) C.A. Mey. 213
baccata (Forssk.) Moq. 214
hortensis (non (Forssk.) Moq.) Boiss. 214
Schoberia C.A. Mey. 210
Scleranthus L. 29
annus L. var. *uncinatus* Boutigny 29
hamatus Hausskn. 40
uncinatus Schur 29, fig. 13
Sclerocephalus Boiss. 14
arabicus Boiss. 14, fig. 5
Seidlitzia Bunge ex Boiss. 236
cinerea (Moq.) Bunge ex Botsch. 238
lanigera Post 245
rosmarinus Ehrenb. ex Boiss. **236**, 255, fig. 94
Senniella spongiosa (F. Muell.) Aellen var. *holocarpa* (F. Muell.) Aellen 189
Silene L. 96
sect. Ampullatae Boiss. 105
sect. Atocion Otth 117
sect. Auriculatae (Boiss.) Schischk. 106
sect. Bipartitae Boiss. 119
sect. Compactae Boiss. 113
sect. Conoimorpha Otth 122
sect. Fimbriatae Boiss. 106
sect. Lasiocalycinae Boiss. 115
sect. Lasiostemones (Boiss.) Schischk. 98
sect. Rigidulae Boiss. 120
sect. Sclerocalycinae (Boiss.) Chowdh. 100
sect. Spergulifoliae (Boiss.) Chowdh. 102
aegyptiaca (L.) L. f. 117, fig. 47
affines (non Godr.) Boiss. 119
ampullata Boiss. 105
apetala Willd. 115, fig. 46
arabica Boiss. 119
arguta Fenzl 104
arenosa C. Koch 120, fig. 46
armeniaca Rohrb. 102
atocioides Boiss. 117
atocion Jacq. 117
aucheriana (non Boiss.) Boiss. 104
avromana Boiss. & Hausskn. 102
bornmuelleriana Freyn 104
brevicaulis Boiss. 110
var. *latifolia* Boiss. 110
caricifolia (Boiss.) Bornm.
caryophylloides (Poir.) Otth 113
subsp. caryophylloides 113
subsp. **subulata** (Boiss.) Coode & Cullen 113
var. *nardifolia* Boiss. ex Rohrb. 113
chaetodonta Boiss. 121
chlorifolia *Sm.* 100
subsp. **chlorifolia** 100, fig. 40
subsp. **swertiifolia** (Boiss.) Ghazanfar 102
var. *swertifolia* Rohrb. 102
colorata Poiret 119
subsp. **oliveriana** (Otth) Rohrb. 119
commelinifolia Boiss. 108
var. *erimicana* (Stapf) Bullock 110
var. *heterophylla* Fenzl 108
var. *isophylla* Bornm. 108
var. *ruscifolia* Hub.-Mor. & Reese 108
compacta Fisch. 113, fig. 45
coniflora Nees ex Otth 122
conoidea L. 122, fig. 46
debilis Stapf 121
dichotoma Ehrh. 115
subsp. *sibthorpiana* (Reichenb.) Rech. f. 115
diversifolia Otth 119
erimicana Stapf 110, fig. 44
eriocalycina Boiss. 105
hohenackeri Boiss. 104
ispirensis Boiss. & Huet 104
lagenocalyx Fenzl ex Boiss. 117
leyseroides Boiss. 120
linearis DC. 121
longipetala Vent. 98, fig. 39
lycia Boiss. 106
microphylla Boiss. 106
monantha Boiss. & Hausskn. ex Boiss. 108
montbretiana Boiss. 104
muscipula L. 123
nardifolia Boiss. et Huet 113
odontopetala Fenzl 112
subsp. *congesta* (Boiss.) Melzh. 112
var. *cerastifolia* Boiss. 112
var. *congesta* Boiss. 112
var. *glabrifolia* Blakelock 112
oliveriana Otth 119
orchidea L. f. 117
oreophila Boiss. 105
var. *latifolia* Chowdh. 105
paniculata Ehrenb. ex Rohrb. 102
porringens Gouan ex L. 85
pruinosa Boiss. 104
pungens Boiss. 110
racemosa Otth 115
retinervis Ghazanfar 112
rhynchocarpa Boiss. 104, **106**, fig. 43
var. *lycia* Boiss. 106
rubella L. 119
ruscifolia (Hub.-Mor. & Reese) Hub.-Mor. 108
salsa Boiss. 120
saxatilis *Sims* 100
schizopetala Bornm. 106, fig. 42
sibthorpiana Reichenb. 115
spergulifolia Boiss.
spergulifolia (Desf.) Bieb. 102, fig. 41
var. *elongata* Boiss. 102
var. *arbuscula* Boiss. 102
stenobotrys Boiss. & Hausskn. 102
subulata (Boiss.) Coode & Cullen 113
swertiifolia Boiss. 100, **102**
var. stenophylla Boiss. 102
trinervis Banks & Sol. 115
villosa Forssk. 120, fig. 46
Spergula L. 23
arvensis *L.* 23
diandra (Guss.) Heldr. et Sint. 27
fallax (Lowe) E.H.L. Krause 23, fig. 10
flaccida (Roxb.) Aschers. 23
media Wahl. 25

pentandra L. var. *intermedia* Boiss. 23
saginoides L. 57
vulgaris Boenn. 23
Spergularia (Pers.) J. & C. Presl 24
bocconii (Scheele) Aschers. & Graebn. 27
diandra (Guss.) Boiss. 27
fallax Lowe 23
marginata (DC.) Kittel 25
marina (L.) Griseb. 25
media (L.) C. Presl 25, fig. 11
salina J. & C. Presl 25
SPHENOCLEACEAE (Lindl.) Mart. ex DC. 273
Sphenoclea Gaertn. 273
pongatium DC. 273
zeylanicum Gaertn. 273, fig. 106
Spinacia L. 190
oleracea L. 192
var. **oleracea** 192
var. **inermis** (Moench) Metzg. 192
tetrandra Stev. 192, fig. 74
Stellaria L. 44
apetala auct., non Ucria 46
cerastioides L. 47
kotschyana Fenzl ex Boiss. 46
media (*L.*) *Vill.* 44, fig. 18
media (L.) Vill. subsp. *pallida* (Dumort) Asch. & Graebner 46
pallida (Dumort) Piré 46
serpyllifolia (L.) Scop. 33
Suaeda Forssk. ex J.F. Gmel.
acuminata Moq. 214
aegyptiaca (Hasselq.) Zohary 214, fig. 84
altissima (L.) Pall. 213
anatolica (Aellen) Sukhor. 213
baccata Forssk. ex J.F. Gmel. 214
birandii Aellen 214
carnosissima Post 214
catenulata Bornm. & Gauba 211
cochlearifolia Wol. 214
fruticosa Forssk. ex J.F. Gmel. 211
guestii Aellen 214, 217
maritima auct. non L. 215
maritima auct. non (L.) Dum. 213
mesopotamica Eig 211
monoica Forssk. ex J.F. Gmel. 213, fig. 84
prostrata Pall. subsp. *anatolica* Aellen **213**, 215
rawii Aellen (ined.) 211
salsa auct. non (L.) Pall. 215
setigera (non (DC.) Moq.) Boiss. 214
vermiculata Forssk. ex J.F. Gmel. 211, fig. 84

Telanthera bettzickiana Regel 269
Telephium L. 4, **27**
imperati L. subsp. **orientale** (Boiss.) Nyman 29
var. *orientale* (Boiss.) Boiss. 29
oligospermum Steud. ex Boiss. 29, fig. 12
orientale Boiss. 29
THELIGONACEAE 271
Theligonum L. 271
alsinoideum Lam. 271
cynocrambe L. 271, fig. 105
Traganum Del. 255
brachypetala Jaub. & Spach
nudatum Delile 255
Tunica brachypetala Jaub. & Spach 74
cretica (L.) Fisch. & Mey. 74
gracilis Williams 74
macra Boiss. & Hausskn. ex Boiss. 74
pachygona Fisch. & Mey. 74
stricta (Ledeb.) Fisch. & C. A. Mey. 74

Vaccaria Wolf 94
grandiflora (Fisch. ex DC.) Jaub. & Spach 94
hispanica (Miller) Rauschert 94, fig. 38
var. *oxydonta* (Boiss.) Léonard 94
oxyodonta Boiss. 94
pyramidata Medik. 94
Velezia L. 75
rigida L. 75, fig. 30

Willemetia lanata Maerkl. 199

INDEX TO VERNACULAR NAMES

These are cited in the text along with their source, and in some cases are translated into English. Arabic and Kurdish names are capitalized in this list; English names are given in lower case letters.

'AIRUM 243
'AIWA 233
ALAICH-AL-GHAZAL 199
ALDŌZ 130
'ARAD 249
ARID 200
AWAISJA 238
BAQLA 130
BAQLATU 'L-HAMQA 130
BARBÎN 130
Batchelor's Buttons 270
Beetroot 169
BŪ'AIDH 233
Burning Bush 201
CHIBCHĀB 252
CHUKUNDAR 169
Cock's-comb 257
CRNAROKA 105
DAMM AL-'ĀSHIQ 230
DAMM AL-ĀSHAQ 270
DHUMRĀN 255
DUMRĀN 255
GA SANGOAL 104
GA SURIK 240
GHIDHĀM 218, 221
GIAH SURIK 240
Globe Amaranth 230
GLUNCA SPIA 105
GŌGALLA 212
HAITAM 203
HALŪLĀN 199
HĀMIDH 203
HAMTH 134, 207
HAMUDH 212, 215
HAMUTH 134
HUMATA AL-GHANAM 263
HUMMAIDH 200, 210
HUMMAIR 210
IJRŪM 243
ITHNE 249
Joy weed 268
JURSHKA 134
KHANAIQ 200
KHATHRAF 199
KHEIYES 263
KHIDHRĀF 221, 228
KHIDRĀF 221
KHUDHRADH 199
Leaf-Beet 169
Love-lies-bleeding 259
NETOON 249
PANJAR 169
PARPÎNA 130
Pigweed 130
Potager 130
Pouprier 130
Purslane 130
Pursley 130
QUTAINAH 199
RAIWAS 142
REMITH 249
RIJLA 230
RIMITH 218, 233, 243, 249
RIMTH 207, 249
RIMTH ABU RAIDA 234
RŪTHA 219
SARMAQ 265
SHA'RĀN 252, 255
SHAWATĀN 215
SHINAN 233
SHIRĀN 255
SHNAN 238
SHUATAN 215
SHUWANDAR 169
SILIQ 169
Spinach-Beet 169
SUĀD 200
SUKARI 169
TAHĀMA 212
TAKHME 212
TARTAI' 215
THELETH 207
THELLAITH 207
THILAIDH 207, 210, 249
'UBAIRAH 252
URF AD-DÎK 257, 265
WARD AD-DUKMA 230
WARDA 234
YALDIZ 130
YALDIZ CHICHAJI 130
YATHNAH 249
YITNA 249
ZILIQ 169
ZIRR HABASHI 230